SUSTAINABLE DEVELOPMENT – AUTHORITATIVE AND LEADING EDGE CONTENT FOR ENVIRONMENTAL MANAGEMENT

Edited by **Sime Curkovic**

Sustainable Development
– Authoritative and Leading Edge Content for Environmental Management
http://dx.doi.org/10.5772/2562

Edited by Sime Curkovic

Contributors

A.S. Abdulkareem, A.S. Afolabi, J. Abdulfatai, H. Uthman, J.O. Odigure, Ekaterina Anyanova, Brigitte Leblon, Laura Bourgeau-Chavez, Jesús San-Miguel-Ayanz, Koushen Douglas Loh, Sasathorn Tapaneeyakul, Maria J. Masanet-Llodra, Paulo Morgado, Marina Toger, Patricia Abrantes, Jérémy Fiegel, Martin Oteng-Ababio, Jorge Rocha, António Perdigão, Raquel Melo, Cristina Henriques, Gilberto Santos, Manuel Rebelo, Síria Barros, Martinha Pereira, Amir Hamzah Sharaai, Noor Zalina Mahmood, Abdul Halim Sulaiman, Cristian Teodorescu, Margareta Nicolau, Adriana Cuciureanu, Marinela Petrescu, Doina Guta, Ciprian Tetiu, Rodica Baciu, Cristian Teodorescu, Lucian Constantin, Margareta Nicolau, Aurelia Ballo, Cristiana Cosma, Massimo Battaglia, Tiberio Daddi, Francesco Testa, Francesco Testa, Tiberio Daddi, Fabio Iraldo, Marco Frey, Ray-Yeng Yang, Ying-Chih Wu, Hwung-Hweng Hwung, Antonio Zanin, Fabiano Marcos Bagatini, Mansoor Zoveidavianpoor, Ariffin Samsuri, Seyed Reza Shadizadeh, Felipe de Jesús Escalona-Alcázar, Bianney Escobedo-Arellano, Brenda Castillo-Félix, Perla García-Sandoval, Luz Leticia Gurrola-Menchaca, Carlos Carrillo-Castillo, Ernesto-Patricio Núñez-Peña, Jorge Bluhm-Gutiérrez, Alicia Esparza-Martínez, Danica Fazekašová, Beatriz Gómez-Muñoz, David J. Hatch, Roland Bol, Roberto García-Ruiz, Nils Max McCune, Francisco Guevara-Hernández, Jose Nahed-Toral, Paula Mendoza-Nazar, Jesus Ovando-Cruz, Benigno Ruiz-Sesma, Leopoldo Medina-Sanson, Boško Josimović, Igor Marić, Richard Y.M. Kangalawe, Stella Kyvelou, Maria Sinou, Isabelle Baer, Toni Papadopoulos

Published by InTech

Janeza Trdine 9, 51000 Rijeka, Croatia

Copyright © 2012 InTech

Notice

Publishing Process Manager Silvia Vlase

Typesetting InTech Prepress, Novi Sad

Cover InTech Design Team

Additional hard copies can be obtained from orders@intechopen.com
Sustainable Development – Authoritative and Leading Edge Content for Environmental Management, Edited by Sime Curkovic
p. cm.
ISBN 978-953-51-0682-1

For EU product safety concerns:
IN TECH d.o.o., Prolaz Marije Krucifikse Kozulić 3, 51000 Rijeka, Croatia,
info@intechopen.com or visit our website at intechopen.com.

Meet the editor

Dr Sime Curkovic is a Professor of Supply Chain Management at Western Michigan University. He received his undergraduate degree in Management Systems from Kettering University and his Ph.D. degree from Michigan State University. His research interests include environmentally responsible manufacturing, green purchasing, total quality management, supply chain management, supply chain risk management, and integrated global strategic sourcing. Dr Curkovic's publications have appeared in the Journal of Supply Chain Management, the IEEE Transactions on Engineering Management, the Decision Sciences Journal, the International Journal of Operations and Production Management, the International Journal of Production Research, the Journal of Quality Management, the Journal of Operations Management, and the International Journal of Production Economics. His dissertation on the relationship between Total Quality Management and Environmentally Responsible Manufacturing has been given the following awards: 1) The APICS Educational Foundation Edward & Marion Plossl Fellowship; 2) The Richard J. Lewis Quality of Excellence Award; and 3) The National Decision Sciences Institute Buffa Doctoral Dissertation.

Contents

Preface

Welcome to the Authoritative and Leading Edge Content for Sustainable Development. The practice of sustainable development is becoming widespread in all industries around the globe today, and both small and large firms are realizing the benefits provided by sustainability. The World Commission on Environment and Development defines sustainable development as "development that meets the needs of the present without compromising the ability of future generations to meet their own needs." Sustainability implies efforts directed towards preserving the environment, doing more with less (cost savings, profit increasing), reducing dependence on foreign resources such as oil, using alternate forms of energy, etc. The critical issues in sustainable development are energy consumption, air pollution, and climate change to name just a few. Confronted with the prospect of non-availability of fossil fuels within the foreseeable future, scarcity of natural resources, an increasing frequency of natural calamities and the worst economic downturn in nearly a century, the sustainability landscape has changed drastically in the last five years.

Researchers are trying to meet the challenge posed by the rapidly evolving landscape of sustainability. All in all, a lot has been done but a lot more needs to be done on the part of researchers to meet the pace of the sustainability movement. In order to address the sustainability deficit in education and research, *InTech* has invited submissions of manuscripts for publication in its focused issue on sustainability. All chapters are high-quality, high-impact manuscripts addressing integration of sustainability efforts with clear lessons and insights for the stake holders (faculty, students, administrators, communities, and industries).

We think this book is unique in that it uses a novel and logical approach to present discussions of this topic from three general perspectives: scientific, social, and economic, and the integration within these important areas. We think this book is somewhat different than other environmental management books since we offer a more balanced view of the topic – many of the books available today concentrate primarily on just one of these three perspectives. A primary objective of this book is to help readers think about how sustainable development impacts all areas and processes of the firm and the society that it functions within, and to show how managers, researchers, and policy makers can improve their position by employing the practices described throughout the book.

This book addresses the strongly related environmental issues (e.g., energy production, pollution, climate change, etc.) by considering them in an integrated manner and linking them to economic, social, and technical issues and environmental policies. These chapters collectively can help transform the values and principles that will directly influence the development of future sustainability strategies and our lifestyle. Likewise, this book will help decision makers consider both the economic, social, and technical issues with environmental concerns, at which point sustainable development becomes more logical and ultimately sustainable.

Sime Curkovic
Western Michigan University, Haworth College of Business,
Department of Management, Kalamazoo,
Michigan,
USA

Chapter 1

Oil Exploration and Climate Change: A Case Study of Heat Radiation from Gas Flaring in the Niger Delta Area of Nigeria

A.S. Abdulkareem, A.S. Afolabi, J. Abdulfatai, H. Uthman and J.O. Odigure

Additional information is available at the end of the chapter

http://dx.doi.org/10.5772/47730

1. Introduction

Nigeria like most other developing countries in the early part of the 70s was engaged in intensive natural resource exploitation as a way of stimulating economic growth (Ajayi, 1999; Abdulkareem et al, 2009). As at 1976, about 10 years from the start of oil exploration in the Niger-Delta area of Nigeria, figures available from the office of federal statistics stated that the oil has come to account for about 14% of the nation's gross domestic product (GDP) of Nigeria, 95% of the total export and over 80% of government's annual revenue (Olukoga, 2002; Tolulope, 2004; Abdulkareem and Odigure, 2006). Also total export peaked at 2 millions barrels per day of crude oil with price range of 18-22US dollars per barrel (Odigure and Abdulkareem, 2006). It is worth of mentioning that the current price per barrel of oil is now in the range of 80-110 US dollars (Ufarana, 2004; Akpan, 2009). This has created more opportunities for the development of new fields and increase granting of mining licenses and intensive exploration of oil mineral resources in the country (Ajayi, 1999; Abenege, 2004). The crude oil comes from reservoir containing gas, which is produced along with the oil. This associated natural gas is separated from the oil at a flow station. However in Nigeria, about 75% of the associated gas is flared due to the underdeveloped local market for gas in the country. The gas currently flared is estimated at two billion cubic feet per day (scf/d) or 56,600 m^3, the highest in any member nation of OPEC (Ageh et al, 2009; Ajayi, 2009). The quantity of gas flared in Nigeria is equivalent to the total annual power generation in the sub-Saharan Africa 'according to World Bank'(Odigure, 2001; UNDP/World bank report, 2004; Akpan, 2009). Nigeria is said to have natural gas reserves of 100 trillion standard cubic feet (about 2.832 trillion cubic meters), with about 45 trillion standard cubic reserves in the Niger-Delta Area of Nigeria (Ufarana, 2004; Tolulope, 2004). In energy terms, the quantity of natural gas in Nigeria is said to be more than twice the quantity of crude oil (Agbalino and Eyinal, 1997; Ajayi, 1999; Abdulkareem and Odigure, 2006) .It is estimated that the country's reserve-production ratio is about 125 years compared to that of crude oil of less than 30 years (Akpan, 2009). Consequently, petroleum experts often describe Nigeria as a natural gas province with some oil in it; this put the country in the ten top nations in the world in terms of natural gas reserves (Abowei et al., 1997, Abdulkareem, 1999; Oni, 2011). Though, there is no proper record of natural gas production per year in Nigeria, however the total Nigeria natural gas production is put in the range of 22-25 billion cubic meters per year (Abdulkareem and Odigure, 2006). Around 18 billion cubic meters of the total gas produced is associated gas, most of which is flared , with small amount re-injected into the sandstone sponge while the remaining is sold to electricity generating stations and industries (Abdulkareem and Odigure, 2010). It has been reported that about 3 billion cubic meters per year of non associated gas is currently trapped for industrial consumptions, this is an indication that the level of gas utilization on Nigeria is very low (Akpan, 2009; Abdulkareem et al., 2010). Gas utilization is for now limited to some small quantities being used as fuel for petroleum operations for enhancement of oil recovery project, for pressure maintenance in some industrial processes on a relatively modest scale and for power generation (Reymond, 2007; Udetal et al., 2007). Gas flaring is therefore not only wastes valuable resources, but is also a major cause of environment pollution in the Niger-Delta, where most of Nigeria's oil output is produced (Abdulkareem and Odigure, 2002; Odigure et al., 2003; Ufarana, 2004).

The Niger-Delta oil fields of Nigeria covers about 70,000 square kilometer and is one of the world's largest wetlands, which houses Nigeria's proven gas reserves, estimated to be 120 trillion cubic feet (Abdulkareem, 1999; Uyigue and Agho, 2005; Onyiah, 2005). However, while the exploitation and exploration of oil has created some fortunes and contributed positively to the economic and technological advancement of Nigeria as a whole, the accompanying socio-economic and ecological fallout remain problematic (Alakpodia, 1980; Akpan, 2009). The public considers the oil-producing companies operating in the Niger-Delta oil fields responsible for polluting the environment by way or relentless flaring and venting of gas in to the environment, heat radiation, noise radiation, oil spillage, water pollution, site clearing, deforestation and destruction of the flora and fauna and consequences disturbances of the ecosystem in the 70,000 square kilometers Niger-Delta wetland (Ifeanyichukwu, 2002; Odigure and Abdulkareem, 2002; Abdulkareem et al., 2011). The situation that led to growing anger among the local peoples on the damages caused to their health and ecosystem by oil exploration activities, especially gas flaring and crude oil spillage (Oyekunle, 1995; Onosode, 1996; Abdulkareem et al., 2011). It has also been noted that there are currently 100 gas flaring sites, some of which have been burning ceaselessly for 40 years. Each one of these bonfires as shown in Fig 1 has an adverse effect on the inhabitants and the natural environment (Ifeanyichukwu, 2002; Oyekunle, 1995). The extent of human damage attributable to gas flaring is unclear, but doctors have found an unusual high incidence of asthma, bronchitis, skin and breathing problems in communities in oil producing areas (Abdulkareem and Odigure, 2006). Moreover, flaring is a global source of green house gas emissions, contributing to global warming. The World Bank estimates that gas flaring in the Niger-Delta releases some 35 million tonnes of carbon dioxide annually into the air (FEPA Report, 1998; UNDP/World Bank report, 2004). Flared gas also releases hazardous substances into the environment that heighten the problem of the depletion of the ozone layer (Chimaroke, 2004). The attendant "green house effect" is one of the most frightening environmental problems of our time. Ozone layers that serve as blanket for regulating the earth's temperature are stripped as a resulting of gas flaring thereby causing global warming (Ikelegbe, 1993). In spite of advances in technology and the potential to convert the flared gas into a source of enormous nation revenue, the practices has continued in Nigeria, ostensibly underscoring the problems of our national development (Oyekunle, 1999; Akpan, 2009). Combustion of associated gas during gas flaring also releases heat into the environment. The heat radiation from gas flaring greatly affects the surrounding environment and particular crops planted within the vicinity of gas flare stations (Abdulkareem and Odigure, 2002). It also has a devastating effect on microorganisms and aquatic life. Heat radiation from gas flaring also causes an increase in heat waves hence there is the possibility that habitants of Niger-Delta Area, where the gas flaring stations are located will suffer heart stroke, heart attacks and other ailments aggravated by the heat (Odigure et al., 2003). For instance, it has been reported that heat wave killed more than 700 people in Chicago area alone and if this is happening already from heat, what would occur in the future with global warming (Aduku, 1997; Olukoga, 2002; Abenege, 2004) . Heat radiation from gas flaring also contributed to the increment in the soil temperature, which destroys the plant thereby affecting the ecosystem since plants absorb CO_2 in the atmosphere (Kearns et al., 2000; Tolulope, 2004; Tzimas et al., 2007). The world health organisation advised the Nigerian government to address the problem of gas flaring by paying close attention to the activities of companies engaged in gas flaring and the environmental problems associated with their exploratory methods and to invite experts from developed countries to work with Nigerian professionals and environmentalists to proffer remedy (Global gas reduction initiative, 2002; UNDP/World bank report, 2004; Ufarana, 2004). Critics of the flares in the Niger-Delta have said that the Nigerian government puts profit ahead of the environmental safety and the welfare of its citizens. The harmful effects of gas flaring and inability of the oil companies and government to quantify the resultants effects of gas flaring on the environment has led to strain relationship between the oil producing companies and the people of Niger-Delta Area of Nigeria. To control the activities of oil companies in the Niger-Delta area of Nigeria, there is the need to concentrate on environment management as a tool of liberation in improving the quality of life and to make the environment friendly for human beings. This however brings about models and simulation, which is now applied generally to look into the inter-relationship between the parameters and its resultant effects on the environments (Abdulkareem et al., 2011). In this work, mathematical modelling that can be used to predict the quantity of heat radiation from gas flaring station will be developed. The developed model will be simulated and find interaction between various parameters such as distance, volume of gas flared, flared stack efficiency that influence the rate of heat radiation from gas flaring station. Mathematical modeling is a simplified image of processes taking place in a system. These could include heat propagation, concentration of dispersion of gases from combustion and generation of heat and propagation e.t.c. Models retain the most essential properties of the actual process but presents them in mathematical forms. According to Luyben (1995)." Mathematical modeling is very much an art. It takes experience, practice and brain power to be good mathematical modelers". Mathematical model of a system must be sufficiently simple, easy to grasp and give a clear idea about all the qualitative aspects of the phenomenon of interest. On the other hand, it must be sufficiently accurate in bringing the quantitative aspects of the process. Simulation represent the application of modelling techniques to real system, thus enabling information on plant characteristics to be gained without either constructing or operating the full scale plant or system under consideration. Simulation methods come in two type viz. Digital simulation and Analogue simulation of these two types, Digital simulation which involve the use of codes and programme are more in use since they can be implemented on modern computer with exceptional speed (William, 1995)

Figure 1. Gas flaring station in the Niger-Delta area of Nigeria

1.1. Oil exploitation and climate change

Pollution is a term that defines any environmental state or manifestation which is harmful or unpleasant to life, it can also be defined as the introduction of natural and artificial particulate contaminants into the atmosphere (Cemak, 1985; Odigure, 1998). This can be caused due to man's failure to achieve control over the chemical, physical or biological consequence or side effects of his scientific industrial and social habit (Drake and Hubacek, 2007). Industrial source of pollution is as a result of release of hazardous waste and huge mass of unclear gases and other airborne particles produced as by- products in process industries (Odigure, 1998). The degree of the increase of process industries is increasing drastically, this improvement has changed man's way of life remarkably, however the advantages derived have not being without a price which is basically the effect of pollution on the environment (Carvalho et al., 2005; Baroutian, 2006). Pollutants are emitted into the atmosphere as either gases or particles, and are eventually removed by natural self cleansing processes (Ajayi, 1999; Alameddine and El-Fadel, 2005). The waste mostly originates from the burning of fossil fuels and the processing of materials by industries (Abdulkareem et al., 2011). Other sources include waste from burning engines in cars, fuel use in domestic sectors, oil boom, gas flaring, agricultural processes, but of all these the process industries have been recognized as the major source of air pollutants. The self-cleansing ability of the atmosphere which involves dispersion and dilution, is used as a grant channel, is presently too small and cannot match up to the rate of introduction of pollutants into the atmosphere, this is due to meteorological influences (Held et al., 2005; Potocnik et al, 2007; Soylu, 2007). During the 20th century contamination of the environment as a result of human activities has risen drastically (Hussain et al., 2005; Abdulkareem et al, 2009). Pollution problem have risen in all industrialized areas as well as in various inland, coastal water and stretches of ocean (Appleby, 1992; Kinnee et al., 2004). The capacity of the biosphere to disperse, degrade and assimilate human waste is in serious question, waste due to human activities now overweigh natural forces of putting toxic elements into the atmosphere (Ufarana, 2004). From colloid chemistry atmospheric air belongs to the group of incoherent, coarse dispersal system. More specifically, it is a system of colloid particles dispersed in gas in which solid and liquid component can be found in a mixture of gases, with this knowledge the term "pure air" cannot be defined because that will depend on the content of the air expected. Stern (1962 – 1968) said the composition of the air found near the soil level (homosphere) composes of various amount of substances additional to its constant component, this additional component are the air pollutants, thereby making pure air never to be in existence (Ufarana, 2004, Onyiah, 2005). The activities of the oil companies in the Niger Delta area of Nigeria are the major sources of the environmental pollution in the country. The situation that makes the public to consider the oil producing companies operating in the Niger-Delta oil fields responsible for polluting the environment by way of relentless flaring and venting of gas in the environment. Gas flaring in Nigeria today has posses an environmental hazard to the nation at large. So much damage is being done to the environment through gas flaring, that if nothing is done in a few years from now, serious environmental and health problems such as premature death and diseases will emerge. With respect to gas flaring, its effect on vegetation, health and the micro-climate are equal searing. Apart from the noise produced from the fire at the flare sites, the thick smoke that bellows into the sky contains poisonous gases which give rise to acid rain and eventually poison streams, lakes, lagoons and rivers thereby destroying aquatic organism and making the water unhealthy for drinking. It has been proven that gas flaring generates heat which is felt over an average radius of 0.5 kilometres thereby causing thermal pollution (Ikelegbe, 1993). Gas flaring cause green house effect, thereby producing global warming and green house gases which include: water vapour, carbon dioxide, Methane, Nitrous oxide; Ozone, Carbon monoxide and Nitrogen oxide (Indriani, 2005; Nwaichi and Uzabona, 2011).

1.1.1. Global warming and green house effect

Man's expanding activity has reached a level at which their effects are global in nature. The natural system i.e. the atmosphere, land and sea as well as life's and plants therein are clearly being disturbed (Cermak, 1985; Odigure, 1998) .It is obvious that some trace of CO_2, N_2O, CH_4 and O_3 gases have increased during the last century (Nigeria natural gas stategy, 2002; Nwaichi and Uzabona, 2011). In addition, other gases are being emitted that are not naturally part of the global eco-system, notably Chloro Fluoro Carbons (CFCS) (Odjugo, 2010; Abdulkareem and Odigure, 2010). These trace gases absorb and emit radiation and are thus able to influence the earth's climate. They are refereed to collectively as green house gases. The green house effect is described as a warming of the earth's surface and lower atmosphere that tends to intensify with an increase in atmospheric carbon dioxide. The atmosphere allows a large percentage of the rays of visible light from the sun to reach the earth's surface and heat it (Nyong, 2004; Odjugo, 2010). A part of this energy is radiated by the earth's surface in the form of long wave infrared radiation, much of which is absorbed by molecules of carbon dioxide and water vapour in the atmosphere and which is reflected back to the surface as heat (Abdulkareem and Odigure, 2002). This is roughly analogous to the effect produced by glass panes of a green house, which transmit sunlight in the visible range but hold heat. The trapping of this infrared radiation causes earth's surface and lower atmospheric layers to warm to a higher temperature than would otherwise be the case. Without this green house heating, the earth's average temperature would be only -73°C, even the ocean would be frozen under such conditions (Odjugo, 2010). Owing to the rise in atmospheric carbon dioxide caused by the modern industrial societies, widespread combustion of fossil fuels (coal, oil and natural gas), the green house effect on earth may be intensified. An increase in concentration of the atmospheric concentrations of other trace gases such as Chlorofluorocarbons (freons), nitrous oxide and methane due human activity may also aggravated greenhouse condition (Nwaichi and Uzabona, 2011).

The realization that climate might change, as a result of emission of CO_2 into the atmosphere is not new. Arrhenius (1896) pointed out that burning of fossil fuel might cause an increase of atmospheric CO_2 and thereby changing the heat balance of the earth. Calendar (1938) convincingly showed that the atmospheric CO_2 concentration was increasing. The observation which began in 1958 has clearly shown that the concentration of CO_2 in the atmosphere has increased from about 315ppm to about343ppm in 1984 (Nyong, 2007). We know today that approximately, the amounts of CO_2 that have been emitted into the atmosphere by fossil fuel combustion and changing land use (deforestation and expanding agriculture) can related to the observed increase of atmospheric CO_2 to these human activities. Since a continued increase of atmospheric CO_2 concentration might lead to changes of global climate, it is essential to be able to project the likely future concentration that may occur due to various possible rate of CO_2 emissions .As far as the expected climatic change is concerned, it can be prognoses that a doubling of the CO_2 concentration would lead to an increase of the globally averaged temperature by 1.5-4.5 ^{o}C (Ufarana, 2004). The prediction that climate change due to human activities began with a prediction made by the Swedish Chemist, Svant Arrhenius, in 1896. Arrhenius took note of the industrial revolution and realized that the amount of carbon dioxide being released into the atmosphere was increasing. He further believed that carbon dioxide concentrations will continue to increase as the world's consumption of fossil fuel particularly coal and fossil fuel increased even more rapidly. His understanding of the role of carbon dioxide in heating the earth led him to predict that if the atmospheric carbon dioxide is doubled, earth would become several degrees warmer (Ufarana, 2004).

1.1.2. Acid rain formation

Acid rain is commonly used to describe the deposition of acidic components in rain, snow, fog, dew or dry particles .(Ufarana, 2004). The primary sources are sulphur dioxide, carbon dioxide and oxide of oxides of nitrogen (Uyigue and Agho, 2007). A variety of industrial processes, such as the production of iron and steel, utility factories, oil producing companies are responsible for the emissions of obnoxious gases that resulted into acid rain. This report focused on the contribution of oil exploration and exploitation in the Niger-Delta area of Nigeria to the formation of acid rain. It has been reported that acid rain is one of the most discussed modes of environmental pollution in recent years (Odigure and Abdulkareem, 2001). This is caused by the considerable quantities of Sulphur dioxide and Nitrogen oxide released into the atmosphere as fossil fuel is burnt (SPDC Report, 1981). These pollutants combine with water vapour contained in the air to form dilute solution of sulphuric acid, which are subsequently washed out of the atmosphere by rain or other type of precipitation such as fog and snow (Thompson, 1991). The carbon (iv) oxide (CO_2) gas released during gas flaring also dissolves in water to form carbonic acid (H_2CO_3), the concentration of the gas in water depends on the extent of dissolution of gas in water (Abenege, 2004). Although, acid rain usually consists of relatively mild acids, they are sufficiently caustic to do great harm over time to certain natural ecosystems. Already there is much evidence that deposition of acid leads to lakes and stream acidification, stunted growth of flora while acid sensitive crops will die in affected areas (Abdulkareem, 2000). In cities, the corrosions of buildings and monuments are both exacerbated and accelerated (Odigure and Abdulkareem, 2001). To some extent acid rain has always been present in certain humid environment, originating from natural events as volcanic eruption, forest fires and even the bacterial decomposition of dead organism (Odigure and Abdulkareem, 2001). However, as the worldwide industrial revolution spread globally, the destructive capabilities of natural acid rain have been enhanced by human actions. The phenomenon of acid rain due to the dissolution of these acid gases in the atmospheric water in the sky so that it becomes acidic thereby leading to formation of rain with pH below 5.0 as shown in Equations 1-3 (Abdulkareem, 2000). Its effect is usually by its corrosion of rooftops, discoloration of paints on building, premature rusting of metallic object, damage to flora and fauna (Plant productive parts).

$$CO_2(g) + H_2O\ (l) \longrightarrow H_2CO_3(aq) \tag{1}$$

$$SO_3(g) + H_2O\ (l) \longrightarrow H_2SO_4(aq) \tag{2}$$

$$NO_2(g) + H_2O(l) \longrightarrow HNO_2(aq) + HNO_3(aq) \quad (3)$$

1.1.3. Ozone layer depletion

Increasing concentration of the synthetic chemicals known as Hallon (Bromine fluoro carbons) and Chlorine Fluoro Carbons (CFC'S) are known to enhance the breaking down of the ozone layer, allowing more of the ultraviolet rays to penetrate to the earth surface (Ifeanyichukwu, 2002, Abenege, 2004; Ufarana, 2004). Ultraviolet rays can break apart important biological molecules including Deoxyribonuclectic acid (DNA). Increased ultraviolet radiation can lead to greater incidence of skin cancer and immune deficiencies as well as decreased crop yield and reduced population of certain fish larvae Pluto Plankton and Zooplankton that are vital to the food chain. Increased ultraviolet radiation can also lead to smog and reduced the useful life of paints and plastics. Stratospheric ozone protects oxygen at lower altitude from being broken up by ultraviolet light and keeps most of those harmful rays from penetrating to the earth's surface (Abdulkareem, 2000) . Hallons are an industrial group of chemical that contains bromine, which reacts in a manner similar to chlorine by catalytically destroying ozone. Halons are used primarily in fire extinguishing foam. Chloro Fluoro Carbons (CFC'S) are compounds that consist of chlorine, fluorine and carbon, first introduced in the late 1920s, these gases have been used as coolant for refrigeration and air conditioners, propellant for aerosol sprays, agents for producing plastic foam and cleaners for electrical parts (Abdulkareem, 2000; Odjugo, 2011). CFC'S do not degrade easily in the troposphere as a result they raise into the stratosphere where they are broken down by ultraviolet light. The chlorine when liberated reacts with ozone to produce two molecules of oxygen. In the upper atmosphere, ultraviolet light breaks off a chlorine atom from a CFC'S molecule. The chlorine radicals attack ozone molecules and breaking it into an ordinary oxygen molecule and a molecule of chlorine monoxide, while the free oxygen radicals breaks up the chlorine monoxide as shown in Equations 4-7 (Abdulkareem, 2000). The freed chlorine radical is again available to repeat the process. Chlorine acts as catalyst and is unchanged during the process. Each chlorine atom can destroy as many as 10,000-ozone molecules before it returned to the troposphere.

$$CFCl_3 \rightarrow FCl_2 + Cl + C \quad (4)$$

$$Cl + O_3 \rightarrow ClO + O_2 \quad (5)$$

$$O_2 \rightarrow 2O \quad (6)$$

$$ClO + O \rightarrow Cl_2 + O_2 \quad (7)$$

1.2. Effect of heat radiation from gas flaring

There is no doubt that the Nigerian oil industry has affected the country in a variety of ways at the same time. It has fashioned a remarkable economic landscape for the country but on the other hand, ever since the discovery of oil in Nigeria in the 1950s, the country has been suffering the negative environment consequences of oil development, these negative impacts precipitated by the introduction of its own unwanted by-products into the environment may be catastrophic if allowed to build-up and unattended to (Nyong et al., 2007). The growth of the country's oil industry, combined with a population explosion and a lack of environmental regulation has led to substantial damage to Nigeria's environment predominantly the Niger-Delta region, the centre of the country's oil industry (Odjugo, 2010). For example, the rampant flaring of natural gas in the Niger Delta during oil production is the main culprit making natural gas the main source of carbon emissions in Nigeria (Abdulkareem et al., 2009). The people in most oil communities have to live with gas stacks that flare gas 24 hours a day at a temperature of 13 – 14,000°C. In 1994, these gases flared according to World Bank Report produced 35 million tons of CO_2 and 12 million tons of methane more than the rest of the world (UNDP/World Bank, 2004). This makes the oil industry in Nigeria the single biggest source of global warming in the world. Therefore the impact of gas flaring in Nigeria is of local and global concern (Uyigue and Agho, 2007). Even in the immediate environment of these flares; amidst conflicting claims, field evidence seems to support the widespread postulation that flaring apart from human impacts has a direct relationship with heat radiation and elevation of temperature (Odigure et al, 2003). Typical gas flare in the Nigerian oil field are located at the ground level as shown in Fig 1 and surrounded by thick vegetation, farmland and villages huts 20-30m from the flare station (Odigure et al., 2003). The heat radiation from the flare station is a function of the flare temperature, gas flow rate and the geometrical design of flare stack (Odigure et al., 2003).

The combustion of gaseous hydrocarbons contained in the natural gas is an exothermic process, which result in the evolution of heat to the atmosphere (Abdulkareem et al., 2009). This endangers both the plant and animal life around the vicinity gas flaring stations. Although the mechanism of radiant energy transfer is not completely understood, however the associated phenomenon is explained in terms of dualistic theory (Abdulkareem and Odigure, 2002). This theory deals separately with the emission and reception of radiation and with its transmission, radiation is emitted and received in discrete particle called photon. The geometric relationships between emitting and receiving surfaces have been kept very simple by arranging that emitting surface sees only the receiving surfaces. In practice it is not possible for heat radiated to strike the receiver completely. This necessitated the introduction of geometrical factor to relate the radiant energy striking a surface to the total radiant energy emitted (Abdulkareem and Odigure, 2002). While the concept of heat as a pollutant may seem impossible on a cold winter day, it is important to note that at any time of the year, an increase in water temperature has effect on aquatic life (Odigure et al., 2003). Heat pollution is a consequence of the rising in energy demand by man's activities. For instance the power plants burns fossil fuel or nuclei fuel to provide the energy needed for industrial consumptions, they released considerable amounts of heat (Abdulkareem, 2000). The power plants that release these huge amounts of energy are located near bodies of water, which the plant used for heat dissipation purposes. Living things especially cold blooded animal i.e. fish are very sensitive to small change in the average temperature. It has been reported that fish hatch its eggs before the hatching period due to change in water temperature. It may also prevent fish eggs from hatching at all (Abdulkareem, 2009; Odigure et al., 2003). In addition, a small rise in average temperature could produce a profound climatic change. Some experts' belief a small rise in temperature would cause the Greenland and Antarctic ice to melts, raising ocean levels and inundating large areas of land (Abdulkareem., 2009). The average worldwide temperature can be affected when the products of combustion such as carbon monoxide, water vapour and carbon dioxide are emitted in large quantities into the atmosphere (Tzimas et al., 2007). Although, solar energy on its way to the earth surface easily pass through the layers of carbon dioxide, some of the heat escaping through the earth will be absorbed by the increase amount of carbon emitted to the atmosphere by the process industries and releases back to the earth with negative impact on the soil (Abdulkareem and Odigure, 2002). For instace, the heat from gas flared falls on the soil thereby heating it up, increases in heat deposition on soil reduces diseases generally, it may not be suitable for some plants and crops to survive hence rendering such land unsuitable for cultivation (Aduku, 1997). There is therefore, a great physiological impact on crops planted in the vicinity of the gas flares station. It has been reported that there could be about 100% loss in yield of crops cultivated 200m away from the flares stations, 45% loss in yield of crops at 600m away and 10% loss in yield of crops cultivated 1000m away (Oyekunle, 199).

2. Development of mathematical modelling

The burning process of natural gas also referred to as combustion is described as a rapid oxidation or burning of substances with simultaneous evolution of heat. In the case of common fuels, the process is one of the chemical combinations with atmospheric oxygen to produce as the principal product. Gas flaring of produced gas i.e the process of burning-off surplus combustible vapours from a well, either as a means of disposal or as a safety measure to relieve well pressure - is the most significant source of air emissions from offshore oil and gas installations. Hence gas flaring activity in the Niger- Delta area, and the pollutants released to the atmosphere is causing a lot of damage to the area. It is on this basis that a mathematical model that can quantify the quantity of heat discharged from gas flaring stations into the environment will be developed. The following assumptions were made in order to develop the mathematical model for the heat radiation from gas flaring:

i. The area is assumed to be a bed of soil i.e. of constant heat capacity.
ii. The intensity of the sun is uniform for a given area at a given time
iii. Heats from flares are used in vapourising water, retained by the soil and the remaining reflected back
iv. Combustion is incomplete in air
v. The area is a tropical forest

Below is the schematic diagram of heat radiation in a flare station.

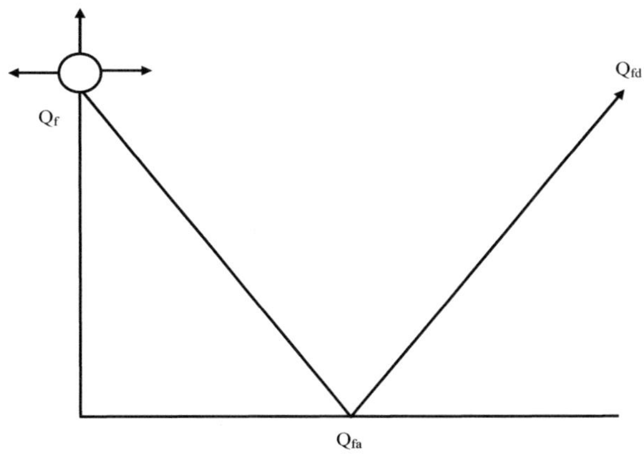

Figure 2. Schematic of heat radiation from gas flared station

Heat balance

Taking heat balance from Fig 2

$$Q_f = Q_{fd} + Q_{fa} \tag{8}$$

Where

- Q_f = heat from flare gases
- Q_{fa} = heat absorbed by earth from flare gas
- Q_{fd} = heat from flare reflected

From assumption (iii) i.e. heat from flares are used in vaporizing water, retained by the soil and the remaining reflected i.e.

$$Q_{fa} = Q_s + Q_v \tag{9}$$

Where

- Q_s = heat retained by soil
- Q_v = heat used in vapoursing water

Substituting equation (9) into (8) gives

$$Q_f = Q_{fd} + Q_s + Q_v \tag{10}$$

Rearranging the variables to make Q_{fd} the subject of the formula gives

$$Q_{fd} = Q_f - Q_s - Q_v \tag{11}$$

Where

$$Q_s = M_s C_s \int_{T_{soil}}^{T_s} dT \tag{12}$$

From equation (12)

$$Q_s = M_s C_s [T_s - T_{soil}] \tag{13}$$

Where

- M_s = mass of soil
- C_s = specific heat capacity of soil
- T_s = temperature of flare gas
- T_{soil} = temperature of soil

$$Q_v = M_w C_w \int_{T_{soil}}^{T_s} dT + M_w \lambda_v \tag{14}$$

From equation (14)

$$Q_v = M_w C_w [T_s - T_{soil}] + M_w \lambda_v \tag{15}$$

Where

M_w = mass of water, C_w = specific heat capacity of water, λ_v = latent heat of vapourisation of water

According to Albedo, a fraction of the heat radiated from the source strikes the receiving surface. (Andy, 2003). Therefore,

$$Q_c = \alpha Q_f (1 - a) \tag{16}$$

Where

α = absorptive factor which varies with distance, a = Alhedo constant, Q_c = fraction of the heat which strikes the receiving surface

Hence,

$$\alpha Q_f (1 - a) = M_w C_w (T_s - T_{soil}) + M_w \lambda_v + M_s C_s [T_s - T_{soil}] \tag{17}$$

Substituting equations (13), (15) and (17) into equation (11) gives

$$Q_{fd} = Q_f - [M_s C_s [T_s - T_{soil}] + M_w C_w [T_s - T_{soil}] + M_w \lambda_v] \tag{18}$$

Similarly, from Equations 11 and 16;

$$Q_{fd} = Q_f - \alpha Q_f (1 - a) \tag{19}$$

Rearrange Equation 19 to obtain

$$Q_{fd} = Q_f (1 - \alpha (1 - a)) \tag{20}$$

Evaluation of Q_f

Ufarana (2004) suggest that that the flame from is titled at 45°, hence

$$h_{fv} = L(\sin 45^{o}) = 0.707L \tag{21}$$

Where h_{fv} is the vertical height vector of a flare stack

And L is the flame length

From Equation 21

$$L = \frac{h_{fv}}{0.707} \tag{22}$$

The vertical height vector of the flare stack (hfv) can also be calculated from (Ufarana, 2004)

$$h_{fv} = 0.0042Q_f^{\,0.478} \tag{23}$$

From Equations 22 and 23;

$$L = \frac{0.0042}{0.707} Q_f 0.478 \tag{24}$$

$$\therefore\ L = 0.00594Q_f^{\,0.478} \tag{25}$$

From Steward's correlating equation (Ufarana, 2004)

$$L = 0.8632Q_f^{\,0.4}N' \tag{26}$$

Where,

$$N' = \text{a combustion parameter} = \frac{\left(\dfrac{r + w\rho_a}{\rho}\right)^{0.4}}{[(NHV)^{0.4}(1 - w)]} \tag{27}$$

Where

$$w = \text{combustion parameter} = \frac{rC_pT_a}{(rC_pT_a + NHV)}$$

NHV = flared gas net heating value, Btu/lb

r = stiochiometric air fuel ratio of flared gas, T_a = air temperature, ρ_a = ambient air density
ρ = fuel density

Equating (25) and (26) gives

$$0.00594Q_f^{\,0.478} = 0.8632Q_f^{\,0.4}\,N' \tag{28}$$

From Equation (28)

$$Q_f^{\,0.078} = 145.32N' \tag{29}$$

From Equation 27

$$N' = \frac{\left(\frac{r + \left(\frac{rC_pT_a\rho_a}{rC_pT_a + NHV} \right)}{\rho} \right)^{0.4}}{(NHV)^{0.4}\left(1 - \frac{rC_pT_a}{rC_pT_a + NHV} \right)} \quad (30)$$

Rearrange Equation 30 to obtain;

$$N' = \left(\frac{(rC_pT_a(r\rho + \rho_a) + r\rho NHV)^{0.4}}{\rho^{0.4}(NHV)^{1.4}} \right)(rC_pT_a + NHV)^{0.6} \quad (31)$$

Substituting Equation 31 into Equation 29 gives

$$Q_f^{0.078} = 145.32\left(\frac{(rC_pT_a(r\rho + \rho_a) + r\rho NHV)^{0.4}}{\rho^{0.4}(NHV)^{1.4}} \right)(rC_pT_a + NHV)^{0.6} \quad (32)$$

From Equation (32)

$$Q_f = \sqrt[0.078]{145.32\left(\frac{(rC_pT_a(r\rho + \rho_a) + r\rho NHV)^{0.4}}{\rho^{0.4}(NHV)^{1.4}} \right)(rC_pT_a + NHV)^{0.6}} \quad (33)$$

Substituting Equation 33 into Equation 20 gives

$$Q_{fd} = \left(\sqrt[0.078]{145.32\left(\frac{(rC_pT_a(r\rho + \rho_a) + r\rho NHV)^{0.4}}{\rho^{0.4}(NHV)^{1.4}} \right)(rC_pT_a + NHV)^{0.6}} \right) \times (1 - \alpha(l - a)) \quad (34)$$

Relationship between net heating value and distance from flare point is given as (Ufarana, 2004)

$$NHV = \frac{mc\theta}{x} \quad (35)$$

Wherem= mass of flared gas, c = heat capacity of flared gas, θ = temperature of flared gas, x = distance

Substituting Equation 35 into Equation 34 to obtain

$$Q_{fd} = \left(\sqrt[0.078]{145.32 \left(\frac{\left(rC_pT_a x(r\rho + \rho_a) + \rho mc\theta \right)^{0.4}}{\rho^{0.4}(mc\theta)^{1.4}} \right) (rC_pT_a + mc\theta)^{0.6} x^{0.4}} \right) \times \quad (36)$$

$$(1 - \alpha(l - a))$$

But

$$m = \rho_T V_T \quad (37)$$

- Wherem = mass of flared gas
- V = volume of flared gas
- ρ_T = density of gas produced by gas flaring
- V_T = volume of gas produced by gas flaring which can be calculated as follows:

The volume of obnoxious gases produced such as CO, CO_2, NO_2, SO_2 and THC when Vm^3 of gas is flared was estimated on the assumption that combustion of the associated gas is incomplete in air as shown in Equations 38-44. The calculation was also based on the compositions of associated gas in crude oil at the flare stations (Table 1).

Component of gas flared	Percentage of Composition
CH_4	47
C_2H_4	18
C_3H_8	20
C_4H_{10}	5
C_5H_{12}	9
H_2S	0.03
N_2	0.022
Others	0.068

Table 1. Component of gas flared (Abdulkareem, 2000)

Evaluation of volume of gas flared was estimated on the assumption that combustion is incomplete in air with the following reactions take place during the process of combustion.

$$2CH_4 + \frac{7}{2}O_2 \rightarrow CO_2 + CO + 4H_2O \quad (38)$$

$$C_2H_6 + 3O_2 \rightarrow CO_2 + CO + 3H_2O \quad (39)$$

$$C_3H_8 + 4O_2 \rightarrow CO_2 + 2CO + 4H_2O \quad (40)$$

$$C_4H_{10} + 5O_2 \rightarrow CO_2 + 3CO + 5H_2O \quad (41)$$

$$C_5H_{12} + \frac{13}{2}O_2 \rightarrow 2CO_2 + 3CO + 6H_2O \quad (42)$$

$$H_2S + O_2 \rightarrow SO_2 + H_2 \quad (43)$$

$$N_2 + 2O_2 \rightarrow 2NO_2 \quad (44)$$

Basis: $1m^3$ of flared gas

Let S_E = stack energy

The total volume of gas produced by flaring $1m^3$ of gas = volume of CO_2 + volume of CO + volume of NO_2 + volume of SO_2 + volume of THC

$$= 0.00845S_E + 0.1235S_E + 0.000003S_E + 0.00004S_E + 0.99 - 0.00990S_E =$$
$$= (0.99 + 0.0109074S_E)m^3 = V_T \quad (45)$$

Equation (45) represents the total volume of gas produced by flaring $1m^3$ of gas. But when Vm^3 of gas is flared equation (45) becomes

$$(0.99 + 0.0109074S_E)Vm^3 = V_T \quad (46)$$

From Equation 37 i.e. $m = \rho_T V_T$

Substituting Equation (46) into Equation (37) gives

$$m = \rho_T(0.99 + 0.0109074S_E)V \quad (47)$$

Substituting Equation 47 into Equation 36 gives

$$Q_{fd} = 0.078 \sqrt{145.32 \left(\frac{[rC_pT_ax(r\rho + \rho_a) + \rho\rho_T(0.99 + 0.0109074S_E)VC\theta]^{0.4}}{\rho^{0.4}\left(\rho_T(0.99 + 0.0109074S_E)VC\theta\right)^{1.4}} \left(rxC_pT_a + \rho_T(0.99 + 0.0109074S_E)VC\theta\right)^{0.6} x^{0.4} \right)}$$

$$(1 - \alpha(1 - a)) \quad (48)$$

Where $\theta = T_s - T_a$

$$\rho = \frac{\rho_a T_a}{T_s}$$

Equation (48) is the model equation for the heat reflected due to gas flaring.

3. Results and discussion of results

Environmental pollution has transcended natural boundaries; stratospheric ozone depletion, global warming, the green house effect, deforestation, acid train and mega disaster are some of the various environmental problems attributed to pollution. The potential effects of global pollution have necessitated global cooperation in other to secure and maintain a live able global environment (Odjugo, 2010). It has been reported that pollutants emitted from one country can easily cross political boundaries. People are beginning to recognize that pollutants can affect not just a region but the entire planets. Modern industrial society creates far more carbon (IV) oxide (CO_2) that what the planet vegetation can consume (Odigure and Abdulkareem, 2001). As the excess CO_2 rises into the atmosphere, it acts as absorptive body, which trap heat reflected from the earth surface. Scientists accept that green house effect from increased level of CO_2 and other heat trapping gases eventually will cause an increase in global temperature. Some predicted that the temperature will rise significantly within the next century and that global pattern could be drastically disrupted. Air pollution is not restricted to outdoor air, although relatively little attention is given to the hazards of many substances found in indoor air most especially in the developing nations where there is no proper regulation in place to combat air pollution. But it is however well established that people may spend as much as 80-90% of their time indoor. The sources of indoor pollution are different for developing and industrialized country (Odigure and Abdulkareem, 2001). In developing countries, indoor pollution comes mainly from using biomass fuels (Wood, agricultural waste, dung etc.) for cooking and heating. The majority of the world's population depends on biomass for most of their energy supply. It is estimated that as many as 400-500 million people mainly in the rural area of developing countries and primarily women and children may be adversely affected by indoor pollution (FEPA Report, 1998). The fuel is burnt inefficiently in rooms that poorly ventilated. Biomass smoke contains numerous substances, the most hazardous of which include suspended particulate matter, nitrogen dioxide, carbon dioxide and sulphur dioxide (Gwendolyn et al, 1993). It also releases a number of aldehyde. While the key indoor pollutants in an industrialized countries are nitrogen dioxide, carbon monoxide, radon (from building material), formaldehyde (from insulator, asbestos, mercury, manmade fibers etc. Also polluting the environment is the heat radiated as result of these indoor activities and process industries. The released may be harmful to plant and animal. In Nigerian context, gas flaring from the oil exploitation and exploration in the Niger-Delta area of the country has been considered as the major sources of environmental pollution. Every day in southern Nigeria, almost 2million cubic feet of natural gas is burnt (flared) during crude oil production, more than is flared anywhere else in the world (Oni and Oyewo, 2011). Hence, gas flaring is not only wastes of valuable resources, but is also a major cause of environment pollution in the Niger-Delta, where most of Nigeria's oil output is produced. Nigeria has a population of over 170million people and an abundance of natural resources especially hydrocarbons. The Nigerian economy is largely dependent on its oil sector which supplies 95% of its foreign earnings. While the exploitation and exploration of oil has created some fortunes and contributed positively to the economic and technological advancement of Nigeria as a country, the accompanying socio-economic and ecological fallouts remain problematic. The public considers the oil producing companies operating in the Niger-Delta oil fields responsible as major environmental pollutants by way of relentless flaring and venting of gas in the environment, oil spillages, site clearing, deforestation and destruction of flora and fauna, and disturbances of the ecosystem in the 70,000 square kilometres Niger-Delta wetland (Oguejifor, 1993). Gas flaring in Nigeria today has posses an environmental hazard to the nation at large. So much damage is being done to the environment through gas flaring, that if nothing is done in a few years from now, serious environmental and health problems such as premature death and diseases will emerge. It is therefore, on this ground that a mathematical model that can quantify the amount of heat radiation from a flare stack is developed in this work. The model will assist in estimating the quantity of heat migration from gas flaring as a function of flare temperature, gas flow rate and geometric design (Efficiency) of the flare stack. Data gathered on the rate of flaring of gas and the measured quantities of heat radiation are presented in Tables 1 and 2. While the simulated results at different conditions for a period of one year for gas flare stations 1 and 2 are presented in Figs. 2-7

Table 2 present the average volume of gas flared from two flare stations per month for a period of one year. Results as presented indicate non uniformity in the volume of gas flared by the flare station per months. For instance, the volume of gas flared in station 1 for the month of May was $2.03m^3/sec$ while the volume of gas flared in the same station in the month of June was $1.65m^3/sec$. Results also reveal variation in the rate of gas flaring by the two stations investigated, with average rate of gas flaring by station 2 higher than that of station 1. The variation in the rate of gas flaring as presented can be attributing to the variation in the rate of production of crude by the two stations. The crude oil obtained from wells is a mixture of crude oil itself, water and natural gas. At the flow stations, the components are separated and the gas that not be contained is flared. Hence when the rate of crude oil produced increases, the quantity of gas flared will also increases.

	Volume of gas flare (m^3/s)	
Months	Station 1	Station 2
Jan	1.78	2.79
Feb	2.10	3.16
Mar	1.52	2.78
April	1.56	3.21
May	2.03	2.91
June	1.65	2.66
July	1.74	2.64
Aug	1.39	2.87
Sept	1.92	3.01
Oct	1.30	2.56
Nov	0.99	2.81
Dec	1.75	2.97

Table 2. Flow rate of gas flared per month

Presesnted in Table 3 are the measured values of heat radiation per month by station 1 for a period of eight months. The measured values showed that 0.805×10^{-3} kW/m^2 of heat was radiated at a distance of 100m for the month January, while 0.050×10^{-3}kW/m^2 of heat was radiated at the same distance for the month of February. Values of heat radiation presented in Table 3 do not show any distribution pattern with seasons. The un-pattern nature of heat radiation per season could be attributed to the volume of gas flared, distances from flared point and wind speed. However, the measured heat radiation to large extent conforms to the physical law of pollutant dispersion from the generating sources.

	Heat Radiation (kW/m^2) ×10^{-3}							
Months	Jan	Feb	Mar	April	May	June	Sept	Dec
Distances (m)								
100	0.805	0.553	0.050	3.016	2.672	2.735	1.527	0.837
150	0.716	0.514	0.076	2.465	2.429	1.994	1.668	0.725
200	0.620	0.440	0.075	1.010	3.034	1.697	2.708	0.620
250	0.330	0.332	0.075	3.916	2.175	1.682	2.262	0.522
300	0.319	0373	0.094	0.312	1.675	1.520	1.929	0.457
500	0.558	0.237	0.091	0.979	1.262	1.519	1.849	0.305

Table 3. Heat radiation from gas flaring station 1

Simulation of the model is the use of computer code to show the operation and behavior of the system. The model equation developed for the heat radiation from gas flaring was simulated using Q-basic programme. The results obtained are presented in Figs 3-8. The simulated results values are obtained at various distances ranging from 25m to 1500m for different volume of gas flared and at different stack efficiencies of 65%, 75% and 85%. The choice of these 3 stack is as favoured by the world bank report that flare stack efficiency in the flare station of Niger-Delta area of Nigeria is 75%. It can be seen from the simulation results presented that the heat radiation from gas flaring for different stations increases with increase in volume of gas flared and stack efficiency, while the quantity of heat radiated reduces with increase in distance from the flare point. For instance in the month of May at a distance of 100m the quantity of heat radiated is 0.00262287 kW/m^2 while at a distance of 200m the quantity of heat radiated is 0.0023729 kW/m^2 for station 1. Results as presented also indicate that at a stack efficiency of 64% in the month of May, the quantity of heat radiated at a distance of 100m is 0.00262287 kW/m^2, while for that same month at the same distance in the same station when the stack efficiency is 74%, the quantity of heat radiated is 0.00279044 kW/m^2. Simulation results presented also indicates that the highest quantities of heat is radiated in the month of June for station1 and this as a result of the fact that highest amount of gas flared in this station during this month. Results as presented in Figs for station2 indicates same pattern of results as obtained in station1. It could be observed from the experimental and simulated results that the habitants of the Niger-Delta area of Nigeria are exposed to serious environmental risk based on the quantities of heat released into the environment from the gas flare stations. Results obtained support the claims by the researchers that the fire form gas flaring stations generate constant heat, which in turn evaporate water produced around the flare point, thus increasing the salinity of the pool water. There is also an evidence including the observation by the local farmers that the flare is considerably diminishes the value of agriculture productivity. Other observation made in the Niger-Delta area as a negative consequence of heat radiation from gas flaring is that the game animals were scared away by the fire.

Comparison of experimental results with simulated results showed that there is variation between them. The variation between experimental and modeling simulation results could be attributed to the following factors.

1. The un-patterned nature of the experimental data, which could be attributed to the fact that the weather conditions and rate of rainfall are not constant throughout the season. As a result, the heat will accumulate in the air during no rain period. However, when it rains part of the accumulated heat pollutant will be washed away
2. Experimental values are a measure of the extent of atmospheric pollution because of possibility of accumulation as stated above. While the simulation results are an instantaneous value i.e. it measured the possible amount of heat that could be released during flaring at a given time.
3. Atmospheric conditions such as wind speed, humidity, temperature e.t.c affect the dispersion and dilution of heat radiation from flare stack as a function distance.
4. The variation in experimental and simulations values could also be attributed to some assumption made at the initial stage of the modeling, such as wind speed, weather condition, volume of gas flared e.t.c. These assumptions may not conform to prevailing atmospheric condition.

Despite the variation between the experimental and simulated results, the dispersion pattern of the obtained values from experimental and simulation showed that the model and experimental results to a large extent conforms to the modified physical law proposed by (Gwendolyn, 1993).

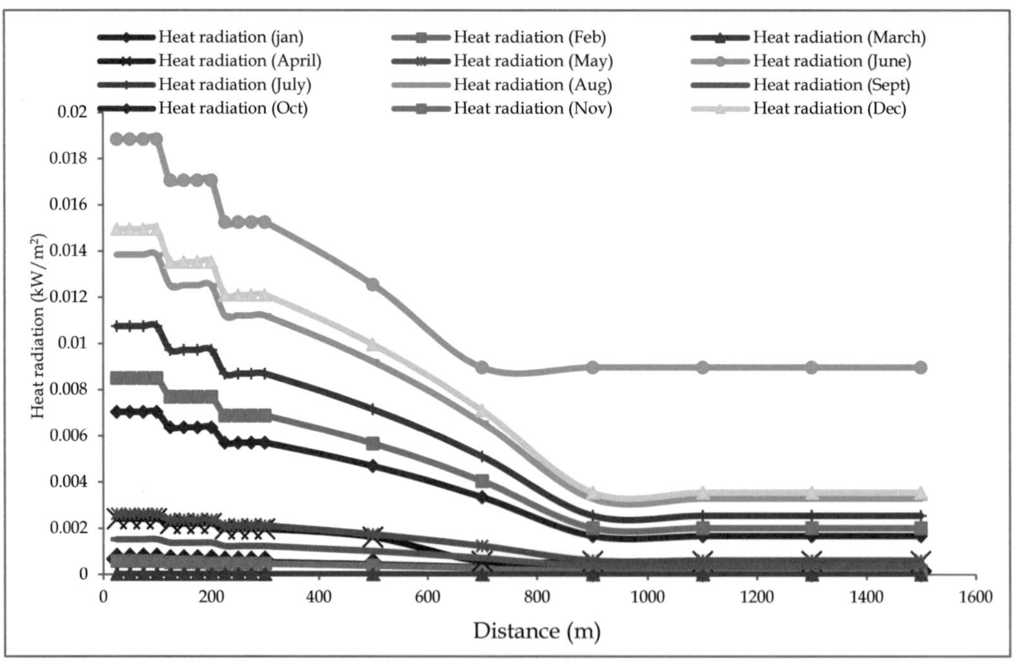

Figure 3. Heat radiation from gas flaring (Station 1) with a stack efficiency of 65%

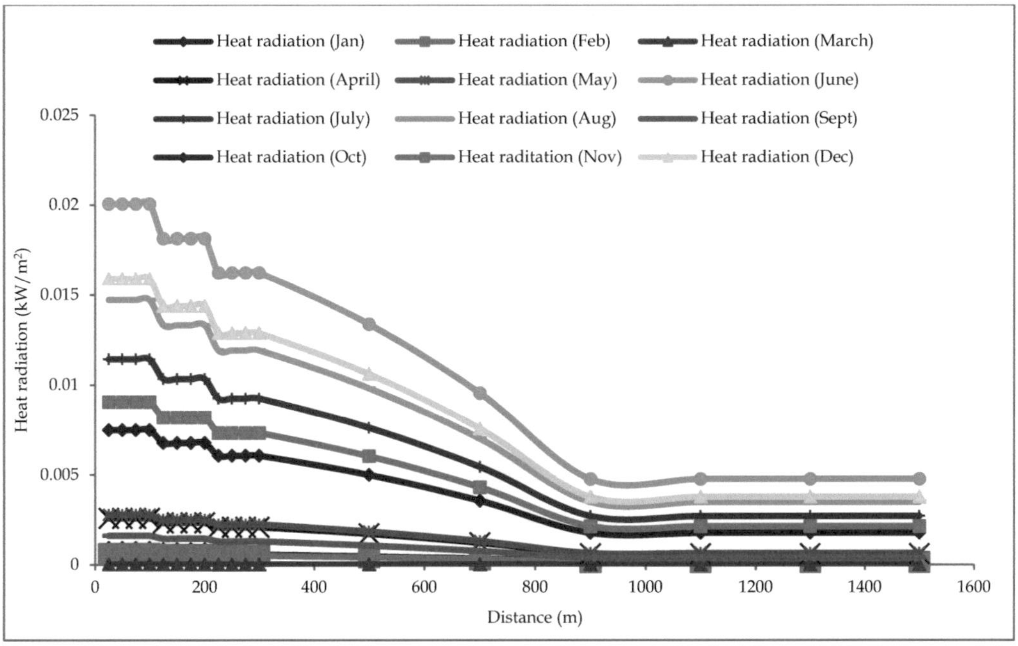

Figure 4. Heat radiation from gas flaring (Station 1) with a stack efficiency of 75%

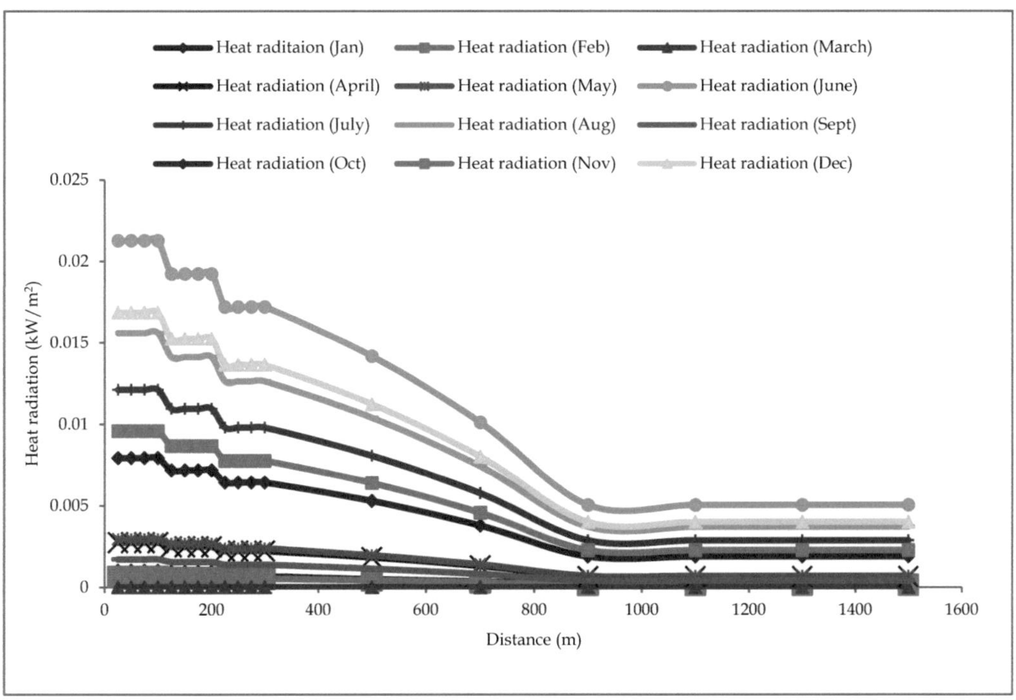

Figure 5. Heat radiation from gas flaring (Station 1) with a stack efficiency of 85%

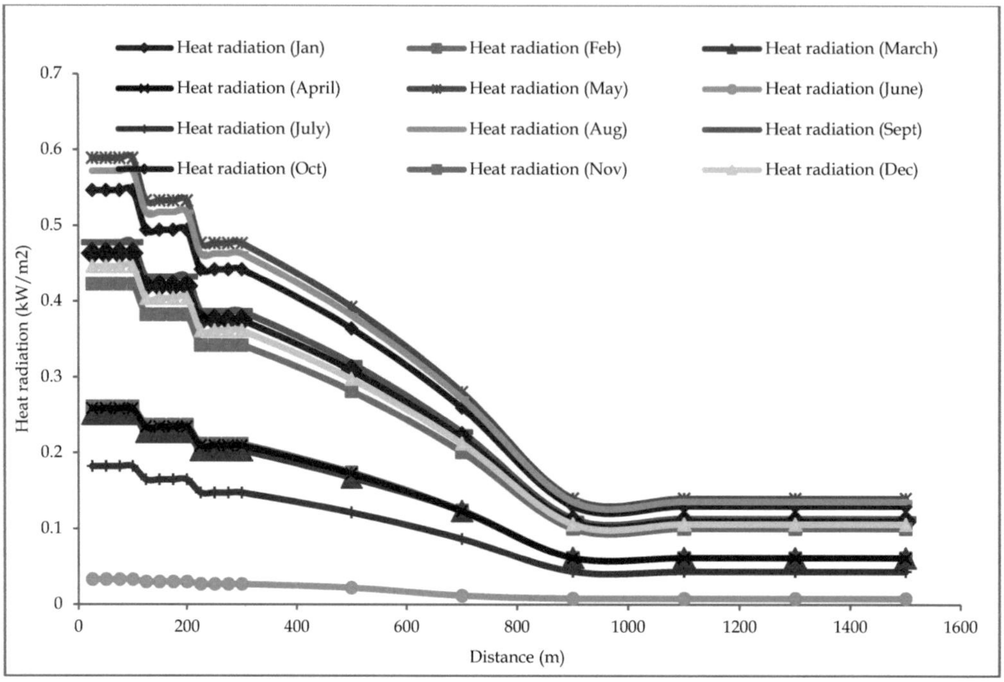

Figure 6. Heat radiation from gas flaring (Station 2) with a stack efficiency of 65%

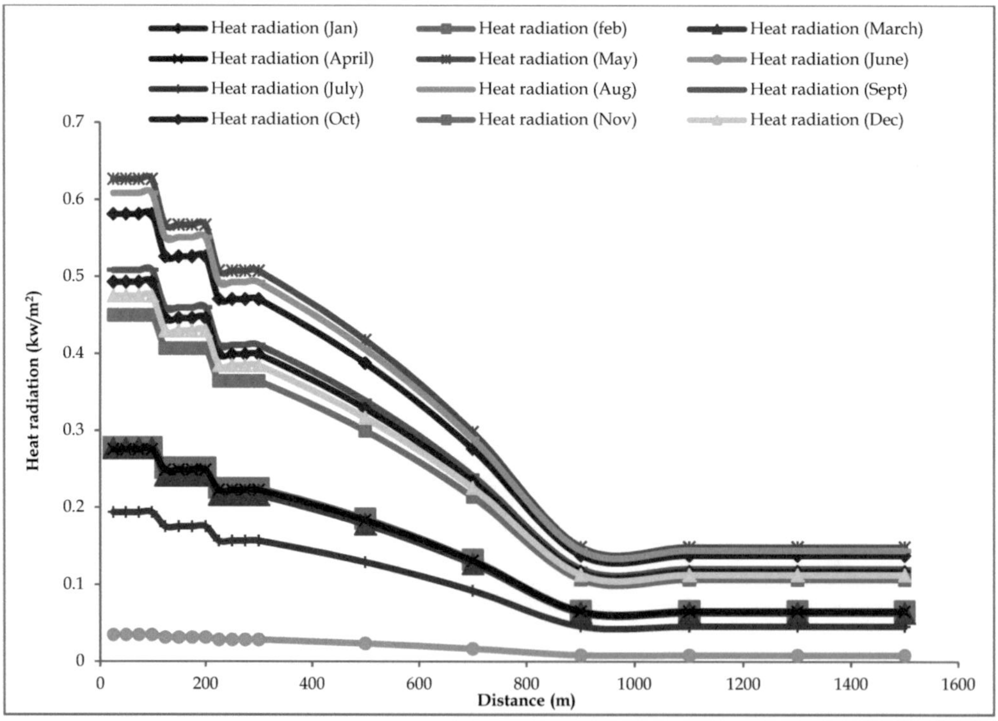

Figure 7. Heat radiation from gas flaring (Station 2) with a stack efficiency of 75%

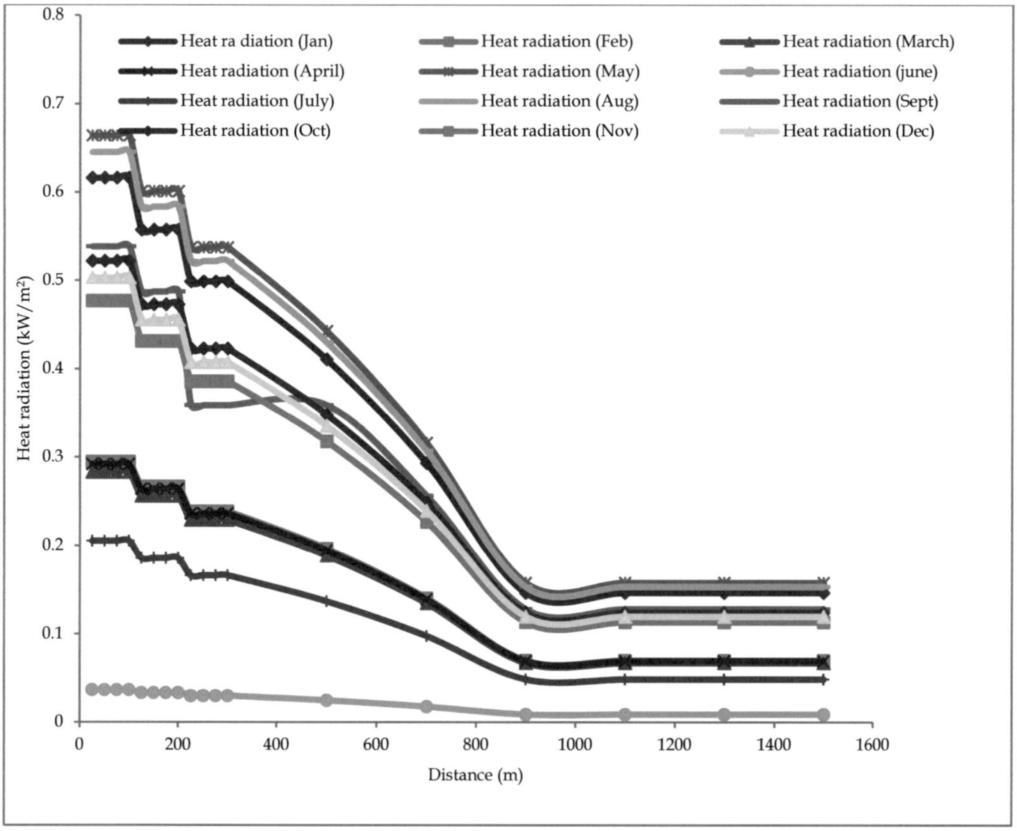

Figure 8. Heat radiation from gas flaring (Station 2) with a stack efficiency of 85%

4. Conclusion

Environmental pollution due to heat radiation from gas flaring stations in the Niger-Delta area of Nigeria has been identified as one of the major causes of strives, demonstration and sometimes-violent protest between the oil exploration companies and habitants of Niger-Delta area of Nigeria. Experimental analysis of heat radiation from gas flaring has been conducted. Attempt at modeling heat radiation from flare station using q basic program is hereby presented. It can be inferred from the simulation results of the developed model that volume of gas flared considerably affects the quantity of heat radiation from gas flaring in a direct proportionate manner. Also influence the heat radiation from gas is the distances of measurement from the point of flare, as the distance increases the quantity of heat radiated decreases. It can be inferred from the results that the effect of heat radiated will be felt mostly felt at distances of 25-100m within the point if flare. The result also clearly show that show that continuous gas flaring irrespective of the quality deposited in the immediate environment will in the long run lead to change in the physicochemical properties of environment due to the quantity of heat radiated. From this research, the following conclusions can be deduced:

1. It was observed that the result of simulation of model developed based on the modified principles of pollutants dispersion agreed with the experimental results.
2. The dispersion pattern of heat radiation based on the simulation results showed that the extent of spread heat from the flare point is dependent on nearness to source of flaring, volume of gas flared and stacks efficiency.
3. Model equation that best represents pollutant dispersion pattern is:

$$Q_{fd} = \left(\sqrt[0.078]{145.32 \left(\frac{\left(rC_pT_a x(r\rho + \rho_a) + \rho mc\theta \right)^{0.4}}{\rho^{0.4}(mc\theta)^{1.4}} \right) (rC_pT_a + mc\theta)^{0.6} x^{0.4}} \right) \times (1 - \alpha(l - a))$$

Author details

Abdulkareem A.S.
Department of Chemical Engneering, School of Engineering and Engineering Technology, Federal University of Technology, Minna, Niger State Nigeria

Department of Civil and Chemical Engineering, College of Science, Engineering and Technology, University of South Africa, Private Bag X6, Florida, Johannesburg, South Africa

Afolabi A.S.
Department of Civil and Chemical Engineering, College of Science, Engineering and Technology, University of South Africa, Private Bag X6, Florida, Johannesburg, South Africa

Abdulfatai J.
Department of Chemical Engneering, School of Engineering and Engineering Technology, Federal University of Technology, Minna, Niger State Nigeria

Uthman H.
Department of Chemical Engneering, School of Engineering and Engineering Technology, Federal University of Technology, Minna, Niger State Nigeria

Membrane Research Unit (MRU), Block L-01, Universiti Teknologi Malaysia (UTM), International Campus, Jalan Semarak, WP, Kuala Lampur Malaysia

Odigure J.O.
Department of Chemical Engneering, School of Engineering and Engineering Technology, Federal University of Technology, Minna, Niger State Nigeria

5. Acknowledgement

We wish to acknowledge the support received from Federal University of Technology Minna, for successful completion of this research. National research foundation (NRF), South Africa and Faculty of Science, Engineering and Technology University of South Africa are also appreciated for their support.

6. References

Abdulakreem, A. S. & Odigure, J.O. (2002). Radiative Heat Evaluation from Gas Flaring By Computer Simulation. Journal of Association for the advancement of Modelling and simulation in enterprises, Lyon France. Vol.71, No 2, pp 19 – 35, ISSN 0761 – 2516.

Abdulkareem A.S. (2000). Mathematical modelling of pollutant migration from gas flaring in the Niger Delta Area. MsC thesis, Department of Chemical Engineering, Federal University of Technology, Minna. Nigeria (unpublished), 1-85.

Abdulkareem, A. S. (2005). Evaluation of ground level concentration of pollutant due to gas flaring by computer simulation: A case study of Niger – Delta area of Nigeria. Leonardo Electronic Journal of Practices and Technologies, Technical University of Cluj - Napoca Romania. Issue 6, pp 29 – 42, ISSN 1583 - 1078.

Abdulkareem, A. S.(2005). Urban Air Pollution Evaluation by Computer Simulation: A Case study of Petroleum Refining Company, Nigeria. Leonardo Journal of Science Technical University of Cluj - Napoca Romania. Issue 6, pp 17 – 28, ISSN: 1583 - 0233.

Abdulkareem, A.S & Odigure, J.O. (2006). Deterministic Model for Noise Dispersion from gas Flaring: A case study of Niger – Delta area of Nigeria. Journal of Chemical and Biochemical Engineering, Croatia Q 20, No 2, pp 139 – 146, ISSN 0352 – 9568.

Abdulkareem, A.S,; Odigure, J.O & Abenege, S (2010): Predictive model for pollutant dispersion from gas flaring: A case study of oil producing area in Nigeria. Journal of energy sources, Part A, 1004-1015. ISSN 1556-7036.

Abdulkareem, A.S. & Odigure. J.O. (2010): Economic Benefit of Natural Gas Utilization in Nigeria: A case study of Food Processing Industry. Journal of Energy Source Part B, Vol 5 Pp 106-114.

Abdulkareem, A.S.; Idibie, C.A.; Afolabi, A.S.; Pienaar, H.C.vZ. & Iyuke S.E. (2010): Kinetics of sulphonation of polystyrene butadiene rubber in sulphuric acid medium. Journal of International Review of Chemical Engineering, vol. 2, No7, pp 832-839, ISSN 2035-1755

Abowei, M.F.N.; Susu, A.A & Onyeme, J.O (1997). Oil spill in marine enviroment. CJC press (Nigeria) ltd, Lagos, Nigeria.1-50.

Adeniyi, O.D.; Kovo, A.S.; Abdulkareem. A.S & Chukwudozie. C. (2007): Ethanol Fuel Production from Cassava as a Substitute for Gasoline. Journal of Dispersion and Technology, vol. 28, No4, pp 501-504, ISSN 0193-2691.

Aghalino, S.O & Eyinal, B (2009). Oil exploitation and marine pollution: Evidence from Niger-Delta, Nigeria. Journal of human ecology, vol.28(3), 177-182.

Ajayi, O.M (1999): Impact of gas flaring on the Niger-Delta Area: A case study of Port-Harcourt. B.Eng degree project, department of Chemical Enginnering, Federal University of Technology, Minn. Nigeria (unpublished) 10-26.

Akpan, S (2009). The production and utilisation of natural gas resources in Nigeria: A review. Proceedings of annual international conference and exhibition of the Nigerian Society of Petroleum Engineers, Abuja, Nigeria.

Alakpodia, I.J (1980): The effect of gas flaring on the microclimate and adjacent vegetation in Isoko area. Master of Science thesis, University of Lagos, Nigeria (unpublished), 1-50.

Alameddine, I & El-Fadel, M (2005). Stack emission from desalination plant: A parametric senstivity analysis for exposure assessment. Journal of desalination, vol. 177, 15-29.

Baroutian, S.; Mohebbi, A & Goharrizi, S.A (2006). Measuring and modelling particulate dispersion: A case study of Kerman cement plant. Journal of hazrdous materials, vol. A36, 468-474.

Carvalho, J.C& De Vihema, T.M.B (2005): Pollutant dispersion simulation fro low wind speed condition by the ILS method. Journal of atmospheric enviroment, vol.39, 6182-6288.

Cermark, J.E (1985): Physical modeling of urban air pollutant transport. Journal of wind and industrial aerodynamics, vol.59, 51-67.

Chimaroke, A.C (2004): Urban air pollution by computer simulation. B.Eng degree project, department of Chemical Enginnering, Federal University of Technology, Minn. Nigeria (unpublished) 1-56.

Davidson, O.R (1998). The climate convetion and Kyoto agreements: Opprotunities for Africa. Proceedings of an international conference on enviroment, Elephant Hills, Victoria Falls. Zimbabwe.

Davies, S & Ebbe, K (1995). Traditional knowledge and sustainable development. Proceedings of international conference on enviromentally suistainable development, world bank, Washigton, DC.

Drake, B & Hubacek, K (2007). What to expect from greater geographic dispersion of wind farms?-A risk portfolio approach. Journal of energy policy, vol.35, 3999-4008.

Eludinni, O.I (2004): A review of eceonomic benefits of natural gas utilization in Nigeria. B.Eng degree project, department of Chemical Enginnering, Federal University of Technology, Minn. Nigeria (unpublished), 1-51.

Global gas flaring reduction initiative (2002). Report on consultaion with stakeholders, world bank group on collaboration with goverment of Norway. 27-30.

Ifeayichukwu, G. P (2001). Modelin of air pollution in industrial cities: Case study of Lagos, Nigeria, B.Eng degree project, department of Chemical Enginnering, Federal University of Technology, Minn. Nigeria (unpublished) 1-60.

Ikelegbe, O.O (1999). Pollution in Nigeria: Cause effect and control. Proceddings of Nigerian Geographical Association conference, Minna, Niger State, Nigeria. 18-22.

Indriani, G (2005). Gas flaring reduction in the indonesian oil and gas sector –Technical and economic potential of clean development mechanism (CDM) projects. Hamburg instituite of international economics, report 253, 1-141. ISSN 0179-2253.

Janes, J.; Armstrong, K,; Shirvill, L,; Emannuel, G & Monoplis, J(2000). Flaring & venting in the oil & gas exploration & production industry: An overview of purpose, quantities, issues, practices and trends. International association of oil & gas producers, report no:2.79/288, 1-18.

Joint UNDP/World bank energy programme sector management programme (ESMAP). Strategic gas plan for Nigeria, February, 2004.

Kinee, E.J.; Touma, J.S.; Mason, R.; Therman, J.; Belder, A.; Bailey, C & Cook, R (2004). Allocation of on road mobile emissions to road segments for air toxics modelling in an urban area. Journal of transportation research, partB9, 139-150.

Luyben, W.L (1999). Process modeling, simulation and control for chemical engineers. 2nd edition, McGraw-Hill, Singapore. 15-91.

Nigerian natural gas stratergy under supervison of teh world bank (2002). IPA Energy consulting and Co. 12-25.

Nwaichi, E.O & Uzazobona, M.A (2011). Estimation of CO_2 level due to gas flaring in the Niger-Delta. Research journal of enviromental sciences, vol. 5(6), 565-572. ISSN 1819-3412.

Nyelong, P.N (2004). Global warming and global water. Journal of energy and enviroment, vol. 17 (1), 79-90.

Nyong, A.; Adesina, F & Osman Elasha, B (2007). The value of indigeneous knowledge in climate change mitigation and adaptation strategies in the African Sahel. Journal of mitigation, adaptation and startegtic global change, vol.12, 787-789.

Odigure, J. O & Abdulkareem, A. S. (2001). Modelling of Pollutant's migration from gas flaring: A case study of Niger – Delta area of Nigeria. Journal of Association for the advancement of Modelling and simulation in enterprises, Lyon France. Vol.62, No 3, p 57 – 68, ISSN 0761 – 2524.

Odigure, J. O.; Abdulkareem, A.S. & O.D Adeniyi, O.D (2003). Computer simulation of soil temperature due to heat radiation from gas flaring. Association for the advancement of Modelling and simulation in enterprises, Lyon France. Vol. 72, No 6, pp 1 – 10, ISSN 1259 – 5969.

Odigure, J.O.; Abdulkareem, A.S& Adeniyi, O.D (2004): Mathematical modelling and computer simulation of noise radiation by generator. Assumption university journal of technology, vol 7(3), 111-119., ISSN 1513-0886.

Odjugo, A.O.P (2009). Quantfying the cost of climate change impact in Nigeria: Emphsis on wind and rainstorms. Journal of humna ecology, vol.28 (2), 93-101.

Odjugo, A.O.P (2010). Regional evidence of climate change in Nigeria. Journal of geography and regional planning, vol. 3(6), 142-150. ISSN 2070-1845.

Odjugo, A.O.P (2011). Climate change and global warming. The Nigeria perspective. Journal of suistainable development and enviromental protection, vol.1(1), 1-12.

Olukoga, E.A (2002). Natural gas as a vehicular fuel. Proceedings of Nigeria gas association, international gas conference exhibition, Abuja, Nigeria. 6.

Oni, I.S & Oyewo, A.M (2011). Gas flaring, transportation and susitanable energy development in the Niger-Delta, Nigeria. Journal of human ecology, vol. 33(1), 21-28.

Onyiah, I.M (2005): Modificcation and verification of Milton equation for measuring gaseous pollutant concentration. Post graduate diploma thesis, Department of Chemical Engineering , Federal University of Technology, Minna, Nigeria (unpublished), 1-63.

Oyekunle, L.O (1999): Effect of gas flaring in Niger-Delta area. Proceddings of Nigerian Society of Chemical Engineers, Port-Harcourt, Nigeria. 13.

Potocnik, P.; Thaler, M.; Gavekar, E.; Grabec, I & Puredoz, A (2007). Forecasting risks of natural gas consumption in Slovenia. Journal of energy policy, vol.35, 4271-4282.

Reymond, M (2007). European key issues concerning natural gas: Development and vulnerability. Journal of energy policy, vol. 35, 4169-4176, ISSN 0301-4215

Soylu, S (2007). Estimation of Turkish road transport emssions. Journal of energy policy, vol.35, 4088-4094, ISSN 0301-4215

Tolulope, O.A (2004). Exploration and enviromental degredation: The Nigerian experience. Journal of enviromental informatics archive, vol.2, 387-393.

Tsimas, E.; Mercier, A.; Cormos, C.C & Peteves, S.D (2007). Trade-off emissions of acid gas pollutants and of carbon dioxide in fossil fuel power plant with carbon captive. Journal of energy policy, vol.35. 3991-3997, ISSN 0301-4215

Udeta, M.E.M.; Burani, G.E.; Maure, J.O.A & Oliva, R.C (2007). Economics of secondary energy from GTL regarding natural gas reserviors of Bolivia. Journal of energy policy, vol.35, 4095-4106, ISSN 0301-4215

Ufarana, O.R (2004). Dispersion of heat radiation from gas flaring in Niger-Delta. B.Eng degree project, department of Chemical Enginnering, Federal University of Technology, Minn. Nigeria (unpublished), 1-85.

Uyigue, E & Agu, M (2007). Coping with climate change and enviromental degradation in the Niger Delta of Southern Nigeria. Report of community research and development center (CREDC), Lagos, Nigeria, 1-31.

Yassin, M.F.; Kato, S.; Ooka, R.; Takahashi, T & Kouno, R (2005). Field and wind tunnel study of pollutant dispersion in a built area under various meterological conditions. Journal of wind engineering and industrial aerodynamic, vol. 93: 361-382.

Chapter 2

Oil Pollution and International Marine Environmental Law

Ekaterina Anyanova

Additional information is available at the end of the chapter

http://dx.doi.org/10.5772/37399

1. Introduction

After the Second World War not only the public interest in the environment increased in general. Concerns of coastal states about increasing ship-source marine pollution and oil spills started to grow as well. Some of the occurred incidents with tankers clearly demonstrated that oil spills in an environmentally or economically sensitive area could cause irreparable damage (Gold, 1998).

Oil pollution of the ocean comes from shipping activity and offshore oil production. Sea-bed activities on oil exploration and production constitute a relatively small part in the general amount of the pollution of marine environment with oil. The principal cause of marine pollution with oil is shipping. Traditionally shipping is considered to be "a polluting industry". The world's tanker fleet counts approximately 7 000 vessels with cargo capacities between 76 000 and 175 000 tons (Gennaro, 2004). Usual shipping operations, especially transportation of oil by tankers and accidents, result in the dumping of around 600 000 – 1 750 000 tons of oil into the ocean per year (Brubaker, 1993).

Due to the use of pipelines for petroleum products, oil transportation with tankers decreased significantly (Gennaro, 2004). However, the incidents with this type of vessels and the occurred oil spills occur constantly. The last oil pollution incident, which gained publicity and attention of the mass media, happened in October 2011 off the New Zealand's coast. The grounding off of the tanker "Rena" and the followed oil leaking caused the environmental disaster. This oil spill seriously damaged wildlife, including penguins, seals, dolphins, whales and rare sea birds (New Zealand oil spill ship captain charged, 2011).

It must be stressed here that the oil spills and individual catastrophes are very spectacular, but the scientific research demonstrates that pollution from other sources damages the marine environment more. Furthermore, it should be noted that a small amount of oil is constantly seeping in the seas being assimilated into the ocean environment (Brubaker,

1993). Many chemicals carried at sea are intrinsically far more harmful to the marine environment. Although the impact of the oil pollution constitutes only a small part of a general pollution to the maritime environment, the consequences of oil spills and oil wastes are extremely damaging for marine landscape and ocean's inhabitants.

Spilled oil is very toxic. It can be lethal to adult animals even in relatively low concentrations. It may also cause physiological or behavioral disruptions of species. Oil spills also cause death through the prevention of normal feeding, respiration and movement functions not only of ocean wildlife, but also of marine life at the sea shore. Particularly dangerous oil spills are for birds. Oil spill can lead sometime to the tainting of fish and shellfish. Sometimes one can feel the consequences of the oil spills through the oily taste or smell to the seafood. An oil spill directly damages not only animals, plants and corals, fisheries, but also affects human activity in the area of fisheries through damaging of fishing boats, fishing gear, floating fishing equipment.

Oil spills affect not only the ocean space around them, but also shorelines, open waters and the seabed; wetlands; corals. They also damage fisheries and coastal amenities. Especially vulnerable for the potential damage is the area of shorelines. The caused damage is unpredictable and does not depend on the size of the oil spill. It depends rather on the closeness to the shoreline and vulnerability of the area. For example, a 9 000-ton diesel fuel spill from the "Tampico Maru", in the Baja California in 1957, damaged over 10 km of coastline. On the other hand, 10 000 tons of crude oil discharged by the "Argea Prima", in Puerto Rico in 1962, caused very little actual damage. The oil spill of 476 000 tonnes of crude oil, caused by the Ixtoc I oil platform blowout in the Gulf of Mexico, had caused relatively little damage. The damage from the "Argo Merchant" grounding in 1976 and oil spill of 50 000 tons were very serious. The oil spill of 40 000 tons by the VLCC "Exxon Valdez" in especially vulnerable area of Prince William Sound in Alaska, in 1989, resulted in an ecological disaster and very long and costly clean-up operations. The same phenomena were observed during the Iran-Iraq and Iraq-Kuwait military actions and resulted oil spills. The oil spill with "Atlantic Empress" with loss of almost 300 000 tons of crude oil in 1978 in the Atlantic Ocean did not cause any significant impact on economy, but seriously damaged an offshore ecosystem around the site of the catastrophe.

Under the right conditions the marine environment recovery natural process is incredibly quick and "painless", however, the internal mechanisms of the nature are not endless and marine environment needs proper treatment and protection. The new oil and gas development projects also raise more and more serious concerns of the environmentalists. For example, the recent decision to start the drilling in the Arctic seriously worried the environmentalists especially in light of the climate change issue, which have been widely discussed in the mass media. On 29 August 2011 Exxon Mobil Corp and Rosneft signed an agreement on the development of oil and gas in the Russian sector of the Arctic (Korsunskaya & Reddall, 2011). The region presumably obtains around 13% of the undiscovered oil resources and 30% of its natural gas. Although this project is considered to be highly beneficial for both sides, it is stressed by both sides that environmental safety is very important in this area (Howard, 2011), since this area is considered to be ecologically fragile. Partly the concerns address the transportation of oil and possible oil incidents.

In the international law in the course of time a comprehensive regulatory regime on prevention of marine oil pollution (particularly oil spills) was developed. Special attention was paid to the regulation of marine oil pollution by shipping (Salter & Ford, 2001), so the existing rules cover mostly vessel-source pollution. The most effective instruments in the marine environment protection are regional treaties. Almost all regional treaties include a general obligation for signatory states to prevent, reduce and control all forms of maritime pollution. In the Helsinki convention[1] and the Convention for the Protection of the Marine Environment of the North-East Atlantic (OSPAR Convention)[2] one can find more concrete clauses like the precautionary concept, polluter pays concept, best available technology, and best environmental practice.

However, the elaborated rules need to be enforced and complied with. A closer co-operation and sharing of informational resources within the international community is urgently required, especially in the cases of conventions and their amendments ratification.

This chapter is devoted to the existing rules of international law and certain unilateral legislation on the issue of marine environment pollution with oil as well as their development in the XX-XXI centuries. The liability and compensation schemes in cases of occurred oil pollution are also analyzed. The chapter also deals with the existing regional conventions on marine oil pollution and makes certain proposals on the improvement of the existing at present legislation.

2. International law documents on marine environment pollution

The first international convention on oil pollution was adopted in 1926 by the International Maritime Conference in Washington. This document however was not ratified.

Because of the significant pollution especially of the Atlantic Ocean during the World War II (military operations with submarines, torpedoes etc.), since 1945 the issue of oil pollution became very acute and more and more important.

Marine pollution particularly with oil is not clearly regulated in any particular global environmental convention. This form of pollution is considered in some of the international legal documents. The provisions of the international conventions on this issue are, however, relatively limited.

2.1. Stockholm declaration

The Declaration on the Human Environment (Stockholm Declaration)[3] and Action Plan[4] were adopted at the United Nations Conference on the Human Environment (UNCHE),

---

[1] Convention on the Protection of the Marine Environment of the Baltic Sea Area, 1992. http://www.helcom.fi/Convention/en_GB/text/

[2] Convention for the Protection of the Marine Environment of the North-East Atlantic, 1992. http://www.ospar.org/html_documents/ospar/html/OSPAR_Convention_e_updated_text_2007.pdf

[3] Declaration of the United Nations Conference on the Human Environment, 1972. http://www.unep.org/Documents.Multilingual/Default.asp?documentid=97&articleid=1503

held in Stockholm in June 1972. Both documents have special sections on marine pollution. This conference was one of the first attempts of the integrated approach to the global environmental issues.

It was stressed in the Principle 7 of the Stockholm Declaration on the Human Environment, 1972 that states shall take all possible steps to prevent pollution of the seas by substances that create hazards to human health, harm living resources and marine life, damage amenities or interfere with other legitimate uses of the sea (Sokolova, 2005).

Principle 22 addresses the issue of liability and compensation for marine pollution damage requiring from states further cooperation in order to develop rules of international law regarding this issue.

Action Plan consisting of 109 recommendations proposes to address pollution by means of the environmental assessment, environmental management and supporting measures.

One of the most important achievements during the UNCHE was the establishment of a new United Nations institution - the United Nations Environment Programme (UNEP). As for the issues of marine environment protection, the UNEP adopted "regional seas action plans". The organization also monitors pollution in some of the regional seas areas. This shift to the regional perspective was caused by the idea that the transboundary problems of the oceans or environmental protection of any particular sea could be better managed from a regional basis. The first covered region was the Mediterranean (Suarez, 2006).

2.2. Global conventions on the law of the sea

As a separate issue oil pollution is not regulated in the global conventions on the law of the sea. However, the Geneva Conventions of 1958 contain the provisions on environmental protection of the ocean against oil pollution through oil pipelines or continental shelf development (Art. 5(1) and 5(7) of the Geneva Convention on the Continental Shelf[5] and Art. 24 of the Geneva Convention on the High Seas[6]) (Gennaro, 2004). These provisions are, however, rather superficial. Geneva Convention on the High Seas in its Art. 24 proclaims the obligation of states to draft national legislation on pollution prevention from ships or pipelines or sea-bed activities. Art. 5(1) and 5(7) of the Geneva Convention on the Continental Shelf concern the exploration and exploitation of the continental shelf and its natural resources. The coastal state has to ensure that there is no unjustifiable interference with navigation, fishing or the conservation of the living resources of the sea, oceanographic or other scientific research. The coastal states shall also establish safety zones around the offshore installations and take measures for the protection of the living resources of the sea from harmful agents.

Another comprehensive document for the different aspects of ocean matters including marine environment protection is the United Nations Convention on the Law of the Sea

[4] Action Plan for the Human Environment, 1972.
http://www.unep.org/Documents.Multilingual/Default.asp?DocumentID=97&ArticleID=1492&l=en

[5] Convention on the Continental Shelf, 1958. United Nations, Treaty Series. Vol. 499. P. 311.

[6] Convention on the High Seas, 1958. United Nations, Treaty Series. Vol. 450. P. 11, p. 82.

(UNCLOS)[7] adopted on 10 December 1982 at Montego Bay (came into force on 16 November 1994). This document became global. This "umbrella convention" does not contain detailed rules for the protection of the marine environment, only general provisions. These rules shall be implemented by means of the further regulations of the international law (Zhu, 2006).

The provisions of UNCLOS on maritime protection are of great importance world-wide. They are contained in Part XII of the Convention. These regulations are general. The convention proclaims general obligation of states to protect the marine and coastal environment and its resources (Art. 192). Art. 193 grants to the states the right to develop their natural resources under the consideration of their natural environmental policy. Art. 193 stresses as well the duty of states to protect and preserve the marine environment.

The general provisions of the mentioned Art. 192 and Art. 194 on measures to prevent, reduce and control pollution of the marine environment are considered to be a part of the international customary law.

Art. 195 and 197 require from state not to transfer damage or hazards or transform one form of pollution into another and to cooperate with each other on global or regional basis (Dahm et al., 2002).

The convention also contains the enforcement procedure for the rules on marine pollution prevention by the flag state (Art. 217) and by the coastal states over the vessels in its ports or off-shore terminals (Art. 220).

The protection of marine environment during the offshore development of oil is reflected in Art. 207, regulating the protection of marine environment against pollution from land-based sources. Art. 208 regulates protection of the marine environment from sea-bed activities under their jurisdiction. Art. 208 (4) stresses the need to reach a compromise in this respect on a regional level, what should be considered as a recognition of necessity to solve this problem on the regional level. Art. 213-214 contain the enforcement rules for the mentioned provisions.

Art. 235 proclaims liability of the states for their international obligations concerning the preservation and protection of marine environment. Art. 235 (2) requires the states to ensure the possibility to obtain compensation or other relief in case of the damage caused by the pollution.

2.3. Agenda 21 – Programme of action for sustainable development

Another international conference – The Earth Summit, which took place in Rio de Janeiro, Brazil from 3-14 June 1992 – was also very important for the environmental and development issues. This meeting was prepared by the United Nations Conference on Environment and Development (UNCED).

---

[7] United Nations Convention on the Law of the Sea, 1982.
http://www.un.org/depts/los/convention_agreements/texts/unclos/unclos_e.pdf

The outcome of this conference was the adoption of several non-binding legal instruments, including Agenda 21[8]. Agenda 21 is a programme of action for sustainable action world-wide.

Chapter 17 of this document is devoted to the oceans "Protection of the Oceans, All Kind of Seas, including Enclosed and Semi-enclosed Seas, and Coastal Areas and the Protection, Rational Use and Development of their Living Resources". This chapter links the activities of the conference with the UNCLOS. The chapter stresses the need of the marine pollution protection (Gold, 1998). The danger of offshore oil and gas operations for the marine environment and the need to minimize this danger are also in this chapter. Agenda 21 stresses the need to prevent especially the marine pollution from vessels including illegal discharges and pollution caused by ships in particularly sensitive areas. The need to reduce the risks of accidents is also stressed. Shipping has to comply with the provisions of the International Convention for the Prevention of Pollution From Ships, 1973 as modified by the Protocol of 1978 (MARPOL)[9] and the UNCLOS. International cooperation with regional or global international organizations and, where appropriate, with industry-based organizations shall be improved.

This document did not establish a new legal framework on ocean governance, but confirmed the importance and fundamental role of the UNCLOS in this respect especially in the protection and sustainable development of the marine and coastal environment and its resources (Suarez, 2006).

3. International legal documents on the oil pollution at sea

3.1. OILPOL'54

International Convention for the Prevention of Pollution of the Sea by Oil (OILPOL)[10] was adopted in London on 12 May 1954. This convention became the first international treaty dealing with the oil pollution.

It addressed the discharge of oil and oily wastes into the water. OILPOL`54 prohibited the intentional discharge of oil and oily mixtures from certain vessels in specified ocean areas. The ballast discharges have to be made in the permitted areas with a special record in an oil record book. This book shall be inspected at regular intervals. The enforcement of the convention had to be fulfilled by the flag state.

The Convention became a significant achievement at that time. In the preamble of the later adopted MARPOL convention it is stressed that OILPOL was the first multilateral instrument to be concluded with the prime objective of protecting the environment. The preamble of the MARPOL also appreciates the significant contribution, which the OILPOL has made in preserving the seas and coastal environment from pollution.

---

[8] Agenda 21. http://www.un.org/esa/dsd/agenda21/res_agenda21_00.shtml

[9] International Convention for the Prevention of Marine Pollution from Ships, 1973, United Nations, Treaty Series. Vol. 1340. P. 184 (as modified by Protocol, 1978, United Nations, Treaty Series. Vol. 1340. P. 61).

[10] International Convention for the Prevention of Pollution of the Sea by Oil, 1954. United Nations, Treaty Series. Vol. 327. P. 3.

3.2. "Torrey Canyon" disaster and intervention convention

On 18 March 1967 in the English Channel the accident with the oil supertanker "Torrey Canyon" occurred. The grounding of the vessel was caused by human error. The entire cargo of the vessel - 120 000 tonnes of crude oil – was spilt. Around 15 000 sea birds died because of the spill. Damage claims in Great Britain amounted to GBP 6 million and to FRF 40 million in France. States quickly recognized the danger of a major oil spill to the coastlines.

The "Torrey Canyon" incident demonstrated that there was no internationally agreed means of responding to accidents that had environmental implications. The regulation of compensation to be paid also did not exist on the international level (Özçayir, 2004).

This incident was the first major oil pollution incident. Due to its strong impact on the development of the international law it became historical. This oil spill clearly demonstrated that the development of the marine environmental legislation is closely connected with the occurring incidents at sea.

The International Maritime Organization established a Legal Committee to deal with the deficiencies in the international system for assessing liability and compensation for oil-spill damage, and a new subcommittee of the Maritime Safety Committee (MSC) to deal with environmental issues (Özçayir, 2004).

The occurred incident accelerated the formation of MARPOL convention and in 1973 the *International Convention for the Prevention of Pollution from Ships (MARPOL)* was adopted to cover pollution by oil, chemicals, harmful substances in packaged form, sewage and garbage (Özçayir, 2004).

As another consequence of this incident the International Maritime Organization adopted the *International convention relating to intervention on the high seas in cases of oil pollution casualties* done at Brussels on 29 November 1969 (Intervention Convention)[11], enabling a government to take action, if an accident in international waters threatened its coastline with pollution (Özçayir, 2004).

In its preamble it stresses the need to protect the interest of people against the consequences of a maritime casualties resulting in oil pollution. Measures of an exceptional character taken in order to protect the environment on the high seas are admissible. These measures do not affect the principle of freedom of the high seas.

In Article I the Convention grants a permission to take such measures on the high seas as may be necessary to prevent, mitigate or eliminate grave and imminent danger to their coastline or related interests from pollution or threat of pollution of the sea by oil in cases of a maritime casualty or acts related to it against vessels which pose a threat to their coastlines.

Article VI declares the obligation of a state caused damage to others to pay compensation in case, if the measures exceeded those reasonably necessary.

---

[11] International Convention Relating to Intervention on the High Seas in Cases of Oil Pollution Casualties, 1969. United Nations, Treaty Series. Vol. 970. P. 211.

It provides coastal states with limited rights to take preventive measures on the high seas against foreign vessels which are considered to present a grave and imminent danger to coastlines and other coastal interests from oil pollution as a result of a maritime casualty.

The Convention has been criticized both for allowing too much discretion to coastal states and for limiting the rights of such states to take action. In 1973 a Protocol[12] covering substances other than oil was added.

4. International liability regime for oil pollution

4.1. Liability for Oil Pollution in the international conventions

The "Torrey Canyon" incident demonstrated that in case of the oil pollution of the ocean there were no rules of international law making the polluter liable. OILPOL'54 left the issue of liability for pollution to the national law.

It was decided to develop international legal scheme with the liability regime for oil spills.

The present international regime of compensation for damage caused by oil pollution is based on two conventions: *International Convention on Civil Liability for Oil Pollution Damage, 1969 (CLC 1969)*[13] and *International Convention on the Establishment of an International Fund for Compensation for Oil Pollution Damage, 1971 (FUND 1971)*[14].

The CLC 1969 was elaborated within the Inter-Governmental Maritime Consultative Organization and signed on 29 November 1969 in Brussels. It ensures the compensation to be paid. The general principle provided in the convention is that those causing oil pollution should pay compensation.

The convention aims to ensure the adequate compensation to victims of oil pollution damage resulting from maritime casualties involving oil-carrying ships. The convention applies to the pollution damage caused on the territory of the Member States to the Convention and related preventive measures (Art. II). The CLC does not apply to ships or vessels owned or operated by a State and used for non-commercial service. The CLC applies to State-owned merchant fleets.

Art. III of the Convention makes the owner of a ship strictly liable for the pollution damage caused by the discharge from the ship. The shipowner is liable even in the absence of any fault, for any damage by pollution caused by the oil. However, the shipowner can normally limit his financial liability up to an amount established according to the tonnage of the ship. This amount is guaranteed by his liability insurer. The liability insurance is compulsory.

Traditional liability exemption concept of fault and negligence did not apply under the CLC. The admissible exceptions out of "strict liability"-rule are damage resulting from an act of

[12] Protocol Relating to Intervention on the High Seas in Cases of Pollution by Substances Other Than Oil, 1973. United Nations, Treaty Series. Vol. 1313. P. 4.

[13] International Convention on Civil Liability for Oil Pollution Damage, 1969. United Nations, Treaty Series. Vol. 973. P. 3.

[14] International Convention on the Establishment of an International Fund for Compensation for Oil Pollution Damage, 1971. http://www.iopcfund.org/npdf/Text%20of%20Conventions_e.pdf

war or grave natural disaster or damage wholly attributable to sabotage by a third party or wholly caused by the failure of authorities to maintain navigational aids.

The Amendments of the Protocol of 1992[15] extended the sphere of application of the Convention to the exclusive economic zone. The amendments adopted in the Protocol in 2000[16] raised liability rates.

The convention uses the standard of the Special Drawing Rights (SDR) of the International Monetary Fund.

This convention provided for a uniform set of international rules and procedures for determining liability and compensation. The introduction of the CLC convention significantly facilitated the recovery of compensation for oil pollution damage.

This convention is widely accepted, however, the US refused to adopt the CLC. The CLC's provisions do not apply to the waters of states not accepted the CLC. In such a way, oil spills in the waters of the non-CLC states, such as the United States, remain uncovered.

During the 1969 conference at the Inter-Governmental Maritime Consultative Organization, which developed the CLC, it was recognized that there would be incidents which would require compensation limits in excess of the available under the CLC.

To provide for compensation for the circumstances not covered by this convention, an international fund was established under the terms of the 1971 Convention on the Establishment of an International Fund for Compensation for Oil Pollution Damage to administer this compensation system.

The FUND Convention is expertly administered by the International Oil Pollution Compensation Fund Secretariat in London. This fund is an intergovernmental organization established by States. Any state which accepts the FUND Convention automatically becomes a member of the International Oil Pollution Compensation (IOPC) Fund. The FUND is financed by a levy applied to individuals and corporations dealing with the import and export of oil in contracting states.

The convention also introduced a compulsory liability insurance requirement for ship owners.

Only oil (cargo or bunkers) carried in bulk by vessels is covered. Parties to the FUND must also be parties to the CLC, and the flag state of the vessel, which caused the damage, must also be a party to the FUND, if the shipowner is also seeking compensation.

The IOPC Fund becomes involved by providing supplementary compensation when the amount payable by the shipowner and his insurer is insufficient to cover all of the damage.

---

[15] Amendments of the Limitation Amounts in the Protocol of 1992 to Amend the International Convention on Civil Liability for Oil Pollution Damage, 1969.
http://www.iopcfund.org/npdf/Text%20of%20Conventions_e.pdf

[16] Resolution LEG.1(82), 18 October 2000. Amendments of the Limitation Amounts in the Protocol of 1992 to Amend the International Convention on Civil Liability for Oil Pollution Damage, 1969.
http://www.austlii.edu.au/au/legis/cth/consol_act/potsla1981357/sch3.html

Any person or company, which has suffered pollution damage in a Member State of the IOPC Fund 1992 caused by oil transported by ship, can claim compensation from the shipowner, his insurer and the Fund. This applies to individuals, business, local communities or States.

To be entitled to compensation, the damage must result from pollution and have caused a quantifiable economic loss. The claimant must substantiate the amount of his loss or damage by producing accounting records or other appropriate evidence. Compensation may be claimed for property damage, clean-up operations, economic losses of fishermen or those engaged in mariculture and tourism sector (ITOPF, 2002).

Since its establishment the IOPC Fund has been involved in some 120 incidents of different graveness in around 20 countries. Over US$ 630 million were paid as a compensation.

The FUND pays compensation to any claimant who has suffered pollution damage in cases where no liability arises under the CLC as the shipowner is protected by one of the CLC exemptions; or the shipowner is financially unable to meet the CLC obligations and the available insurance coverage is insufficient; or the damage exceeds the shipowner's CLC liability. Most cases fall within the third category.

Claims for the compensation shall be brought in the applicable courts of contracting states.

During the 1984 Diplomatic Conference at the International Maritime Organization it was decided to revise completely both the CLC and the FUND instruments. This decision was strongly lobbied by the USA. After the "Amoco Cadiz" incident it was obvious that the CLC and FUND limits were not sufficient.

These protocols were some sort of the compromise between oil and shipping industries and coastal states for the protection of the marine and coastal environment. The Protocol of 1984 to Amend the International Convention on Civil Liability for Oil Pollution Damage, 1969; and the Protocol of 1984 to Amend the International Convention on the Establishment of an International Fund for Compensation for Oil Pollution Damage, 1971 raised the liability limits and extended jurisdiction to the exclusive economic zone. However, both documents met the opposition of the USA after the "Exxon Valdez" grounding and did not enter into force. The 1992 Protocols[17] raised the liability limitation and based the limitation on the vessel's gross tonnage. The 1992 CLC and FUND Protocols repeated contents of the 1984 protocols. The Protocols raised the compensation limitations and extended the jurisdiction of the conventions to the exclusive economic zones of contracting states. The amendment procedures for the conventions were also simplified.

The 2000 amendments[18] raised again the liability limitations up to 4,51 – 89,77 million SDR depending on the gross tonnage of the vessel.

---

[17] Protocol of 1992 to Amend the International Convention on Civil Liability for Oil Pollution Damage, 1969. http://www.iopcfund.org/npdf/Text%20of%20Conventions_e.pdf; Protocol of 1992 to Amend the International Convention on the Establishment of an International Fund for Compensation for Oil Pollution Damage, 1971. http://www.iopcfund.org/npdf/Text%20of%20Conventions_e.pdf

[18] IMO Resolution, 18 October 2000. Adoption of Amendments of the Limits of Compensation in the Protocol of 1992 to Amend the International Convention on the Establishment of an International Fund for Compensation for Oil Pollution Damage, 1971. http://www.iopcfund.org/npdf/Text%20of%20Conventions_e.pdf

4.2. Industry Initiatives

Simultaneously with the development of the CLC and the FUND Conventions the shipping and oil industries elaborated compensation regime schemes. It was made under the pressure of the publicity after the Torrey Canyon incident. *Tanker Owners' Voluntary Agreement Concerning Liability for Oil Pollution, 1969, as amended (TOVALOP)*[19] and *Contract Regarding an Interim Supplement to Tanker Liability for Oil Pollution, 1971, as amended (CRISTAL)*[20] were developed. The introduction of these schemes took place, since oil tanker industries were aware that the CLC and FUND conventions would take too long to enter into force and coastal states could take unilateral measures.

TOVALOP was a voluntary agreement amongst tanker owners, including bareboat charterers. The main aim of this agreement was to encourage tanker owners to clean up spills, regardless of fault, with the assurance of recovering their costs from their P&I club and to compensate governments for the clean-up costs. TOVALOP members were required to carry sufficient insurance to cover their obligations under the scheme, including the clean-up costs. TOVALOP scheme was made as close as possible to the CLC provisions. After the elaboration of the Protocol to the CLC in 1984 there were even proposals to make in the TOVALOP the tanker owner liable for all cases of oil spills, even covered by the FUND. This risk had to be additionally insured. However, these proposals have been rejected.

TOVALOP was managed by the International Tanker Owners' Pollution Federation (ITOPF) in London. ITOPF was charged with interpreting the scheme and handling claims directly.

Another voluntary interim agreement introduced by the oil industry was Contract Regarding an Interim Supplement to Tanker Liability for Oil Pollution (CRISTAL). This initiative commenced in April 1971 (the FUND convention entered into force seven years later).

The scheme was administered by the Oil Companies Institute for Marine Pollution Compensation Ltd. in Bermuda. Marine Pollution Compensation Services Ltd. (CRISTAL) in London was responsible for its everyday activities.

The both initiatives clearly demonstrated that the oil industry under the environmental pressure was ready to take the responsibility for pollution claims.

Both schemes have been amended several times according to the amendments of the CLC and FUND conventions.

From the beginning the both schemes were developed as the interim measures. However, CLC and FUND schemes became widely accepted and both TOVALOP and CRISTAL have been terminated on 20 February 1997.

---

[19] Tanker Owners Voluntary Agreement Concerning Liability for Oil Pollution , 1969. International Legal Materials. Vol. 8, 1969. P. 497.

[20] Contract Regarding an Interim Supplement to Tanker Liability for Oil Pollution, 1971. International Legal Materials. Vol. 10, No. 1, 1971. P. 137.

Another proposal from the oil industry sector was a *Pollution Agreement among Tanker Owners (PLATO)* in 1985. This initiative reduced the liability of oil sector trying to put a greater burden on the shipping sector, since the level of the safety of shipping was still not sufficient. The initiative was not supported.

5. Further international documents on oil pollution of the marine environment

5.1. International convention on the prevention of marine pollution by the dumping of wastes or other matter (London Convention), 1972

Further international document on the oil pollution was the International Convention on the Prevention of Marine Pollution by the Dumping of Wastes or Other Matter (London Convention), 1972[21] with Protocol of 1996[22]. This document was adopted under the influence of the Stockholm Conference. The adoption of this convention served as a demonstration of the readiness of states to protect marine environment against oil pollution. The Convention deals with 'dumping', i.e. the deliberate disposal of wastes and other matter (other than operational discharges) from vessels and aircraft. The provisions of the convention prohibit dumping except for the wastes listed in Annex 1. Dumping of some wastes requires a prior special or general permit (Art. IV). Convention stresses that the capacity of the sea to assimilate wastes and render them harmless, and its ability to regenerate natural resources, are not unlimited.

The effectiveness of the London convention was demonstrated in the incident with the dumping of high-level radioactive waste in the Arctic by the USSR (Stokke, 1998). The dumping was documented in the Yablokov Report in the 1990s. Certain efforts were made to enforce the provisions of this convention through the Russian-Norwegian Environmental Committee and the International Arctic Sea Assessment Program. The behavior of the USSR was condemned during the Consultative Meeting under the London Convention. Further treatment and storage projects for the hazardous wastes for the former countries USSR under the framework of the London Convention.

5.2. International Convention for the prevention of pollution from ships (MARPOL), 1973/1978

The international community was dissatisfied with the OILPOL'54 convention and its environmental and technical requirements. The International Maritime Organization adopted another convention on the prevention of vessel-source pollution on 2 November 1973 - the International Convention for the Prevention of Pollution from Ships (MARPOL). It covered pollution by oil, chemicals, and harmful substances in packaged form, sewage and

[21] Convention on the Prevention of Marine Pollution by Dumping of Wastes and Other Matter, 1972. United Nations, Treaty Series. Vol. 1046. P. 120.

[22] Protocol of 1996 to the 1972 Convention on Prevention of Marine Pollution by Dumping of Wastes and Other Matter. 2006 Australian Treaty Series 11.

garbage. The convention aimed to prevent both international and negligent pollution and minimize accidental spills. The Convention represents the compromise between coastal and shipping interests. The Convention introduced new anti-pollution regime and more modern operating requirements.

MARPOL effectively superseded OILPOL'54. However, in its preamble MARPOL Convention recognizes the importance of the International Convention for the Prevention of Pollution of the Sea by Oil, 1954.

Tanker accidents which took place in 1977/78 led the Inter-Governmental Maritime Consultative Organization to convene an International Conference on Tanker Safety and Pollution Prevention (TSPP Conference) in February 1978. This conference adopted MARPOL Protocol of 1978 in order to speed up the adoption of MARPOL convention. This 1978 Protocol is also known as the Tanker Safety and Pollution Protocol.

A few weeks after this conference the VLCC "Amoco Cadiz" spilt 230 000 tons of crude oil (Gold, 1998). This incident stressed the importance and urgency of the oil pollution prevention for the international community.

The combined instrument is referred to as the International Convention for the Prevention of Marine Pollution from Ships, 1973, as modified by the Protocol of 1978 relating thereto (MARPOL 73/78), and it entered into force on 2 October 1983 (Annexes I and II). Later further Annexes (III-VI) were adopted to this document.

Annex I with regulations for the prevention of pollution by oil covers prevention of pollution by oil from operational measures as well as from accidental discharges. It also contains the mandatory requirements for new oil tankers to have double hulls and a phase-in schedule for existing tankers on double hulls (Gold, 1998).

Annex II with regulation for the control of pollution by noxious liquid substances - regulates the discharge of noxious substances, and prevents their discharge within 12 miles of the nearest land.

Annex III on prevention of pollution by harmful substances carried by sea in packaged form requires the packaging, labeling and documenting of harmful substances.

Annex IV is devoted to the prevention of pollution by spillage from ships.

Annex V on prevention of pollution by garbage from ships specifies distances from land and manner of disposal of garbage from ships and completely prohibits disposal of plastic.

Annex VI on prevention of air pollution from ships regulates discharge of sulphur oxide and nitrous oxide. The discharge of ozone depleting gases is prohibited.

Since the definition of "ship" in the convention includes vessels of any type operating in the marine environment (hydrofoil boats, air-cushion vehicles, submersibles, floating craft and fixed or floating platforms, oil-drilling vessels and platforms), all of their discharges of oil into the sea are also covered by the scope of this convention.

The "oil" in the convention is defined as petroleum in any form, including crude oil, fuel oil, sludge, oil refuse and refined products (other than petro-chemicals).

The discharge of oil is completely prohibited in a number of 'special areas' which are considered particularly vulnerable to pollution, for example, Mediterranean, Baltic, Black Sea, Red Sea, Gulf and Antarctic. In these areas for technical oceanographic and ecological reasons and the particular character of traffic, special mandatory methods for the prevention of sea pollution by oil are required.

MARPOL laid down the mechanism to check the seaworthiness of a ship by providing a framework for certification of ships with respect to safety and pollution compliance. Powers to inspect, detain and prosecute have been given to flag states and port states (Gautam, 2010).

The convention requires from vessels to have on board Oil Record Book (for tankers and other vessels over a certain tonnage) and Cargo Record Book (for ships carrying noxious substances other than oil). These books have to contain records on on-board operations.

MARPOL introduces a number of certificates to be kept on board: International Oil Pollution Certificate 1973 (IOPC); International Pollution Prevention Certificate for the Carriage of Noxious Liquid Substances in Bulk 1973; International Sewage Pollution Certificate 1973. These certificates could be demanded by surveyors and inspectors in flag and port states. Lack of or improper certification may result in sailing permission being withheld.

The convention introduces a system of communication between states. All relevant information of interest shall be communicated to other state(s) or organizations. For information on accidents and results of investigations reports are required. The International Maritime Organization shall be informed on such issues as reception facilities, inspection/survey authorities, specimens of certifications, texts of laws and regulations, annual reports and statistical index of violations and fines imposed.

MARPOL concerns mainly pollution from vessels. Exploration and other relevant offshore resources development activity is excluded.

The MARPOL Convention became the main international convention covering prevention of pollution of the marine environment by ships from operational or accidental causes. For the first time the whole issue of marine pollution was addressed. This document became an innovation. If earlier anti-pollution conventions had been limited to pollution by oil, MARPOL aimed at all kinds of sea-borne pollution: oil, chemicals, sewage, garbage, and other harmful materials (Rosenne, 1999).

In order to maintain its impact considering the constantly developing technical innovations the convention is being updated by amendments.

This convention serves as a certain "environmental code" for the shipping industry.

5.3. Other relevant documents

One can find certain provisions on marine pollution with oil in other relevant international legal documents.

Convention for the prevention of marine pollution from land-based sources[23] done at Paris on 4 June 1974 in Art. 1 puts an obligation on the Contracting Parties to take all possible steps to prevent and combat pollution of the sea from land-based sources. The pollution from land-based sources also covers the maritime pollution from installations under the jurisdiction of the member states to the convention including offshore installations and structures.

International Convention Relating to the Limitation of the Liability of Owners of Sea-going Ships, 1957[24] was developed by the Comité Maritime Internacional. It includes the principle of limitation of liability. Besides the liability is limited for shipowners in cases of death, personal injury and property damage claims depending on the tonnage of the vessel. This provision of the convention was often overruled by the courts, so the convention was replaced by the *Convention on Limitation of Liability for Maritime Claims, 1976 (LLMC 1976)*[25]. This document sets general limitation of liability. The limitation-rule does not apply in cases of intentional or reckless personal act or omission. By the limitation rule shipowner, charterer, manager, operator, salvors and insurers are covered.

Convention on Civil Liability for Oil Pollution Damage Resulting from Exploration of Seabed Mineral Resources, 1977 (CLEE 1977)[26] is a liability convention for offshore oil and gas operations. The convention did not enter into force since there is a developed liability regime for oil industry under the bilateral agreements with the involved coastal states.

The issue of oil pollution offshore drilling and exploration and exploitation activities is also concerned in a voluntary agreement amongst oil companies operating in northwestern Europe *the Offshore Pollution Liability Agreement (OPOL)*[27]. According to its provisions operators accept strict liability for pollution damage and remedial measures.

In aftermath of the "Exxon Valdez" catastrophe under the pressure of the USA the *International Convention on Oil Pollution Preparedness, Response and Co-Operation (OPRC 1990)*[28] was adopted in London on 30 November 1990 addressing the issues of response and preparedness of the international community to the oil spills. By means of this convention the International Maritime Organization developed a framework for the international cooperation in combating major oil pollution incidents.

The convention stresses in its preamble the serious threat posed to the marine environment by oil pollution incidents and reminds that in case of the oil pollution incident, prompt and

---

[23] Convention for the Prevention of Marine Pollution from Land-Based Sources, 1974. http://www.opcw.org/chemical-weapons-convention/related-international-agreements/toxic-chemicals-and-the-environment/marine-pollution-from-land-based-sources/

[24] International Convention relating to the Limitation of the Liability of Owners of Sea-Going Ships, 1957. http://www.admiraltylawguide.com/conven/limitation1957.html

[25] Convention on Limitation of Liability for Maritime Claims, 1976. http://www.admiraltylawguide.com/conven/limitation1976.html

[26] Convention on Civil Liability for Oil Pollution Damage Resulting from Exploration and Exploitation of Seabed Mineral Resources, 1977. http://www.dipublico.com.ar/english/convention-on-civil-liability-for-oil-pollution-damage-resulting-from-exploration-and-exploitation-of-seabed-mineral-resources/

[27] Offshore Pollution Liability Agreement. http://www.opol.org.uk/agreement.htm

[28] International Convention on Oil Pollution Preparedness, Response and Co-operation, 1990. http://www.admiraltylawguide.com/conven/oilpolresponse1990.html

effective action is essential in order to minimize the damage. In Art. 6 Convention puts an obligation upon Party States to establish a national system addressing the oil pollution incidents. The convention recognizes the importance of mutual assistance and international cooperation and establishes the basis for the exchange of information respecting the capabilities of states to respond to oil pollution incidents, the preparation of oil pollution contingency plans, the exchange of reports of incidents of significance which may affect the marine environment or the coastline and related interests of states, as well as research and development respecting means of combating oil pollution in the marine environment. The convention also sets a requirement for vessels and offshore units to have on board oil pollution emergency plans.

This convention is a framework treaty containing only general obligations without concrete ones. However, it should be stressed that a convention provides for legal basis for better cooperation between states in responding to oil pollution incidents (Gold, 1998). Besides the convention granted to the International Maritime Organization a central role in organizing an immediate response to polluting incidents, providing the basis for International Maritime Organization coordination of technical support and financial assistance for governments.

International Convention on Civil Liability for Bunker Oil Pollution Damage (BUNKER), 2001[29] was adopted to reduce a number of gaps in the CLC regime. This document provides for prompt compensation system for the damage caused by oil spills, when oil was carried as fuel in ships' bunkers. This convention is applied to the territorial seas and exclusive economic zones of the States Parties. The registered owner of the vessel is under the obligation to maintain compulsory insurance cover. A claim for pollution damage could be brought directly against an insurer.

6. Unilateral "Solutions"

6.1. The US Oil Pollution Act, 1990

The international measures for the prevention and protection of the marine environment pollution are often considered to be slow and ineffective. Sometimes it makes the states to adopt the unilateral (or regional) measures for the protection of marine environment. However, it should be stressed that the number of states with a unilateral approach remains modest (Dahm et al., 1989).

As examples one could mention the reaction of the European Commission on the "Erika» oil spill or the Oil Pollution Act (OPA)[30] of the US in 1990 after the «Exxon Valdez» grounding on 24 March 1989. The vessel ran into the Bligh Reef in King William Sound in Alaska. 10,8 million gallons (40 000 tons (Brubaker, 1993) of crude oil were spilt. Over 3 400 square kilometres were damaged by oil. Because of this oil spill approximately 250 000 sea birds and 2 000 sea otters died (Graham, 2003).

---

[29] International Convention on Civil Liability for Bunker Oil Pollution Damage, 2001. http://www.official-documents.gov.uk/document/cm66/6693/6693.asp

[30] U.S. Oil Pollution Act, 1990, Public Law 101-380, 104 U.S. Statutes at Large 484 (1990).

The company "Exxon Mobile" spent over $2 billion in order to clean up the Prince William Sound area after the Exxon Valdez oil spill (Gennaro, 2004).

It was the largest spill in the US waters. Being dissatisfied with the international community's environmental protection legislation and common international standards (Pamborides, 1999), the US started its own new policy towards vessels calling at the US ports. Looking at the statistics one should say that after this legislation the volume of spilled oil in the US waters dropped by 70%.

The OPA, promulgated by the US Congress, introduced ahead of the International Maritime Organization double hull standards for oil tankers. This act was called "draconian legislation" (Gold, 1998). It allows to use single-hull tanker vessels of 5 000 gt or more for trade with the US until 2015, depending on their age, only if they were equipped with a double bottom or double sides. The vessels not complying with this requirement could be refused access to US waters after 2010.

This idea concerning the tank of the vessel was not particularly new. As far back as 1971 the amendments to the OILPOL have been elaborated. The proposed rules regulated the tank sizes of tankships in order to reduce the pollution. However, at that time this amendment proposal met a considerable opposition from the ship construction industry. Amendments did not enter into force (Gold, 1998).

The OPA applies to the navigable waters of the US, including the exclusive economic zone up to 200 miles from the baseline. 'Oil' is defined broadly to include petroleum, fuel oil, sludge, oil refuse and oily wastes. Hazardous substances are not included.

The OPA introduces liability provisions and establishes a supplemental fund to be used as compensation for losses not covered by polluters.

The OPA'90 allows to the US states to enact their own laws on issues of pollution liability and response.

The behavior of the US was condemned by the international shipping industry, especially tanker operators. In 1992 the International Maritime Organization adopted the amendment to MARPOL requiring mandatory double hulls for new tankers delivered on or after 6 July 1996. Existing tankers shall be fitted with double hulls within 30 years.

The revised measures on the phasing-out of single-hull oil tankers were adopted in December 2003 after incident with the "Prestige" tanker. Regulation 13G of Annex I of MARPOL concerning the final phasing-out dates was revised. The dates were pushed forward from 2007 to 2005 for "pre-MARPOL tankers" and from 2015 to 2010 for MARPOL tankers and smaller tankers. The Condition Assessment Scheme was made applicable to all single-hull tankers older than 14 years (Anianova, 2006).

6.2. Other national legislation on oil pollution prevention

The Arctic Waters Pollution Act of 1970 of Canada was the result of the beginning of the use of the Alaska territory for the transportation of oil. The Canadian law prohibits the dumping of hazardous wastes from vessels and from the land. The legal act grants to the Governor in Council the possibility to declare a certain area of the Arctic waters a safety zone. For such safety zone it is possible to enact separate regulations concerning the technical requirements to the vessels, their loading. The navigation of vessels through such zones can also be prohibited (Dahm et al., 1989). This act raised the protests of the USA. The attempt of the Canada to lobby the introduction of the environmental safety zones into the UNCLOS during the conference was not successful.

On December 12, 1999 the oil tanker "Erika" broke off the coast of Brittany, France. 14 000 tonnes of oil were spilled damaging 400 km of Atlantic coast. The "Erika" incident accelerated the amendments to MARPOL adopted in April 2001 on the phase-out of single-hull tankers by 2010. The case was brought to the court. The court found guilty the energy company Total, the owner and manager of the tanker, and the Italian classification society "RINA", of negligence and criminal liability. They were ordered to pay 192 million Euros in damages. The Paris Court of Appeal increased the sum of damages to 200 million Euros.

The incident with the tanker "Prestige" in November 2002 off the coast of Spain when the vessel broke in two was followed by the collision of the Turkish tanker "VICKY" with a sunken carrier off the coast of France and the spill of 70 000 tons of flammable kerosene.

These accidents were reflected in the legislation of the European Union: the transport of heavy fuels have been prohibited in single-hulled tankers in EU waters and a phase-out plan for single-hulled tankers was introduced. For example, in Spain a ban against the transportation of heavy fuel in single hull tankers was introduced.

These unilateral measures concerning the international navigation could seem to be more effective than the international ones. However, they do not consider the interests of the international shipping. Such "competition" between unilateral initiatives and International Maritime Organization standards could be detrimental to international shipping and lead to the "selected ports" approach: the ports with the higher standards could not be simply ignored by some ships, while the ports with weaker port state control would be chosen. These ships with lower standards could threaten the environment in those regions (Anianova, 2006).

7. Regional conventions

As for unilateral or national anti-pollution measures, one could say, that the effective marine environment protection is not possible without the cooperation with other states. However, the regional conventions dealing with pollution of certain seas or waters by oil could be more effective than the global ones with the general requirements. In particular it concerns regional seas and the need to protect their environment and regulate maintenance, exploration and exploitation of their resources (Gelberg, 1979). It is easier to consider the particularities of the regional seas with the help of the coastal states cooperation.

Most regional conventions on protection of particular regional seas include in their texts a common general obligation for the parties to take "appropriate" measures to prevent and control pollution arising from the exploration and exploitation of their seabed mineral resources:

- Art. 8 of the Convention for Co-operation in the Protection and Development of the Marine and Coastal Environment of the West and Central African Region (Abidjan Convention), 1981[31];
- Art. 4 of the Convention for the Protection of the Marine Environment and Coastal Area of the South-East Pacific (Lima Convention), 1981[32];
- Art. VII of the Regional Convention for the Conservation of the Red Sea and Gulf of Aden Environment (Jeddah Convention), 1982[33];
- Art. 8 of the Convention for the Protection and Development of the Marine Environment of the Wider Caribbean Region (Cartagena de Indias Convention), 1983[34];
- Art. 8 of the Nairobi Convention for the Protection, Management and Development of the Marine and Coastal Environment of the Eastern African Region (Nairobi Convention)[35];
- Art. 8 of the Convention for the Protection of the Natural Resources and Environment of the South Pacific Region and related Protocols (Nouméa Convention), 1986[36];
- Art. 7 of the Convention for the Protection of the Mediterranean Sea against Pollution (Barcelona Convention), 1976[37] (Gavouneli, 1995).

Most of the regional conventions on the seas protections contain an obligation to take decisive actions to protect their marine environment. Most of these conventions refer to the national legislation and the need to adopt the corresponding measures.

In some of the conventions (the Abidjan, Nairobi, Cartagena and Nouméa Conventions) the requirement on obligatory environmental management is introduced.

Most regional conventions also include cooperation clauses in their texts requiring the proper communication between concerned states and international organizations. Regional conventions contain mostly general rules and principles of environmental protection of the seas, however, the more detailed requirements and standards fail.

The anti-pollution legislation for all regional seas is at present still being developed and formed. There is already a number of regional arrangements for cooperation in combating pollution.

The first convention relating to offshore oil spills was *the 1969 Bonn Agreement for Cooperation in Dealing with Pollution of the North Sea by Oil*[38]. Originally it covered only oil spills

[31] Convention for Co-operation in the Protection and Development of the Marine and Coastal Environment of the West and Central African Region, 1981.
http://www.unep.org/AbidjanConvention/docs/Abidjan%20Convention%20English.pdf

[32] Convention for the Protection of the Marine Environment and Coastal Area of the South-East Pacific, 1981. http://sedac.ciesin.org/entri/texts/marine.environment.coastal.south.east.pacific.1981.html

[33] Regional Convention for the Conservation of the Red Sea and Gulf of Aden Environment, 1982.
http://www.unep.ch/regionalseas/main/persga/redconv.html

[34] Convention for the Protection and Development of the Marine Environment of the Wider Caribbean Region, 1983.
http://www.offshore-environment.com/regionalconventions.html

[35] Convention for the Protection, Management and Development of the Marine and Coastal Environment of the Eastern African Region, 1985.
http://www.offshore-environment.com/regionalconventions.html

[36] Convention for the Protection of the Natural Resources and Environment of the South Pacific Region, 1986. http://www.offshore-environment.com/regionalconventions.html

[37] Convention for the Protection of the Mediterranean Sea Against Pollution, 1976.
http://www.unep.ch/regionalseas/regions/med/t_barcel.htm

emanating from tankers but after the "Ekofisk" disaster in 1977 the parties agreed that it covered spillages from offshore installations as well. It has been replaced as of 1 September 1989 by the *1983 Bonn Agreement for Cooperation in Dealing with Pollution of the North Sea by Oil and other Harmful Substances*[39].

Mediterranean sea is protected by means of the *Convention for the Protection Of The Mediterranean Sea Against Pollution (Barcelona convention), 1976*. The provisions of this convention contain a non-binding obligation to undertake all possible steps in order to avoid the marine pollution of the Mediterranean sea by means of the seabed exploitation. Another obligation of coastal states concerns the adoption of the national legislation regarding the disposal of the offshore installations not in use in accordance with international guidelines and standards.

Kuwait Regional Convention for Co-operation on the Protection of the Marine Environment from Pollution, 1978[40] devotes some of its provisions to the marine pollution from oil development activity.

The 1992 Convention for the Protection or the Marine Environment of the North-East Atlantic (OSPAR) also mentions the issue of marine pollution from offshore installations (Art. 5 of the Annex III). Member states to the convention are obliged to avoid marine pollution caused by the activity on offshore installations. Article 3 of the Annex VI requires to organize a special environment examination before the exploration and exploitation of the sea-bed. The convention requires to apply the standards of the MARPOL 73/78. Besides the convention requires to cover all not used drilling holes and to dispose offshore installations not in use (Art. 8 Annex VI).

The 1972 Convention for the Prevention of Marine Pollution by Dumping from Ships and Aircraft (the Oslo Dumping Convention)[41] covers the Northeast Atlantic, Arctic Ocean and North Sea, Baltic and Mediterranean Seas (Art. 2) and sets out strict standards of dumping of industrial wastes and redundant materials at sea from vessels.

It is stressed in the preamble of this Convention that the ecological equilibrium and the legitimate uses of the sea are increasingly threatened by pollution. Art. 1 of the Convention contains the general obligation of the Contracting Parties to take all possible steps to prevent the pollution of the sea by hazardous substances.

The list of the prohibited substances is in the Annex I. Annex II contains the list of substances and materials requiring special care.

The Convention On The Protection Of The Maritime Environment Of The Baltic Sea Area (Helsinki Convention)[42] came into force in 1980. The document concerns the protection of the Baltic

---

[38] Agreement for Cooperation in Dealing with Pollution of the North Sea by Oil, 1969. http://sedac.ciesin.org/entri/texts/pollution.north.sea.by.oil.1969.html

[39] Agreement for Cooperation in Dealing with Pollution of the North Sea by Oil and Other Harmful Substances, 1983. Official Journal L188 (16.07.1984). P. 9.

[40] Kuwait Regional Convention for Co-operation on the Protection of the Marine Environment from Pollution, 1978. http://www.offshore-environment.com/regionalconventions.html

[41] Convention for the Prevention of Marine Pollution by Dumping from Ships and Aircraft, 1972. http://sedac.ciesin.org/entri/texts/marine.pollution.dumping.ships.aircraft.1972.html

[42] Convention on the Protection pf the Marine Environment of the Baltic Sea Area, 1974. http://www.offshore-environment.com/regionalconventions.html

Sea. Its member states are all eight Baltic states. The Helsinki Commission regulates discharges and dumping from ships and direct discharges from land, atmospheric inputs, emergency action against oil spills, the use of antifouling paints containing tributyl tin, co-ordinating monitoring programmes, and seal conservation. Regularly meetings of experts evaluate the scientific evidence for the further recommendations (Clark, 1989).

All decisions of the Helsinki Commission are only recommendations. In 1992 the area of responsibility was extended to include the inland waters landward of the baselines of the Baltic Sea states (Valencia, 2001).

The Helsinki Convention is a legal basis for permanent co-operation of Baltic states, especially through the Helsinki Commission (HELCOM) (Zhu, 2006).

The Regional Convention for the Conservation of the Red Sea and Gulf of Aden Environment[43] was signed in Jeddah in 1982. It has a separate protocol on combating oil pollution. An independent regional intergovernmental organization, the Programme of Environment for the Red Sea and Gulf of Aden (PERSGA), was established to implement the Convention. Egypt, Jordan, the Palestinian Authority, Saudi Arabia, Somalia, Sudan, and Yemen participate in PERSGA, which is headquartered in Jeddah, Saudi Arabia.

For the Red Sea and Gulf of Aden a special Strategic Action Programme was adopted in 1982 and initiated in 1995 in order to develop a regional framework for protection of the environment and sustainable development of coastal and marine resources (Dzurek & Schofield, 2001).

Abidjan Convention for Co-Operation in the Protection and Development of the Marine and Coastal Environment of the West and Central African Region, 1981 applies to the West African Atlantic Ocean. Its provisions concern first of all offshore oil producers in West and Central Africa. A general obligation is placed on contracting parties to prevent and relieve pollution caused by sea dumping and exploitation activities. Such activities should conform with recognized rules and practice in the London Convention, 1972. The Convention stresses in its preamble the need for sustainable, environmentally sound practices and inter-generational justice.

Another industry scheme for the North-Western European waters was adopted in 1975 - the *Offshore Pollution Liability Agreement (OPOL)*. Its parties are oil companies operating in the mentioned area (Gavouneli, 1995).

One could also mention an attempt to develop a convention on liability for damage resulted from seabed pollution. One adopted the *1976 London Convention on Civil Liability for Oil Pollution Damage Resulting from Exploration for and Exploitation of Seabed Mineral Resources*[44]. It is still not in force.

---

[43] Regional Convention for the Conservation of the Red Sea and the Gulf of Aden Environment, 1982. http://www.offshore-environment.com/regionalconventions.html

[44] Convention on Civil Liability for Oil Pollution Damage Resulting from Exploration and Exploitation of Seabed Mineral Resources, 1976. http://www.dipublico.com.ar/english/convention-on-civil-liability-for-oil-pollution-damage-resulting-from-exploration-and-exploitation-of-seabed-mineral-resources/

8. Conclusion

Although it was scientifically proved that many chemicals carried at sea are intrinsically far more harmful to the marine environment, the impact of oil upon the ocean and its ecosystem is very dangerous. The spillage of even few tons of oil into sea causes a thin film on the water surface, what is deadly for marine life (Gautam, 2010).

Since the middle of the XX century not only numerous international legislative measures were adopted in the area of oil pollution prevention for the marine environment, but also national laws and regulations. This new legislation reflected not only the development of the legal position on the certain issues, but also the new developments in construction technology like, for example, improved tank stripping pumps, the load-on-top system, and other technological advances. All these preventive measures considerably reduced both vessel-source and offshore oil development pollution.

Beside the main legal documents on oil pollution and marine environment protection, general principles of international environmental law are also applicable to the cases of oil pollution. Such soft concepts as the «precautionary principle» and «polluter pays principle» could be applied (Salter & Ford, 2001). Besides these principles being a substantive element of sustainable development are reflected in conventions on liability and compensation in case of pollution (e.g. CLC, FUND etc.)

In comparison to the oil pollution prevention during the offshore oil development pollution, measures against the vessel-source oil pollution represent the better and more detailed regulated area of marine environmental law (the 1954 Brussels Convention for the Prevention of Pollution of the Sea by Oil (OILPOL) was superseded from 2 October 1983 by the 1978 protocol relating to the 1973 International Convention for the Prevention of Pollution from Ships (MARPOL 73/78)).

The statistics demonstrates that since the beginning of the international legislation on the oceans protection against the oil pollution there had been considerable improvements in the prevention of ship-generated oil pollution. It is not surprising, since the environmental regulation of the industry is becoming wider in its scope and tougher in its implementation (Salter & Ford, 2001).

Tanker incident at sea especially close to the coasts always raises the significant attention of the publicity (Mehr Sicherheit fuer Tanker, 2003). It should be stressed that 99% of the transported oil (about 1,9 billion tons of oil by some 3 000 tankers) is delivered safely (Gold, 1998). However, even this tiny amount of the spilled oil is sufficient to cause the irreparable damage.

Damage to coastal amenities, beaches, tourist and recreational areas, harbors, offshore installations depends on the geographical location of each spill. For example, a relatively small spill, due to the holing of the tanker "American trader" off the coast of California in 1990, caused serious damage. Claims for damage, clean-up costs and fines amounted to over USD 25 million. In the case of the oil spill of the VLCC "Haven" off Genoa in 1991, the French, Italian and even Spanish Mediterranean coasts were damaged. 1 300 Italian claims alone amounted to GBP 705 million.

There exist very good means and instruments to combat the oil pollution, what was demonstrated by clean-up operations after the "Exxon-Valdez" oil spill. There are four major options of responding to marine spills: mechanical containment and collection; use of chemical dispersants; physical shoreline clean-up; and natural removal, requiring no clean-up action. Other counter-measures that are less frequently used due to their limitations are burning, sinking, gelling and enhanced biodegradation. A decision, which clean-up action shall be applied, depends upon a given situation (Yoder, 1985). However, the best clean-up operations won't recover the existed ecosystem. So let the oil pollution never had happened.

On the other hand, the demands of the maritime transportation system increase. For example, the Red Sea is still one of the world's least ecologically damaged bodies of water. However, the vessel traffic through the Suez Canal and from oil terms along the Red Sea coast increases, what raises the concerns on environment, living resources, and tourism, but the vessels traffic continues to grow.

In such a way, there is still certain resistance of the oil industry and slow development of the anti-pollution legislation. It is quite understandable. For example, the move of the tankers from one-hull to the double hulls is just uneconomical (Mehr Sicherheit fuer Tanker, 2003). The other aspect concerns the quality and technical level of the vessels transporting oil (Mehr Sicherheit fuer Tanker, 2003). Often they are built in the developing countries with low loans and escape of technical innovations. Besides, it is now often discussed that only those ships which are insured by the shipping insurance companies or certified by "classification societies" shall be allowed to sail in international waters (Gautam, 2010).

It can be easily noted that maritime catastrophes of large scale lead to the development of the international law. For example, the "Titanic" (1912), "Torrey Canyon" (1967), "Amoco Cadiz" (1978), "Exxon Valdez" (1987) accidents served as a reason for the adoption of new safety and anti-pollution rules (Rosenne, 1998). As for oil pollution legislation, a series of tanker accidents occurred off the coast of North America in 1976-1977 (Özçayir, 2004) with "Sansinena"; "Oswego Peace"; "Olympic Games"; "Daphne"; "Grand Zenith"; "Barcola; Mary Ann"; "Universe Leader", and several other tankers (although almost all of these disasters were caused by human error or negligence) accelerated adoption of the MARPOL convention and Protocol. The stranding of the "Argo Merchant" (Anianova, 2006), in December 1976 and an oil spill of 27 000 tonnes of the coast of Massachusetts resulted in the Conference of the International Maritime Organization on Tanker Safety and Pollution Prevention in February 1978 under the US lobby and adopted amendments on tanker design and operation incorporated in the Protocol of 1978 to the SOLAS Convention[45] (Özçayir, 2004). The accident with the "Amoco Cadiz" on March 16, 1978 off the coast of France, which resulted in the oil spill of 221 000 tonnes of crude oil, served as an accelerating factor for the entry into force of the MARPOL convention.

On the basis of the made research a conclusion should be made that hydrocarbons are also the main pollutants from offshore installations used for the exploitation of the natural

[45] Protocol of 1978 relating to the International Convention for the Prevention of Pollution from Ships, 1973, as amended, United Nations, Treaty Series. Vol. 1340. P. 61,62

resources of the seabed (Gavouneli, 1995). Offshore resources are used more and more. 20% of the today's oil production comes from offshore wells.

Although the liability regime for oil rigs is well-established in the private sector and adequate pollution compensation for most incidents is available, one could imagine that the need for a widely accepted international regime, covering all aspects of oil rigs, will become necessary in future.

The subject is at present on the agenda of the International Maritime Organization. There is some opposition to a more comprehensive international treaty from some sectors of the oil industry, especially offshore operators who, at present, are able to conclude simple bilateral agreements with coastal states.

Intentional pollution occurs not too often because any loss of oil and gas goes against the commercial interests of the operator. The cases of accidental pollution from blowouts or tanker spillage and collisions are still numerous. The statistics shows that 75-90% of the oil released at sea comes from discharges of the oily water produced during drilling and the dumping of oil-based drilling muds and chemicals. The reported oil spills represent only about 1% of the oil released into the sea. For example, in the North Sea the prevailing geological formations tend to react with water-based muds. It results in the instability (Gavouneli, 1995), so it is necessary to use for the oil development the highly toxic oil-based muds.

The international legislation regarding the marine environmental protection during exploration and exploitation of seabed mineral resources is subject to surprisingly few international regulations. It is not well developed. Operations in the Area are under the control of the International Seabed Authority according to the rules of the United Nations Convention on the Law of the Sea, 1982, but oil and gas drilling operations are conducted in the continental shelf under the direct control of the coastal state, usually through a state-controlled oil company.

Another critical issue in this topic remains, whether unilateral measures in the environmental protection should be admissible. Although the USA unilaterally introduced their own liability system for the cases of oil pollution, it was clearly demonstrated that this approach could be damaging to the international economy. Effective environmental protection is not possible with the efforts of only one state. However, regional conventions on cooperation in this area could be very effective. Geographical differences between various regions make regional efforts in the marine environmental protection more useful. The pollution problems are better tackled by regional agreements.

As the recent accidents show the established legal mechanisms (the CLC, MARPOL etc.) do not help totally to prevent oil spills in the ocean. The proposals to launch a system with the stronger advance cooperation with the shipping industry still remain only proposals. The incidents with "Prestige" and "VICKY" oil spills also demonstrated the inability of the existing liability system to compensate the victims of the oil pollution. They are also criticized for low liability limits (Gennaro, 2004).

In this research it was demonstrated by means of the overview of the existing international legal documents, that there exists a regime for oil pollution liability and compensation with more or less adequate coverage for the most serious incidents. One can make a conclusion that the industry is well controlled by legislation (Salter & Ford, 2001). But the question arises whether one can foresee everything in advance and to prevent it?

It was scientifically confirmed that the marine environment may eventually recover from very serious oil pollution incidents. However, it does not mean that there is no shorter-term damage to the marine environment, coasts, people and property. Besides we should think not only about the today's interests, but consider "the interests of future generations" (Birnie & Boyle, 1992).

One shouldn't forget that in such cases as oil pollution prevention on the level of the international legislation the most important aspect is a quick response in its time (Anianova, 2006).

Author details

Ekaterina Anyanova
The Institute of State and Law, Russian Academy of Science & OOO "LUKOIL-KMN"
Russian Federation

9. References

Anianova, E. (2006). The International Maritime Organization – Tanker or Speedboat?, In: *International Maritime Organizations and their Contribution towards a Sustainable Marine Development*, Ehlers, P., Lagoni, R. (Eds.), pp. 77-103. LIT Verlag, ISBN 3-8258-9296-4, Hamburg

Birnie, P., Boyle, E. (1992). *International Law and the Environment*, Clarendon Press, ISBN 0-19-876282-8, Oxford

Brexendorff, A. (2006). *Rohstoffe im Kaspischen Becken. Voelkerrechtliche Fragen der Foerderung und des Transports von Erdoel und Erdgas*, Peter Lang GmbH, ISBN 3-631-54968-7*Pb., Frankfurt am Main

Brubaker, D. (1993). *Marine Pollution and international law: principles and practice*, Belhaven, ISBN 1-85293-273-2, London

Clark, R.B. (1989). *Marine Pollution* (2nd ed.), Clarendon Pr., ISBN 0-19-854263-1, Oxford

Dahm, G., Delbrueck, J., & Wolfrum, R. (2002). *Voelkerrecht. Band I/2* (2nd ed.), Walter der Gruyter, ISBN 3-89949-023-1, Berlin

Dzurek, D.J., Schofield, C. (2001). *Parting the Red Sea: Boundaries Offshore Resources and Transit*, IBRU, ISBN 1-89764-346-2, Durham

Gautam, D. (2010). Trans-Boundary Marine Oil Pollution and Its International Legal Aspects, In: *Private Law: Rights, Duties and Conflicts*, Kierkegaard, S.M. (Ed.), pp. 980-988. IAITL, ISBN 978-87-991385-8-6, Copenhagen

Gavouneli, M. (1995). *Pollution from Offshore Installations*, Graham and Trotman, ISBN 1-85966-186-6, London

Gelberg, L. (1979). *Rechtsprobleme der Ostsee*, Sample, ISBN 3-921654-06-8, Hamburg

Gennaro, M. (2004). Oil Pollution Liability and Control under International Maritime Law: Market Incentives as an Alternative to Government Regulation. *Vanderbilt Journal of Transnational Law*, Vol. 37:265, No. 1, (January 2004), pp. 265 – 298, ISSN 0090-2594

Gold, E. (1998). *Handbook on Marine Pollution* (2nd ed.), Assuranceforeningen Gard, ISBN 82-90344-11-2, Arendal

Graham S. (19 December 2003). Environmental Effects of Exxon Valdez Spill Still Being Felt, In: *Scientific American*, 31.10.2011, Available from: <www.scientificamerican.com/article.cfm?id=environmental-effects-of>

Howard, R. (2 September 2011). How Arctic oil could break new ground, In: *The Guardian*, 31.10.2011, Available from <http://www.guardian.co.uk/commentisfree/2011/sep/02/arctic-oil-exxonmobil-russian-deal>

ITOPF (Eds.) (2002). *International Oil Pollution Compensation Fund 1992*, ITOPF, London

Korsunskaya, D. & Reddall, B. (30 August 2011). Exxon, Rosneft tie up in Russian Arctic, In: *U.S. (Reuters)*, 31.10.2011, Available from <http://www.reuters.com/article/2011/08/30/us-rosneft-exxon-idUSTRE77T2OM20110830>

Valencia, M. (Ed.) (2001). *Maritime Regime Building*, Martinus Nijhoff Publishers, ISBN 90-411-1580-3*hc, The Hague

Mehr Sicherheit fuer Tanker (2003). *HANSA*, No. 9, pp. 15 – 18

New Zealand oil spill ship captain charged (12 October 2011), In: *BBC News*, 31.10.2011, Available from <http://www.bbc.co.uk/news/world-asia-pacific-15268314>

Özçayir, O. (2004). *Port State Control* (2nd ed.), LLP, ISBN 1-8431-1328-7, London

Pamborides, G. (1999). *International Shipping Law*. Sakkoulas, ISBN 960-150014-6, Athens

Rosenne, S. (1999). The International Maritime Organization Interface with the Law of the Sea Convention, In *Current maritime issues and the international maritime organization*, Nordquist, M. & Moore, J. (Eds.), pp. 251-265, Martinus Nijhoff Publishers, ISBN 90-411-1293-6*alk., The Hague

Salter, E. & Ford, J. (2001). Holistic Environmental Assessment and Offshore Oil Field Exploration and Production. *Marine Pollution Bulletin*, Vol. 42, No. 1., pp. 45-58, ISSN 0025-326x

Stokke O. (1998). Beyond Dumping? The Effectiveness of the London Convention, In: *Yearbook of International Co-operation on Environment and Development*, Bergesen, H., Parmann, G., & and Oystein, T. (Eds.), pp. 39-49, Earthscan, ISBN 1853835269, London

Suarez, S. (2006) Towards sustainable Ocean Development: The Participation and Contribution of Non-governmental Organizations to the Work of International Organizations in the Ocean Sector, In: *International Maritime Organizations and their Contribution towards a Sustainable Marine Development*, Ehlers, P., Lagoni, R. (Eds.), pp. 53-75. LIT Verlag, ISBN 3-8258-9296-4, Hamburg

Yoder, C. (1985). *The Canadian regulation of offshore installations*, The University of Calgary, ISBN 0919269184, Calgary, Canada

Zhu, L. (2006). Do we need a Global Organization for the Protection of the Marine Environment, In: *International Maritime Organizations and their Contribution towards a Sustainable Marine Development*, Ehlers, P., Lagoni, R. (Eds.), pp. 157-180. LIT Verlag, ISBN 3-8258-9296-4, Hamburg

Sokolova, N. (2005). Zaschita i sohranenie morskoj sredy kak uslovie osuschestvleniya svobod otkrytogo morya. *Morskoe pravo i praktika*, Vol. 1, pp. 20 – 32

Chapter 3

Use of Remote Sensing in Wildfire Management

Brigitte Leblon, Laura Bourgeau-Chavez and Jesús San-Miguel-Ayanz

Additional information is available at the end of the chapter

http://dx.doi.org/10.5772/45829

1. Introduction

Wildfire is the one of the prominent disturbance factor in most vegetation zones throughout the world, like forests and grasslands. Wildfires present a challenge for ecosystem management, because they have the potential to be at once beneficial and harmful. On the one hand, wildfires are a natural part of several ecosystems for maintaining their health and diversity in numerous ways, such as regulating plant succession and fuel accumulations, controlling age, structure and species composition of vegetation, affecting insect and disease populations, influencing nutrient cycles and energy flows, regulating biotic productivity, diversity and stability and determining habitats for wildlife.

On the other hand, wildfires can also become a threat to property, human life and economy, particularly in ecosystems where fires are an uncommon or even unnatural process. Despite the prominence of fire events, current estimates of the extent and impact of vegetation fires globally are still a challenge. Several hundred million hectares of forest and other vegetation types are estimated to burn annually throughout the world, consuming several billion tons of dry matter and releasing emission compounds that affect the composition and functioning of the global atmosphere and human health. According to the FAO (FAO 2012), wildfires are important climate forcing factors as they release aerosol between 25-35% of the total CO_2 net emissions to the atmosphere. Over the last decade in Canada, wildfires have consumed an average of 1.9 million ha/year and induced fire suppression costs ranging from about $500 million to $1 billion a year (Canadian Forest Service, 2012). In Europe, wildfires burn more than half a million ha of forested areas every year. Over 95% of the burnt areas are located in the Mediterranean region, in which critical fire events have taken place in recent years (http://effis.jrc.it).

Because of the threat that fires represent, operational systems have been developed for use in fire management that includes fire danger prediction, fire detection and fire control. Given expected increases in fires across the world due to climate changes, better prediction of fire danger and fire detection will have significant benefits both from the economical and the human safety points of views across the world. There is also the need to accurate assessment of burnt areas because they are related to greenhouse gas emissions into the atmosphere that need to be accounted for following Kyoto's protocol requirements as well as for managing post-fire environmental impacts, such as regeneration and erosion. Space-borne remotely sensed imagery can play an important role in these systems. Indeed, satellite imagery offers the advantages of extensive regional coverage, zero disturbances of the area to be viewed, as well as a method for acquiring data in less accessible areas on a regular and cost effective basis. In the first part of this chapter, we will present the use of remote sensing in pre-fire conditions management. The second part of the section will deal with the use of remote sensing for detecting fires and burn scar mapping.

2. Pre-fire conditions management

Ignition and spread of wildfires depends on fuel moisture and weather conditions as well as on fuel types and topography. These parameters are as inputs into fire danger predicting systems that have been developed for fire management, among others for fire suppression. These systems are among others the *National Fire Danger Rating System* (*NFDRS*) in USA (US Forest Service, 2012) and the *Canadian Forest Fire Danger Rating System* (*CFFDRS*) in Canada (Canadian Forest Service, 1992). The *CFFDRS* is also used in Alaska and in some other parts of the world, including Europe and Asia. Both systems are based primarily on weather parameters that are point source data which are often acquired in a sparse network of weather stations. The availability of satellite images coupled with the development of geostatistics and spatial analyses using geographic information technology allows moving fire danger rating from point-based estimates from weather stations to spatially-explicit estimates. Indeed, satellite images have the advantages of larger sampling areas, lack of destruction of the studied resource, gathering data on less accessible areas and are measuring, in essence, the integrated response of vegetation (including fuel) to environmental influences (including drought).

Several pre-fire conditions can be monitored using remote sensing. The first one is related to the fuel type, which can be mapped, like classical vegetation mapping, from high spatial resolution optical or radar images (e.g., Chuvieco and Martin, 1994; Burgan et al., 1998; Chuvieco et al., 1999a). These maps can then be linked, within a wildfire threat analysis system, to other pre-fire conditions variables, such as topography, proximity to roads and to urban areas, etc... (Burgan et al., 1998; Chuvieco et al., 1999a; Chuvieco et al. 2010). Another pre-fire condition, which can be estimated by remote sensing, is the fuel moisture condition. We will focus here on live fuel moisture conditions, which are in current fire prediction systems, either directly measured (Pinol et al., 1998) or broadly estimated (Canadian Forest Service, 1992). Dead fuel moisture conditions will also be considered, although they can be more easily computed from weather data and fuel characteristics, because dead fuel moisture is in balance with that of the surrounding atmosphere (Burgan et al., 1998; Pinol et al., 1998; Chuvieco et al., 1999b). In most of the remote sensing studies on live fuel moisture estimation, live fuel moisture conditions have been quantified as an absolute measurement of plant water content, through the Fuel Moisture Content (FMC) or the Equivalent Water

Thickness (EWT). FMC is defined as the ratio between the quantity of water (fresh weight–dry weight) and either the fresh weight or the dry weight (see the review of Ceccato et al., 2001). EWT is the leaf water content per unit leaf area which is defined as the ratio between the quantity of water and the leaf area (see the review of Ceccato et al., 2001). Live fuel moisture conditions have been also quantified indirectly, through the degree of water stress which is expressed in terms of evapotranspiration rates (Vidal et al., 1994). In the present study, the term "optical" is used to describe wavelengths between 400 and 2500 nm, in contrast to the thermal infrared bands, which range from 3000 to 15000 nm. Both types of wavelengths are recorded by optical sensors. The present study will primarily focus on satellite data, although the theory may also be applied to airborne sensors, which are currently used during fire suppression activities rather than as fire danger prediction tools.

2.1. Optical remote sensing

The first remote sensing studies on fuel moisture conditions monitoring used optical data, mainly NOAA-AVHRR NDVI images (e.g., Paltridge and Barber, 1988; Burgan et al., 1998; Chuvieco et al., 1999b; Hardy and Burgan, 1999). This supposes that timing and extent of drought can be assessed from vegetation greenness, as retrieved from satellite data. NDVI data were also correlated to simulated forest evapotranspiration (e.g., Deblonde and Cihlar, 1993), to FWI codes and indices (Dominguez et al., 1994; Camia et al., 1999; Leblon et al., 2001; Oldford et al., 2006; Leblon et al., 2007), to fuel moisture content of grasslands (Yebra et al., 2008), and to fire occurrences (e.g., Lopez et al., 1991; Illera et al., 1996; Burgan et al., 1998). NDVI-based operational systems have been proposed to assess fire potentials (Figure 1) (Burgan et al., 1998) and crop droughts or fire dangers (Kogan, 2001).

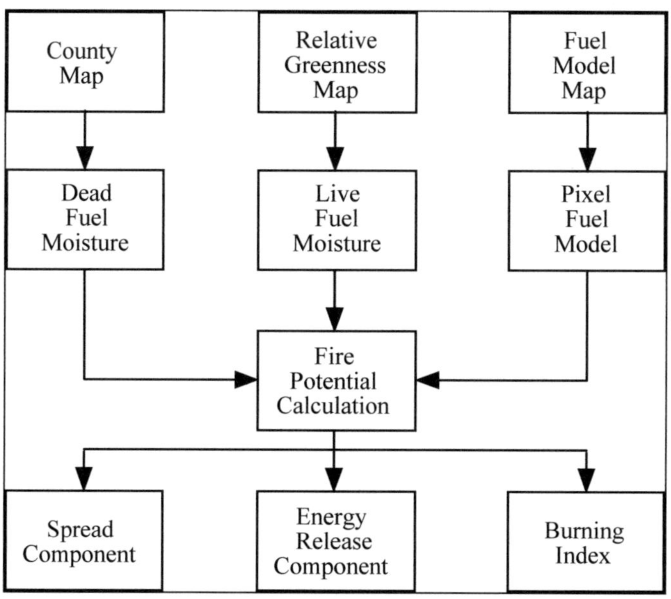

Figure 1. An operational system to compute fire potential maps from NOAA-AVHRR NDVI images (adapted from Burgan et al., 1998)

These studies listed several problems related to the use of NDVI images in fuel moisture mapping, namely the saturation of relationships (Paltridge and Barber, 1988), the influence of site wetness on relationships (Deblonde and Cihlar, 1993) and the difficulty of using NDVI over forests, due to the spectral mixture of the overstory with the understory, both being different in nature and in moisture content (e.g., Hardy and Burgan, 1999; Leblon et al., 2001). In fact, NDVI and associated vegetation indices are only indirectly related to fuel moisture conditions, because it rather measures the greenness and the chlorophyllous activity of the vegetation (Ceccato et al., 2001; Leblon, 2005). In a study on pre-fire conditions using NOAA-AVHRR over Northwest Territories boreal forests, Oldford et al. (2003) showed that high FWI areas correspond to high surface temperature areas on the surface temperature NOAA-AVHRR image, indicating water stress, but to high NDVI areas over the NOAA-AVHRR NDVI image, indicating no drought conditions (Figure 2).

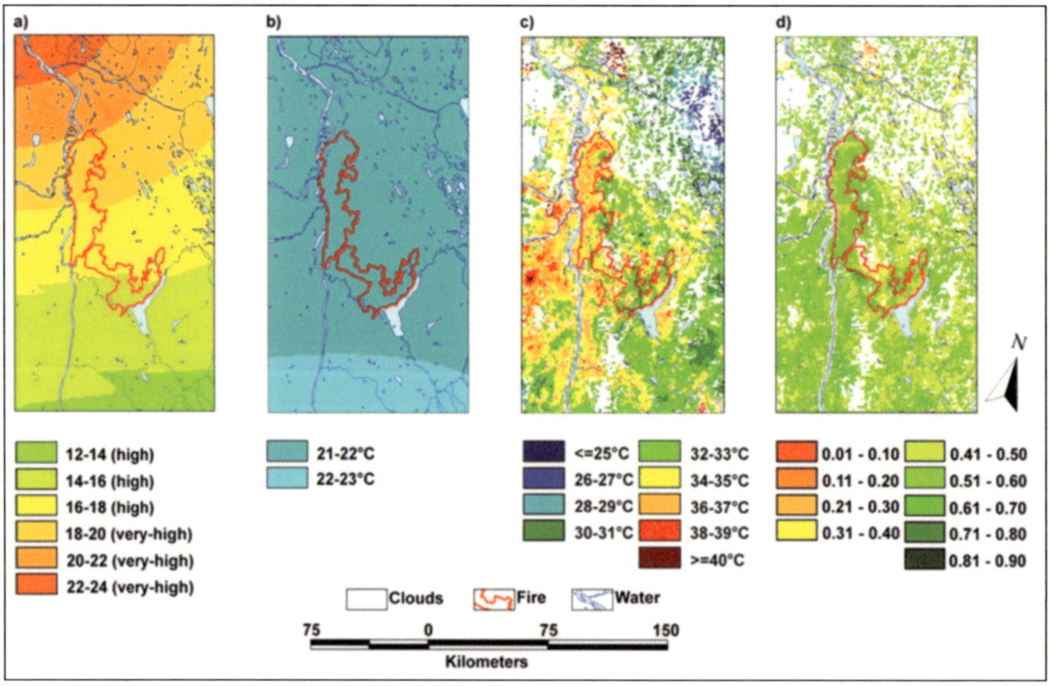

Figure 2. Map of (a) interpolated FWI, (b) interpolated T_a, and corresponding (c) T_s NOAA-AVHRR image and (d) NDVI NOAA-AVHRR image for the area, where the fire "*vq0332*" starts. The image was acquired 3 days before fire ignition (after Oldford et al., 2003)

Greenness and the chlorophyll activity of the vegetation explained the positive correlations between ΣNDVI and FWI codes and indices found by Leblon et al. (2001, 2007) over Canadian northern boreal forests, since both types of variables increase in parallel throughout the fire season, but for two different reasons: FWI codes and indices, because of drought, and ΣNDVI because of vegetation growth. In addition, reduction in NDVI could be induced by factors other than drought, like disease or senescence (Leblon, 2005) and shadowing or penumbra (Chuvieco et al., 1999b). For all these reasons, a better use of NDVI images over forests will be to map timing of deciduous leaf flushing, which is critical in fire management, because of its relationship to fire occurrence in mixed-deciduous forests.

Fuel moisture is theoretically better related to another optical band, the shortwave infrared (1300-2500 nm) (e.g., Pierce et al., 1990; Pinol et al., 1998; Chuvieco et al., 1999b, Ceccato et al., 2001, Yebra et al., 2008). Relationships were significant only when the water stress was already well developed (Pierce et al., 1990; Pinol et al., 1998). Reflectance variations associated with water changes were smaller than those associated with leaf structure (Pierce et al., 1990; Ceccato et al., 2001). In addition, shortwave bands are highly disturbed by atmospheric effects. Fuel moisture is also probably estimated better using hyperspectral data. Indeed, hyperspectral data allow derivative analysis which is useful to remove, on reflectance, the effect of leaf structure, of background and of atmosphere as well as to resolve overlapping spectra to better separate components of the global spectrum (see the review of Leblon, 2005). Hyperspectral data were related to plant water content through empirical relationships (e.g., Pinol et al., 1998) or analytical models (e.g., Ustin et al., 1998; Ceccato et al., 2001). Multispectral data of the MODIS sensor were used into analytical model to retrieve fuel moisture content of shrublands (Yebra and Chuvieco, 2009).

However, from the operational point of view, both hyperspectral data are, up to now, only provided by airborne sensors and shortwave infrared data are acquired by only a few numbers of spaceborne sensors, among others LANDSAT-TM, SPOT-VEGETATION, NOAA-16 and MODIS. While the oldest ones like LANDSAT-TM have a long revisit period, the newest ones, like SPOT-VEGETATION and the new series of the AVHRR sensor, on board NOAA-16, or MODIS, have the advantage to allow daily image acquisition. This temporal scale may be longer on cloudy periods. The performance of these new sensors is still under evaluation. By contrast, for many years, thermal infrared data are provided more often and mostly at the same time as the optical visible and near-infrared ones, by several existing spaceborne sensors, e.g., NOAA-AVHRR, LANDSAT-TM, ATSR-2, RESURS-01, METEOSAT, GOES or MODIS.

2.2. Thermal infrared remote sensing

Surface temperatures (T_s) were better correlated than NDVI to FWI codes and indices (Dominguez et al., 1994; Camia et al., 1999; Aguado et al., 2003; Oldford et al., 2003; Oldford et al., 2006, Leblon et al., 2007), to foliar moisture content (Chuvieco et al., 1999b) and to shrub water potentials (Gouyet et al., 1991). They were also useful to detect water-stressed coniferous stands, when extreme differences in canopy water content occurred (Pierce et al., 1990). In fact, the difference between surface and air temperatures is a better spectral index to monitor plant water status than the surface temperature solely, the last being too sensitive to weather conditions (Camia et al., 1999; Duchemin et al., 1999). In addition, according to the energy budget equation, plants respond to water stress by stomata closure, thereby decreasing latent heat transfer from leaf surface to the air and causing an increase in leaf surface temperature (Pierce et al., 1990). Solving the energy budget equation, in which the sensible heat flux (H) is inferred from the difference between surface and air temperatures

(T_s-T_a), as a function of the latent heat flux (LE) leads to an analytical relationship between actual evapotranspiration rate (AET) and T_s-T_a. Cumulative T_s-T_a data were well related to monthly fire start numbers throughout the fire season over Mediterranean forests (Prosper-Laget et al., 1995). For the same ecosystem, Vidal et al. (1994) used the energy budget equation to compute the ratio between actual and potential evapotranspirations (AET/PET) from daily NOAA-AVHRR surface temperatures and synoptic air temperatures. The ratio was related to fire occurrences (Vidal et al., 1994) and to two shrub flammability variables (Desbois and Vidal, 1996). The ratio was used to operationally monitor fire danger over Mediterranean forests in 1994 (Desbois and Vidal, 1995) and was correlated to FWI codes and indices over Canadian northern boreal forests (Strickland et al., 2001).

However, these studies also showed that estimating AET from T_s-T_a using the energy budget equation is more problematic over forest canopies than over crop canopies (Leblon, 2005). First, canopy height makes forests different from a thin leaf surface, as supposed by the energy budget equation, because of an additional level of radiation absorption and convective heat exchange between the ground and the superior stratum. Second, the measured surface radiative surface temperature is different from the aerodynamic surface temperature (T_{rad}) required by the equation, because of an additional excess resistance (known as the kB^{-1} factor) to heat transfer from leaves, which increases with the canopy height. Third, the aerodynamic resistance (r_a) is lower than the canopy resistance (r_c) and T_s-T_a is thus less sensitive to moisture fluctuations. This lower sensitivity is compensated by the sensitivity of satellite signals to ground vegetation patches which are an important fire danger parameter. Also, the clumped nature of canopy elements in tree crowns reduces wind speed near leaves and allows sunlit leaves to have temperatures elevated well above T_a. Wind can affect temporal fluctuations of T_s-T_a, but these fluctuations on the 1 km pixel basis of NOAA-AVHRR may be very small, because eddies near the surface are on a scale of about 10 m and because of the spatial integration over the pixel.

The energy budget equation requires an estimate for T_a. If synoptic T_a measurements are used, they should be corrected for shelter and tree height effects (Prosper-Laget et al., 1995). They can also be estimated as the radiative surface temperature of nearby well-watered canopies (Duchemin et al., 1999), since for not well-watered canopies, a systematical bias has been observed because the difference between surface and air temperatures is an indicator of water stress. T_a was also estimated as the radiative surface temperature corresponding to the extrapolation of the NDVI/T_s relationship to an NDVI of an infinitely thick vegetation canopy (e.g., Goward et al., 1994). However, such an estimate requires first that the range of variation in NDVI and T_s is enough to accurately define the slope (Pierce et al., 1990). Second, the images should not be contaminated by clouds, snow or standing water, because the slope can then be positive (e.g., Goward et al., 1994). Third, the canopy should be well-watered because the NDVI/T_s slope can changed as a function of the moisture of the canopy. Indeed, the slope was related to several moisture-related variables which are listed in Section 4.

The energy budget equation also requires the knowledge of the aerodynamic and canopy resistances which are difficult to estimate (Vidal et al., 1994; Vidal and Devaux-Ros, 1995;

Strickland et al., 2001) and whose estimates are valid only for small areas. Thereby, other analytical models for computing AET from T_s have been proposed. The first one is the Surface Energy Balance Algorithm for Land (SEBAL) (Bastiaanssen et al., 1998). It computes LE as a residual quantity of the energy budget equation, but H is derived from the vertical difference in air temperature (δT_a) between the surface roughness length to heat transport (z_{oh}) and the reference height (z_h), δT_a being directly inferred from T_s. SEBAL has been validated on both short and tall vegetation (Bastiaanssen et al., 1998).

The second one does not compute LE as a residual of the energy budget equation. It uses the Penman-Monteith approach, in which the vapour pressure deficit of the air (VPD) is estimated from the saturation vapour pressure at the mean daily surface temperature (VP^*_{Ts}) (Granger, 1997). Indeed, according to the feedback theory, feedback links between the surface and the overlying air are such that the observed surface temperature is a good indicator of the air humidity over the surface (Granger, 1997). It is applicable to both short and tall canopies, because the VPD-VP^*_{Ts} relationship does not depend on the cover type. However, it does not distinguish between vegetated and non vegetated surfaces having the same surface roughness, temperature and air humidity, unless they have a different albedo leading to a different R_n. Its operational use thereby requires a careful land use mapping. There are other more sophisticated approaches to estimate AET from T_s, like soil-vegetation-atmosphere transfer (SVAT) models (see the review in Olioso et al. (1999)). SVAT models usually require a high number of input variables and thereby have little operational potentials in fire management.

2.3. Synergisms between optical and thermal infrared remote sensing

Several empirical studies already showed that inclusion of thermal infrared data improved correlations between NDVI-related indices and drought-related variables (Dominguez et al., 1994; Chuvieco et al., 1999b, 2003; Aguado et al., 2003; Oldford et al., 2003; Oldford et al., 2006; Leblon et al., 2007). Oldford et al. (2006) showed that for slow-drying fuel moisture code (DC) mapping, compared with weather station data interpolation, improved spatial resolution can be achieved at the pixel level when DC is computed using a regression model which has surface temperature and NDVI NOAA-AVHRR images sensing data as independent variables (Figure 3). The fire shown in the center of the 15x15 pixel area was classified by the Sustainable Resource Development Department of Alberta as a surface fire, caused by lightning. It is interesting to observe that the fire burned in a closed coniferous forest cover type which was classified as having a high DC danger rating, when the NOAA-AVHRR image was used, but it was classified as having a moderate DC danger rating in the weather station-based map.

Combining optical vegetation indices with surface temperature data helps account for the influence on the ground cover rate over the composite surface temperature measured by the sensor. This led to defining several drought indices, like the Vegetation and Temperature Condition Index (VT) (Kogan, 2001), an empirical index (Chuvieco et al., 2003), the Water Deficit Index (WDI) (Vidal and Devaux-Ros, 1995), and the Temperature-Vegetation Wetness

Index (TVWI) (Akther and Hassan, 2011). WDI was related to the number of fires and the area burned in the case of Mediterranean forests (Vidal and Devaux-Ros, 1995). TVWI together with the surface temperature and the normalized multiband drought index were related to fire occurrence maps in the case of boreal forests (Akther and Hassan, 2011). The inverse relationship between NDVI and T_s was related to fire occurrences in Mediterranean forests (Prosper-Laget et al., 1994) and to moisture-related variables, such as canopy resistance (Nemani and Running, 1989), sensible and latent heat flux (Nemani and Running, 1989; Olioso et al., 1999), leaf water potential (Goward et al., 1994), accumulated rainfall (Duchemin et al., 1999), FWI codes and indices (Dominguez et al., 1994; Aguado et al., 2003, Oldford et al., 2006; Leblon et al., 2007), and foliar moisture content (Chuvieco et al., 1999b, 2003).

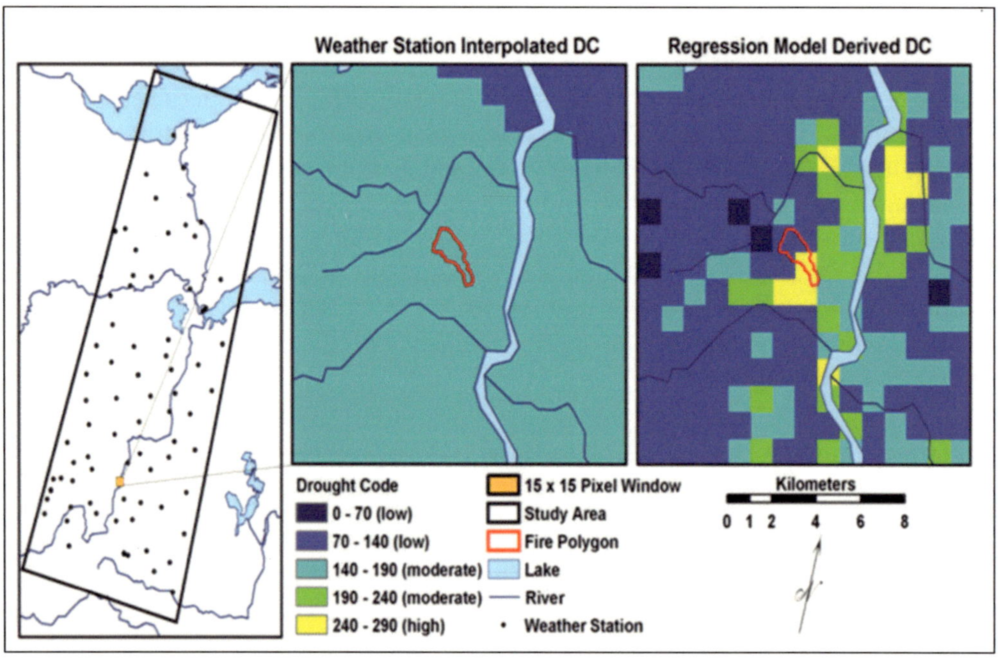

Figure 3. Comparison between a slow-drying fuel moisture code (DC) mapped by weather station interpolation and the one computed by stepwise multiple regression models from NOAA-AVHRR images for the June 1- June 10 1995 compositing period. The fire polygon corresponds to a 62 ha area burned between 2 and 17 June 1995 (after Oldford et al., 2006)

Other synergisms between optical and thermal infrared data can also be considered when estimating AET using the energy budget equation. Indeed, the required net radiation flux (R_n) can be computed from the solar irradiance at the surface or from the surface albedo, both variables being inferred from optical data (e.g., Granger, 1997; Bastiaanssen et al., 1998). Also, the ratio between the soil heat flux (G) and R_n can be analytically derived from optical vegetation indices (e.g., Bastiaanssen et al., 1998; Leblon, 2005).

Using both thermal infrared and NDVI images improve the correlation with fuel moisture variables, but these images have the same operational inconvenience of limited image availability during cloudy days. As reviewed in Leblon (2005), many strategies can be applied to overcome the problem of cloudy days, like the interpolation of evaporation fractions for the cloudy days, or the use of images acquired by passive or active microwave sensors, which are able to penetrate cloud cover. Currently, only the SSM/I sensor provides images acquired in passive microwaves, but at a coarser spatial resolution than NOAA-AVHRR images. For all these reasons, this paper has no further discussion of the use of passive microwaves in fuel moisture monitoring. By contrast, active microwave (or radar) images can be acquired by several existing satellites, i.e., ERS-1/2, ENVISAT, and RADARSAT-1/2, ALOS-PALSAR. In addition to acquiring images under all illumination and weather conditions, these satellites provide data at a finer spatial resolution than NOAA-AVHRR.

2.4. Radar remote sensing

Studies reviewed in Leblon et al. (2002) and in Abbott et al. (2007) have shown that radar backscatter ($\sigma°$) measurements over forested areas depend on (i) vegetation type, species, and structure, (ii) vegetation biomass, (iii) topography and surface roughness and canopy height; (iv) flooding and the presence/absence of standing water, and (iv) moisture. Three sources of moisture variation may contribute to the forest radar backscatter: the forest floor, the canopy (including its woody elements) and the environmental conditions (rain events). Over boreal forests, positive relationships between radar backscatters and rainfall amounts were found with ERS-1 C-VV SAR images (Bourgeau-Chavez et al., 1999; Leblon et al., 2002) and with RADARSAT-1 C-HH SAR images (Abbott et al., 2007). The good correlation between $\sigma°$and weather variables, which are used to compute the various FWI codes and indices, may expect that these indices and codes are also well related to $\sigma°$. FWI codes and indices were correlated to $\sigma°$ derived from ERS-1 C-VV and RADARSAT-1 C-HH SAR images acquired over burned and unburned boreal forests located in Alaska (Bourgeau-Chavez et al., 1999; 2001, 2006, 2007) and in the Northwest Territories, Canada (Leblon et al., 2002; Abbott et al., 2007) (Figure 4).

While these studies produced encouraging results, they also showed that single channel C-band SAR images are restricted in their applicability across the landscape primarily due to variations in surface roughness and biomass which act as confounding factors. Recently, fully polarimetric X-, C- and L-band SAR sensors have been launched into orbit (ALOS-PALSAR in 2006 and TerraSAR-X and RADARSAT-2 in 2007) allowing for decomposition of the backscattered energy into dominant scattering mechanisms which may prove useful for reducing the confounding factors and allowing improved extraction of the variable of interest in the absence of ancillary information.

Bourgeau-Chavez et al. (2012) compared RADARSAT-2 polarimetric SAR images acquired under the same incidence angle and during an extreme dry date and a wet date over a chronosequence of Alaskan boreal black spruce ecosystems (recent burns, regenerating forests dominated by shrubs, open canopied forests, moderately dense forest cover). They found that there was a significant difference between the wet and the dry dates for all backscatter polarizations and for the Freeman-Durden (Freeman and Durden, 1998) and van Zyl decomposition (van Zyl et al., 2011) parameters particularly for the parameter corresponding to odd bounce or surface scatters (Table 1). However, none of the Cloude-Pottier decomposition (Cloude and Pottier, 1997) parameters exhibited significant differences between the wet and dry dates. Indeed, the Cloude-Pottier decomposition works with the polarimetric state only, and does not consider the span information (i.e., radar intensity) in contrast to the two other decompositions. Both use intensity information implicitly and therefore more information from the imaged area. These polarimetric decomposition parameters are currently under investigation in empirical algorithm development for a multi-date dataset (across a range of soil moisture conditions) over the Alaska boreal test area.

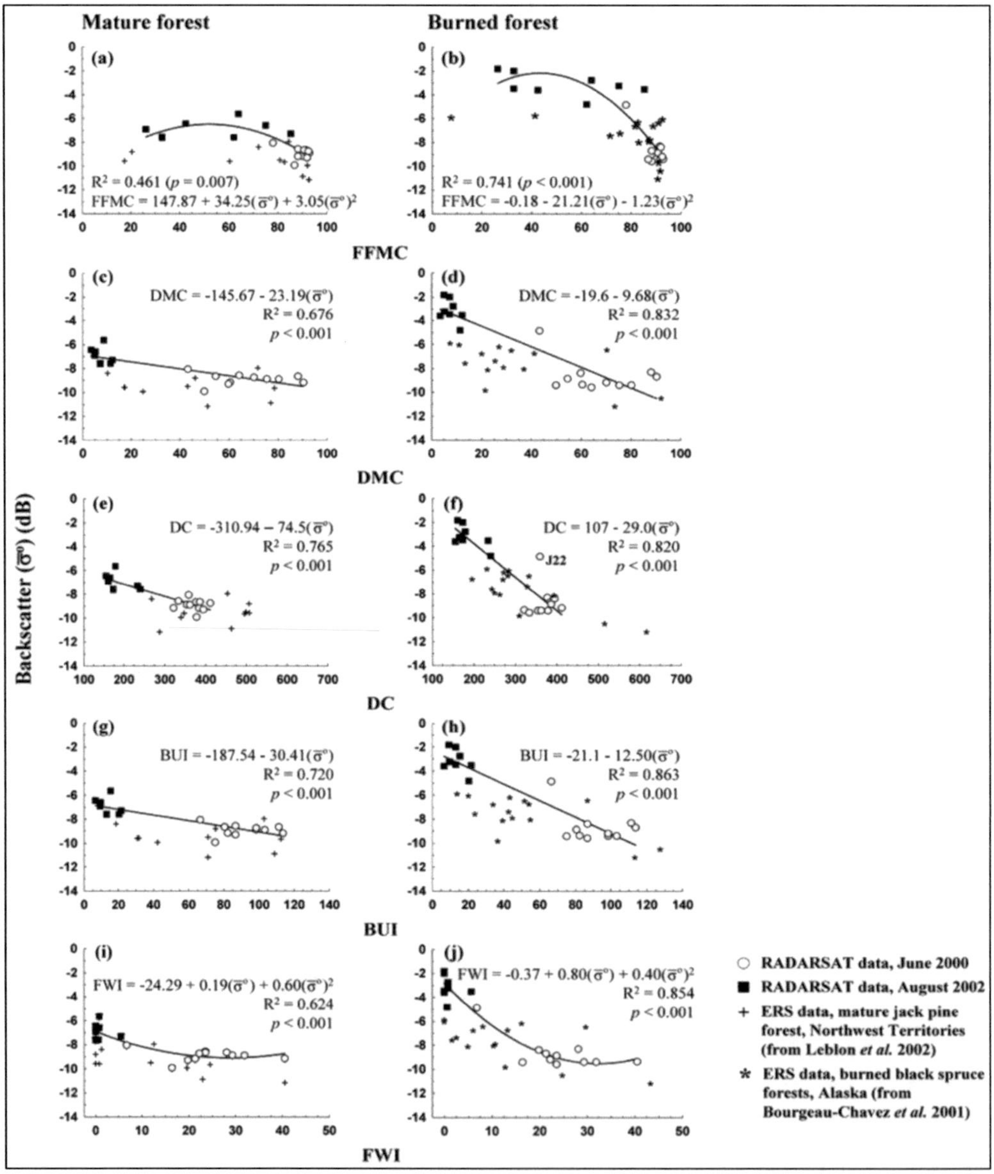

Figure 4. Relationship between ERS-1 C-VV SAR and RADARSAT-1 C-HH SAR radar backscatters (σ^o) and Fire Weather Index (FWI) codes and indices over boreal forest sites in the Northwest Territories, Canada and in Alaska, USA (data from Bourgeau-Chavez et al., 2001; Leblon et al., 2002; Abbott et al., 2007)

Parameter	*p-value*
C-HH backscatter	0.000
C-HV backscatter	0.031
C-VV backscatter	0.000
C-RR backscatter	0.000
C-LR backscatter	0.005
C-LL backscatter	0.020
Cloude-Pottier Alpha	0.698
Cloude-Pottier Anisotropy	0.577
Cloude-Pottier Entropy	0.609
Freeman Durden Double Bounce	0.052
Freeman Durden Odd Bounce	0.005
Freeman Durden Volume Scatter	0.020
van Zyl Double Bounce	0.003

Table 1. *P-value* of the one way ANOVA test for wet vs. dry conditions by SAR parameter measured over several sites of a chronosequence of Alaskan boreal black spruce ecosystems (recent burns, regenerating forests dominated by shrubs, open canopied forests, moderately dense forest cover) (after Bourgeau-Chavez et al. 2012)

Although radar images are theoretically available independently of the weather conditions, their availability could be limited because of the longer repeat cycle of the satellites. For example, ERS-1/2 had a repeat cycle of 35 days. Fortunately the revisit period is shorter for the Canadian radar satellites (RADARSAT-1/2), which has a possible quasi-daily coverage due to its pointing capability (Abbott et al., 2007). In addition, often the radar images have a finer spatial resolution than optical or thermal infrared images, while covering a smaller area. Thus, radar data represent a data source that is complementary to optical or thermal infrared data. Consequently, synergisms between optical or thermal infrared bands and radar bands should be investigated.

3. Fire detection and burnt area mapping

Fire detection is one critical stage of wildfire management, which is aimed at either fighting or monitoring the fire. For fire fighting the early detection is essential; so far, fire detection for fire fighting is based on human observation, the use of fixed optical cameras to monitor the surrounding environment, or aerial survey. The revisit time provided by current satellite sensors is not considered sufficient for fire fighting operations by forest fire mangers. However, the monitoring of wildfires and wildfire effects for large territories is mainly based on satellite remote sensing. Mapping of burnt areas and assessment of wildfire effects is one of the most successful applications of satellite remote sensing. Satellite remote sensing provides the means for acquiring comprehensive and harmonized information on wildfire effects for large territories at low cost. For this purpose, burnt area mapping is performed with a wide variety of remote sensors and techniques. A wide variety of optical and radar sensors have been used for fire detection and burnt area mapping, from local to global scales. This section reviews the application of remote sensing in active fire detection and the assessment of fire damages through the mapping of the extent of burnt areas.

3.1. Fire detection

Fires produce anomalies that are detectable in many different parts of the electromagnetic spectrum and are therefore suitable for detection with the use of remote sensing techniques. Although fire detection is possible in the microwave range of the spectrum, these techniques are not used operational because of the high cost of the sensors and the low nominal achievable spatial resolution of the detection (Kempka et al., 2006). Therefore the focus of this section is on the detection of fires from optical remote sensors.

Firstly, active fires can be detected from the light they emit in the visible part of the spectrum; however, the discrimination of the fire-emitted light is only possible at night (Cahoon et al., 1992, Elvidge, 2001). Since most of the fires occur and have their highest intensity during the day, the detection of them solely at night is not of high interest for operational fire management.

Secondly, fires can be detected by the smoke plume they produce (Figure 5). This detection method is widely used at local scale, as an alternative to visual detection by human operators. Image processing algorithms can be used to single out this smoke plume in contrast to its background, and associate it to a fire. Although these systems eliminate false alarms produced by overheating of ground areas, they also present some limitations. The limitations arise from two facts; first, the smoke plume can only be detectable some time after the fire has started; and second, smoke is often conducted along the surface and emerges in an area different from that where the fire started. Ground automatic detection systems can make use of cameras mounted in towers, buildings or masts with good visibility of the surveyed terrain. The cameras can be fixed (attached to the structure) or mounted on a positioning system to vary the azimuth and elevation angles. A positioning system can be used to survey the entire environment by varying automatically the scanning angles. The detection delay depends on the scan velocity given by the motors and the optical system in the camera, noting that the image processing requirements for automatic detection are higher when using mobile sensors. The sensor technologies used in today's automatic ground detection systems are mainly infrared and visual cameras (San-Miguel-Ayanz et al., 2005).

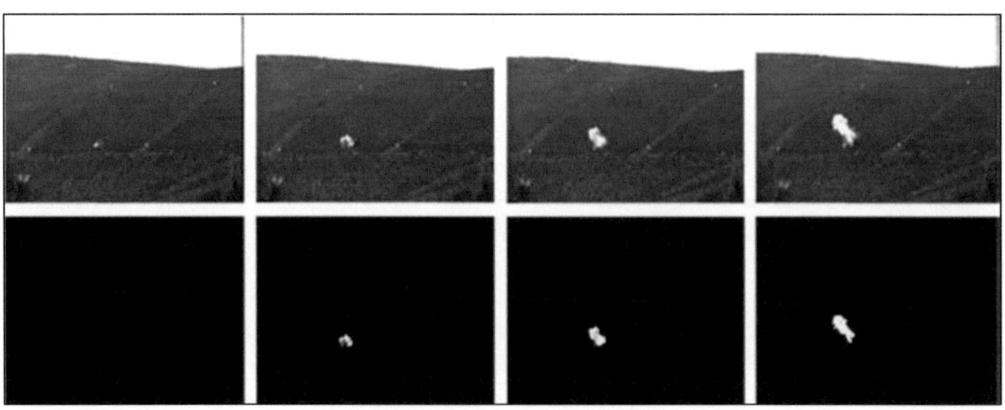

Figure 5. Smoke plume identification for fire detection with optical cameras

Lastly, and most commonly, fires are detected due to the distinct high temperature they produce, which results in a high reflection signal in the mid-infrared and thermal electromagnetic spectra. Active fires produce temperatures ranging between 800 K and 1200 K, although they can reach up to 1800 K. These temperatures are easily detectable in the mid-infrared part of the spectrum (Matson and Dozier 1981). This mid-infrared spectral window is suitable for fire detection because it is far from the peak of the Earth and Solar radiations at 0.5 and 9.7 μm, respectively (Figure 6). Fires also radiate in the thermal part of the spectrum, i.e. between 8 μm and 12 μm; however, the peak radiation at these wavelengths corresponds to a normal environmental temperature of 300 K. Fires can be detected as local or absolute maximum in the mid-infrared and thermal spectra. An absolute (or regional) maximum is used in the so-called thresholding algorithms. Any area above a given threshold temperature is considered a fire. However, differences in fire characteristics among regions in the world lead to problems of false alarms and/or missed fires using this method. Although fixed thresholding algorithms were used in the past most current techniques fore fire detection make use of the so-called contextual algorithms. Contextual algorithms detect local maxima. Multispectral criteria are aimed at detecting the difference between a fire pixel (active fire) and the background temperature (environmental temperature in the proximity of the fire pixel (Flasse and Ceccato, 1996, Giglio et al., 2003).

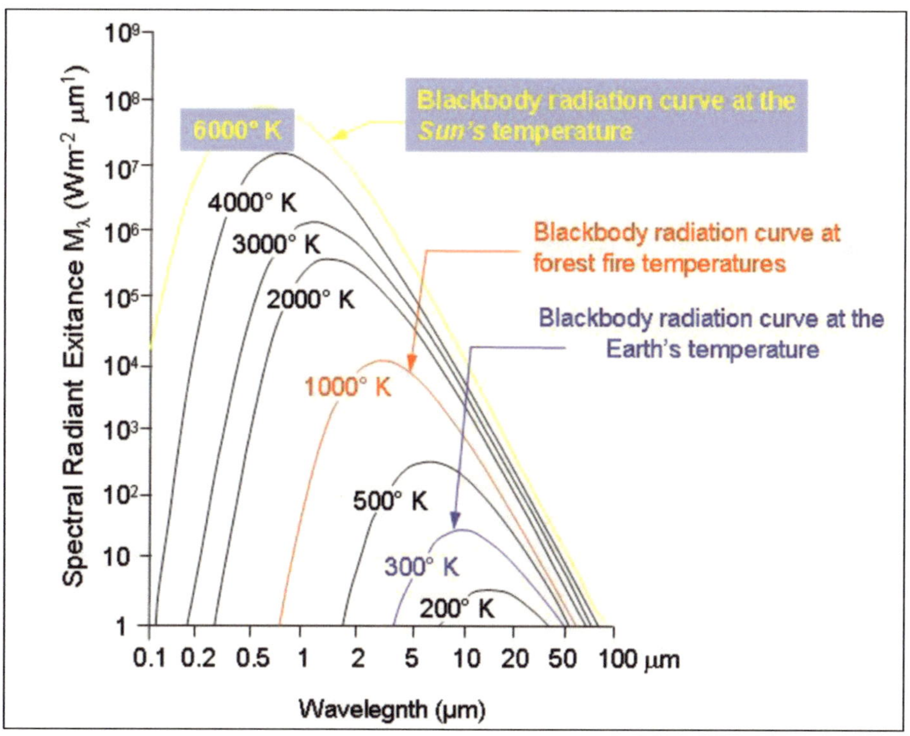

Figure 6. Spectral radiant exitance as a function of the temperature of the black body. The figure shows that forest fires being hotter than the Earth's surface exhibit a peak in their spectral exitance at a shorter wavelength than the Earth's surface

Active wildfire monitoring is performed through the use of geo-stationary satellite sensors such as GOES (Geostationary Operational Environmental Satellite) or SEVIRI on board of the Meteosat Second Generation (MSG) satellite, or geo-synchronous satellite sensors such as the AVHRR on board of the NOAA meteorological satellite, the ATSR (Along Track Scanning Radiometer) on board the ERS-1 and 2 and the Envisat, and the MODIS (Moderate Resolution Imaging Spectroradiometer) on board of the Terra and Aqua satellites.

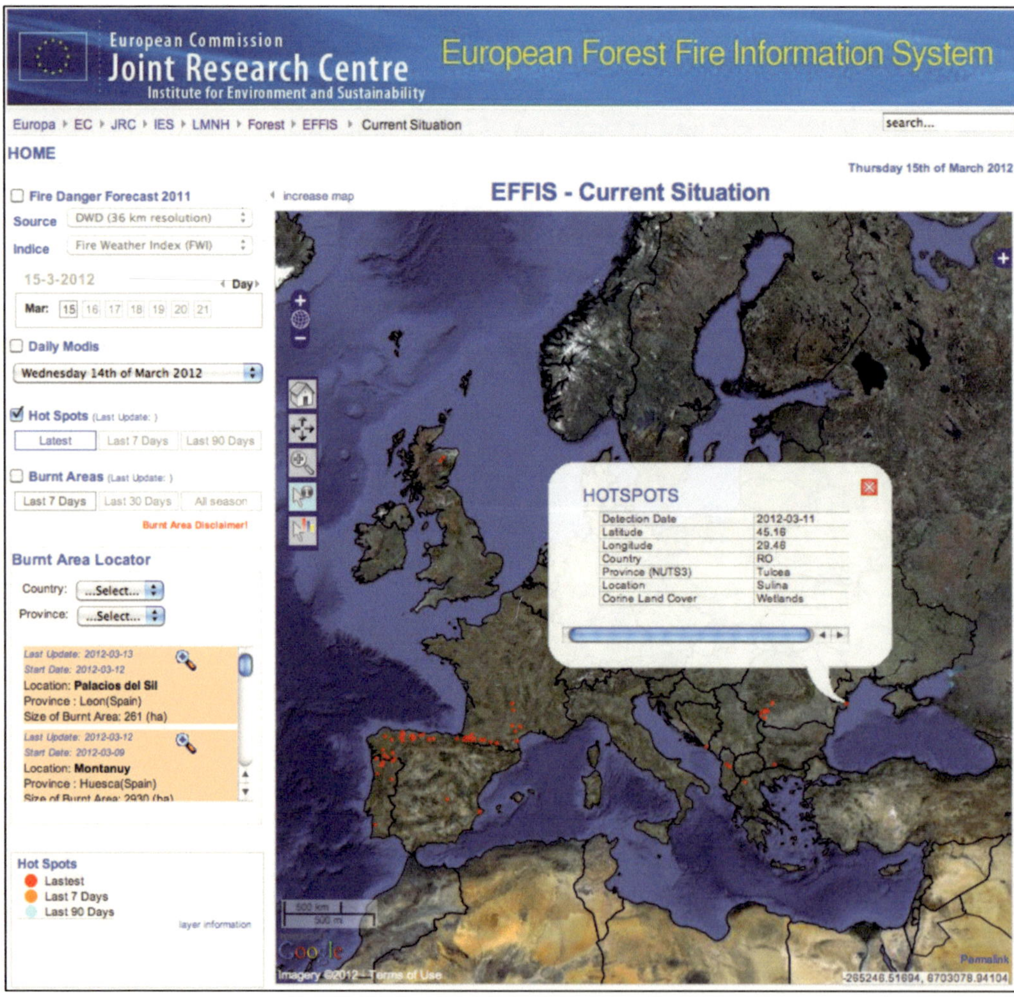

Figure 7. MODIS-based active fire detection in the European Forest Fire Information System (EFFIS) (http://effis.jrc.ec.europa.eu)

GOES and SEVIRI provide high frequency coverage in the order of 30 minutes and 15 minutes, respectively. They are thus suitable for the monitoring of most wildfire processes

(Prins and Menzel, 1992; Prins et al., 1998). This is a relative advantage for the monitoring of fire activities, as compared to ATSR, AVHRR that provide a maximum of 1 daily pass, or MODIS, which provides 2 daily passes. However, due to its high spatial resolution, its good fire detection capabilities and its global coverage, MODIS has become the standard sensor for active fire monitoring at regional to global scales. The Aqua MODIS instrument acquires data twice daily (1:30 PM and AM), as does the Terra MODIS (10:30 AM and PM). These four daily MODIS fire observations serve to advance global monitoring of fire processes and their effects on ecosystems, the atmosphere, and climate (Giglio et al., 2006a and 2006b). Some operational fire monitoring systems using MODIS active fire detection include the Canadian Wildland Fire Information System (CWFIS) (http://cwfis.cfs.nrcan.gc.ca), the USA Active Fire Mapping Service, or the European Forest Fire Information System (EFFIS) (http://effis.jrc.ec.europa.eu). In the case of EFFIS, post-processing filters based on landcover ancillary data are applied to the MODIS product to reduce the number of false alarms produced by non-fire hot surfaces (e.g. industrial areas, hot ground soils) and therefore increase the reliability of the active fire detection (San-Miguel-Ayanz et al., 2012). Figure 7 shows operational fire detection monitoring in the European region within EFFIS.

3.2. Burnt area mapping

Remotely sensed data have been extensively used for burnt area mapping. Fires produce a significant change in the structure and the reflectance of vegetation and the soil properties within the burnt area that are noticeable in the microwave, visible and especially the infra-red part of the electromagnetic spectrum.

At the global scale, NOAA-AVHRR data were extensively tested in the 1990s. Studies differed mainly on the use of diverse spectral indices, although most commonly, burn scar areas were discriminated from a multi-temporal comparison of NDVI (Kasischke and French, 1993; Martin and Chuvieco, 1995; Pereira, 1999). More recently other global burnt area datasets were derived from SPOT Vegetation and the ATSR-2 on board of Envisat (Gregorie et al., 2003, Tansey et al., 2004; Simon et al., 2004). Although these data provide gross estimates of burnt areas at the global level, the lack of extensive validation and agreement between them limit their use at regional or national levels. Nevertheless, partial validations of the global burnt area products were performed by Roy et al. (2005) and Boschetti et al. (2007). Pereira et al. (1999) showed that the accuracy of the results for mapping burnt areas with AVHRR data in the Mediterranean region of Europe was about 80% for large fires. The methods were considered suitable only for fires larger than 1000 ha, and reliable for fires larger than 2000 ha. However, the mapping of those fires would correspond only to approximately 30% and 21%, respectively, of the total yearly burnt area in the European Mediterranean region.

With the launch of the MODIS sensor on board of the TERRA and AQUA satellites, a new capability for regional mapping of burnt areas was put in place. The availability of free data of medium spatial resolution from the MODIS sensors since 2000 provided a definite impulse for the use of remote sensing at the regional and global scales (Justice et al, 2002). Better radiometry and higher spectral information of the MODIS sensor provided the right data for the discrimination of burnt areas at these scales. The simultaneity in the operation of both satellites provided higher frequency in data acquisition and enough revisit time for accurate mapping of burnt areas. At the global scale, the MODIS program has released a standard product on burned areas that is based on a multitemporal change detection approach to analyze differences between modeled and actual reflectance, and to take into account Bidirectional Reflectance Distribution Function (BRDF) corrections (Roy et al., 2002, 2005).

At regional scale, MODIS is operationally used in systems such as Canadian CWFIS and the European EFFIS, mentioned above. Two full mosaics of MODIS data are received and processed daily in EFFIS to provide near-real time monitoring of wildfires and map burnt areas. The systems is thus updated up to two times daily, providing accurate information of fire impacts in Europe ((San-Miguel-Ayanz et al. 2009). The use of higher spatial resolution imagery from Advanced Wide Field Sensor (AWiFS) for regional coverage in Europe was recently tested. However, results of this exercise showed that the benefits derived from the use of high spatial imagery in term of detailed mapping of fire perimeters are obscured by the limitations in the revisit time of the sensor. These results did not enhance those of the standard Rapid Damage Assessment module of EFFIS based on MODIS imagery (Sedano et al., 2012). Figure 8 shows the extent of burnt area as they were mapped from MODIS and AWIFS imagery.

At national to local scales, the wide variety of remotely sensed products at medium to high resolution (10 m to 30 meter ground spatial resolution), make it possible the accurate mapping of burnt areas. However, the increase in spatial resolution is often accompanied by a decrease in revisit time of the sensor, which prevents the acquisition of this imagery for extensive areas.

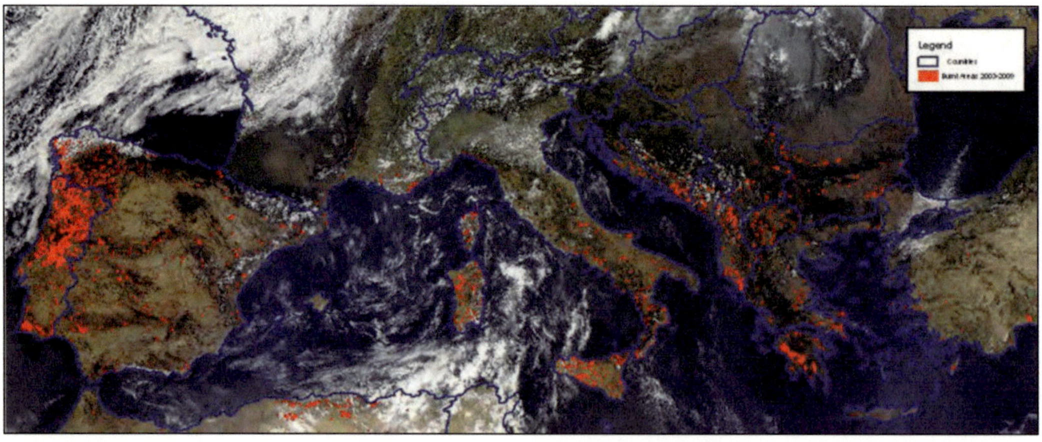

Figure 8. Burnt areas in the European Mediterranean region between 2000 and 2009 (http://effis.jrc.ec.europa.eu)

High-spatial burnt area mapping has been performed with Landsat Thematic Mapper imagery (Michalek et al. 2000, Pereira and Setzer, 1993, Chuvieco and Congalton, 1998) complemented in some cases by the SPOT and ASTER sensors. Some analyses made use of the LISS-3 sensor of the IRS Indian satellite, and the RESURS MSU-K (San-Miguel-Ayanz et al, 1998). A variety of indices computed from the original spectral bands were used to enhance the mapping of burnt areas (Pereira et al., 1997, Li et al. 2000, Chuvieco et al, 2002). However, this exercise was, in most cases, limited to the mapping of burnt areas at local and sub-national scale. An exception to this is the case of Portugal, where an operational system capable of processing Landsat TM scenes for mapping of burnt areas was set up (Pereira et al., 1993).

The use of high resolution remote sensing in the management of critical wildfires has improved dramatically in the last decade. The variety of remote sensing imagery of high and very-high spatial resolutions such QUICKBIRD, IKONOS, FORMOSAT, EARLYBIRD, RAPIDEYE has permitted the rapid coverage of critical fire events. The processing of this imagery provides a great level of spatial detailed that is needed for the accurate analysis of fire damages and the sound planning of restoration measures. Data provision for critical fire events has been supported by the agreement of the space agencies in the so-called International Space Charter, which allows the rapid provision available remotely sensed data from a series of satellites, including RADARSAT, ERS, ENVISAT, SPOT, IRS, SAC-C, NOAA satellites, LANDSAT, ALOS, DMC supporting crisis management.

Although most of the studies on burnt area mapping were based on the use of optical imagery, there are a series of examples in which data from active sensors such as the Synthetic Aperture Radar (SAR) were used. Most of the studies were carried out in boreal forest (Bourgeau-Chavez et al. 1997, 2002, Kasische et al, 1994, French et al. 1999, Siegert and Ruecker, 2000, Menges et al, 2004), but some examples for the Mediterranean area exist (Gimeno and San-Miguel-Ayanz, 2004, Gimeno et al. 2005). Rather than the changes in vegetation condition and structure, the detection of burnt areas from SAR is based on the changes on moisture content in the burnt surface with respect to the unburned areas. Burnt areas tend to have higher moisture content than unburned areas, which reduces the backscatter. Thus, burnt areas appear as dark objects in relation to the surrounding non-affected areas.

Similarly to the studies on pre-fire conditions (see Section 2), polarimetric SAR images have been recently tested. They were more efficient for fire detection and burnt area mapping than single channel C-band SAR images. Figure 9 compares RADARSAT-2 single-polarized and polarimetric SAR images that were acquired at 32.4°-34° incidence angle and using a east-looking direction, during a dry day, between 1 to 2 months after fires that occur in rough terrains (Calabria peninsula in Southern Italy). Similar to the pre-fire conditions study (see Table 1), fire scars were more visible on the Freeman-Durden decomposition images than on the Cloude-Pottier decomposition, probably because the first decomposition works with the polarimetric state only, and does not consider the span information (i.e., radar intensity) in contrast to the second one. The Freeman-Durden decomposition uses more information from the imaged area because it implicitly considers the intensity information.

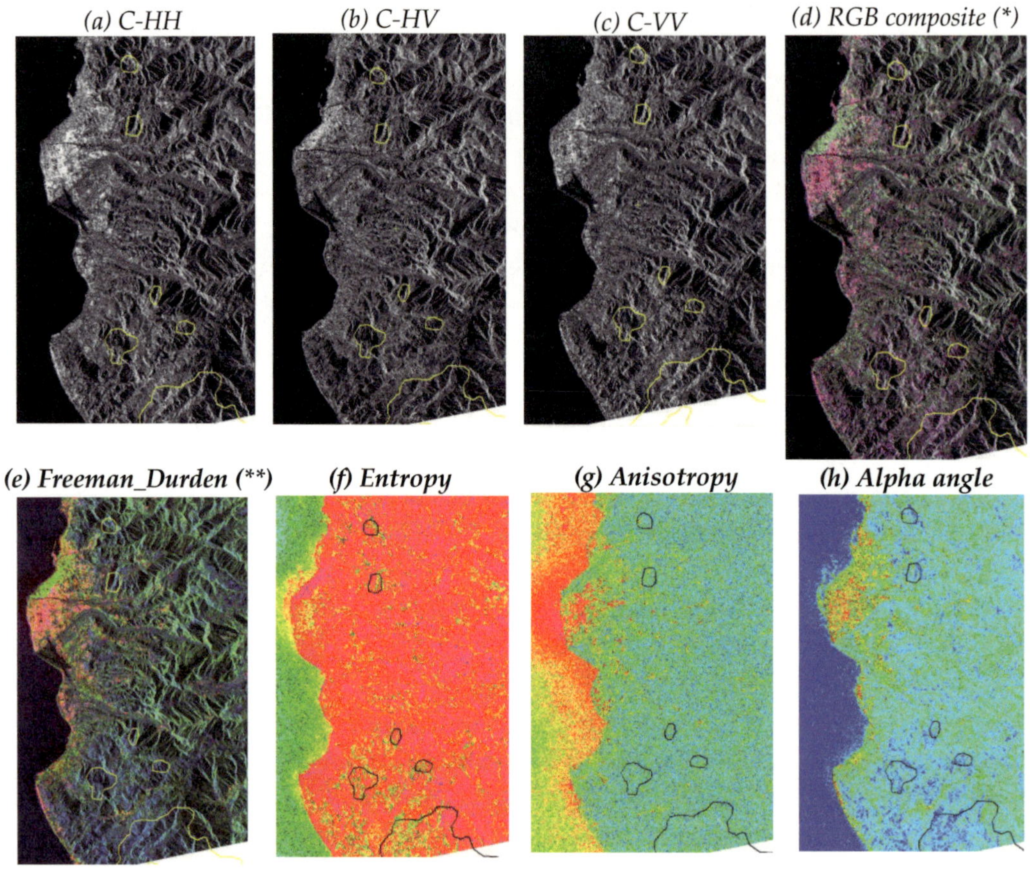

(*)HH in red, HV in green and VV in blue; (**) double-bounce in red, volume in green and surface in blue

Figure 9. Fire scars over various RADARSAT-2 products made from a RADARSAT-2 that was acquired over Calabria peninsula in Southern Italy 1 to 2 months after the fires during a dry date using an ascending pass (east-looking direction) with a FQ13 beam mode (32.4°-34° incidence angle). The fire scar limits are displayed in yellow or black.

4. Conclusions

We reviewed studies using optical, thermal infrared and radar images for pre-fire and post-fire conditions monitoring. For the pre-fire conditions, our review has a particular emphasis on the studies using satellite data to monitor fuel moisture. Remote sensing of fuel moisture was first done with NDVI images (mainly from NOAA-AVHRR), based on the assumption that the greenness of the scene is a good indicator of fuel moisture and fire danger. NDVI images are used operationally to map fire potentials, but the reviewed studies have shown that NDVI is sensitive to the chlorophyll activity of the vegetation rather than to its actual changes in moisture content. By contrast, thermal infrared images allow the computation of surface temperatures which are analytically related to surface moisture-related variables, like evapotranspiration, through the energy budget equation. This analytical approach was used to compute the ratio between actual and potential evapotranspiration (AET/PET) from daily surface temperatures and synoptic air temperatures. The ratio was used in an operational fire danger monitoring system over Mediterranean forests. The same ratio was computed from optical and thermal infrared images acquired over Canadian northern boreal forests and related to FWI codes and indices. In the AET/PET computation, NDVI images were also used, in the calculation of soil heat flux and aerodynamic resistances.

More recent studies on the use of remote sensing in fuel moisture monitoring use both NDVI and surface temperature images. This is done at no additional cost of data acquisition, since both kinds of images are provided in the same time by numerous existing satellites, like NOAA-AVHRR, LANDSAT-TM, ATSR-2, RESURS-01, METEOSAT, GOES, EOS-ASTER, and EOS-MODIS. An important operational limitation of using optical and thermal infrared data is image availability, which still depends on weather and illumination conditions. For this reason, an operational system to monitor fuel moisture using satellite images should probably also include radar images, which can be acquired during cloudy days. These images have also the advantage of having a finer spatial resolution.Good relationships were found between ERS-1 and RADARSAT-1 radar backscatters and FWI codes and indices over northern boreal forests, but these studies also showed that radar backscatters are affected by several confounding factors other than those related to moisture, such as surface roughness and biomass. More recently, the availability of polarimetric SAR images allows for decomposition of the backscattered energy into dominant scattering mechanisms which may prove useful for reducing the confounding factors. Statistically significant differences between wet and dry dates were observed in the case of several polarimetric variables extracted from RADARSAT-2 C-band polarimetric SAR images, such as the Freeman-Durden and van Zyl decomposition parameters particularly for the parameters corresponding to odd bounce or surface scatters. Further studies are required to establish models that can use such data for estimating fuel moisture.

One limitation of the operational use of SAR images in fire danger monitoring is image availability that is limited by the long revisit periods of most existing radar satellites and by the commercial operating mode of some new radar satellites, like RADARSAT-2. However, the availability in the near future of SAR satellite constellations such as the planned RADARSAT-3 mission, will decrease the revisit period. Eventually, further studies are needed to assess the combination of optical and thermal infrared images to radar images for monitoring pre fire conditions.

Most of the reviewed remote sensing studies for pre-fire conditions management are based on the estimation of fuel moisture which is one of the canopy factors which influences fire danger. However, further research is needed to see whether or not current fire danger rating systems can account for canopy variables, like evapotranspiration or moisture content, which are more closely related to spectral variables. On the other hand, all required input variables of an operational fire danger system, like wind parameters, will surely not be derived from remote sensing data and additional ground-based and weather information will always be required to effectively monitor fire danger. In this regard, *in situ* sensing systems, as described in Teillet et al. (2001), will be useful.

Regarding fire detection, remote sensing techniques can be considered fully operational. At local scale they are mainly based on the use of visible and infra-red cameras for the detection of active fires or smoke plumes. Fire detection at this scale is focused on support to forest fire fighting operations. At large scale, information is provided by geo-stationary satellite sensors (GOES, SEVIRI) or geo-syncronous sensors (AVHRR, ATSR, MODIS). The high revisit time of the geostationary satellites provide frequent information (15 to 30 minutes) that is indicated for monitoring fire processes and fire effects. However, although geo-stationary satellites provide a lower revisit time (1 to 2 daily passes), they provide global fire information that is essential for the monitoring of wildfire processes and their effects on ecosystems, the atmosphere, and climate.

Burnt area mapping from remote sensing has been on-going for nearly 30 years. Most of these applications are based on passive optical remote sensing imagery at global and regional scales. Global burn area datasets were derived from AVHRR, ATRS, Vegetation, and recently from MODIS. At local scale, actives sensors such as the ERS SAR and RADARSAT have proven their capacity for monitoring fires under all-weather conditions. Currently, the data acquired by the MODIS sensor has become the standard for fire monitoring at regional to global scales and is used for environmental policy and decision-making. At local level, numerous examples on the use of high-spatial resolution imagery exist. However, the lack of operational routines for the processing of satellite imagery and the difficulties in acquiring cloud free imagery due to the low revisit time of the sensors has prevented the full operationalization of remote sensing. Agreements have been recently established among the Space Agencies in the International Space Charter for the provision of remote sensing data for wildfire crisis management, which permits the rapid monitoring of critical fire events.

Author details

Brigitte Leblon and Laura Bourgeau-Chavez
Faculty of Forestry and Environment Management,
University of New Brunswick, Fredericton, New Brunswick, Canada

Jesús San-Miguel-Ayanz
Forest Resources and Climate Unit, Institute for Environment and Sustainability, Ispra, Italy

Acknowledgement

The pre-fire conditions study is a compendium of the works from the following students: Lisa Gallant, Shannon White, Mark Doyle, Melissa Abbott, Guy Strickland, Keith Abbott, Steven Oldford and P.A Fernandez-Garcia. ERS-1 and RADARSAT-1 SAR image processing were helped by N. French and Gordon Staples Field support for the RADARSAT-2 study was provided by Aditi Shenoy, Eric Kasischke, Kevin Riordan, Kristen Manies, and Nancy French. The whole study was funded by CIFFC, MacDonald, Dettwiler and Associates Ltd. (MDA)and by a NSERC Discovery grant awarded to B. Leblon. Additional funding for this research was provided by NASA grants NNX09AM15G and NNG04GR24G. NOAA-AVHRR images were provided under the EODS program, thanks to J. Cihlar and J. Chen,

Canada Centre of Remote Sensing, Ottawa. ERS-1 SAR images were provided by E.S. Kasischke. RADARSAT-1 images were acquired under the ADRO-2 program of the Canadian Space Agency. RADARSAT-2 polarimetric images were provided by a Canadian Space Agency SOAR grant (SOAR#445). Weather data were obtained from Marty Alexander, Canadian Forest Service, Edmonton and Rick Lanoville, Government of Northwest Territories, Fort Smith. Examples for the sections on active wildfire detection and burnt area mapping were extracted from the European Forest Fire Information System (EFFIS). This is the result of a team of scientists working in different fire-related disciplines at the at the European Commission Joint Research Centre (http://effis.jrc.ec.europa.eu)

5. References

Abbott, K., Leblon, B., Staples, G., Alexander, M.E., & MacLean, D. (2007). Fire danger monitoring in a northern boreal forest region using RADARSAT-1 imagery. *International Journal of Remote Sensing,* Vol. 28, No 5-6, (March 2007), pp. 1317-1338, ISSN 0143-1161

Aguado, I., Chuvieco, E., & Salas, J. (2003). Assessment of forest fire danger conditions in southern Spain from NOAA images and meteorological indices. *International Journal of Remote Sensing,* Vol. 24, No 8, (January 2003), pp. 1653-1668, ISSN 0143-1161

Akther, M. S., & Hassan, Q. K. (2011). Remote sensing-based assessment of fire danger conditions over boreal forest. *IEEE Journal of Selected Topics in Applied Earth Observations and Remote Sensing,* Vol. 4, No. 4, (December 2011), pp. 992-999, ISSN 1939-1404.

Bastiaanssen, W.G.M., Menenti, M., Fedded, R.A., & Holtslag, A.A.M. (1998). A remote sensing surface energy balance algorithm for land (SEBAL). 1. Formulation. *Journal of Hydrology,* Vol. 212-213, (December 1998), pp. 198-212, ISSN 0022-1694.

Boschetti, L., Roy, D., Barbosa, P., Boca, R., & Justice, C. (2008). A MODIS assessment of the summer 2007 extent burned in Greece. *International Journal of Remote Sensing,* Vol. 29, No 8, (April 2008), pp. 2433-2436, ISSN 0143-1161

Bourgeau-Chavez, L.L., Harrell, P.A., Kasischke, E.S. & French, N.H.F. (1997). The detection and mapping of Alaskan wildfires using a spaceborne imaging radar system. *International Journal of Remote Sensing,* Vol. 18, No 2, (January 1997), pp. 355-373, ISSN 0143-1161

Bourgeau-Chavez, L.L., Kasischke, E.S., & Rutherford, M.D. (1999). Evaluation of ERS SAR data for prediction of fire danger in a boreal region. *International Journal of Wildland Fire,* Vol. 9, No 3, pp. 183-194, ISSN 1049-8001.

Bourgeau-Chavez, L. L., Brunzell, S., Nolan, M., & Hyer, E. (2001). Analysis of SAR data for fire danger prediction in boreal Alaska. *Final Report, ASF-IARC Grant NAS-98-129,* 59 pages.

Bourgeau-Chavez, L.L., Kasischke, E.S., Brunzell, S.M., Tukman, M. & Mudd, J.P. (2002) Mapping fire scars in global boreal forests using imaging radar data. *International Journal of Remote Sensing,* Vol. 22, No 18, (January 2002), pp. 3665-3687, ISSN 0143-1161.

Bourgeau-Chavez, L.L, Riordan, K., Garwood, G., Cella, B., Alden, S., Kwart, M., Murphy, K., Ferguson, S., Slawski, J., Medvecz, M., Ames, S.& Walters, T. (2006) Benchmark

Report on the use of Satellite radar imagery for monitoring fuel moisture in Alaska. Final report on NASA SENH project NAS5-03113. 96 pp. Available from http://aiwg.gsfc.nasa.gov/esappdocs/benchmarks/NASASENHFuelMoisture_Bench_sept06.pdf

Bourgeau-Chavez, L.L., Kasischke, E.S., Riordan, K., Brunzell, S.M., Hyer, E., Nolan, M., Medvecz, M. & Ames, S. (2007). Remote monitoring of spatial and temporal surface soil moisture in fire disturbed boreal forest ecosystems with ERS SAR imagery. *International Journal of Remote Sensing,* Vol. 28, No 10, (May 2007), pp. 2133-2162, ISSN 0143-1161

Bourgeau-Chavez, L.L., Leblon, B., Buckley, J., & Charbonneau, F. (2012). Assessment of polarimetric SAR data for fuel moisture estimation: Analysis of wet *versus* dry conditions. *International Journal of Remote Sensing (accepted),* ISSN 0143-1161

Burgan, R.E., Klaver, R. W., & Klaver, J.M. (1998). Fuel models and fire potential from satellite and surface observations. *International Journal of Wildland Fire,* Vol. 8, No 3, pp. 159-170, ISSN 1049-8001.

Cahoon, D. R., Jr., Stocks, B. J., Levine, J. L., Cofer III, W. R., & O'Neill, K. P. (1992). Seasonal distribution of African savanna fires. *Nature,* Vol. 359, (October 1992), pp. 812– 815, ISSN 0028-0836

Camia, A., Bovio, G., Aguado, I., & Stach, N. (1999). Meteorological fire danger indices and remote sensing. In *Remote Sensing of Large Wildfires in the European Mediterranean Basin,* E. Chuvieco Ed., , pp. 39-59, Springer-Verlag, ISBN 978-3540657675, Berlin.

Canadian Forest Service. (1992). Development and Structure of the Canadian Forest Fire Behaviour Prediction System. Canadian Forest Service, Information Report ST-X-3, Ottawa, ONT., 63 p.

Canadian Forest Service. (2012). *Forest fire facts and questions.* Available from http://cfs.nrcan.gc.ca/pages/153

Ceccato, P., Flasse, S., Tarntola, S., Jacquemoud, S., & Grégoire, J.M. (2001). Detecting vegetation leaf water content using reflectance in the optical domain. *Remote Sensing of Environment,* Vol. 77, No 1, (July 2001), pp. 22-33, ISSN 0034-4257

Chuvieco, E., & Martin, M. P. (1994). Global fire mapping and fire danger estimation using AVHRR images. *Photogrammetry Engineering and Remote Sensing,* Vol. 60, No 5, (May 1994), pp. 563-570.

Chuvieco, E., Aguado, I., Cocero, D., & Riaño, D. (2003). Design of an empirical index to estimate fuel moisture content from NOAA-AVHRR analysis in forest fire danger studies. *International Journal of Remote Sensing,* Vol. 24, No 8, (January 2003), pp. 1621-1637, ISSN 0143-1161.

Chuvieco, E., Aguado, I., Yebra, M., Nieto, H., Salas, J., Pilar, M., Vilar, L., Martínez, J., Martín, S., Ibarra, P., de la Riva, J., Baeza, J., Rodríguez, F., Molina, J. M., Herrera, M.A., & Zamora, R. (2010). Development of a framework for fire risk assessment using remote sensing and geographic information system technologies. *Ecological Modelling,* Vol. 221, No 1, (January 2010), pp. 46–58, ISSN 0304-3800.

Chuvieco, E., Aguado, I., Yebra, M., Nieto, H., Salas, J., Pilar, M., Vilar, L., Martínez, J., Martín, S., Ibarra, P., de la Riva, J., Baeza, J., Rodríguez, F., Molina, J. M., Herrera, M.A., & Zamora, R. (2010). Development of a framework for fire risk assessment using remote sensing and geographic information system technologies. *Ecological Modelling*, Vol. 221, No 1, (January 2010), pp. 46–58, ISSN 0304-3800.

Chuvieco, E., Deshayes, M., Stach, N., Cocero, D., & Riaño, D. (1999b). Short-term fire risk foliage moisture content estimation from satellite data. In *Remote Sensing of Large Wildfires in the European Mediterranean Basin*, E. Chuvieco Ed., pp. 17-38, Springer-Verlag, ISBN 978-3540657675, Berlin.

Chuvieco, E., Deshayes, M., Stach, N., Cocero, D., & Riaño, D. (1999b). Short-term fire risk foliage moisture content estimation from satellite data. In *Remote Sensing of Large Wildfires in the European Mediterranean Basin*, E. Chuvieco Ed., pp. 17-38, Springer-Verlag, ISBN 978-3540657675, Berlin.

Chuvieco, E., Salas, F.J., Carvacho, L., & Rodirguez-Silva, F. (1999a). Integrated fire risk mapping, In *Remote Sensing of Large Wildfires in the European Mediterranean Basin*, E. Chuvieco Ed., pp. 61-84, Springer-Verlag, ISBN 978-3540657675, Berlin.

Cloude, S.R., & Pottier, E. (1997). An entropy based classification scheme for land applications of polarimetric SAR. *IEEE Transactions Geoscience and Remote Sensing*, Vol. 35, No 1, (January 1997), pp. 68-78, ISSN 0196-2892.

Deblonde, G., & Cihlar, J. (1993). A multiyear analysis of the relatonsip between surface environmental variables and NDVI over the Canadian landmass, *Remote Sensing of Environment*, Vol. 7, pp. 151-177, ISSN 0034-4257.

Desbois, N., & Vidal, A. (1995). La télédétection dans la prévision des incendies de forêt, *Ingénieries- EAT*, Vol. 1, (March 1995), pp. 21-29.

Desbois, N., & Vidal, A. (1996). Real-time monitoring of vegetation flammability using NOAA-AVHRR thermal infrared data, *EARSeL Advanced in Remote Sensing*, Vol. 4, No 4, (November 1996), pp. 25-32.

Dominguez, L., Lee, B.S., Chuvieco, E., & Cihlar, J. (1994). Fire danger estimation using AVHRR images in the Prairie provinces of Canada. *Proceedings 2nd International Conference on Forest Fire Research*, Vol. 2, No 17, pp. 679-690, Coimbra, Portugal, November 1994.

Duchemin, B., Guyon, D., & Lagouarde, J. P. (1999). Potential and limits of NOAA-AVHRR temporal composite data for phenology and water stress monitoring of temperate forest ecosystems, *International Journal of Remote Sensing*, Vol. 20, No 5, (January 1999), pp. 895-917, ISSN 0143-1161.

Elvidge, C. D. (2001), DMSP-OLS estimation of tropical forest area impacted by surface fires in Roraima, Brazil: 1995 versus 1998. *International Journal of Remote Sensing*, Vol. 22, No 14, (January 2001), pp. 2661–2673, ISSN 0143-1161

Flasse S. P., & Ceccato, P. (1996). A contextual algorithm for AVHRR fire detection. *International Journal of Remote Sensing*, Vol. 17, No 2,(January 1996), pp. 419-424, ISSN 0143-1161.

Food and Agriculture Organization (FAO) (2012). *About the Global Fire Information Management System (GFIMS)*. Available from http://www.fao.org/nr/gfims/about/en/

Freeman, A., & Durden, S.L. (1998). A three-component scattering model for polarimetric SAR data. *IEEE Transactions Geoscience and Remote Sensing*, Vol. 36, No 3, (May 1998), pp. 963- 973, ISSN 0196-2892

French, N. H. F., Bourgeau-Chavez, L. L., Wang, Y. & Kasischke, E. S. (1999). Initial observations of Radarsat imagery at fire-disturbed sites in interior Alaska. *Remote Sensing of Environment*, Vol. 68, No 1, (April 1999), pp. 89-94, ISSN 0034-4257

Giglio, L, Csiszar, I, & Justice, C. O. (2006a). Global distribution and seasonality of active fires as observed with the Terra and Aqua Moderate Resolution Imaging Spectroradiometer (MODIS) sensors. *Journal of Geophysical Research-Biogeosciences*, Vol. 111, No G2, (June 2006), p. G02016, ISSN 0148–0227

Giglio, L, van der Werf, GR, Randerson, JT, Collatz, GJ, & Kasibhatla, P (2006b). Global estimation of burned area using MODIS active fire observations. *Atmospheric Chemistry and Physics*, Vol. 6, No 4, (March 2006), pp. 957-974, ISSN 1680-7316.

Gimeno, M. & San-Miguel-Ayanz, J. (2004). Evaluation of RADARSAT-1 data for identification of burnt areas in Southern Europe. *Remote Sensing of Environment*, Vol. 92, No 3, (August 2004), pp. 370-375, ISSN 0034-4257

Gimeno, M., San-Miguel-Ayanz, J. & Schmuck, G. (2004). Identification of burnt areas in Mediterranean forest environments from ERS-2 SAR time series. *International Journal of Remote Sensing*, Vol. 25, No 22, (November 2004), pp. 4873-4888, ISSN 0143-1161

Gouyet, J.F., King, C., Le Gleau, H., Malon, J.F., Phulpin, T., & Valette, J.C. (1991). Apport des données satellitaires NOAA-AVHRR dans le suivi de la végétation forestière. *Proceedings 5th International Colloquium on Physical Measurements and Signatures in Remote Sensing*, pp. 625-629, Courchevel, France, January 1991

Goward, S.N., Waring, R.H., Dye, D.G., & Yang, J. (1994). Ecological remote sensing of OTTER satellite macroscale observations. *Ecological Applications,* Vol. 4, No 2, (May 1994), pp. 322-343, ISSN 1051-0761.

Granger, R.J. (1997). Comparison of surface and satellite-derived estimates of evapotranspiration using a feedback algorithm. *Proceeding. 3rd International Workshop on Application of Remote Sensing in Hydrology*, pp. 71-81, Goddard Space Flight Center, Washington, DC, USA, October, 1996.

Grégoire, J.-M. Tansey K., and Silva, J. M. N. (2003). The GBA2000 initiative: Developing a global burnt area database from SPOT-Vegetation imagery, *International Journal of Remote Sensing*, Vol. 24, No 6, (January 2003), pp. 1369-1376, ISSN 0143-1161

Hardy, C.C., and Burgan, R.E. (1999). Evaluation of NDVI for monitoring live moisture in three vegetation types of the Western U.S. *Photogrammetry Engineering and Remote Sensing,* Vol. 65, pp. 603-610.

Illera, P., Fernandez, A., & Delgado, J.A. (1996). Temporal evolution of the NDVI as an indicator of forest fire danger. *International Journal of Remote Sensing,* Vol. 17, No 6, (April 1996), pp. 1093-1105, ISSN 0143-1161.

Justice, C., Giglio, L., Korontzi, S., Owens, J., Morisette, J. T., Roy, D., Descloitres, J., Alleaume, S., Petitcolin, F., & Kaufman, Y. (2002). The MODIS fire products, *Remote Sensing of Environment*, Vol. 83, No 1–2, (November 2002), pp. 244-262, ISSN 0034-4257

Kasischke, E. S., Bourgeau-Chavez, L. L., & French, N. H. F. (1994). Observations of variations in ERS-1 SAR image intensity associated with forest fires in Alaska. *IEEE Transactions on Geoscience and Remote Sensing*, Vol. 32, No 1, (January 1994), pp. 206-210, ISSN 0196-2892

Kasischke, E. S., French, N. H. F., Harrell P., Christensen N. L. Jr., Ustin, S. L., & Barry, D. (1993). Monitoring of wildfires in boreal forests using large area AVHRR NDVI composite image data. *Remote Sensing of Environment,* Vol. 45, No 1, (July 1993), pp. 61–67, ISSN 0034-4257

Kempka, T., Kaiser, T., and Solbach, K., 2006, Microwaves in fire detection. *Fire Safety Journal*, Vol. 41, No 4, (June 2006), pp. 327-333, ISSN 0379-7112

Kennedy, P. J., Belward, A. S., & Gregoire, J. M. (1994). An improved approach to fire monitoring in West Africa using AVHRR data. *International Journal of Remote Sensing,* Vol. 15, No 11, (July 1994), pp. 2235–2255, ISSN 0143-1161

Kogan, F.N. (2001). Operational space technology for global vegetation assessment. *Bulletin of the American Meteorological Society,* Vol. 82, No 9, (September 2001), pp. 1949-1964.

Leblon, B. (2005). Using remote sensing for fire danger monitoring. *Natural Hazards,* Vol. 35, No 3, (July 2005), pp. 343-359, ISSN 0921-030X.

Leblon, B., Chen, J., Alexander, M.E., & White, S. (2001). Fire danger monitoring using NOAA-AVHRR NDVI images in the case of northern boreal forests. *International Journal of Remote Sensing,* Vol. 22, No 14, (January 2001), pp. 2839-2846, ISSN 0143-1161.

Leblon, B., Kaschike, E.S., Alexander, M.E., Doyle, M, & Abbott M. (2002). Fire danger monitoring using ERS-1 SAR images over northern boreal forests. *Natural Hazards ,* Vol. 27, No 3, (November 2002), pp. 231-255, ISSN 0921-030X.

Leblon, B., Fernandez-Garcia, P.A., Oldford, S. MacLean, D., & Flannigan, M. (2007). Using NOAA-AVHRR cumulative indices for estimating fire danger codes in northern boreal forests. *International Journal of Applied Earth Observations and Geoinformation*, Vol. 9, No 3, (August 2007), pp. 335-342, ISSN 0303-2434

Lopez, S., Gonzalez, F., Llop, R., & Cuevas, J.M. (1991). An evaluation of the utility of NOAA-AVHRR images for monitoring forest fire risk in Spain, *International Journal of Remote Sensing,* Vol. 12, No 9, (September 1991), pp. 1841-1851, ISSN 0143-1161.

Martin, M. P., & E. Chuvieco (1995), Mapping and evaluation of burned land from multitemporal analysis of AVHRR NDVI images, *EARSeL Advanced in Remote Sensing,*Vol. 4, No 3, pp. 7– 13.

Matson, M., & Dozier, J. (1981). Identification of subresolution high temperature sources using a thermal IR sensor. *Photogrammetry Engineering and Remote Sensing,* Vol. 47, No 9, (September 1981), pp. 1311-1318.

Menges, C. H., Bartolo, R. E., Bell, D. & Hill, G. J. E. (2004). The effect of savanna fires on SAR backscatter in northern Australia. *International Journal of Remote Sensing,* Vol. 25, No 22, (November 2004), pp. 4857-4871, ISSN 0143-1161

Michalek, J.L., French, N.H.F., Kasischke, E.S., Johnson, R.D., & Colwell, J.E. (2000). Using Landsat TM data to estimate carbon release from burned biomass in an Alaskan spruce complex. *International Journal of Remote Sensing,* Vol. 21, No 2, (January 2000), pp. 323–338, ISSN 0143-1161

Nemani, R.R., & Running, S.W. (1989). Estimation of regional surface resistance to evapotranspiration from NDVI and thermal-IR AVHRR data. *Journal of Applied Meteorology,* Vol. 28, No 4, pp. 276-284, ISSN 1558-8424.

Oldford, S., Leblon, B., Gallant, L., & Alexander, M. (2003). Mapping pre-fire conditions in the Northwest Territories, Canada, using NOAA-AVHRR images. *Geocarto International,* Vol. 18, No 4, (December 2003), pp. 21-32, ISSN 1010-6049

Oldford, S., Leblon, B., MacLean, D., & Flannigan, M. (2006). Predicting slow drying fuel moisture codes using NOAA-AVHRR images. *International Journal of Remote Sensing,* Vol. 27, No 18, (September 2006), pp. 3881-3902, ISSN 0143-1161.

Olioso, A., Chauki, H., Courault, D., & Wigneron, J.P. (1999). Estimation of evapotranspiration and photosynthesis by assimilation of remote sensing data into SVAT models. *Remote Sensing of Environment,* Vol. 68, pp. 341-356, ISSN 0034-4257.

Paltridge, G.W., & Barber, J. (1988) Monitoring grassland dryness and fire potential in Australia with NOAA-AVHRR data. *Remote Sensing of Environment,* Vol. 25, No 3, (August 1988), pp. 381-394, ISSN 0034-4257.

Pereira, J.M.C. (1992). Burned area mapping with conventional and selective principal component analysis, *Finisterra* Vo. 27, No 53–54, pp. 63–78.

Pereira, J.M.C., Sousa, A. M. O., Sà, A. C. L., Martin, P. & Chuvieco, E. (1999). Regional-scale burnt area mapping in Southern Europe using NOAA-AVHRR 1 Km data, In *Remote Sensing of Large Wildfires in the European Mediterranean Basin,* E. Chuvieco Ed., pp. 139-155, Springer-Verlag, ISBN 978-3540657675, Berlin.

Pierce, L.L., Running, S. W., & Riggs, G.A. (1990). Remote detection of canopy water stress in coniferous forests using the NS001 Thematic Mapper Simulator and the Thermal Infrared Multispectral Scanner. *Photogrammetry Engineering and Remote Sensing,* Vol. 56, No 5, pp. 579-586.

Pinol, J., Filella, I., Ogaya, R., & Penuelas, J. (1998). Ground-based spectroradiometric estimation of live fine fuel moisture of Mediterranean plants. *Agricultural and Forest Meteorology,* Vol. 90, *No 3, (April 1998),* pp. 173-186, ISSN 0168-1923.

Prins, E. M., & Menzel, W. P. (1992). Geostationary satellite detection of biomass burning in South America, *International Journal of Remote Sensing,* Vol. 13, No 15, (October 1992), pp. 2783–2799, ISSN 0143-1161

Prins, E. M., Feltz, J. M., Menzel, W. P. & Ward, D. E. (1998). An overview of GOES-8 diurnal fire and smoke results for SCAR-B and 1995 fire season in South America, *Journal of Geophysical Research-Atmospheres,* Vol. 103, N0 D24, (December 1998), pp. 31,821– 31,835. ISSN 0148–0227

Prosper-Laget, V., Douguedroit, A., & Guinot, J.P. (1994). A satellite index of forest fire occurence risk in summer in the Mediterranean area. In *Proceedings 2nd International Conference on Forest Fire Research,* vol. 2, pp. 637-646, Coimbra, Portugal, November 1994.

Prosper-Laget, V., Wigneron, J.P., Guinot, J.P., & Seguin, B. (1995). Utilisation du satellite NOAA pour la détection des risques d'incendies de forêts. *La Météorologie,* Vol. 8, No 10, pp. 28-38, ISSN 0026-1181.

Roy, D., P. E. Lewis, & C. O. Justice (2002), Burned area mapping using multi-temporal moderate spatial resolution data: A bi-directional reflectance model-based expectation approach, *Remote Sensing of Environment,* Vol. 83, No 1– 2, (November 2002), pp. 263–286, ISSN 0034-4257

Roy, D., Y. Jin, P. Lewis, & C. Justice (2005), Prototyping a global algorithm for systematic fire-affected area mapping using MODIS time series data, *Remote Sensing of Environment,* Vol. 97, No 2, (July 2005), pp. 137–162, ISSN 0034-4257

San-Miguel-Ayanz, J., Annoni, A., & Schmuck, G. (1998). The use of satellite imagery for retrieval of information on wildfire damage in Mediterranean landscapes. *Proceedings of ERIM'98, 27th International Symposium on Remote Sensing of Environment: Information for sustainability*, pp. 758-762, Tromsö, Norway, June, 1998.

San-Miguel-Ayanz, J., Ravail, N., Kelha, V. & Ollero, A. (2005). Active fire detection for emergency management: Potentials and limitations for the operational use of remote sensing. *Natural Hazards,* Vol. 35, No 3, (July 2005), pp. 361-376, ISSN 0921-030X.

San-Miguel-Ayanz J., Pereira, J., Boca, R., Strobl, P., Kucera, J., & Pekkarinen, A. (2009). Forest fires in the European Mediterranean region: mapping and analysis of burned areas. In *Earth Observation of Wildland Fires in Mediterranean Ecosystems*, Chuvieco Emilio (Ed.), pp. 189-204. Springer-Verlag ISBN 978-3-642-01753-7, Berlin Heidelberg.

San-Miguel-Ayanz, J., Schulte, E., Schmuck, G., Camia, A., Strobl, P., Liberta, G., Giovando, C., Boca, R., Sedano, F., Kempeneers, P., McInerney, D., Withmore, C., Santos de Oliveira, S., Rodrigues, M., Durrant, T., Corti, P., Oehler, F., Vilar L., & Amatulli, G. (2012). Comprehensive monitoring of wildfires in europe: the European Forest Fire Information System (EFFIS), in *Approaches to Managing Disaster - Assessing Hazards, Emergencies and Disaster Impacts,* John Tiefenbacher (Ed.), pp. 87-105, InTech, ISBN 978-953-51-0294-6.

Sedano, F., Kempeneers, P., Strobl, P., McInerney D., & San-Miguel-Ayanz, J. (2012). Increasing spatial detail of burned scar maps using irs-awifs data for Mediterranean Europe. *Remote Sensing,* Vol. 4, No 3, pp. 726-744, ISSN 2072-4292

Sedano, F., Kempeners, P., San-Miguel-Ayanz, J., Strobl, P., & Vogt, P. (2012). Towards a pan-European burn scar mapping methodology based on single-date medium resolution optical remote sensing data. *International Journal of Applied Earth Observation and Geoinformation,* (in press), ISSN 0303-2434.

Siegert, F. & Ruecker, G., (2000). Use of multitemporal ERS-2 SAR images for identification of burned scars in south-east Asian tropical rainforest. *International Journal of Remote Sensing,* Vol. 21, No 4, (January 2000), pp. 831-837, ISSN 0143-1161

Simon, M., S. Plummer, F. Fierens, J. J. Hoelzemann, and O. Arino (2004), Burnt area detection at global scale using ATSR-2: The GLOBSCAR products and their qualification. *Journal of Geophysical Research-Atmospheres,* Vol. 109, (July 2004), p. D14S02, ISSN 0148–0227

Strickland, G., Leblon, B., Gallant, L., & Alexander, M.E. (2001). Monitoring fire danger of northern boreal forests from optical and thermal infrared NOAA-AVHRR images. *Proceedings 23th Canadian Remote Sensing Symposium,* pp. 667-676, Ste-Foy, Canada, June 2001.

Tansey, K., Grégoire, J.M., Stroppiana, D., Sousa, A., Silva, J., Pereira, J. M. C., Boschetti, L., Maggi, M., Brivio, P. A., Flasse, S., Ershov, D., Binaghi, E., Graetz, D., & Peduzzi, P. (2004), Vegetation burning in the year 2000: Global burned area estimates from SPOT VEGETATION data, *Journal of Geophysical Research-Atmospheres*, Vol. 109, (June 2004), p. D14S03, ISSN 0148–0227.

Teillet, P.M., Dudelzak, A.E., Pultz, T.J., McNairn, H., & Chichagov, A. (2001). A framework for *in-situ* sensor measurement assimilation in remote sensing applications. *Proceedings 23th Canadian Remote Sensing Symposium*, pp. 111-118, Ste-Foy, Canada, June 2001.

US Forest Service 2012. *The Wildland Fire Assessment System*. Available from http://www.wfas.net/

Ustin, S., Roberts, D.A., Pinzon, J., Jacquemoud, S., Gardner, M., Scheer, G., Castaneda, C.M., & Palacios-Orueta, A. (1998). Estimating canopy water content of chaparral shrubs using optical methods. *Remote Sensing of Environment,* Vol. 65, No 3, (September 1998), pp. 280-291, ISSN 0034-4257.

van Zyl, J.J., Arii, M., & Kim, Y. (2011). Model-based decomposition of polarimetric sar covariance matrices constrained for nonnegative eigenvalues. *IEEE Transactions Geoscience and Remote Sensing,* Vol. 49, No 9, (September 2011), pp. 3452-3459, ISSN 0196-2892

Vidal, A., & Devaux-Ros, C. (1995) Evaluating forest fire hazard with a Landsat TM derived water stress index. *Agricultural and Forest Meteorology,* Vol. 77, *No 3–4,* (*December 1995),* pp. 207-224, ISSN 0168-1923.

Vidal, A., Pinglo, F., Durand, H., Devaux-Ros, C., & Maillet, A. (1994). Evaluation of a temporal fire risk index in Mediterranean forests from NOAA thermal IR. *Remote Sensing of Environment,* Vol. 49, No 3, (September 1994), pp. 296-303, ISSN 0034-4257.

Yebra, M., & Chuvieco, E., 2009. Linking ecological information and radiative transfer models to estimate fuel moisture content in the Mediterranean region of Spain Solving the ill-posed inverse problem. *Remote Sensing of Environment,* Vol. 113, No 11, (November 2009), pp. 2403–2411, ISSN 0034-4257.

Yebra, M., Chuvieco, E., & Riaño, D., 2008. Estimation of live Fuel Moisture Content from MODIS images for fire risk assessment. *Agricultural and Forest Meteorology,* Vol. 148, No 4, (April 2008), pp. 523–536., ISSN 0168-1923.

Chapter 4

GIS for Environmental Problem Solving

Koushen Douglas Loh and Sasathorn Tapaneeyakul

Additional information is available at the end of the chapter

http://dx.doi.org/10.5772/50098

1. Introduction

The authors are affiliated with the Laboratory of Systems Technology Applications in Renewable Resources (The STARR LAB) at Texas A&M University. The purpose of this chapter is to provide a synopsis of the cumulative research and teaching work for the past twenty years from the STARR LAB. The aim of chapter is to demonstrate holistic understandings of what key environmental issues and problems people are facing and how their concerns may be addressed with the help of geographic information systems (GIS).

We are the environments, and the environments are us. There are many environmental issues and problems the society is facing. Some major categories include environmental disasters, ecological services, and perceptions of environments by people, just to name a few. In terms of environmental disasters, hurricanes, earthquake and wildfires are some examples that exert enormous direct impacts on people's lives. Their increasing recurrences have elevated public awareness on the vulnerability and risks of the environments we live in. An awareness of environmental issues leads to an increase in people's perceptions regarding the surrounding environments. There are many factors contributing to such perceptions. Combined considerations of pertinent factors result in an overall perception. One plausible combined index is called quality of life (QOL). QOL is a practical measurement of the state of an environment. Environmental awareness also raises people's concerns on the sustainability of the ecological services. Ecological services refer to public goods, tangible or intangible, rendered to us by environments and ecosystems. Air, water, food, fiber, and fuel we consume are good examples. Sustaining these services is of great importance to all environmental stakeholders.

There are many ways to help stakeholders gain insights to environmental issues and problems. One handy approach is the use of GIS. GIS are systems of hardware, software, data, people, organizations and institutional arrangements for collecting, storing, analyzing and disseminating information about areas of the Earth [1]. Such technologies enable analyses of spatial-temporal patterns for a geographic span of interest and generations of easy-to-comprehend reports such as maps and images. GIS are maturing and proliferating rapidly in parallel to the quantum leap of personal computer (PC) platforms. It greatly enhances people's ability to know about their environments. Given the advantages, GIS have emerged as a popular subject matter among interested learners on college campuses as well as in environmental fields. A good indicator of this assertion is the sustaining popularity of Environmental GIS courses the authors teach at Texas A&M University. Other institutions are reporting a similar phenomenon.

All things considered, it is timely to provide a rundown of GIS for Environmental Problem-Solving as a chapter of this book. Main thrusts of our presentation consist of four parts. They are: 1) Introduction (this section); 2) Research method; 3) Illustrations of GIS for environmental problem solving applications; and 4) Concluding remarks.

2. Research method

Systems approach is a key research method to incorporate GIS into problem-solving process in addressing environmental issues and problems. The essence of this approach is to envision and to enact relevant endeavors into a cohesive sequence of steps. The whole process is called developing and implementing a GIS project. A typical sequence of steps in a GIS project includes framing the problem, defining a project area, identifying and acquiring data, extracting and preparing data, editing spatial data, geospatial analysis, and generating maps and reports.

2.1. Framing the problem

The first step in solving any problem is to frame the problem. The purpose of this step is to help narrow down the scope and identify the problem to make it easier to solve. This helps address the questions you want to answer. Specifically, what do you want to accomplish from looking at this problem? What are the goal and objectives you are planning to address from the problem?

Then, the next question is what is the potential information associated with the problem? Pertinent information includes:

- *Scope:* To lay out tasks, data, and time frame to solve a problem, a scope needs to be defined so that you know how much information you are dealing with. The scope varies depending upon the nature and objectives of the problem. Questions on whether the problem is looking at a specific region, a particular group of population, or a particular phenomenon are worth investigating. Also, is the problem asking for information, maps, or more in-depth analysis of the problem?
- *Scale:* Is the problem focusing on an institutional scale (individual, family, municipal, state, national, or international) and/or ecological scale (plant, plot, ecosystem, landscape, biome, or global)? As addressed in [2], stakeholders at different spatial scales can (and should) assign different values to environment and ecosystem under interest.

- *Type of information:* two distinctive types of information are quantitative and qualitative. You need to specify if the problem is looking for quantitative and/or qualitative information. Quantitative information focuses on some sort of value or measurable information. Number of population affected by a hurricane or the amount of oil spilled into an ocean are quantifiable. Qualitative information, on the other hand, represents some sort of status that needs to be stated. Wildlife species affected by a hurricane or types of chemical released into a river are some of the examples.

It is also helpful to construct an outline or diagram of the problem so that it is easy for you and/or stakeholders to determine necessary steps, to better organize the tasks, and to be able to comprehend the problem at hand.

Consider the following real world examples using the above criteria:

Example 1: The 2005 Hurricane Katrina

The scope is the Hurricane Katrina in New Orleans, Louisiana. This pertains to the Greater New Orleans Region. Information of interest includes population affected, infrastructural damage, hazardous materials, and situations that might arise afterward. Given this information, one possible answer is the number of population affected as the quantitative information. Quantitative information includes, but not limited to, current stage of hazardous waste, groups of population, animal species, and housing.

Example 2: Bastrop County Complex Fire

The scope is a major wildfire in Bastrop County, Texas in 2011. Information of interest is effects on both human and animal, economics losses, effects on land and environments, and infrastructural damage. With the defined information, possible answers include the number of affected people and animals, income losses from the incident, and the loss of species' habitats, which are accounted as quantitative information. Households and habitats affected by the fire, problem of land degradation and fragmentation, time frame for recovering, and preventive plans are some of the qualitative information that seeks answers.

Example 3: West Nile Virus in Brazos County

Brazos County, Texas and the surrounding areas is the scope of interest. Became widespread in the recent years (with the highest number of 7 severe cases in human being accounted for in 2006 [3]), West Nile virus has been under surveillance for residents in the County. Critical information that needs to be asked include: What causes the West Nile virus?; How can you track the spread of the West Nile virus?; and Where has West Nile virus been found in this location? Quantitative answers are the current number of infected individuals and the past records. Possible locations and trends that may be associated with the spread of the West Nile virus serve as the qualitative answer to the problem.

2.2. Defining a project area

With an identified problem, you can proceed to define a project area. This step delineates a confined boundary of an area of interest. The information from *Step 2.1* helps specify the proper location where the problem occurred and address the possible questions and answers under interest. The process pinpoints the focus of the problem while eliminate unnecessary areas or secondary scope of interest from the picture. Not only that this can help save time, but it also allows you to pay closer attention to the essence of the project. At this stage, the conceptual project area should be carefully thought out before attempting to acquire data, i.e., map layers, in the next step.

GIS enable a variety of ways for convenient delineation of a project's boundary that might not be made possible with other applications. ArcGIS®[1], a worldwide used GIS software, allows users to work with geographic information data by inputting and manipulating map layers in a comprehensive manner. In this chapter, we use ArcGIS 9.3.1 for all GIS applications.

For example, one may select an administrative boundary of a local jurisdiction from a base map layer as the project area as in the cases of Bastrop County Complex Fire and West Nile Virus in Brazos County. One may also "union" multiple local jurisdictions into a broader geographic span for addressing issues that are of cross-boundary nature. The project boundary resulted from one way or another serves as the "cookie cutter" for clipping data from relevant layers and tables in the ensuing steps to expedite problem-solving. As in the case of Hurricane Katrina, at least five parishes (Louisiana's equivalent of counties in other states) should be included as the project area of the Hurricane analysis.

2.3. Identifying and acquiring data

Once the project area is defined, the next step is to locate and acquire needed data. Before looking for data, the methodology needs to be analyzed to establish what data is needed. The most important question that needs to be answered is: Why do I need this data? If the data is truly needed, then this question is easily answered. If not, then the data is most likely not necessary to solve the problem.

To be able to work with data in GIS, you need to understand the nature and procedural steps of working with data in GIS as follows:

2.3.1. GIS datasets formats

Typical formats of datasets, which allow you to conveniently work with multiple information or map layers, include spatial and attribute data.

Spatial data comes in the forms of raster and vector and is generally organized into so-called layers or thematic maps.

- *Raster* data is digital image composed by rectangular grids or cells that contain numeric information from a defined range to characterize geographic features. *Digital Elevation Model or DEM* is a form of raster data important in depicting a terrain. It provides crucial information on the topologies of a geographic span.

---

[1] ArcGIS is a registered trade mark of Environmental System Research Institute (ESRI), Inc.

- *Vector or shapefile* data is constructed as points, lines, and polygons to represent geographical features.

Attribute data is information used to describe characteristics of a locale. The data is organized in a table containing information linked to a spatial feature by a common identifier. This gives you details or certain types of information associated with each specific feature.

2.3.2. GIS data sources

GIS data is vastly available from many sources, including those in public domains at local, state and federal agencies; international non-governmental organizations or NGOs; and private sector providers. In the U.S., the Federal Government generates and provides public access for easily found and downloadable geographic datasets[2]. Each agency supplies relevant datasets pertaining to their line of work. Some of the main providers include: National Oceanic and Atmospheric Administration (NOAA) for extensive National Weather Service datasets; United States Geological Survey (USGS) for DEMs, administrative boundaries, land cover, soil, and water data; United States Environmental Protection Agency (EPA) for toxic and cleanup sites; and Census Bureau for demographic and socioeconomic data, which contains a wide-range of attributes detailed to the block level. State agencies, such as Texas[3], have established a central clearinghouse for state-level datasets. With increasing demand for GIS in solving various problems, many counties and cities have initiated GIS departments, which oversee and provide relevant geographic data to inquirers.

2.3.3. Map projections and coordinate systems

Each map layer contains a coordinate system, which allows one to identify the location of the map and to be able to display, manipulate, and integrate the map layer with other layers for further applications and analysis. It is therefore imperative to understand the fundamentals of map projections and coordinate systems.

A coordinate system is a grid that may be used to define where a particular location is. Two common types of coordinate system are:

- *Geographic Coordinate System:* This uses 3D spherical surface to define locations. Often incorrectly referred to as datum, geographic coordinate system includes not only datum, but also angular unit of measure and prime meridian. Points on Earth's surface are referenced by latitude and longitude, while angles are measured by degree.
- *Projected Coordinate System:* Commonly referred to as map projections, projected coordinate system is defined on flat, 2D surface with constant lengths, angles, and area. X, Y coordinates are presented on grid. It is based on geographic coordinate system.

Often, input maps will be in different projections, requiring transformation of one or all maps to make coordinates compatible. Since monitor screens are analogous to a flat sheet of paper, there is a need to provide transformations from the curved surface to the plane for displaying

---

[2] See http://geo.data.gov/geoportal/ for more information.

[3] See http://www.tnris.org/get-data/ for more information.

data. In order to do so, mathematical formulas to relate spherical coordinates to planar coordinates are required. Some distortions in the shape, area, distance or direction of data can occur during the transformation; different projections cause different distortions. Therefore, careful consideration of the appropriate map projection is crucial. Proper map projection must consider: the map's subject and purpose; the subject area's size, shape, and location; the audience and general attractiveness; size and shape of page; and appearance of the graticule[4].

2.4. Extracting and manipulating data

The fourth step is data extraction and manipulation. In this step, one is to extract data from a conceivably larger original source file. Reduction of the size of datasets and their consolidation expedite the ensuing data management and processing. The project area defined at the onset (the cookie cutter) dictates the extent and size of data to be extracted and prepared.

Typically, data acquired may exist in various forms and shapes, e.g. different coordinate systems and file formats. It is a MUST to prepare and consolidate all datasets into a commonly operable format. GIS have a database management system component to support the proper management of both spatial and attribute data. It also enables convenient linking and relating of various data records by their locations on a common coordinate system. Some common tasks you will encounter during the data extraction and manipulation steps are as follows:

- *Re-projecting data:* This is a basic essential step in any analysis using GIS. The purpose is to convert a particular piece of data from one coordinate system to another. Working with GIS employs more than one map layer, therefore acquired datasets may contain different projections. Different data projections lead to distortion of data and inaccuracy in the analysis.

  For example, in Figure 1, a residential area polygon (in blue) is projected to Geographic Coordinate System: GCS_North_American_1983. The same residential area polygon (in yellow) is in Projected Coordinate System: NAD_1927_UTM_Zone_16N. As shown in Figure, there are some discrepancies in the map layers with different coordinate systems. If this re-projection step is not taken, any analysis preformed will be inaccurate leading to much larger problem in subsequent analysis with multiple layers.

- *Conversion of raster to vector:* Not only data comes in different coordinate systems, the file formats can also be varied; most commonly in the forms of raster or vector (shapefile). Especially with the growing use of GIS, datasets in shapefile have become more available. Shapefile data usually comes embedded with attribute data, which allows user to easily select and manipulate the information of interest. Therefore, converting a raster file to vector enables user to intersect other data with the available vector data. Suppose you have acquired and managed shapefile layers of affected area by Hurricane Katrina and population layer in the Greater New Orleans Region, by intersecting these two layers, you can extract the areas in which population were affected by the Hurricane.

---

[4]See [4] for further explanations on coordinate systems

- *Reclassification:* To extract specific data from a raster, i.e., specific elevation data, reclassification is performed. Given the Hurricane problem, flooding can be assessed as one major result of the incident. In order to extract only the flooded area resulted from the Hurricane, reclassification is utilized to distinguish a specific range of elevation in which flooding occurred from others. This will allow you to analyze the effects pertaining to the flooded area.
- *Selecting by attributes:* The purpose is to extract desired attribute data for analysis. This can be done through conditional statement imposed in attribute data table to select only specific information of interest. Considering an attribute table of chemical sites located within the Hurricane flooding zone, one can select only specific sites containing particular chemicals of interest for further analysis and map report.
- *Exporting data:* To make a temporary layer permanent in a current map, data resulted from steps such as that of above need to be exported and saved in a current working folder. Otherwise, the file may be lost or difficult to locate when you want to revisit and work on it.

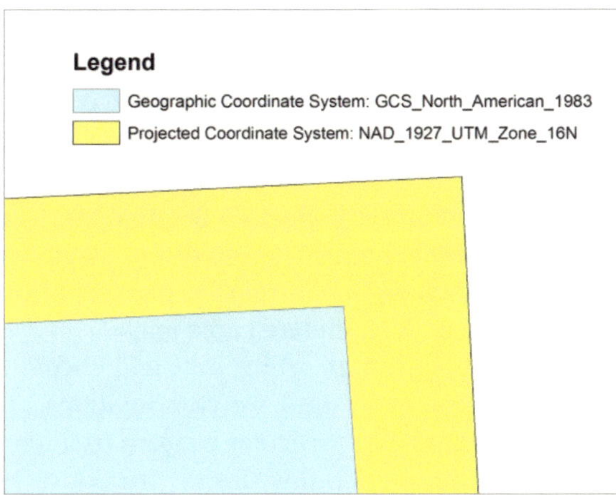

Figure 1. Same layer file with different coordinate systems resulting in 20-meter difference on the map.

2.5. Editing spatial data

Oftentimes, acquired data might not be in the most suitable shape or boundary for problem under consideration. Options to edit spatial data in GIS allow one to manage the data in such a way that is more manageable and ready to be analyzed.

Typical editing tools consist of creating new features, cutting polygons, modifying features, and extending the basic skills to other tasks such as clipping a feature to a desired shape and area.

- *Creating new features:* When creating a new feature, a blank data set is being defined by the editor. A blank data set is like an empty pie shell, while creating a new feature is like filling the pie shell. This task is only used if a new feature is desired or a single part feature is to be converted into a multi-part feature when the second part of the feature does not already exist.

- *Cutting polygon features:* This process is a shortcut to creating a multi-part feature from a single part feature. Simply put, this process is used like a set of scissors to cut an existing feature into multiple parts.
- *Modifying features:* This task is used when an existing feature does not cover the area that is desired. The attribute data will remain the same, while the feature will be modified to suit one's need.
- *Clipping features:* Clipping is a process that is like using a "cookie cutter" to remove a portion of a feature permanently. The attribute data will also be changed due to a permanent removal of the feature.

2.6. Geospatial analysis

Upon data readiness, a project may move on to the sixth step of spatial-temporal analyses. There are many useful procedures for these endeavors. Especially with the versatilities of GIS software, one can utilize extended range of applications available. Some common tools that one should be familiar with and were used specifically for the ensuing applications in this chapter include:

- *Distance analysis:* A suite of tools to produce distance maps are commonly available in GIS. In ArcGIS, distance tools are available under Spatial Analyst option. Euclidean distance tool measures straight-line distance from the center of cell to the nearest object of interest, i.e., your source. Another alternative is the cost distance tool, which incorporates travel cost from different paths into the analysis. The products from these tools are distance maps in raster representing proximity maps with a range of distance values from the source. For instance, one can find proximities from pollution sources at defined interval to any locales within a defined area map.
- *Map algebra:* Another useful application, which you will encounter at certain point of analysis, is map algebra. This can be used for computations of raster data to create spatial patterns that depict locales of a particular concern or interest. Raster calculator, a Spatial Analyst application, allows for this useful procedure by inputting specified mathematical functions and expressions in the calculator. The result will be raster values and layer corresponding to the specified function.

The use of analytic procedures mentioned above and other tools in a proper order results in useful information for a problem under study.

2.7. Generating maps and reports

The final major step is to generate maps and reports. One picture is better than a thousand words. To this end, GIS come handy in presenting information in maps, images, 3D graphs, tables, and other forms. It also expedites the import and export of these presentations between GIS and other software environments, e.g. a word or a graphic processor. With the acceleration of PC powers, the sky is the limit to GIS' capability of generating maps and reports. It is worth noting that you should understand what the readers are looking for when creating the maps and write ups, i.e., what is the focus or message that you want to communicate to others? This should align with the proposed information of interest.

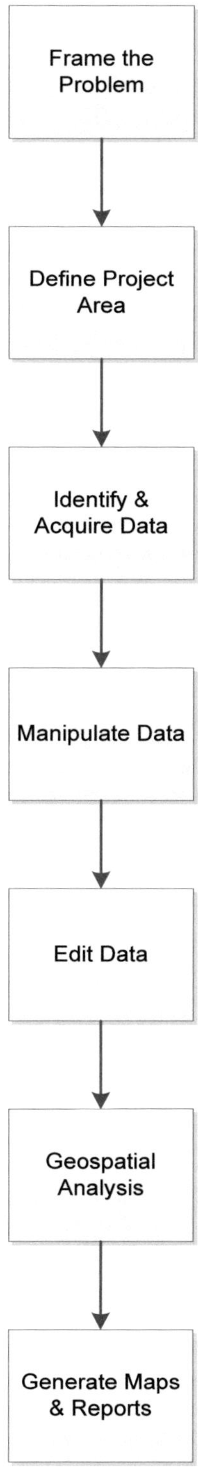

Figure 2. Diagram of problem-solving steps

3. Illustration of GIS for environmental problem solving applications

To illustrate how GIS are used to help address environmental issues and problems, two cases are described herewith in this section. The first one is on flood assessment, and the second is a QOL analysis. The applications help prepare for the building framework of spatial appraisal and valuation of environment and ecosystems (SAVEE), which will be discussed in the following section, tremendously.

3.1. Flood assessment

Considered one of the costliest [5] and most destructive natural disasters in the history of the United States, Hurricane Katrina provides a number of opportunity to understand the risk of nature, and how one could expect to understand and learn from such disastrous effects. The aforementioned problem-solving steps allow us to contemplate the steps as follows:

Step 1: Framing the problem

An analysis of the scenario indicated that Hurricane Katrina occurred in the Greater New Orleans Area. Field measurements and distributions on the majority of victims indicated that roughly those under 1 meter in elevation were initially affected by the flooding [6]. Given this information, the scenario was that every location below 1 meter in elevation was affected and any location that is above this level was unaffected by the flood water. This particular area of impact needed to be delineated. The information of interest included area and population affected by the Hurricane. Additional scenarios of water-rise were then set for 5, 10, and 15 meters to emulate different levels of flooding.

Step 2: Defining the project area

In this case, the City of New Orleans and its five neighboring parishes suffered by the storm were identified as the study area.

Step 3: Identifying and acquiring data

The best type of data for delineating the affected area is the elevation data (DEM). DEMs, Satellite Imagery, and Census datasets were collected from Atlas, the GIS data central from the State of Louisiana [7]. This included DEM, jurisdiction boundaries, street maps of the study area, and Census data.

Step 4: Extracting and preparing data

Initially, the DEM and Census data came projected as GCS_North_American_1983. By assigning a projected coordinate system to the data, further analysis could be proceeded. Given the information, we projected the data to NAD_1927_UTM_Zone_16N. Hillshades of the DEMs were also generated to visually inspect different elevations in the data. Sink holes pervaded in the DEMs were also been filled to prevent erroneous and prepare for proper flow direction process.

Then, the second crucial step was to extract the flooded area from the total area. This employed the reclassification process in which the elevation value was changed to 1 meter to separate the flooded area from the non-flooded area (elevation above 1 meter). In short, the reclassification divided the elevation data into the flooded area and the non-flooded area. The rest of the water-rise scenarios then followed using the same reclassification step as well as the ensuing steps.

Next, this flooded elevation data was converted into vector to prepare for further analysis. The converted flooded layer was the result of the conversion process as well as exporting the data into a new permanent flooded area layer.

In terms of the Census data, the parishes were merged into one layer so that it was more convenient to work with in the subsequent steps for analyzing the total effect on population.

Step 5: Editing spatial data

The acquired data contained certain parts that were irrelevant to the analysis. Lake Ponchartran, for instance, should not be counted toward the flooded area. Therefore, by editing the data, some unnecessary information of interest were taken out. Pertinent steps of editing the lake included: 1) Creating a new blank shapefile; 2) Using the blank shapefile as the base for editing tools to create a new feature around the lake area; and 3) Using the newly created feature as a cookie cutter to clip off the lake area from the flooded area layer. The result was the flooded area without the lake that was ready to be incorporated into other analysis.

Step 6: Geospatial analysis

At this stage, socio-economic analyses were conducted to assess the damage and impact on the livelihoods of residents of the affected areas. Census data developed was used directly for this purpose. Combining census data with the emulated flooded areas, patterns of suffering by which racial stakeholders and by what economic classes were clearly displayed.

Based on the flooded area layer in *Step 5*, we proceeded to calculate the area under the layer's attribute table. Visual Basic Code to calculate the area (available from [8]) or a Calculate Geometry option, an automated tool in ArcGIS, derived the numbers of area affected by the Hurricane. Mathematical formula imposed helped convert the numbers into desirable units such as acres.

Benefited from the above numbers, the population affected was conveniently calculated. Census data contains racial information that represents groups of population in different parishes. By intersecting the flooded area layer with the merged parishes layer, representing population profile in the areas, affected population was extracted. The overall statistics in the attribute table identified the total population affected by the flooding. Figure 3 demonstrates map layers resulting from the above problem-solving steps.

Step 7: Generating maps and report

Upon generating desired information and analysis, each pertinent map was composed as a map report containing a map title, legend (showing values of the map layer), north arrow, and scale bar. Then, the map reports were exported as image files to be included in a report. The report addressed the finding results of effects from the Hurricane as illustrated by the maps and relevant discussions of further applications and analysis that can later be applied based on this project.

Merged DEM layers of project area with sinks filled

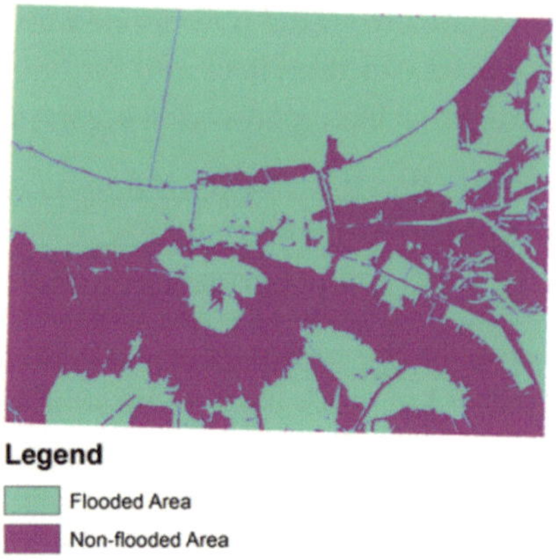

Reclassification distinguishing flooded area (in blue) and non-flooded area (in pink) at 0 meter water

Flooded area layer with lake portion

Flooded area layer without lake portion

Intersection of Census data with flooded area

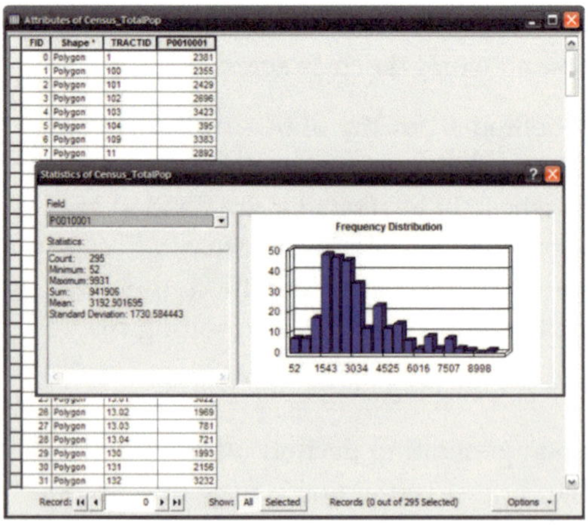

Attribute table and statistics of total population affected

Figure 3. Illustrations of selected problem-solving steps for flood assessment in reference to the 2005 Hurricane Katrina flooding in the Greater New Orleans Region.

3.2. Quality of life assessment

QOL is emerging as a major indicator to monitor citizen's livelihood and wellbeing at the grassroots level. By virtue of its focuses, QOL helps inform local people and organizations of their living environment and optimize the allocations of resources to improve the community development. Canada is perhaps more aggressive in setting up a national framework for QOL [9]. In the U.S., states such as Utah [10]; cities such as San Francisco, California [11]; and organizations, including nonprofit organizations such as the Quality of Life Foundation [12] have been vigorously promoting such term as one of their agendas.

Categories of data to support the development of QOL indicators range from education, environment, economics, social, and justice to transportation/mobility. However, the use of GIS to track QOL progress is still at its infancy stage. City of College Station, Texas, with its advanced GIS installation and rich collection of data, stands to gain a lead role in this area and to provide even superior services to its residents when it embarks on this path.

There are three issues and opportunities in the development of QOL indicators. They are:

- Combining subjective values with objective measurements to create consensus and develop common ground to accommodate multiple perspectives of stakeholders.
- Combining the use of both spatial and attribute information to develop base layer and indices in environment, crimes, recreation, etc. For example:
  - Overlay of census blocks with subdivisions or other neighborhood entities (e.g. apartment complex) to establish the baseline reference (population, its composition, income level, education level, and number of household of an entity)
  - Overlay of crime type, frequency, and location data with entities on the base layer
- Developing a composite score (ranking) of QOL for each neighborhood entity on the base layer

As you set forth to do your research, as in the case of QOL assessment, you are most likely facing with three puzzling situations:

1. Pertinent data/information comes in a variety of forms

It is plausible that the data/information you are facing and plan to collect exists in at least two forms. They are categorical and numeric. Examples of categorical information include "Yes" or "No" on whether a city (or any local jurisdiction) has a neighborhood improvement in place or not program; "Very Good," "Good," "Fair," and "Bad" on how such a program is being regarded by the communities; and "Highly favorable," "Favorable," and "Least Favorable" on how service rendered by the program is perceived by the beneficiaries.

Quite often, information of categorical nature is derived from one's "gut feeling." It may also be convenient to summarize some judgments based on historical data, on some kinds of trends, or on some opinion surveys/polls.

There are two types of numeric information: discrete and continuous. Population of an ethnic group residing in a particular Census unit is an example of discrete type. Example of continuous type is the percentage of an ethnic group versus the total population in such a unit.

2. How to "add" "oranges" and "apples"

When one has data and information of various types in hand, he/she will ask this question:

"How do I add them together?" Indeed, you cannot add oranges and apples together at their original forms. The trick is to convert and normalize all of them into the same numerical scale, say between -1 and +1.

So, what is normalization? Normalization is the act of taking many sets of data that have no clear correlation and placing them under the same quantitative scale. Essentially, normalization allows us to compare apples and oranges. Some decisions must be made prior to normalizing any type of data. The questions include:

1. What are the important factors?
2. Which factors are positive and which factors are negative?
3. How much should each of these factors count in relation to the overall project?

For categorical type of information, what you do is to fix the "best" and the "worst" at +1 and -1 respectively. This is plausible as +1 can represent the best case and -1 the worst. When both ends are fixed, one may logically deduce that a numeric value of "0" represents "Inconclusive." Furthermore, one may come up with a scheme saying that "+0.25" is "somewhat better", "+0.5" is "better" and "+0.75" is "much better." One can also say that "-0.25" is "somewhat worse", "-0.5" is "worse", and "-0.75" is "much worse." As a result, you are converting and normalizing categorical or qualitative data into numeric or quantitative information as illustrated in Figure 4.

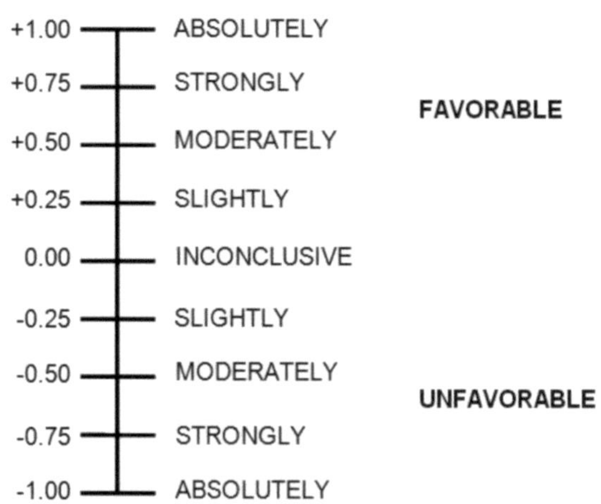

Figure 4. Illustration of the qualitative – quantitative information conversion scheme.

For numeric information, the conversion and normalization is less complicated. Say you deem the ratio of white population in a Census unit at 50% is the best mix (most favorable), in terms of quality of life; 100% or 0% is least favorable. For the best mix, you believe it should be given a score of +1 and for the least favorable a 0. Given this range, you may apply the following equations to convert and normalize the percentage into values in the range (0,+1)

$$F(x) = \begin{cases} 1 - \dfrac{(x-E)^2}{(1-E)^2}, \; E \le x \le 1 & \text{(a)} \\ 1 - \dfrac{(x-E)^2}{E^2}, \; 0 \le x < E & \text{(b)} \end{cases} \tag{1}$$

where E is the expected (best value of x)

Another method is to convert distance to an object from such measurements as miles to the uniform score between (-1 and +1). For example, one may decide that the presence of oil well is bad for quality of life. Evidently, the household right at the oil well would have absolutely unfavorable score of -1. The negative effect most likely would tap off as the distance reaches certain threshold, e.g. 1 mile or 5,280 feet. The tapering effect can then be described by a negative exponential equation as:

$$F(x) = -e^{-x} \tag{2}$$

where x is the distance to oil well(s)

The normalization equation for strictly negative attribute based on the negative exponential equation above becomes:

$$QOL_i = Con\left(\left[Distance_i\right] < X, e^{-\left(\frac{\left[Distance_i\right]+1}{X}\right)}, 0 \right) \tag{3}$$

The translation of the equation is "If Condition < X, True, False." This means if an input value (distance value) falls under the condition (less than X), then the output is negative value. Otherwise, the output is zero. It might be helpful to put this in the oil well scenario above:

$$QOL_{Wells} = Con\left(\left[Distance_{Wells}\right] < 5280, e^{-\left(\frac{\left[Distance_{Wells}\right]+1}{5280}\right)}, 0 \right) \tag{4}$$

The above equation is set so that if the distance to an oil well is less than 1 mile, then the output is negative. As the locales get closer to the actual oil wells path, the more negative they will become (with the minimum at -1). At a distance of 1 mile or greater all the output values are set to zero as shown in Figure 5.

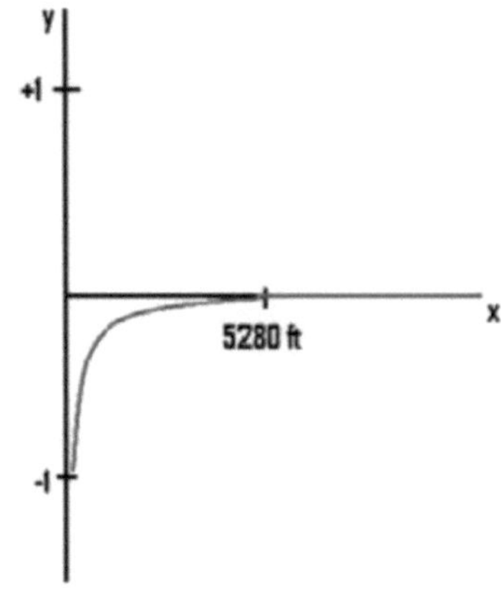

Figure 5. Negative decay graph showing the more negative values as the locales get closer to oil wells.

At any rate, once you have all factors converted and normalized into the scheme of (-1, +1), then the values can be "operated" on to add up their contributions to the overall quality of life assessment of a city. This is done by applying the following formula:

$$Score = \begin{cases} I_A + I_B - (I_A \times I_B), & I_A > 0 \, and \, I_B > 0 \quad (a) \\ I_A + I_B + (I_A \times I_B), & I_A < 0 \, and \, I_B < 0 \quad (b) \\ \dfrac{I_A + I_B}{1 - min\left[\left|I_A\right|, \left|I_B\right|\right]} & Otherwise \quad (c) \end{cases} \tag{5}$$

The method is derived from an expert system algorithm called Emycin [13]. The operations utilize map algebra calculation to integrate two values at a time, i.e., pair-wise calculation, while avoiding the problem of double-counting. The calculations are performed iteratively until all normalized layers are exhausted. As illustrated in Figure 6, the operation calculates the values of two attributes at a time to derive the final score, which is the integration of the values in all attributes. Through fuzzy logic operations, two QOLs (different factor contributing to QOL) can be integrated at a time until all QOLs are exhausted. Iteration 1 integrates QOL_1 and QOL_2 so that only the overlapping portion of both factors values remains. By taking this portion to integrate with another QOL, QOL_3, the final result is the overlapping portion among three factors; QOL_1, QOL_2, and QOL_3.The results can be color-coded as a gradient map of integrated and locale-specific QOL in the range of (-1, +1).

The nicety of the Emycin formula is that:

1. Regardless of the number of factors being used, you always "operate" on two of them in each iteration. This is called pair-wise calculation.
2. Depending on the score values of the two factors, there will be only one of the equations that is applicable.

3. Unlike many "ordinary" algorithms, this formula allows both positive and negative contributions from factors under considerations, which is more realistic.
4. Regardless of the number of factors being considered and operated on, the resulted score will always be bounded between -1 and +1.
5. Regardless of the sequence each factor is put into pair-wise calculation, the result is always the same.
6. Once all factors are exhausted in the calculation, one can always convert the result back to the qualitative scheme to make it more comprehensive to lay persons or people one intend to interpret the results to.

A word of caution: Both -1 and +1 are "singular" points. In other words, if you come up with a score on the contribution of a factor to be either -1 or +1, then other factors' contributions will not matter anymore. This is not a surprise or unreasonable. Because -1 means absolutely "bad" and +1 means absolutely "good." When you have a factor that determines the quality of life to be absolutely bad, then indeed why bother to waste time to assess other factors?

To this end, one may want to adjust or shift the score from a factor that is somewhat different from the absolute values of -1 or +1 so that the pair-wise calculation may proceed logically. Again, this is not unreasonable as there is hardly anything that one can claim that is absolutely good or bad.

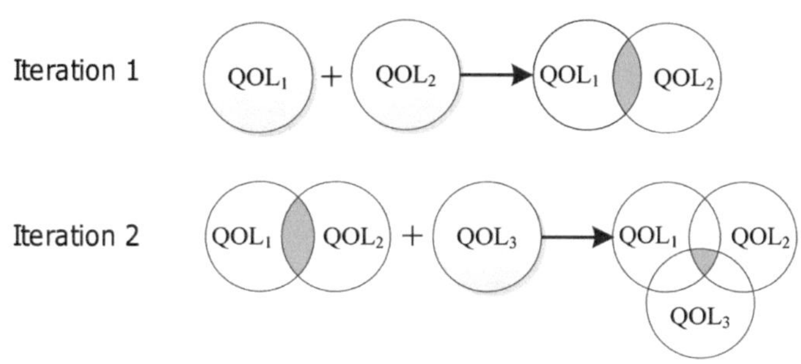

Figure 6. Illustration of pair-wise calculations.

3. Incomplete information

An additional nicety to the above approach is that one can proceed to conduct studies under incomplete information. The condition of incomplete information actually happens quite often in real life. With the kind of flexibility boasted by Emycin, you "add" the contributions from whatever data you are able to get your hand on for a city in determining its quality of life. In the case of comparing multiple cities, you may get this and that for one jurisdiction while not the same categories for all of them. By nature of the conversion, normalization, and pair-wise calculation, you would be able to derive scores on the same scheme and will be able to make comparisons.

With better understanding on the assessment framework, it is time to put such theory into real application.

Step 1: Framing the problem

The information of interest for this case is the factors contributing to the QOL of a city/community. Relevant questions include:

1. What defines a high quality of life?: This depends on who the target audience is: elderly community, students, or married couples.
2. What factors can contribute to the QOL?:
   a. Distance to: hospitals, schools, university, parks, landfill, oil wells, etc.
   b. Census Data Analysis: racial mix, relative income of a population, and number of children per household

Step 2: Defining the project area

In the case of the QOL assessment, the project area was the City of College Station, Texas.

Step 3: Identify and acquiring data

Acquiring data from the City's GIS Department is crucial. From the rich collection of datasets rendered by the City [14], a number of data layers were selected for the ensuing analysis endeavor. They included census data, roads and streets, railroads, parks and green spaces, residence subdivisions, landfills, oil wells, schools, hospitals, flood plains, crime statistics, and many more.

Step 4: Extracting and preparing data

DEMs for College Station were acquired and converted into raster. This represented the base map of College Station for the following steps. Selecting only the areas pertaining to College Station attribute was also another important preparation step since we were looking at the QOL in College Station and nothing else.

Step 5: Editing spatial data

Basemap that contains areas beyond College Station were clipped off, and only the College Station boundary was left for the analysis.

Step 6: Geospatial analysis

Based on the identified QOL factors, proximities to parks, green spaces, schools, hospitals, and some other geographic features were regarded as positive contributing factors. On the other hand, closeness to such factors as landfills, oil wells, railroads, crime occurrences, and flood plains were considered to have negative impacts. The contributions of these factors, positive or negative, were mathematically formulated as distance functions from objects on corresponding data layers. In the ensuing steps, proximity maps encoded with distance functions were generated. The results from each factor layer were then combined with fuzzy logic calculation to form an integrated index between (-1, 1). Any number greater than 0 indicated a good QOL with anything below 0 representing bad index. The index was coded in a color scheme with a gradient from red to green. The color-coded QOL maps displayed clearly the patterns of QOL of the City at every specific neighborhood and locale.

To better illustrate this, four QOL factors, QOL_1, QOL_2, QOL_3, and QOL_4, were used as an example for the calculation (see Figure 7). The first fuzzy operation employed two QOLs, QOL_1 and QOL_2, to derive QOL_{12}. The locales within defined proximity to QOL_1 were color-coded in green representing high QOL with the values approaching 1. On the other hand, those in red represented low QOL with the values approaching -1. Next, QOL_{12} was integrated with QOL_3 resulting in QOL_{123}. The last operation was QOL_{123} and QOL_4 as shown in the final integrated map of QOL_{1234}. The map results in the color gradient reflecting more green in the portion where high QOLs overlap (in the middle of the map) while the outer portion becomes more yellow to orange as a result of integrated low QOLs.

At this stage, twenty sample residential addresses were selected and tabulated in a table. By linking this table to the Address Locator tool in ArcGIS, the residential addresses were shown as a point shapefile on the map. To pinpoint the QOL of each selected address, Identify Tool was used to indicate the QOL index associated with such address.

Step 7: Generating maps and report

Map reports of this project were individually created to reflect normalized layer of each factor contributing to QOL. The normalized values (within -1 to 1 range) were shown in the legend to reflect the results from the analysis. The combined layers resulted from Emycin algorithms were exported into a group of combined layer of strictly positive factors, combined layer of strictly negative factors, and combined layer of the combination of positive and negative factors. The report concludes how QOL assessment was made possible with useful applications of GIS. Future development of applications from the QOL assessment such as the linkage to SAVEE framework was also discussed.

The QOL index illustrated in the above example sheds light on the shape of things to come with SAVEE. First of all, one may acquire the land price and/or real estate information from local authority of a jurisdiction. Using the SAVEE methodology, such information can be converted into the (0, 1) range for services provided by specific environments and ecosystems in an area. Similarly, the QOL index above may also be computed in the same range. Spatial statistical analyses can then be conducted to determine correlation between land prices and QOL. Useful information may be thus generated to pave way for "spatial acres of an environment or ecosystem".

4. Concluding remarks

The chapter depicts the natures and categories of environmental issues people are facing and how GIS can be deployed to help address them. In the context of environmental problem-solving, the systems approach for applying GIS is presented; and a few practical cases are illustrated. This organization casts a holistic view for readers to gain better comprehension of the subject matter.

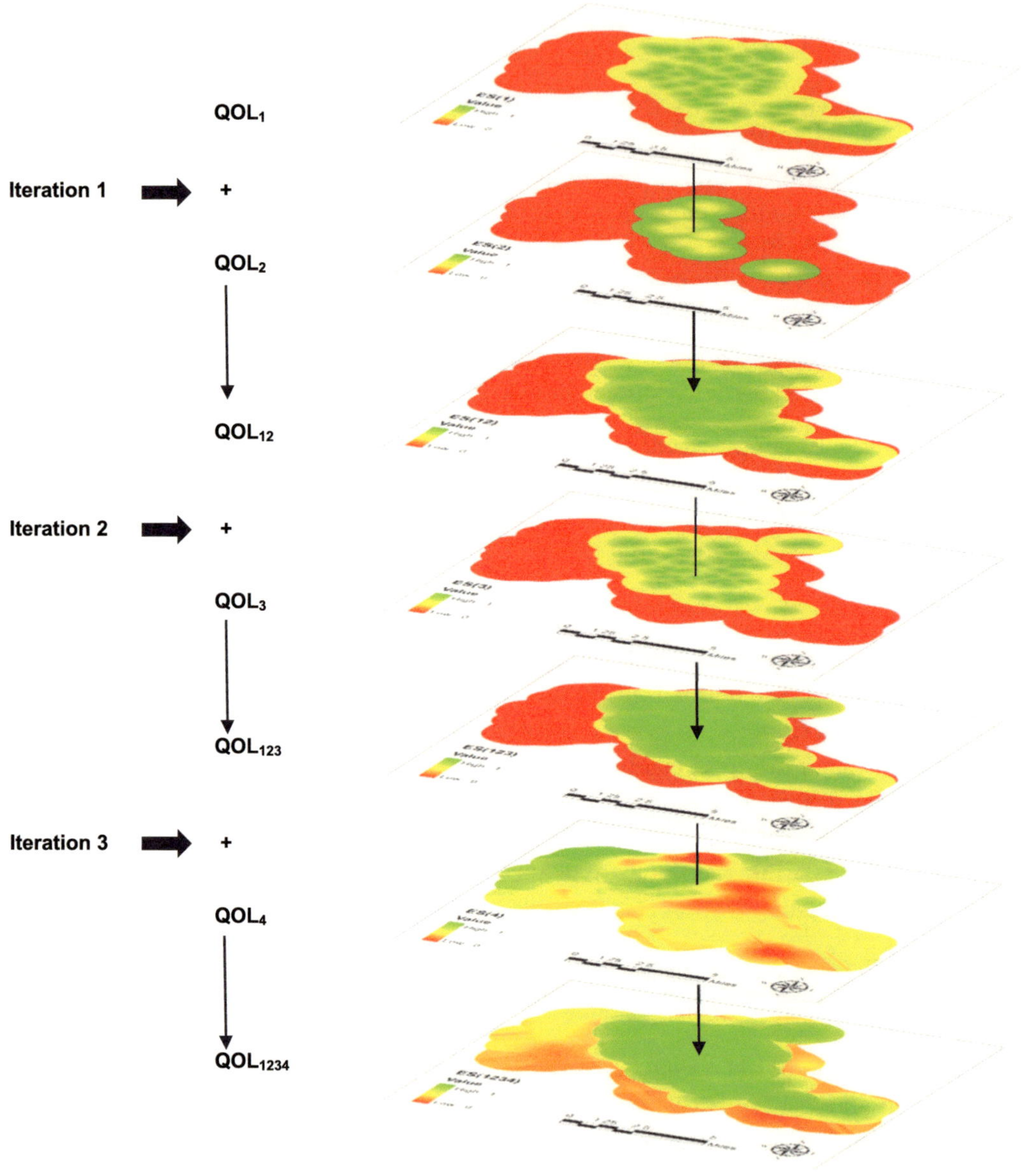

Figure 7. Iteration of map algebra to incorporate fuzzy logic to compute contributions of relevant factors to locale-specific QOL in College Station, Texas.

Problem-solving starts with shaping a mental model on to formulate a solution to the issue at hand. Steps of the solution process are then implemented through the use of appropriate data and tools enabled by GIS. Skills and knowledge to facilitate these endeavors can be best advanced by hands-on practices. For this purpose, interested readers may access the full set of documentation of learning modules at http://starr.tamu.edu/gis2012a/. The materials are from a senior course the authors teach at Texas A&M University. It bears the same title as this chapter, "GIS for Environmental

Problem-Solving." First conceived in the 1990s, the course has gained and maintained its popularity among the student bodies. The learning modules include well-organized step-by-step instructions of applications in ArcGIS presented in this chapter. Being offered online since 2006, this course has proven to be easy yet comprehensive for self-learning, even among students with no prior GIS background.

It is worthwhile mentioning that the approach mentioned above is for a typical GIS project for environment. There is, however, usually one step short. That is asserting monetary values associated with the environment. This issue is emerging as a priority matter in the environmental research community. For example, the monetary losses from the BP Oil Spill in the Gulf of Mexico are yet to be more plausibly determined. Taking on this issue, the authors here at the STARR LAB are developing a new research methodology called Spatial Appraisal and Valuation of Environment and Ecosystems (SAVEE). The aim of this effort is to define "spatial acre" that attaches monetary values to a geographic span of interest.

One main thrust of SAVEE is to cross-reference economic development and ecological sustainability in the framework of Sustainable Development declared in the 1992 Earth Summit and being enhanced continuously ever since. Economic development is tangible and comes with a price tag. It is plausible to assume that the intensity of development of a locale of interest can be reflected in its real estate value, which is generally available. Sampling some locales of their real estate values leads to a price list of real estate values. This price list is then converted into a uniform range between 0 and 1, a well-behaved index representing the intensity of development of locales. On the other hand, ecological sustainability of an area of interest normally does not come with a price tag. However, one may systematically incorporate pertinent ecological services it renders into consideration and develop an index that has the same range of (0, 1). The numbers approaching 1 represent higher sustainability, and the opposites represent lower indices. On the basis of equitability between development and environment, the two index systems may then be mapped. The mapping leads to assigning monetary values associated with development sites to ecological locales with comparable index numbers.

Learning is a life-long process; so are the advances of knowledge and technologies. On the environmental GIS front, asserting monetary values to a system under study has become an imperative. The authors are hopeful that the general framework stipulated in SAVEE shall be advanced to explore this new territory. Only labeling it with dollar signs would make stakeholders appreciate more of our environment of its values. After all, without such dollar values, it is difficult for stakeholders and authorities to understand the magnitude of the environmental problems at hand. It is contended that SAVEE and other similar effort will make a significant contribution to environmental sectors in general and the advancement of GIS.

Author details

Koushen Douglas Loh and Sasathorn Tapaneeyakul
Department of Ecosystem Science and Management, Texas A&M University, USA

5. References

[1] Dueker K.J, Kjerne D (1989) Multipurpose Cadastre: Terms and Definitions. Falls Church: American Society for Photogrammetry and Remote Sensing. 12 p.

[2] Hein L, Van Koppen K, De Groot R.S, Van Ierland E.C (2006) Spatial Scales, Stakeholders and the Valuation of Ecosystem Services. J Ecol econ. 57(2): 209-228.

[3] West Nile Virus Activity in Brazos County [Internet]. College Station (TX): AgriLIFE Extension; Agricultural and Environmental Safety; 2012 [cited 2012 April 21]. Available from: http://www-aes.tamu.edu/public-health-vector-and-mosquito-control/brazos-county-mosquito-borne-disease-surveillance/west-nile-virus-activity-in-brazos-county/

[4] Maling D.H (1991) Coordinate Systems and Map Projections for GIS. In: Maguire D.J, Goodchild M.F, Rhind D.W, editors. Geographical Information Systems: Principles and Applications. London: Longman Group UK. Vol. 1, pp. 135–146.

[5] Knabb R.D, Rhome J.R, Brown D.P (2006) Tropical Cyclone Report: Hurricane Katrina: 23–30 August 2005. National Hurricane Center.

[6] Jonkman S.N, Maaskant B, Boyd E, Levitan M.L (2009) Loss of Life Caused by the Flooding of New Orleans after Hurricane Katrina: Analysis of the Relationship between Flood Characteristics and Mortality. Risk anal. j. 29(5): 676-698.

[7] Atlas: The Louisiana Statewide GIS [Internet]. Baton Rouge (LA): Louisiana State University CADGIS Research Laboratory; 2009 [cited 2012 April 23]. Available from: http://atlas.lsu.edu/

[8] ArcGIS Desktop 9.3 Help [Internet]. Redland (CA): Environmental Systems Research Institute; 2008 [cited 2012 April 23]. Available from: http://webhelp.esri.com/arcgisdesktop/9.3/index.cfm?TopicName=Sample_VBA_code

[9] Atlas of Canada Quality of Life [Internet]. [Place unknown]: National Resources Canada; 2009 [cited 2012 April 23]. Available from: http://atlas.nrcan.gc.ca/auth/english/maps/peopleandsociety/QOL

[10] The 2011 Utah Foundation Quality of Life Index: First Biennial Survey Reveals Strengths, Weaknesses [Internet]. Salt Lake City (UT): Utah Foundation; 2011[cited 2012 April 27]. Available from: http://www.utahfoundation.org/img/pdfs/rr703.pdf

[11] City and County of San Francisco as Successor to the Redevelopment Agency [Internet]. San Francisco: San Francisco Redevelopment Agency; 2012 [cited 2012 April 27]. Available from: http://www.sfredevelopment.org/

[12] The Quality of Life Foundation [Internet]. San Francisco: The Quality of Life Foundation; [Date unknown] [cited 2012 April 27]. Available from: http://www.qualityoflifefoundation.org/

[13] van Melle W, Shortliffe E.H, Buchanan B.G (1984) EMYCIN: A Knowledge Engineer's Tool for Constructing Rule-Based Expert Systems. In: Shortliffe, E.H., editor. Rule-Based Expert Systems: The MYCIN Experiments of the Stanford Heuristic Programming Project. Reading (MA): Addison-Wesley. pp. 302-313.

[14] GIS – Geographic Information Services [Internet]. College Station (TX): City of College Station GIS; 2001 [cited 2012 April 29]. Available from: http://www.cstx.gov/index.aspx?page=3683

Chapter 5

Environmental Management in Businesses: Does It Make Money? An Accounting Perspective

Maria J. Masanet-Llodra

Additional information is available at the end of the chapter

http://dx.doi.org/10.5772/48660

1. Introduction

The growing interest in businesses for conducting environmental management makes us to think about the reasons that justify it. A number of studies (Barth & McNichols, 1994; Power, 1997, Shields & Boer, 1997, Walden & Schwatz, 1997 & Vachon & Klassen, 2008) found that environmental management provided a new challenge for businesses due to the introduction of an environmental perspective in managing them in addition to the economical one. Therefore, the question that arises is whether environmental management makes money in businesses or it is just a new social value with non economic effects. Among the potential benefits attributed to environmental management may be distinguished: energy savings, lower cost of compliance with environmental regulations, waste reductions and more efficient processes. These improvements may be associated to improvements in economic performance through competitiveness (Porter, 1990, 1991; Porter & van der Linde, 1995a, 1995b, McGee, 1998 & Wagner, 2008). Such competitiveness would be translated into positive outcomes for organisations, derived from higher turnover through the "green" customers, lower costs caused by compliance with environmental regulations and better resources management and, finally, total quality management covering the life cycle of products from design to disposal.

Other studies are concerned about the environmental accountability of companies to society (Cormier & Magnan, 2007 & Clarkson et al., 2008). In this sense, the development of environmental accounting can be crucial in sake of comparability of environmental information. Developing an integrated environmental management system into the company's business strategy is needed to articulate an information system that permeated throughout the business area. Accounting, as an analytical tool for decision, is a prominent part of the articulation of the information system; that can translate into financial terms the different business activities (Gray & Bebbington, 2001; Larrinaga-Gonzalez & Bebbington, 2001, Burnett & Hansen, 2007, Vargas et al., 2010 & Gray 2010). Moreover, the accounting system could be employed to seek out, identify and exploit financial savings in resources usage, waste and emissions that would necessarily lead to reductions in the companies' impacts (Ditz et al., 1995, Epstein, 1996, Bennett & James, 1997a, b, Curkovic et al, 2006, Curkovic & Sroufe, 2007, Curkovic et al., 2008 & Abelda, 2011).

From different empirical studies is concluded that there is an ambiguous relationship between environmental performance and financial performance (Shane & Spicer, 1983; Aupperle et al., 1985 & Alexander & Buchholz, 1987). Even though there is a positive correlation among them (Wokutch & Spencer, 1987 & McGuire et al., 1988) it is not known if a stronger financial performance only implies that there are more resources to spend on environmental performance as argued by available funds theory (McGuire et al., 1988, 1990) or if an improvement in environmental performance really improves financial results as advocated by good management theory (Freeman, 1984).

The defenders of stakeholder theory argue that the existence of environmental performance is a requirement for the legitimacy of companies in their environment and that environmental and financial performance tends to be positively associated over time (Freeman, 1984 & Suchman, 1995). Critics of this theory maintain that if managers attend to interests other than those of their investors, there is a breach of their confidence which will harm the wellbeing of shareholders (Friedman, 1962).

The stakeholder theory establishes that companies act in an environmental responsible way for legitimacy with their stakeholders, which means that companies are more concerned about designing of policies and objectives than about achieving results. Therefore, the perceptions of organisations regarding their environmental management are much more optimistic than the achieved results would allow. In accordance to this theory, Moore & Robson (2002) obtained from their study that companies found easier and less costly to design environmental policies and objectives than to implement environmental management that would enable them to achieve and check their results.

Furthermore, other researchers considered the importance of environmental disclosures as a tool of legitimacy of the performances of the companies with their environment according to the tenets of Mouck (1995), Taylor et al. (2001) and Cho & Pattern (2007).

Preston & O'Bannon (1997) distinguished two different approaches to the relationship between environmental performance and financial performance. The first approach is based on the direction of the relationship (positive or negative) while the second one is based on causality (which one leads to the other) and the synergies between them. From these approaches they established 6 hypotheses in their typology:

1. Social impact hypothesis: High (low) levels of environmental performance lead to high (low) levels of financial performance.
2. Trade-off hypothesis: High (low) levels of environmental performance lead to low (high) levels of financial performance.

3. Available funds hypothesis: High (low) levels of financial performance lead to high (low) levels of environmental performance.
4. Management opportunism hypothesis: High (low) levels of financial performance lead to low (high) levels of environmental performance.
5. Positive synergy hypothesis: There is a positive synergy between financial performance and environmental performance.
6. Negative synergy hypothesis: There is a negative synergy between financial performance and environmental performance.

On the other hand, Ullmann (1985) arguments that it is necessary to consider the strategic position of the company towards social demands, because the models should include elements to express the nature of the relationship between environmental performance and financial performance. In this regard, Berman et al. (1999) distinguish three models on the stakeholder theory:

1. The direct effects model: The relationship between the stakeholder and the company strategy has a direct and separate effect on financial performance.
2. The moderation model: The company strategy has a direct effect on financial performance, but moderated by the relationships established between the stakeholders.
3. The stakeholder intrinsic commitment model: The company takes on an intrinsic commitment to the various stakeholders, such that the interests of the stakeholders make up the basis of the strategic decision-making process.

The main objective of our analysis is to understand the development and implementation of environmental management through a particular business sector: the tiles sector in Spain, which is considered as one with high environmental impacts in the United Nations' list and besides, it represents a very dynamic and innovative sector that may suppose flexibility in order to introduce environmental management.

Through three studies carried out over time we were analyzing the above-mentioned sector. This type of analysis allowed us to use the findings of each study as a starting point for the realization of the following study; getting feedback from them. Each of the studies has been used a different methodology: the first has resorted to an exploratory-descriptive survey, the second to a case study research and the third to factor analysis and correlations.

From a survey about the tiles sector in Spain in 1998 an initial diagnosis was established (Masanet-Llodra, 1999). This diagnosis revealed that only a very poor minority of firms were concerned about establishing environmental management systems, although the majority of them proclaimed their environmental awareness. As further studies were required, in 2001 a case study research was undertaken (Masanet-Llodra, 2006). The findings showed the adoption of the most obvious and sensible environmental improvements, which have low costs, together with investments in the most innovative technologies on the market for obtaining competitive advantages that go beyond business image. At present, we carry out an analysis of the relationship between corporate social responsibility (CSR) and financial performance (Masanet-Llodra et al., forthcoming) by using factor analysis for obtaining CSR indicators- environmental management included- in order to look for correlations among those and the financial ones. The obtained results indicate that there is a positive correlation between the CSR self-evaluation and the CSR actions directed to their stakeholders. Not all the CSR indicators have effect on the financial profitability of companies, but direct stakeholders and compliance checks have a positive correlation with return on equity and the debt ratio, which provides evidences of the available funds theory.

All of the three studies corroborate the legitimacy theory, since the perception of firms was much more optimistic than the achieved results would allow.

In the following sections we are going to present the tree studies and their findings as well as the conclusions and implications for future research.

2. Initial diagnosis

The initial diagnosis was carried out by a survey for the whole population -265 companies- of the analysed sector. It had been used an exploratory-descriptive methodology due to there were no previous studies –at that moment- in the tiles sector in Spain related to environmental systems and accounting. This survey was undertaken through a semi-structured questionnaire with 31 items. Previously to design our questionnaire we made a pilot survey with 8 companies randomly selected in order to avoid differences between academic and managerial language.

The survey was made along 1998, and although different efforts -e-mails, telephone-calls, revisiting three or four times- were done to assure the response, the percentage of it was 10.18%. This percentage cannot be representative for inference studies but it did not differ from other similar surveys carried out in Spain. Nevertheless, after analysing our sample it resulted representative of the population regarding to all the analysed categories -size, sub-activity, ways of production, types of management, financial composition, volume of exports, hierarchical organization of the countries to which it is exported, potentially sensitive investments, distribution of the resources for training and respondent position.

From the survey it was obtained an initial diagnosis of the analysed sector from different items.

2.1. Relationship between firms and environment

There were around 20% of firms that considered environment had no relationship with their business. Such percentage even overcame the number of firms that observed environment as an opportunity. This fact together with the high level of no responses reasserted that most of firms saw environment as a threat (figure 1).

However, the majority of firms were aware that they interacted with their environment, what it was very important because the identification of a problem is previous to its solution.

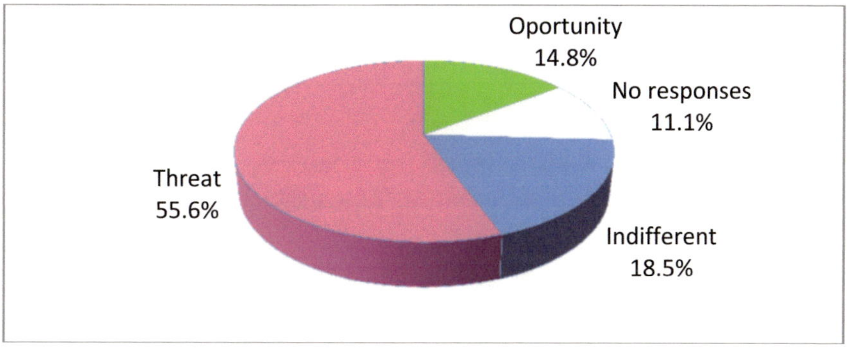

Figure 1. Environmental relationship

2.2. Establishment of environmental policies

The responses showed that the percentage of firms that established environmental policies was superior to the firms that did not (figure 2).

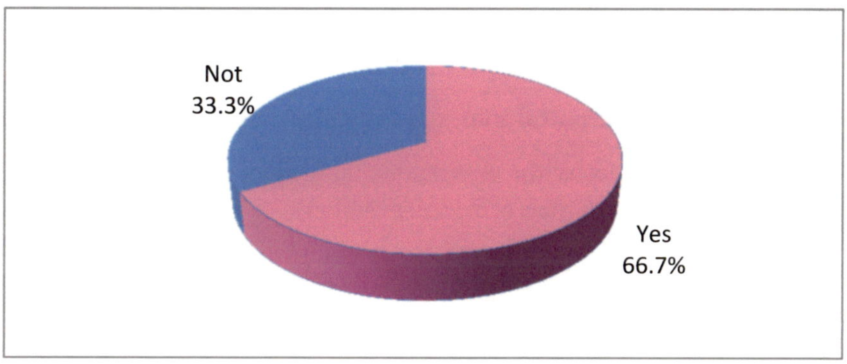

Figure 2. Environmental policies

The majority of firms had established their environmental police in an informal way, but 40% of them had done it in a formal one, so the level of the formality was not so far from the sample mean (figure 3). However, such formality was not reflected in the accounting statements (figure 4).

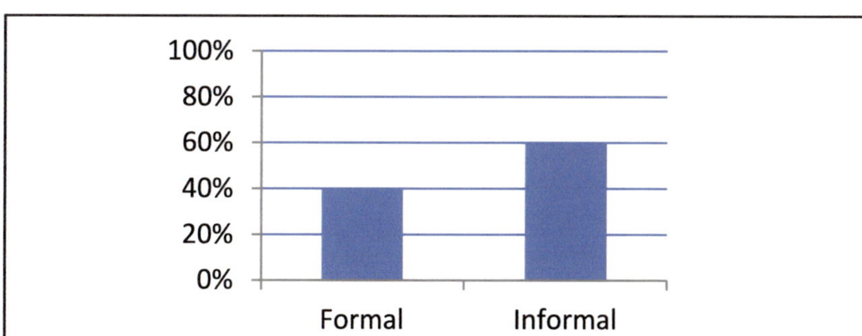

Figure 3. Level of formality

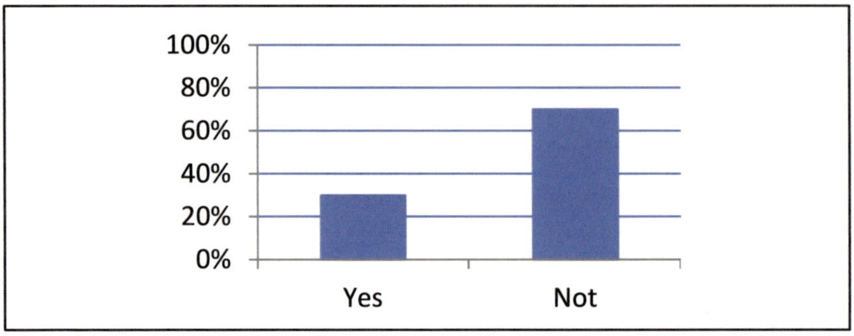

Figure 4. Accounting reflection

The previous hypothesis was that the level of formality of environmental policies should be correlated to the accounting statements reflection, but after calculating Pearson coefficient of correlation such evidence was not found ($r = 0.1889$, p-value = 0.519).

2.3. Knowledge of environmental management standards and adoption level

Although the huge majority knew the environmental management standards, only a small percentage had adopted them in their companies (table 1).

Responses	Knowledge ISO 14001	Adoption ISO 14001
Yes	74.1 %	11.1%
Not	25.9%	85.2%

Table 1. Knowledge and adoption of ISO 14001 (ISO, 1996)

There were a very small percentage of companies that were about obtaining ISO 14001, so the total percentage of adoption did not correspond to 100%.

2.4. Knowledge of European Community regulation, importance and adoption level

There were a very high percentage of companies that did not know the European Community regulation 1836/93 (European Communities, 1993) about voluntary membership in an environmental management and audit system. Nevertheless, the huge majority of companies knew about it (figure 5).

Although 59.3% of companies knew the EC regulation, only 37% considered convenient to adopt it (figure 6) and no one had been adopt it at that moment (figure 7).

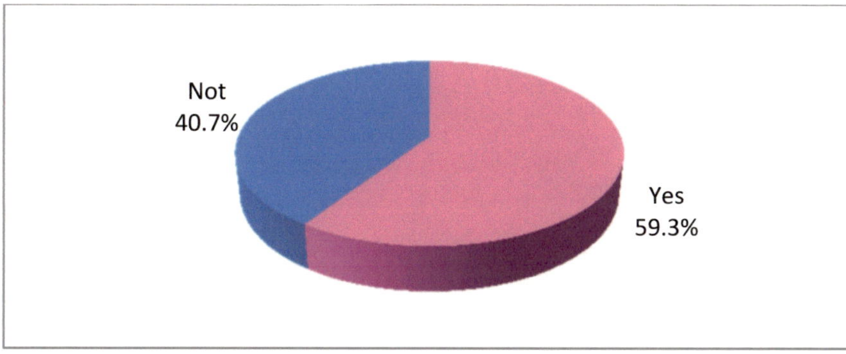

Figure 5. Knowledge of EC regulation 1836/93

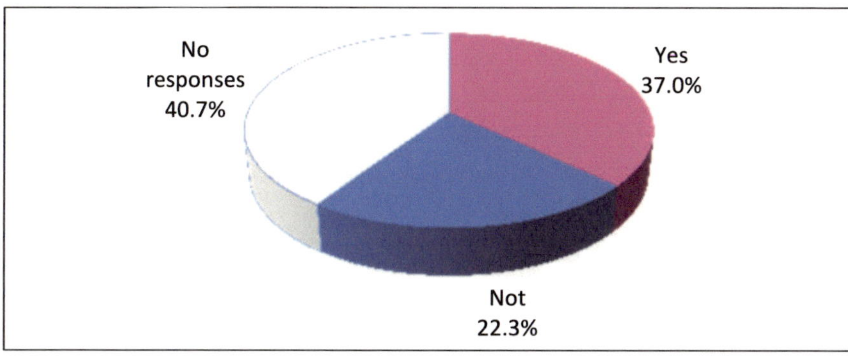

Figure 6. Convenience of adoption

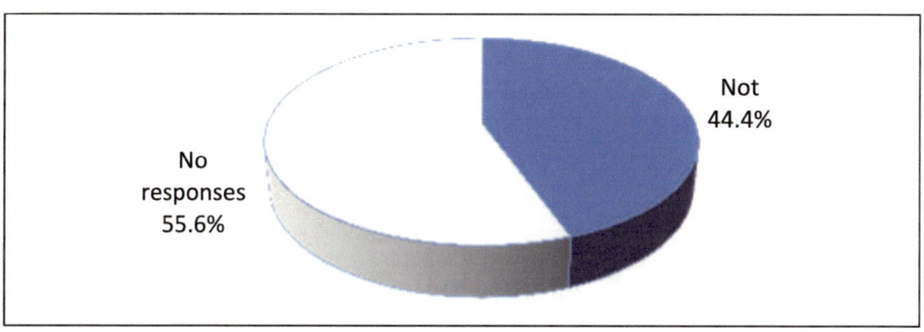

Figure 7. Level of adoption

2.5. Determination of environmental objectives in management strategy

About 50% of companies established environmental objectives in their management strategies (figure 8).

Related to the elaboration of environmental balance-sheets of energy, raw materials... to compile the obtained results, there were around 30% of companies that did not compile environmental results. While among those companies that compiled environmental results there were a higher percentage of those that did it in a formal way (figure 9).

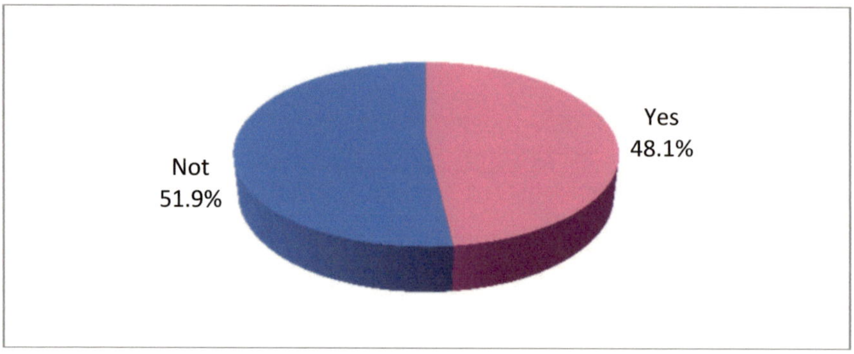

Figure 8. Environmental objectives

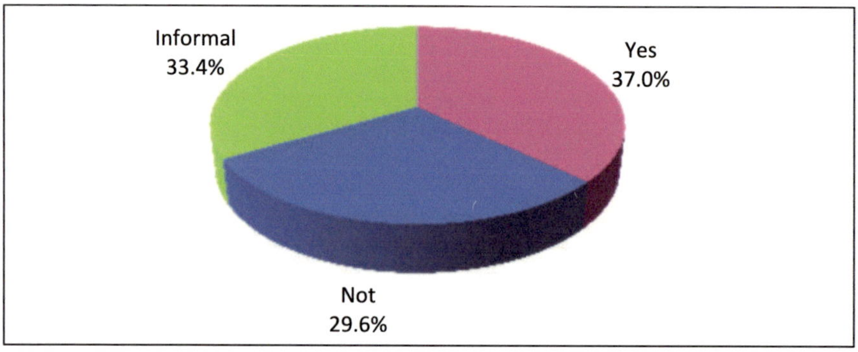

Figure 9. Environmental Balance-Sheets

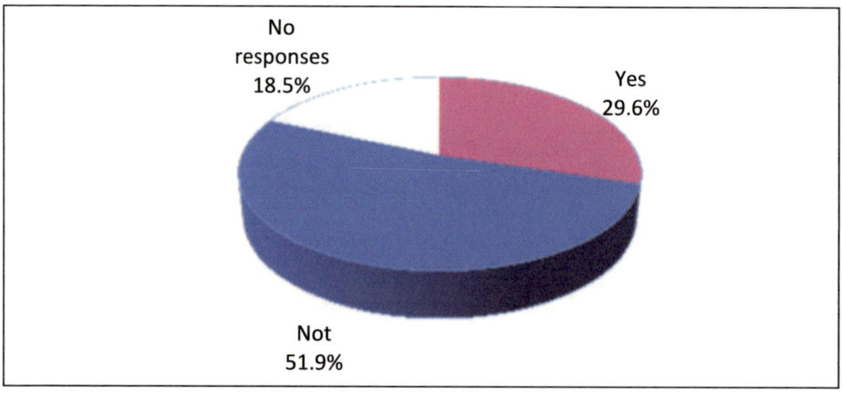

Figure 10. Comparison of objectives and results

However, more than a half of companies did not realise any comparison between environmental objectives and results (figure 10). Jointly with the level of no responses could be explained, in best cases, by the incipient development of environmental awareness so they were in the phase of introducing strategies and far away from controlling results. And in worst cases, such incongruities could be a symptom of an environmental marketing strategy.

2.6. Identification of conditions of environmental improvement

In table 2 are compiled the conditions of environmental improvements mentioned by the firms.

Conditions of environmental improvement	Very agree (5)	Agree (4)	Indifferent (3)	Disagree (2)	Very disagree (1)	Mode
Legislation pressure	11	13	1	1	1	4
Source control	4	12	8	1	1	4
"Green" energies	3	12	7	2	1	4
Organizational culture	7	12	7			4
Subsidies	9	8	7	2		4
Business association	**2**	**8**	**9**	**6**		**3**
Tax incentives	11	10	3	2		5
Export requirements	2	9	6	3		4

Table 2. Conditions of environmental improvement

In order to identify the conditions of environmental improvement the companies showed to be indifferent to business association, while they considered as important higher source control, use of "green" energies, change in organizational culture, subsidies and higher export requirements. However, they stressed tax incentives and an increase of legislation pressure as the necessary conditions for environmental improvement. In conclusion the companies identified a combination of passive and actives techniques for achieving environmental improvement, although giving greater weight to the first.

2.7. Assumption of environmental costs

The huge majority of companies expressed to incur in environmental costs (figure 11).

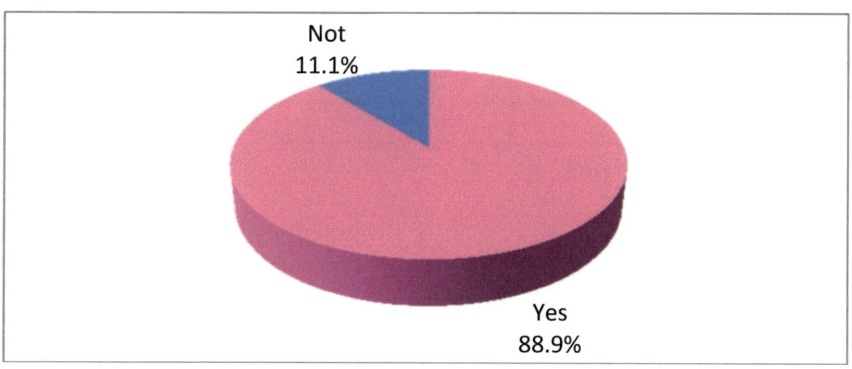

Figure 11. Assumption of environmental costs

2.8. Typology of environmental costs

There were a minority of firms that considered their environmental cost as only restorative, while the restorative and preventive attitude jointly was scarcely higher than the only preventive (figure 12).

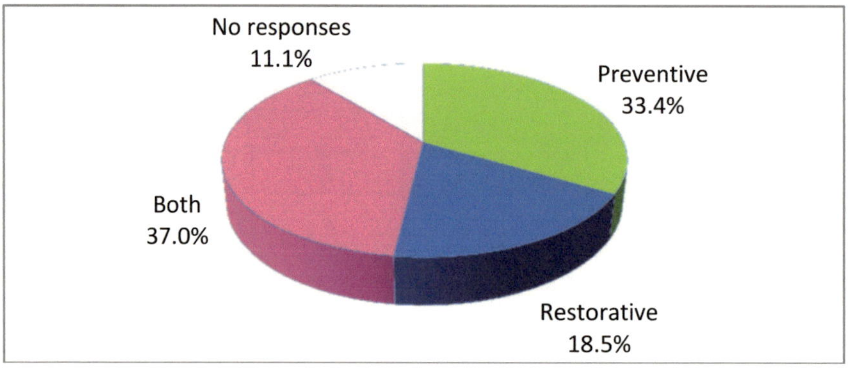

Figure 12. Typology of environmental costs

In order to check the congruity in companies' responses they were asked about the origin of their environmental costs (table 3).

Origin of environmental costs	Very agree (5)	Agree (4)	Indifferent (3)	Disagree (2)	Very disagree (1)	Mode
"Green" technologies	5	5	3	3	5	1,4,5
Redesign of processes	2	5	5	5	3	2,3,4
Ecological products		2	8	6	5	3
Environmental taxes	1	4	8	6	2	3
Treatment of waste	14	8			1	5
Control of discharges and emissions	**8**	**14**	**2**			**4**
Fines and penalties	1		2	4	13	1
Repair of damage	1	1	2	7	9	1
Implementation of EMS[1]		1				4

Table 3. Origin of environmental costs

The origin of environmental costs were mainly attributed to treatment of waste and control of discharges and emissions, so the restorative typology had greater weight of what companies were aware. This situation could be explained by the differences between business and academic language. Thus, the business language considered control of

[1] Environmental Management System

discharges and emissions as well as treatment of waste as preventive costs while it classified as restorative costs repair of damage and payment of fines and penalties. In short, it would be the difference between the academic "ex-ante" and the business "ex-post". Academically is considered as preventive any action to that business activity will not produce environmental impacts, while professionally once has been the activity seeks to control and mitigate its impact.

2.9. Accounting consideration of environmental costs

Environmental costs were considered by the majority of firms according to their nature as expenses or investments. There were about one-third of companies that always considered them as expenses in period and a non-significant minority who considered them always as investment (figure 13).

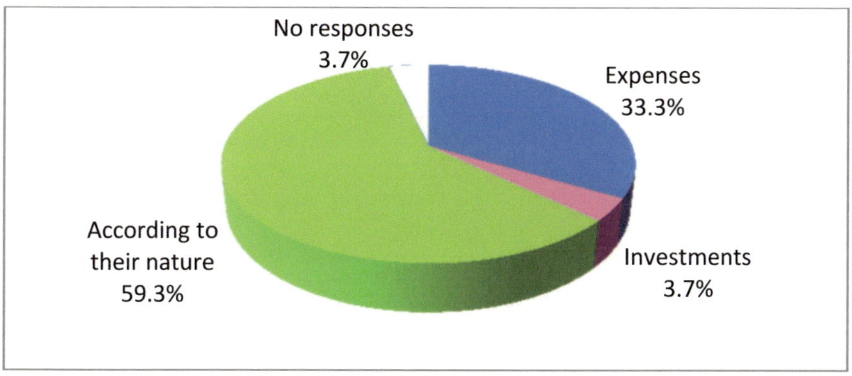

Figure 13. Accounting consideration of environmental costs

2.10. Environmental information in decision-making

More than a-half of companies took in account environmental information in decision-making, although the negative percentage of responses was very high (figure 14).

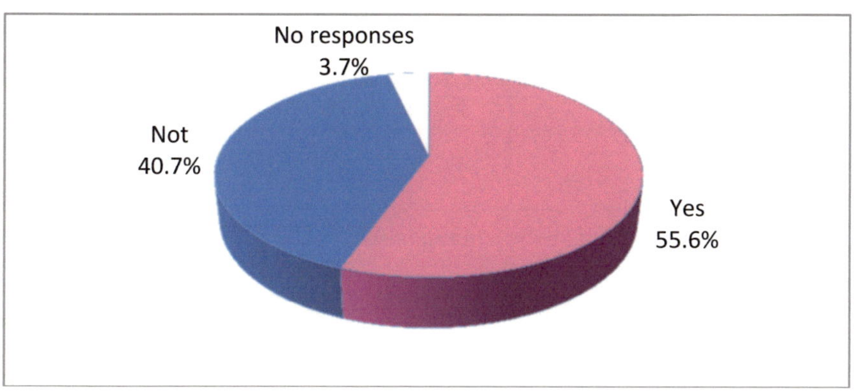

Figure 14. Environmental information in decision-making

2.11. Recognition of environmental responsibilities

The vast majority of companies replied to recognise environmental responsibilities (figure 15). In table 4 the causes of recognition of environmental responsibilities were analysed. The main reason for recognising environmental responsibilities was the compliance with legal regulations, although their environmental commitment and the image to their consumers were also argued causes.

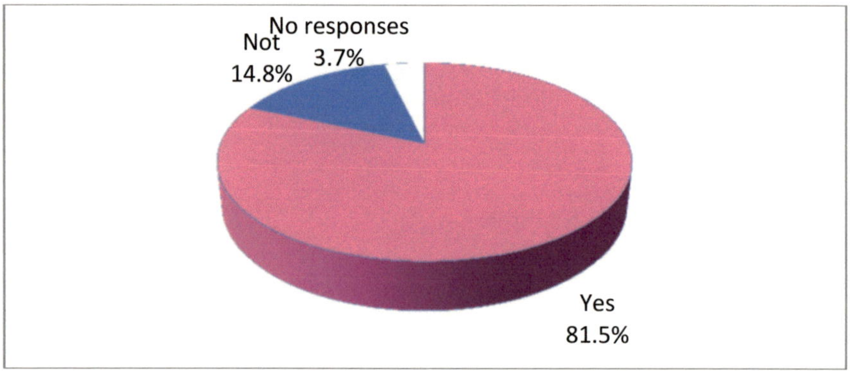

Figure 15. Recognition of environmental responsibilities

Causes of recognition of environmental responsibilities	Very agree (5)	Agree (4)	Indifferent (3)	Disagree (2)	Very disagree (1)	Mode
Legal imposition	13	8	1			5
Image to shareholders	2	7	9	1	2	3
Image to consumers	**5**	**9**	**6**	**1**		**4**
Environmental commitment	9	9	3	1		4 y 5
Accountability	1	5	9	2	1	3
Cost management	1	7	10	1	1	3
Environmental groups' pressure		1	8	4	7	3

Table 4. Causes of recognition of environmental responsibilities

2.12. Environmental provisions

The majority of respondents did not realise environmental provisions in their accounts (figure 16).

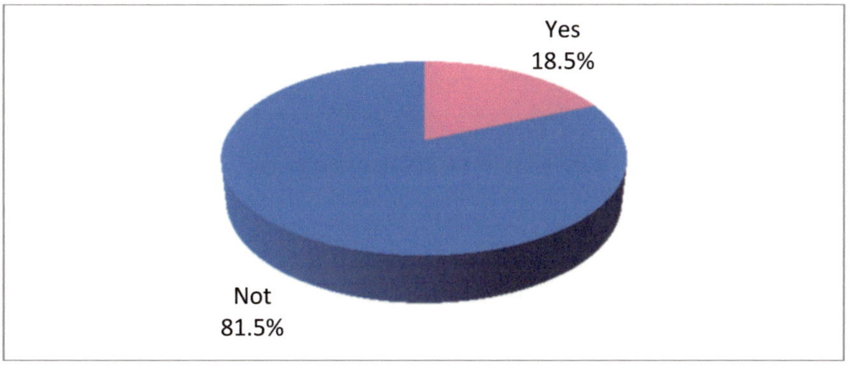

Figure 16. Environmental provisions

2.13. Environmental coverage

A non-significant percentage had contracted environmental coverage, whereas the wide majority of companies did not have them established (figure 17).

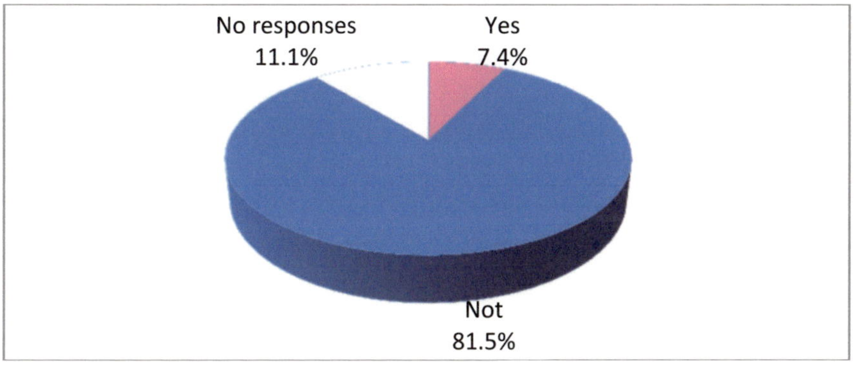

Figure 17. Environmental coverage

2.14. Environmental publication

In table 5 are listed the different documents that companies used to publish environmental information. The documents used to publish environmental information were mainly management report and environmental special reports. Notes was the other most frequently used, followed by the balance sheet document.

The causes argued for publishing environmental publication are described in table 6. The companies recognised as potential users of environmental information detailed in table 7. The potential users of environmental information recognised by companies were top management, non-governmental organization, public administrations, shareholders and executives of companies. They did not identify as potential users consumers, workers and public in general, which could be explained by the fact that those companies sell to middlemen vendors and not to the final consumer.

Documents	Yes	Not
Balance sheet	14.8%	85.2%
Profit and loss account	7.4%	92.6%
Notes	**22.2%**	**77.8%**
Management report	25.9%	74.1%
General report of society	7.4%	92.6%
Letter from the president		100%
Additional information	3.7%	96.3%
Environmental special reports	25.9%	74.1%

Table 5. Environmental publication documents

Causes of environmental publication	Very agree (5)	Agree (4)	Indifferent (3)	Disagree (2)	Very disagree (1)	Mode
Administration requirements	4	2	6	1	3	3
Parent company requirements	1	2	2		8	1
Improving business image	3	5	4	1	2	4
Prior to enforcement regulations	1		8	2	4	3
Increase of market share		2	5	4	3	3
Social responsibility	2	6	5	2		4
Improving management	4	5	4	1	1	4
Environmental awareness	1					5

Table 6. Causes of environmental publication

Potential users of environmental information	Very agree (5)	Agree (4)	Indifferent (3)	Disagree (2)	Very disagree (1)	Mode
Top management	7	10		2	1	4
Shareholders	3	10	2	2	2	4
Executives	2	9	6	2		4
Public administrations	6	8	5	2		4
Non-governmental organizations	8	5	5	2	1	5
Consumers	3	4	10	3	1	3
Workers	2	6	9	3		3
Public in general	2	4	6	6	2	2 y 3

Table 7. Potential users of environmental information

2.15. Findings of initial diagnosis

The environmental management was scarcely implemented and needed to be stimulated so that environmental concerns did not degenerate into a mere accumulation of good purposes. Thus, environmental management systems were rarely taken despite the wide said knowledge of the regulations relating to the environmental management. Unfortunately, the situation was not manifested a tendency to change in the near future, if not provided appropriate tax incentives and an increasing legislation pressure.

The companies considered that environmental variables represent a threat to the development of their activity, which is why established corporate environmental policies in their strategy, generally informal and without perception of accounting reflection.

Nevertheless, the previous hypothesis was that the level of formality of environmental policies should be correlated to the accounting statements reflection, but after calculating Pearson coefficient of correlation such evidence was not found. This result may be due to the conception of what companies meant accounting, reducing its size to a mere bookkeeping technique.

The development of environmental accounting was subject to compliance with legal requirements and to its use as a powerful weapon of business marketing; that in the majority of cases it came to draw up environmental information in an arbitrary manner, avoiding any kind of immediate relations that could set up such as the comparison between obtained results and environmental pursued objectives. In short, environmental accounting was being used as an instrument of legitimacy of the performances of the companies in their environment according to the tenets of Mouck (1995), Taylor et al., (2001) & Cho & Pattern (2007).

Companies recognised environmental responsibilities, which, however, did not seem to have had reflected in environmental provisioning, nor in the employ of another type of coverage by environmental risks. This was due, primarily, to that in a large number of cases the environmental concern was only conceived as a responsibility that attempted, largely, to pass on to third parties.

The inconsistencies expressed between explicit attitudes and actual actions on environmental issues would put revealed differences between the academic and business language. Thus, from the academic perspective the environmental commitment was conceived as a preventive measure while from a business perspective the majority approach is the restorative.

The publication of environmental information was performed, mainly in the management report and the environmental special reports, especially with the aim of improving business image and management as well as for reasons of social responsibility. However, companies not identified as stakeholders in such information to consumers, workers and public in general; which can be explained by the fact that these companies sell to intermediaries and not to the final consumer.

3. Case study research

The perceived incongruities in the initial diagnosis revealed that firms assumed to be highly environmental committed while from facts this commitment was not so high proved. So, it was necessary to introduce case study research methodology to clarify and understand the reasons of these inconsistencies.

The findings of case study research, like experiments, are able to be generalised to theoretical propositions and not to population or universes, because with this methodology the aim is to generalise theories (analytic generalisation) and not to enumerate frequencies (statistical generalisation) (Yin, 1994).

The case study research methodology recalls that a single case is analogous to a single experiment, so its use is accurate under the following circumstances. The first one, when it represents the critical case that can generate a significant contribution to knowledge theory-building, or it can even help to refocus future investigations in an entire field. The second one, in the circumstance of representing an extreme or unique case and the last circumstance is known as a revelatory case because a researcher has the opportunity to observe and analyse a phenomenon previously inaccessible to scientific investigation (Yin, 1994).

The criteria for choosing our case study research' firm were related to three main issues:

Firstly, the identification- in 2001- of a company with a really developed environmental management system and not a "good intentions" firm which had not formalised its environmental procedures. In our case, the ISO 14001 standard was used as a discriminatory criterion because there were no firms in the sector with EMAS verification (Environmental Management and Audit System) (European Communities, 2000).

Secondly, due to the fact that it was an interactive process, it was necessary to detect motivation by the company to participate in it. By using participant observation as a source of evidence in the case study research, the researcher interacts with the firm, so it was necessary to gain access to events or groups that were otherwise, inaccessible to scientific investigation. Unlike other research methodologies the data compilation in case study research demands several fields visits which imply many time dedication from the participating firms.

And finally, there was the importance given by the firm to environmental disclosures. This characteristic was really decisive because the huge majority of companies in the sector considered environmental issues as a matter which was only concerned of their own business, so they were not interested in disclosing environmental information. This particular firm considered that environmental disclosures meant a real commitment with its stakeholders or interested parts; therefore they were keen to disclose environmental information as an essential part of its environmental management system.

According to Yin (1994) we carried out the protocol of our case study research:

- Overview of the case study project. The findings of initial diagnosis served as a point of departure of the present case study.

- Case study questions. The questions were designed by adapting the questionnaire of our previous survey to its obtained conclusions and also, they were redefined for making suitable to the real context.
- Field procedures. Of the six sources recognised for evidence compilation, the following five were mainly used: direct observation, participant observation, documents, archives and interviews. Besides, physical evidence was used for recognising the operativeness of a particular database.
- Compilation of information. This research was conducted over a 7-month period from March to September 2001 and the information was compiled via on-site visits, as well as by e-mail and by telephone.
- Guide for case study report. During the process the firm had full information about its development and the different topics to be discussed in any of the on-site visits were determined beforehand, which allowed the firm to prepare all the documentation related to them.
- Case study draft. The last case study draft was given to the firm for approval in a final discussion meeting. In those issues where concurrence cannot be achieved, the different positions were clearly and objectively stated in the report.

3.1. Findings of case study research

The implementation of an environmental management system could be introduced to avoid the tightening of environmental regulations and to adopt continuous improvement strategies as an assurance of environmental performance. Some kind of environmental management usually existed in firms in a chaotic way and to formalise and to provide connections to that chaos involved a lot of effort and money. But, when there was a specific objective to be achieved, such as obtaining the environmental certification, it helped to make a swift progression from some technical environmental improvements to the configuration of an integrated environmental management system. The environmental certification acts as a guide to formalise the environmental management system and, therefore the willingness to achieve the environmental certification can stimulate the adoption of environmental management in any organization.

In the process of obtaining the environmental certification to transpose previous experience in quality management systems contributes to paperwork to be reduced and to failings and improvements to be detected more easily as Houthuysen (2000) obtained from his experience since later the system itself allows learning from it, as Pedersen & Nielsen (2000) found.

Accounting, as an internal information system, is considered to be very useful and it was recognised that the development of physical and financial indicators to configure a balanced scorecard would have improved the environmental management system. Although other business indicators previously existed in our firm, the development of environmental indicators had favoured the elaboration of a balanced scorecard that gathers economic and environmental ratios, which was clear proof of how the introduction of an integrated environmental management system improves business management.

Nevertheless, such company did not show any interest in disclosing any environmental information in the Annual Report, as evidence of the opacity and secrecy that exist within a sector characterised by a permanent technological innovation.

The strategy of the firm was based on a permanent rationalization process that involved adopting the latest technologies on the market. Thus, it had simultaneously implemented both the first and the second stage in the development of its environmental management system. This consists in applying the most obvious and sensible improvements, which have low costs, together with investments in the most innovative technologies on the market.

From the observation of the two main environmental indicators in the firm, we obtained that the trend of its environmental investments (table 8) corroborated the general strategy of the company since once technological changes has been adopted the financial payments decrease until a new technology is adopted, and the trend of its environmental expenses (table 9) placed our firm in an initial stage of environmental management development, where such expenses progressively increase.

Environmental investments (EI)	1995	1996	1997	1998	1999	2000	2001
(EI /Total investments)	18.05	0.00	37.37	25.08	8.00	4.90	1.98

Table 8. Environmental investments over the Period 1995-2001

Environmental expenses	1999	2000	2001
(expressed in euros)	258,832.65	45,846.76	62,735.99

Table 9. Environmental expenses over the Period 1999-2001

Environmental information had been gathered about the trends of its environmental investments since 1995 but its expenses data were only available from 1999 onwards because, prior so that, environmental expenses were not distinguished from the total expenses of the organisation.

The company considered that its environmental management system had to provide more competitive advantages than those that could be achieved by simply improving its own image. Therefore what the company should do the management of the production system more efficient by spreading the efficiency of the environmental department through the whole organization. Such efficiency would be based on a relevant information system that allowed the different business activities to be coordinated and that relied on the commitment of all human resources.

Whereas from the environmental strategy of the company derived that the environmental information system should be integrated at all business levels and it ought to be useful for decision-making as decision-usefulness theory states, its initial stage of development circumscribed it to specific projects, which are almost exclusively oriented towards obtaining funds such as subsidies or grants.

Again the inconsistencies appeared between the environmental position assumed in strategy and the environmental behaviour reflected in facts. This incongruities revealed that the company assumed to be highly environmental committed while from facts this commitment was no so high proved. Despite most of detected incongruities in our firm could be related to initial stages of implementation of environmental management systems, the one referring to disclosures cannot be explained in such terms. On this matter, the particular firm proclaimed that environmental disclosures meant a real commitment with its stakeholders or interested parts, but, the facts revealed that its environmental disclosures were restricted to particular responses to individual demands -as requested by ISO requirements- instead of providing general information, without considering environmental disclosure in Annual Reports.

Finally, even though our case fulfils the requirements for being considered as a rigorous case study research we are aware of the limitations of such methodology as possible bias in findings due to the researcher perceptions and values, the difficulty in designing case study research and the self-election of the participating company. Furthermore, this methodology enables to generalise findings to theory but not to the whole population.

4. Relationship between corporate social responsibility and financial performance

Following with our in-depth study we tried to analyse the effects of corporate social responsibility (CSR) in the financial results of the companies in the analysed sector.

Data were collected through personal interviews and using factor analysis the dimensions of the CSR were obtained. Subsequently, we analysed whether or not there were correlations between the factors and the financial indicators -collected from their Annual Accounts. Mahoney and Roberts (2007) found significant relationship between some measures of CSR and financial performance in their study, although they did not find any relationship between a unique index of CSR and financial performance.

From the population of the Spanish tile sector -255 companies in 2004- 70 firms were selected by simple random sampling, according to their size by volume of sales. Only 64 interviews were valid, so the analysis was restricted to them, being the size error of +/- 11% and the significance level of 95% ($p = q = 50\%$).

In order to carry out the factor analysis we tested Kaiser-Meyer-Olkin (KMO) and Bartlett's sphericity, -obtaining a KMO of 0.73 and rejection of the null hypothesis of the Bartlett's sphericity test- which allowed us to carry out the analysis. The method of extraction of factors was principal components, specifically, as we were primarily interested in the interpretation of the factors, the Varimax orthogonal rotation method was used. We obtained 6 eigenvalues greater than unit, therefore 6 factors (table 10), with an explained variance of 77.4%.

The analysis of correlation between the dimensions of CSR and the financial performance was conducted using the factors as indicators of the first and the financial performance indicators were used as the following four ratios: return on assets, return on equity, return on sales and debt ratio.

Following with the inconsistencies between the environmental position assumed in strategy and the environmental behaviour reflected in facts by companies, we compared whether the CSR perception-self-evaluated from 1 to 5- of the companies corresponded with their CSR actions. Then, we carried out a correlation analysis between the score that the companies gave themselves and their score in each factor- excluding the question related to self-evaluation to avoid bias.

Factor 1	Socio-environmental objectives	Responsibles for results Socio-environmental results Resources for socio-environmental results Timetabling for implementation Socio-environmental incentives
Factor 2	Direct stakeholders	Suppliers Shareholders Customers Workers
Factor 3	Indirect stakeholders	Non-profit organizations General public Public administrations Consumers Competitors
Factor 4	Selection of suppliers or customer	In relation to compliance with their socio-environmental objectives
Factor 5	Compliance check	Certifications Socio-environmental management system External quality audit Socio-environmental publication Control of socio-environmental results
Factor 6	Impact limitation and self-evaluation	Socio-environmental policy Socio-environmental evaluation Socio-environmental provisions Responsible for product design

Table 10. Dimensions of CSR

4.1. Findings of relationship between CSR and financial performance

From the analysis of correlation between the factors of CSR and the financial ratios three results were obtained:

1. Factor 2 (direct stakeholders) had a positive correlation with return on equity (p-value = 0.26, significant at 5%) and marginally with the debt ratio (p-value = 0.22, significant at 10%).
2. Factor 4 (supplier selection) had a negative correlation with return on assets (p-value = -0.29, significant at 5%) and with return on sales (p-value = -0.27, significant at 5%), and marginally with return on equity (p-value = -0.21, significant at 10%).
3. Factor 5 (compliance check) had a positive correlation with the debt ratio (p-value = 0.25, significant at 5%).

From these results we could extract the hypothesis that both taking into account the consequences of the decisions of companies on their direct stakeholders such as the use of instruments to check compliance of socio-environmental commitments could require additional resources that would be reflected in a higher debt ratio in these companies. These results would validate the available funds theory (McGuire et al., 1988 & Moore, 2001) which postulates that companies with abundant financial resources have more opportunities to carry out CSR initiatives. Besides, taking into consideration the close relationship with their direct stakeholders it would be also rewarded with greater return on equity, which positioned these companies in the moderation model of Berman et al. (1999), where the company strategy has a direct impact on the financial performance although it is moderated by the relationship with their stakeholders.

Furthermore, the positive correlation between factor 5 (compliance check) and the debt ratio could be explained by the fact that companies who need to rely on external financing must meet certain socio-environmental requirements to obtain it, as Roberts (1992) found.

On the other hand, companies that stated a policy of supplier selection by socio-environmental criteria seemed to be penalised with lower return on assets, sales and equity. Due to only 9 of the 64 companies had a high score in factor 4 (supplier selection) the results seemed to indicate rather an absence of penalties for companies who had not such supplier selection.

As factor 2 (direct stakeholders) provided evidence for the hypothesis of positive synergy, while factor 4 (supplier selection) did so in favour of the negative synergy, the behaviour of companies could not be characterised according to the typology of Preston & O'Bannon (1997). Hence, further studies will be required to determine whether the difference in sign is a particular feature of the industrial sector analysed or, on the contrary, it is present in other sectors.

From the analysis of correlation between the score that the companies gave themselves and their score in each factor two results were obtained:

1. Factor 2 (direct stakeholders) had a positive correlation with the CSR self-evaluation ratio (p-value = 0.21, significant at 5%).
2. Factor 6 (impact limitation) had a positive correlation the CSR self-evaluation ratio (p-value = 0.21, significant at 5%).

The positive correlation between the CSR self-evaluation and the CSR actions directed to their stakeholders could be explained by the need for legitimacy of the companies in their environment, corroborating the stakeholder theory (Freeman, 1984 & Suchman, 1995). On the other hand, it seemed sensible that companies taking measures to limit their impacts would have considered that they was making greater efforts in CSR, so they gave themselves higher scores.

5. Conclusions

The environmental management in the Spanish tiles sector needs to be stimulated, due to companies are adopting it for reasons of competitiveness that do not exceed legal imposition, environmental commitment and business image and forgetting the integrated management of resources. Which is why established corporate environmental policies in their strategy, generally informal and without perception of accounting reflection. This result may be due to the conception of what companies meant accounting, reducing its size to a mere bookkeeping technique.

The environmental certification acts in the sector as a guide to formalise the environmental management system and, therefore the willingness to achieve the environmental certification can stimulate the adoption of environmental management in any organization.

The development of environmental accounting was subject to compliance with legal requirements and to its use as a powerful weapon of business marketing; that in the majority of cases it came to draw up environmental information in an arbitrary manner, avoiding any kind of immediate relations that could set up such as the comparison between obtained results and environmental pursued objectives. In short, environmental accounting was being used as an instrument of legitimacy of the performances of the companies in their environment, as it defends the theory of stakeholders.

There is no interest in disclosing any environmental information in the Annual Report, as evidence of the opacity and secrecy that exist within a sector characterised by a permanent technological innovation. Although, internal accounting is considered to be useful and the development of environmental indicators to configure a balanced scorecard have improved the environmental management system and, furthermore it has favoured the elaboration of a balanced scorecard that gathers economic and environmental ratios, which is proof of how the introduction of an integrated environmental management system improves business management.

The strategy of the sector involves adopting the latest technologies on the market. Thus, it had simultaneously implemented both the first and the second stage in the development of its environmental management system, which consists in applying the most obvious and sensible improvements, which have low costs, together with investments in the most innovative technologies on the market.

Although the environmental management system should be integrated at all business levels and it ought to be useful for decision-making as decision-usefulness theory states, the actual stage of development circumscribed it to specific projects, which are almost orientated towards obtaining funds such subsidies or grants.

We can extract the hypothesis that both taking into account the consequences of decisions on their direct stakeholders such as the instruments of checking compliance of socio-environmental commitments could require additional resources that would be reflected in a higher debt ratio, validating the available funds theory. Besides, the closest relationship with their stakeholders would be also rewarded with greater return on equity, as the moderation model of Berman et al. (1999) holds. We also obtained evidences that the instruments of checking compliance were correlated with debt ratio, which could be explained by the fact that companies who need to rely on external financing must meet certain socio-environmental requirements as Roberts (1992) found.

There are some incongruities between the environmental position assumed in strategy and the environmental behaviour reflected in facts in the sector, so from the analysis of correlation between the score that the companies gave themselves and their score in each CSR factor it was obtained a positive correlation between CSR self-evaluation and direct stakeholders as well as impact limitation. The correlation with stakeholders corroborates the stakeholder theory and the correlation with impact limitation could be explained by the fact that companies which are taking measures to limit their impacts generally consider that they are making greater efforts in CSR, so they gave themselves higher scores.

Further studies will be required to determine whether the obtained evidences are a particular feature of the Spanish tiles sector, or on the contrary, some of them could be extrapolated to other sectors.

Author details

Maria J. Masanet-Llodra
Department of Finance and Accounting, Universitat Jaume I, Castellon, Spain

6. References

Alexander, G. J. & Buchholz, R. A. (1987). Corporate Social Responsibility and Stock Market Performance, *Academy of Management Journal*, Vol.21, pp. 479-489

Abelda, E. (2011). The Role of Management Accounting Practices as Facilitators of The Environmental Management. Evidence from EMAS, *Sustainability Accounting, Management and Policy Journal*, Vol.2, No.1, pp. 76-100

Aupperle, K. E.; Carroll, A. B. & Hatfield, J. D. (1985). An Empirical Examination of the Relationship Between Corporate Social Responsibility and Profitability, *Academy of Management Journal*, Vol.28, pp. 446-463

Barth, M. & McNichols, M. (1994). Estimation and Market Valuation of Environmental Liabilities Relating to Superfund Sites, *Journal of Accounting Research*, Vol.32 (Supplement), pp. 177-209

Bennet, M. & James, P. (1997a). Environment Related Management Accounting: Current Practice and Future Trends, *Greener Management International*, Vol.17, pp. 32-51

Bennet, M. & James, P. (1997b). Making Environmental Management Count: Baxter International's Environmental Financial Statement, *Greener Management International*, Vol.17, pp. 114-127

Berman, S. L.; Wicks, A. C.; Kotha, S. & Jones, T. M. (1999). Does Stakeholder Orientation Matter? The Relationship Between Stakeholder Management Models and Firm Financial Performance, *Academy of Management Journal*, Vol.42, pp. 488-506

Burnett, R. D. & Hansen, D. R. (2007). Eco-efficiency: Defining a Role for Environmental Cost Management, *Accounting, Organizations and Society*, Vol.33, No.6, pp. 551-581

Cho, C. & Pattern, M. (2007). The Role of Environmental Disclosures as tool of Legitimacy: A Research Note, *Accounting, Organizations and Society*, Vol.32, pp. 639-647

Clarkson, P.; Li, Y.; Richardson, G. D. & Vasvari, F. (2008). Revisiting the Relationship Between Environmental Performance and Environmental Disclosure: An Empirical Analysis, *Accounting, Organizations and Society*, Vol.33, pp. 303-327

Cormier, D. & Magnan, M. (2007). The Revisited Contribution of Environmental Reporting to Investors' Valuation of a Firm's Earnings: An International Perspective, *Ecological Economics*, Vol.6, No.2, pp. 613-626

Curkovic, S. & Sroufe, R. (2007). Total Quality Environmental Management and Total Cost Assessment: An Exploratory Study, *International Journal of Production Economics*, Vol.105, No.2, pp. 560-579

Curkovic, S., Sroufe, R. & Landeros, R. (2008). Measuring TQEM Returns from the application of Quality Frameworks, *Business Strategy and the Environment*, Vol.17, No.2, pp. 93-106

Curkovic, S., Sroufe, R. & Eckert, J. (2006). A Preliminary Framework for Using Total Cost Assessment in Total Quality Environmental Management, *The International Journal of Environmental, Cultural, Economic and Social Sustainability*, Vol.1, No.3, pp. 81-88

Epstein, M. J. (1996). Improving Environmental Management with Full Environmental Cost Accounting, Environmental Quality Management, Vol.6, pp. 11-22

European Communities. (1993). Regulation (CEE) 1836/93 of June 29, that Permits to Industrial Organizations the Voluntary Adherence to a Community System of Environmental Management and Auditing (Official Diary of European Communities L.168.10.07.93)

European Communities. (2000). Common Position for Future Adoption of the Regulation (CEE), that Permits to Organizations the Voluntary Adherence to a Community System of Environmental Management and Auditing (Official Diary of European Communities C.128.08.05.00)

Freeman, R. (1984). *Strategic Management: A Stakeholder Approach*, Pitman Publishing Inc., Marshfiels: MA

Friedman, M. (1962). *Capitalism and Freedom*, University of Chicago Press, Chicago

Gray, R. & Bebbington, J. (2001). *Accounting for the Environment, Part II*, 2nd ed., Paul Chapman Publishing Ltd, London

Gray, R. (2010). A Re-evaluation of Social, Environmental and Sustainability Accounting: An Exploration of an Emerging Trans-disciplinary Field?, *Sustainability Accounting, Management and Policy Journal*, Vol.1, No.1, pp. 11-32

Houthuysen, S. D. (2000). Deployment and Operation of a Business-Wide EMS, in R. Hillary (eds.), *ISO 14001 Case Studies and Practical Experiences*, Greenleaf Publishing Limited, UK, pp. 18-30

ISO (International Organization for Standardization). (1996). *ISO 14001 Environmental Management Systems: Specification with Guidance for Use*, (ISO/TC207/SC), Geneva

Larrinaga-Gonzalez, C. & Bebbington, J. (2001). Accounting Change or Institutional Appropriation: A Case Study of the Implementation of Environmental Accounting, *Critical Perspectives on Accounting*, Vol.12, pp. 269-292

Mahoney, L. & Roberts R. (2007). Corporate Social Performance, Financial Performance and Institutional Ownership in Canadian Firms, *Accounting Forum*, Vol.31, pp. 233-253

Masanet-Llodra, M. J. (1999). The Development of Environmental Accounting and Managerial Environmental Accounting in the Ceramics sector in Castellon, *I Reunion in Environmental and Social Accounting*, CSEAR, Dundee, UK.

Masanet-Llodra, M. J. (2006). Environmental Management Accounting: A Case Study Research on Innovative Strategy, *Journal of Business Ethics*, Vol.68, pp. 393-408

Masanet-Llodra, M. J., Nieto, L & Barreda, I. J. (forthcoming). Corporate Social Responsibility and Financial Performance: A multidimensional approach for SME

McGee, J. (1998). Commentary on "Corporate Strategies and Environmental Regulations: An Organizing Framework" by A. M. Rugman & A. Verbeke", *Strategic Management Journal*, Vol.19, pp. 377-387

McGuire, J. B.; Schneeweis, T, & Branch, B. (1990). Perceptions of Firm Quality: A Cause or Result of Firm Performance, *Journal of Management*, Vol.16, pp. 167-180

McGuire, J. B.; Sundgren, A. & Schneeweis, T. (1988). Corporate Social Responsibility and Firm Financial Performance, *Academy of Management Journal*, Vol.31, pp. 854-872

Moore, G. & Robson, A. (2002). The UK Supermarket Industry: An Analysis of Corporate Social and Financial Performance, *Business Ethics: A European Review*, Vol.11, No.1, pp. 25-39

Moore, G. (2001). Corporate Social and Financial Performance. An Investigation in the UK Supermarket Industry, *Journal of Business*, Vol.34, pp. 299-315

Mouck, T. (1995). Financial Reporting, Democracy and Environmentalism: A Critique of the Commodification of Information, *Critical Perspectives on Accounting*, pp.535-553

Pedersen, C. & Nielsen, B. (2000). Maintaining the Momentum: EMS after the Certifier has Left, in R. Hillary (eds.), *ISO 14001 Case Studies and Practical Experiences*, Greenleaf Publishing Limited, UK, pp. 31-38

Porter, M. E. & van der Linde, C. (1995a). Green and Competitiveness: Ending the Stalemate, *Harvard Business Review*, Vol.5, No.73, pp. 120-134

Porter, M. E. & van der Linde, C. (1995b). Towards a New Conception of the Environment-Competitiveness Relationship, *Journal of Economic Perspectives*, Vol.4, No.9, pp. 97-118

Porter, M. E. (1990). *The Competitive Advantage of Nations*, Fee Press-Macmillan, New York

Porter, M. E. (1991). America's Green Strategy, *Scientific American*, pp. 168, April

Power, M. (1997). Expertise and the Construction of Relevance, Accounts and Environmental Audit, *Accounting, Organizations and Society*, Vol.2, No.22, pp. 55-65

Preston, L. E. & O'Bannon, D. P. (1997). The Corporate Socio-Financial Performance Relationship, *Business and Society*, Vol.36, pp. 419-430

Roberts, R. (1992). Determinants of Corporate Social Responsibility Disclosure: An Application of Stakeholder Theory, *Accounting, Organization and Society*, Vol.17, pp. 595-612

Shane, P. B. & Spicer, B. H. (1983). Market Response to Environmental Information Procedure Outside the Firm, *The Accounting Review*, Vol.58, pp. 521-538

Shields, D. & Boer, G. (1997). Research in Environmental Accounting, *Journal of Accounting and Public Policy*, Vol.2, No.6, pp. 117-123

Suchman, M. C. (1995). Managing Legitimacy: Strategic and Industrial Approaches, *Academy of Management Review*, Vol.20, pp. 571-611

Taylor, D. W.; Sulaiman, M. & Sheahan, M. (2001). Auditing of Environmental Management Systems: A Legitimacy Theory Perspective, *Managerial Auditing Journal*, Vol.16, No.7, pp. 411-422

Ullmann, A. A. (1985). Data in Search or a Theory: A Critical examination of the Relationships Among Social Performance, Social Disclosure and Economic Performance of US Firms, *Academy of Management Review*, Vol.10, pp. 540-557

Vachon, S. & Klassen, R. (2008). Environmental Management and Manufacturing Performance: The Role of Collaboration in the supply chain, *International Journal of Production Economics*, Vol.111, No.2, pp. 299-315

Vargas-Vargas, M.; Meseguer-Santamaría, M. L.; Mondéjar-Jiménez, J. & Mondéjar-Jiménez, J. A. (2010). Environmental Protection Expenditure for Companies: A Spanish Regional Analysis, *International Journal of Environmental Research*, Vol.4, No.3, pp. 373-378

Wagner, M. (2008). Empirical Influence of Environmental Management on Innovation: Evidence from Europe, *Ecological Economics*, Vol.66, No.2-3, pp. 392-402

Walden, W. D. & Schwartz, B. (1997). Environmental Disclosure and Public Policy Pressure, *Journal of Accounting and Public Policy*, Vol.2, No.6, pp. 125-154

Wokutch, R. E. & Spencer, B. A. (1987). Corporate Saints and Sinners: The Effect of Philanthropic and Illegal Activity on Organizational Performance, *California Management Review*, Vol.29, pp. 62-77

Yin, R. K. (1994). *Case Study Research. Design and Methods. Applied Social Research Methods Series 5,* Sage Publications

Chapter 6

A Bottom Up Approach to Modeling Habitat Connectivity Dynamics Through Networks Analysis

Paulo Morgado, Marina Toger, Patricia Abrantes and Jérémy Fiegel

Additional information is available at the end of the chapter

http://dx.doi.org/10.5772/45939

1. Introduction

Urbanization processes conquer more and more land through sprawl. Cities unite into large functional regions, leaving species other than humans lacking habitat areas. Biodiversity is in decline worldwide [1], but does it has to be? City development doesn't happen monotonously, as an ever spreading wave of sprawl, like it was believed for some time. But rather cities expand by leapfrogging [2] and have a fractal form [3]. This type of process leaves vacant "open spaces" in-between urbanized areas.

Urban planners normally would address urban development from human centered perspective, treating the open spaces as a void, left out by the city. As long as there was a lot of this void, nature could cope with this approach. But as open spaces become scarcer, biodiversity declines due to habitat depletion and fragmentation, and the conservational biodiversity-centered landscape planning can no longer be ignored. A typical process of planning would involve trying to make restrictive policies on development. Do restrictive policies really work if there are strong economic incentives to develop? In Israel the existing planning land-use system has very limited impact on real life land-cover configurations, though the resulting sum area allocations mostly follow the guidelines. [9]. On the other hand, if there is knowledge that allows making educated guidelines for future development, perhaps the sprawl processes can be directed and designed in a less destructive way, the knowledge about nature's spatial configuration requirements in order to thrive as opposed to minimal requirements to survive.

Of course open spaces are becoming more fragmented as a result of human activities and infrastructures. Open spaces fragmentation has an important ecological effect, due to its contribution to habitat depletion and degradation and subsequent biodiversity loss. According to EEA [4], fragmentation results in the inability of different species to access resources, in reduction of the amount and quality of habitat areas, and their isolation, among other effects. Fragmentation is one of the major problems in EU nowadays, thus contradicting the principles of sustainability, especially concerning biodiversity conservation and land use policies [5, 4]. A wide range of literature, regarding connectivity and landscape fragmentation, assumes that increase in fragmentation of open space is associated with decrease in connectivity [5, 6]. Therefore spatial planning and environmental policies should be arguing in favor of the need to increase connectivity between green spaces, as a way to enhance habitat preservation. Unfortunately some of the planning tools and policies have poor scientific basis, neglect or ignore spatial influence and have more aims than the measures to achieve them.

This chapter offers a summarized overview of methods and models, skills and tools required for educated decision making regarding environmental management of open spaces, and then proposes a bottom up approach to model complex phenomena in a landscape involving habitat connectivity. To pursue this, several concepts, technologies and methods are assembled together, such as GIS, Multi-Agent modeling (MA) and Complex Networks analysis (CA). Finally an explicit spatial model is set up in order to analyze the evolution of connectivity habitat and deliver some insights about environmental spatial planning. This model is based on the understanding that different species of animals have different dispersal characteristics and operate in the environment at different scales, meaning that there is a need for suitable linkages between natural spaces at a scale relevant to each species [5, 6]. Hence a generic spatially explicit model, constructed for analyzing habitat fragmentation of a landscape, should be implemented in a range of species specific scales. In order to serve decision makers as a tool, this type of analysis should take into account several focal species (e.g. small and large mammals, small birds, insects and plants) and then compile these results together into a landscape metric. This model is intended for use as a tool for spatial analysis that could be implemented on a sequence of temporal data and thus used for understanding of spatio-temporal dynamics of open space connectivity.

This work's purpose is thus threefold. First we introduce the importance of explicit spatial models for environmental studies, e.g. biodiversity conservation planning, environmental impacts of urbanization and urban planning, and more particularly explanation of the fitness of bottom up models to analyze evolution of connectivity habitat, based on state of the art of referenced papers and works published. Second, a step-by-step explanation of the modeling methods is conducted, so it could be used as a benchmark for others, devoted to this kind of studies. Third, rather than drawing conclusions, we prefer to make final remarks on further developments and define promising applications based on new techniques and methods.

This methodology is developed in following stages: 1) the definition of the conceptual model regarding the integration of GIS, multi agent system and complex networks; 2) model implementation; 3) network analysis. Therefore the methodology contributes to the discussion on the relevance and suitability of multi agent systems, GIS & CN to model landscape fragmentation, as well as to improve the spatial dimension analysis techniques using connectivity algorithms.

2. Explicit spatial models for environmental studies

Longley *et al* say that "There is something special about spatial" [7].

We cannot agree more. Considering that most of events occur in space, geography should be a key variable in the equation of any model. In fact geography is both a constraint and a condition of life throughout territory. Our research is focused on species interactions with the territory as a basis for landscape conservation prioritization and land use decision-making. Of course we could have many different models that ignore geography. But as it can be observed from experience, even the more elegant and simple models are sometimes a poor proxy for the complex spatial reality.

Therefore, it is fundamental to contain in the model the complex spatial reality that includes connectivity and fragmentation (e.g. RAMAS [8]). Location (Where?) is crucial to understand the relationships (What? and How?) and the resulting decisions making support, and help to build better tools for spatial planning.

2.1. Spatial environmental models

Past studies analyzed open space functioning through landscape modeling using network theoretic approach. A complex landscape can be conceptualized as a mosaic of habitat patches, connected by corridors, and surrounded by hostile matrix [12]. Spatial models usually take a snapshot in time and explore spatial aspects of ecosystems: pattern analysis, habitat fragmentation, patch structure, corridors and connectivity, least cost paths through a landscape [11]. Theory of island biogeography [13], together with the meta-population theory [14] [15], was seminal for development of spatial models in ecology. Different patches levels of species richness were compared.

Graph based approaches of landscape can yield estimates of the value of individual patches and corridors for the whole system. The approach is typical when tracking population dynamics and when detailed biological or demographic processes are not needed, or simply when such information is not available in practice [16]. A different type of study developed ranking of dispersal obstruction effect of the land and application of this index as a cost surface to the landscape [17]. The more detailed approach is to consider the land mosaic in its functional landscape heterogeneity and not as binary habitat-matrix. By creating a map of functional land-cover types, the impact of landscape heterogeneity on biodiversity is assessed [18]. Cantwell and Forman [19] explored the topology of graphs defined by the juxtaposition of different patch types. Links could represent potential dispersal routes, functional connections between different patches, trophic or mutualistic interactions, weighted by the strength of interaction or flow rate and direction [20].

Another method utilizes scenario examination modeling. A moderately complex implementation framework consisting of modeling future habitat state, simulating persistence of individual surrogate entities and then projecting it across multiple real entities was developed by Ferrier and Drielsma [21] and has been employed widely in various whole-landscape planning processes throughout New South Wales, Australia.

2.2. GIS-based modeling

Although modern GIS became popular in the 1980's, various lineages can be found in GIS histories articles. Many works were done that point out some of the referenced in GIS functionality even before Roger Tomlinson GIS acronym creation in the 1960's. Some of these pre-GIS works were done by landscape architects and urban planners [22] alongside with geographers and computer scientists, and they highlight the need for well-informed environmental spatial decision concerning land use planning and land resource management. Two of the major GIS software producers in the market in the beginning of GIS computer automation were ESRI (with the ArcInfo software) and INTERGRAPH (with the MGE software). Both of them were built for Environmental spatial modeling purposes, testified by the names adopted, e.g. Environmental System Research Institute (ESRI) and Modular GIS Environment. This only emphasizes the importance of GIS tools and methods for environmental modeling and assessments.

2.3. Biodiversity and land use change

Biodiversity can be measured using different metrics. Species richness (alpha diversity) is the most intuitive way to understand it. Ecosystem (or beta) diversity is another metric. Of course different ecosystems have different species to inhabit them, so the two metrics are correlated in a landscape-wide context. [10]

Land-use changes that follow urbanization processes impact nature in various ways. Agricultural land-use captures space that belonged to nature and had its own biodiversity indices. After the change both diversity metrics tend to decline. Examining the dynamic process it could be said that some species will die out with more and more land captured by agriculture while several new ones will flourish (e.g. the grown species themselves and their parasites and predators). Every agricultural patch of land has some area of ecological influence outside its borders, dependent on type: intensive or extensive agriculture, husbandry, fish or bird farming, greenhouses, etc. Urban areas contain different land-uses. Built areas can be classified by population density, type of activity, etc. Different configurations of urban fabric can have different ecological footprints. One of the ways to classify urbanization is by percentage of impervious surfaces, which is expected to correlate with population density (residential Land Cover) and/or economic activity levels (commercial or industrial LC). Each urbanized patch has ecological influence that extends outside its borders, eg. residential areas have garbage collection points and other sources of food for some species; And industry has pollution effects that have different distance of influence (water, soil, noise and light pollution, magnetic fields etc.), which brings intensification of development, and more and more transport infrastructures are added. Roads, rails and navigable rivers are linear elements of the landscape that function as movement barriers with different impedance dependent on species characteristics and level of traffic. Therefore spatial elements of urbanization processes ecological impacts should be modeled using GIS. Spatial modeling is done as a snapshot in time, which masks the dynamic nature of the processes modeled, for instance, a relevant aspect is the history of each patch. If the patch was a part of a big natural area, but recently became isolated by surrounding urban fabric, its biodiversity probably has not declined yet. This is because there is a time lag in ecosystem response to environmental disturbance. Another patch of similar configuration could have very different characteristics because it was used for agriculture sometime in the past. Land-use legacies of the patches themselves and their neighbors can affect ecological systems for long periods. Thus temporal perspective on patches land-use history is important [11] and should be implemented in spatial modeling using GIS.

2.4. Networks and graph-theory

Graph-theory is a branch of mathematics that is dedicated to network topology. Leonhard Euler created the graph theory in 1736, but until the 1950's it has been almost solely a mathematics and physics research subject. Due to GIS revolution and performance computing improvements between the end 1980's and 1990's, graph theory has been undergoing explosive growth in many disciplines including, geography, and of course landscape ecology and conservation biology [22]. Also, during that time period, we have witnessed the emergence of a new type of science: the science of complex networks [23]. Landscape and geographic networks can be considered as a subclass of complex networks, where nodes are located in an Euclidean bi- or three-dimensional space, and arcs express real physical relationships instead of abstract ones [24].

There is a natural affinity between Graph Theory and Geography. In fact, if we look at the 1960's emergence of the quantitative analysis and mathematics in the Geography, we find that geographic network analysis (GNA) is not new [29]. Topology also echoes Tobler's groundbreaking work such as spatial dependence and interaction models. Therefore, and the fact that topology is about connectivity, the need becomes obvious to incorporate graph modeling and analysis of landscape habitat connectivity.

Graph theory and complex networks metrics are useful not only to give a simplified visualization of system structure, but also to describe and analyze its accessibility, resilience and robustness, and to evaluate and compare its evolution through time. This approach allows us to understand how species interact, what are the implications of their spatial organization and their reaction to human activity inside and around their habitat. [5]

Graph theoretic and complex networks metrics can be divided into two types: connectivity and accessibility metrics. Connectivity metrics allow comparing the structural complexity of the network, and measuring the degree of network connectivity. Accessibility metrics allow to measure individual nodal accessibility and can be considered as advanced network analysis. Using these metrics we can identify individual changes in the network and causes for global structural changes.

2.5. Multi-agent systems

Our conceptual data model uses GIS, Multy-agent systems and Network analysis. Why to combine these three approaches? The aim is to benefit from the advantages of these different methods, that have proved to be fitted for analyze complex systems, namely environmental spatial systems.

3. Data and methods

3.1. Model conceptualization

According to Longley, et al (2001), conceptual model is "…Human oriented, partially structured, model of selected objects and processes that are thought relevant …". Bearing in mind our central issue: how to measure habitat connectivity from one point to another -, our first step is to the design the diagram that portrays this conceptual model. So, this first step is essential to define the data to be used and the subsequent operative processes. The conceptual model is developed in following stages (Fig. 1): 1) data acquisition; 2) GIS integration; 3) computational modeling; 4) multi agent system and network simulation; 5) validation of the model. Here we present a brief review on these topics.

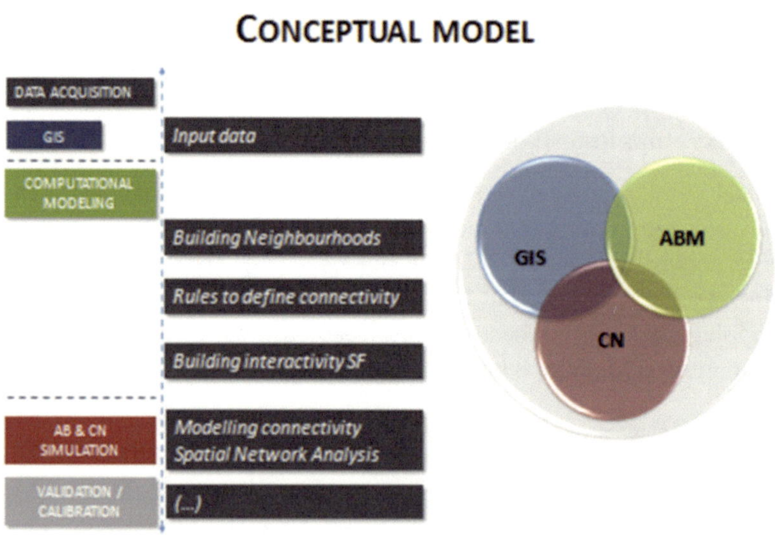

Figure 1. Conceptual model

1) *About data acquisition and validation*. When acquiring data to integrate in a GIS environment we need to consider the following principles: the data is georeferenced and in this case uses a local datum in order to provide spatial accuracy. It has a detailed scale enabling to produce comprehensive information and it preserves topologic relationships. The analysis of habitat connectivity and its evolution considers land use data for a test-region. It uses land use/cover data from the Corine Land Cover (CLC) database of the EEA (European Environment Agency). The CLC data is obtained through satellite images and is available for Europe in vector format for three different periods: 1990, 2000, and 2006. It uses a pixel size of 30 meters and the smallest mapping unit area is 25 hectares. We use the 1990 and 2006 period as a toy example to analyze habitat connectivity evolution and we extract

five types of uses/covers: urban (urb), industry and commerce (wrk), agriculture (agr), forest (gre) and wetlands (wet). The scale of the data (1:100 000) is more appropriate to deal with global/regional phenomena and in this case not so adequate to study habitat fragmentation. Despite that we use this data for our toy model as it is rapidly accessible and enables temporal analysis and thus is suitable for our test purposes. Further testing with other types of data is planned.

2) *About Geographic Information Systems data integration*. GIS are a special class of information systems that keep track of the location of the phenomena [7].The integration of georeferenced information in our computation process uses a specific application of NetLogo, the GIS extension. This extension provides the ability to load vector GIS data and raster GIS data into NetLogo, thus preserving the geographical integrity of the data.

3) *About the computational modeling*. The computational modeling combines multi-agent system and network analysis using both NetLogo and Gephi software. First we model the connectivity graphs between land use patches based on rules that are defined using literature and are species specific parameters. Then we save the results (nodes and edges) with their coordinate system in order to explore them in network analysis software while maintaining the spatial perspective of the data.

4) About multi-agent system (MAS) and Network analysis. The MAS enables to build connectivity graphs between the different land uses. These graphs can then be analyzed trough network analysis. The metrics are of two types: 1) global, i.e. representing the all structure of the networks (e.g. diameter, density, etc.), and 2) individual, i.e. representing the centrality and accessibility of each node (for instance Eigenvector, Betweenness and Closeness centrality metrics)

5) About the validation and calibration of the model. During the development of the model, some problems were take in consideration: the scale is too global thus we need to test the model with data using a more detailed scale; the pixel size is too big so we need to decrease of the pixel size from 30x30m to 5x5m in order have more accuracy. We are only using one data type of generalized land uses. We need to combine other relevant spatial data such as infrastructure networks, urbanization densities, and areas with restriction to urban development.

3.2. Model implementation

The Unicorn model has been implemented in the NetLogo multi-agent system, in four steps.

First, it was necessary to find a way to load the data files from GIS into the NetLogo multi-agent system and, to display them properly (without distortion, etc.) inside the small patches that constitute the world of NetLogo [1]. Then, neighborhood graphs were created between adjacent patches with the same land use. After this step, other connections were made between these neighborhoods, symbolizing all the possible paths between them that have been found according to the specific rules based on the distance neighborhood. Finally, these connections were saved and exported in a format, which can be used by Gephi [2], an open-source software for network visualization and analysis.

3.3. Multi-agent system: NetLogo

NetLogo is a software program and a programming language that is part of the Logo languages family, designed in a process of active learning and constructivist teaching for educational purposes (strong notion of play, testing and experimentation).

It is both:

- An agent-oriented programming language massively used to create, among various other things, virtual worlds in which thousands of entities (agents) can be created and simulated, each with their own processes and their own rules. Because of it's Logo ancestry, this language is extremely easy to understand and learn;
- An application that offers a complete and flexible graphical interface. It has multiple tools to visualize and study models (agent monitors, command center, viewing window, plots, etc.). In NetLogo it is possible to start and stop simulations or even directly manipulate some variables throughout the process (using graphical items such as sliders, buttons, menus, etc.).

The NetLogo language does not include many advanced or complex features in its basic use, compared to other multi-agent systems (such as some widely used pedestrian behavior models or graph creating features), since it must stay easy to learn. However, in addition to the advantages mentioned above, the system has a plethora of useful features that make it an excellent tool in communication and that enable designers to focus on implementing their models. Thus, it is possible:

- To export the results, displayed at any time through the graphical interface, as images, movies or CSV files (Comma-separated values files), useful for further data analysis in spreadsheet software;
- To extend the language by creating extra functions oneself (written in Java or Scala) or just by using extensions created by other programmers;
- To integrate the whole system into another application;
- To replicate the same simulations or even to run entire experimental designs, when the studied model requires more than one try (such as stochastic models), etc.

These possibilities led NetLogo to become a tool used to program models in many different sectors, such as urban planning, biology, chemistry, social sciences, small games, etc. The implemented models belong to a wide range of complexity, from those which are called "simple" (such as the two-body problem) to much more complex systems (such as the simulation of pedestrian behavior during building evacuation).

Finally, it could be noted that the representation of agents in NetLogo is unusual. There are in fact only four different kinds of agents:

- The turtles are entities that are closest to the commonly used definition of agents – moving entities that can change state according to rules;
- The patches are in fact small pieces of the "floor" of the model world. The viewing window is modeled as a grid of patches, each representing a square land area. Turtles are located on patches. As opposed to the mobile turtles, the patches are stationary;
- The links are a type of "linear" agents. They connect two turtles together;
- The observer is a unique agent and is the "master" of the world. It is through the observer that the modeler creates and manipulates other agents. It has no location or visual representation.

In the case of the Unicorn model, all types of agents are used: patches are used to store the land use data from files generated with the GIS application; turtles and links are used to create connectivity graphs, first between adjacent patches of the same land use value, then between non-neighboring patches of the same land use.

3.3.1. Loading land use data into patches

An extension of NetLogo, logically called "GIS extension", can be used to load topological data, raster or vector, contained in files created with a GIS (shape files or ASCII files).

In our case, we imported the land use GIS data into the model. The data was in fact simple numerical values. The map was rasterized and each patch (or each "cell") contained a numeric code. Each one of those represented a particular type of land use. For instance, "111" meant that the current patch was a part of an urban area, "121" corresponded to industrial land-use, "211" to agriculture, "311" to semi-natural and natural landscape elements (woods, grasslands, etc.), "411" to water bodies (lakes, ponds, etc.) and, finally, negative numbers to not specified spaces (patches with no data or outside correspond the studied map).

NetLogo automatically matched the raster pixel grid contained in the imported file to the one formed by the patches. The data was then stored in a *patches-own* variable. The accuracy of the result of this mapping could be decreased or increased by adjusting the number of patches in the model. If there were fewer patches than values in the source file, they would acquire the value of the majority of patches that were inside the square of each pixel.

The translation of these values was then performed in the model: each patch received the land use it represented as a string and a fill color, for visualization (see Figure 1):

- In red, urban areas (label, in the model: "urb");
- In purple, working areas ("wrk");
- In yellow, agricultural areas ("agr");
- In green, grasslands and woods ("gre");
- In blue, water bodies ("wet");
- In gray, not specified areas (empty label).

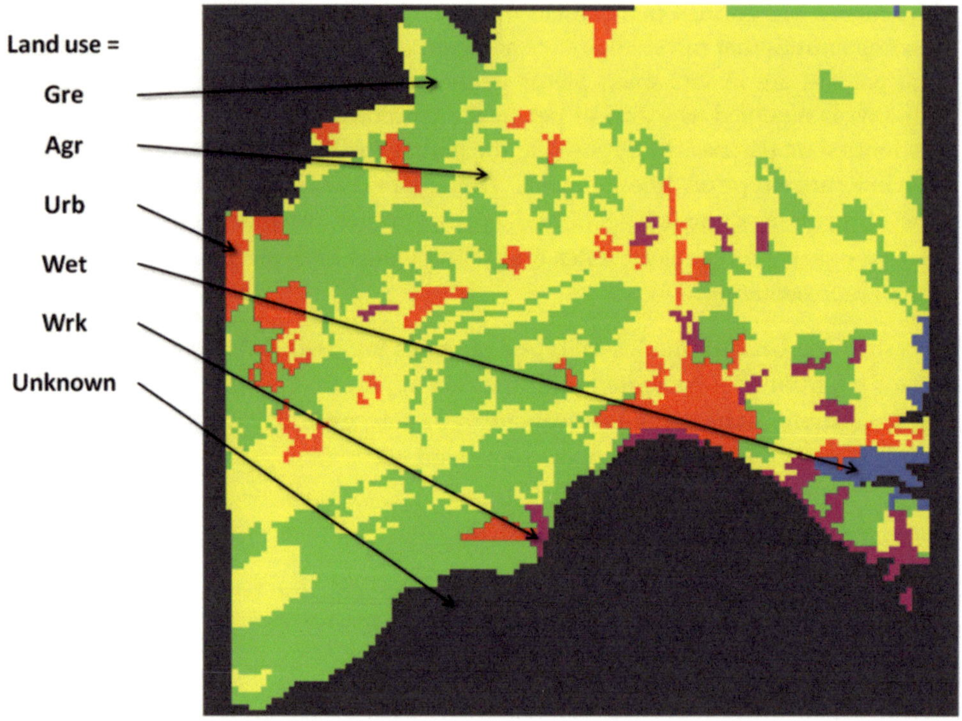

Figure 2. Overview of an imported land use dataset in the Unicorn model

3.3.2. Land use neighborhoods

After each patch received its land use value, it was necessary to classify them in groups belonging to the same area.

An area was defined as a group of patches sharing the same land use value and having at least one of their 4-neighbors (patches with one side in common with the current patch: patches directly to the north, south, to the east and west) following the same rule.

A node (turtle) was created in the center of each patch with a certain land use value. Next each one of the nodes was connected (through a link) to those present on the four neighboring patches with the same land use value. Then, we used a propagation algorithm to retrieve sets of turtles, which were interconnected (see Algorithm 1). Sets of links that belonged to each set of turtles were stored in another list, but in the same order. All interconnected nodes belonged to the same and unique group.

```
PROCEDURE find_areas

LET A = [ ]: sets of turtles in the same area
LET L = [ ]: sets of links between turtles in the same area
LET T: set of turtles which are not member of any item of A

WHILE T is not empty
SELECT one random turtle from T (seed)
LET a = { T }: set of turtles in the current new area
```

```
LET n: connected turtles to turtles ∈ a

WHILE there are turtles in n ∉ a
FOREACH n ∉ a
ADD n to a
END FOREACH
SET n : connected turtles to turtles ∈ a
END WHILE

ADD a to A
ADD to L the set of links between turtles ∈ a
REMOVE all turtles ∈ a to T
END WHILE

END PROCEDURE
```

Scheme 1. Areas detection (propagation through neighborhoods) algorithm.

The fig. 3 is an example of output: 7 areas were detected.

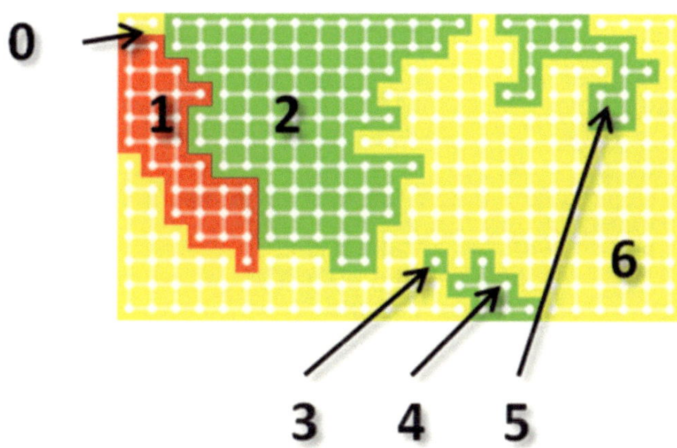

Figure 3. An example of creation of neighborhoods

At the end of the process, we obtained two lists containing the following seven items:

Position in turtles and links lists	Land use label	Number of turtles in the same neighborhood	Number of links between turtles
0	Agr	2	1
1	Urb	30	43
2	Gre	85	141
3	Gre	1	0
4	Gre	7	7
5	Gre	18	20
6	Agr	169	273

Table 1.

3.3.3. Building connections between same land-use neighborhoods

We considered the distance between the neighborhoods with the same land use. The algorithm searched for all existing paths between their patches, according to a certain maximum threshold distance (see scheme 2). Only turtle-to-turtle displacements were allowed, through links, so we used the Manhattan distance (also called the taxicab norm) to find the paths, instead of the Euclidian distance, as animals disperse from one patch to another they have a general destination, but they don't move in a straight line. Rather, they exhibit foraging behavior, exploring the area and searching for food on the way.

```
PROCEDURE find_connections_between_areas

LET distance_max: maximum distance for finding paths between areas with same
land use value
LET A: sets of turtles in the same neighborhood

FOREACH set S in A
LET B: turtles ∈ S, with less than 4 links between them and others turtles ∈
S (i.e. turtles on the edge)
LET list_num_paths: list of numbers of paths between S and other sets of
turtles ∈ A

FOREACH turtle b ∈ B
FOREACH set Si in A (Si ≠ S)
LET i: position of Si in A
LET Bi: set of turtles bi ∈ Si, on the edge AND with manhattan_distance(b,
bi) ≤ distance_max
LET C: paths(b, Bi)
INCREMENT item i of list_num_paths by COUNT C

FOREACH C
CREATE red links between start and end
END FOREACH

END FOREACH
END FOREACH
END FOREACH

END PROCEDURE
```

Scheme 2. Procedure for finding the number of connections between areas with same land use value, within a given maximum distance algorithm.

In scheme 2, the Paths(start, ends) procedure was used to compute all paths between a turtle from the edge of an area (start of the searching) and turtles from the edge of another area which were not too far from the first one (ends). A path is a list of turtles. We used a simple graph search algorithm (see Algorithm 3).

```
PROCEDURE paths(turtle_start, turtles_ends)

LET distance_max: maximum length for a path
LET PF = [ ]: list of paths found
LET Q = [[turtle_start]]: stack of paths found partially
LET CA: land use that can be "crossable"

WHILE Q is not empty
LET C: Pop(Q) (last partial path in Q)

IF last turtle of C ∈ turtles_ends
```

```
THEN
ADD C to PF
ELSE
FOREACH ni ∈ 4-neighbors of last turtle of C
IF (land use value of ni ∈ CA OR ni ∈ turtles_ends) AND
manhattan_distance(last turtle of C, turtle_start) ≤ distance_max AND ni ∉ C
THEN
ADD ni at the end of C
Push(C, Q)
END IF
END FOREACH
END IF
END WHILE

END PROCEDURE
```

With these procedures, the paths were found two times: from an area A to an area B and, in the reverse situation, from the area B to the area A.

Applied with the example shown in Fig. 3 and a maximum distance of 3 patches, we obtained the following result, illustrated in Fig. 4 below:

[2 5 1] [5 2 1] [2 3 1] [3 2 1] [3 4 3] [4 3 3]

Each triplet contains the position of the starting area, the position of the ending area in the list of groups of turtles (with the same land use value created in section 3.3.1) and the number of paths between them.

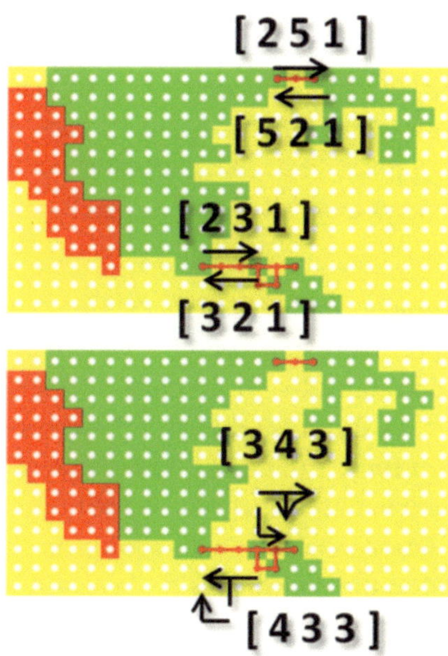

Figure 4. Connections between green areas with a maximum distance of 3 patches

If the maximum distance is 5 patches, then the algorithm produces more paths (see Fig. 5):

[4 5 10] [5 4 10] [3 5 3] [5 3 3] [3 4 28] [4 3 28] [3 6 37] [6 3 37]

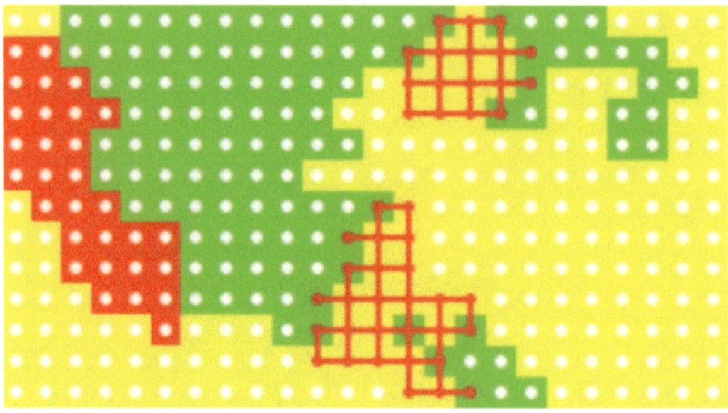

Figure 5. Connections between green areas with a maximum distance of 5 patches

The results were then saved into a file to be read by Gephi. For each area, the model computed and stored the polygon centroid (average of coordinates of the patches) instead of center (average of the coordinates) and the number of patches that composed it, its land use value and the year of the studied data. Then, for each connection between areas, the starting and the ending areas are stored, and the number of paths found between the two.

3.4. Network analysis

Complexity dictates the end of the reductionism, and has explained that for us to comprehend nature we can no longer first decipher its components, but instead we have deciphered the relations taking place. Nowadays, scientists recognize that nothing happens in isolation and most events and phenomena are connected, caused by, interacting with each other [26].

We already have subscribe Longley et al phrase that's something special about spatial, and also we have argue that graph theory and complex networks metrics are useful tools not only for networks visualization, but also to describe and analyze network structure, it's resilience and robustness, well as the accessibility of the components, and to evaluate and compare its evolution, through time. Taking that in account, we aim to have software that allows us to do spatial-temporal network analysis, and to uncover information that has been unrevealed due to classic linear analysis methodologies, that rather considering the phenomena dialectic are focus on the phenomena's. Therefore, we come into Gephi which is an open source software for graph and network analysis that also provides easy and broad access to network data and allows for spatializing, filtering, navigating, manipulating and clustering [28].

The networks metrics for analysis that we consider can be shifted in two: global metrics and individual metrics. Global metrics are the ones that analyze the network as a whole, its connectivity and structure. The density (that tell us how close is a graph to become complete), the diameter (How far apart are the two most distant nodes), modularity (communities detection) and the time-line (that allow us to compare the evolution of connectivity). For individual network metrics, which are the ones that revealing the accessibility of each node of the network and better understand the role and importance of the node in the network. The degree of the node or eigenvector (measures the importance of a node in the network based on node's connections), the betweeness (measure how often a node appears on shortest-path between nodes in the network) and closeness (the average distance from a given starting node to all others nodes in the network) [27, 28].

4. Results and discussion

The global metrics revealed, for both time periods (1990 and 2006), the presence of poor connectivity networks (table 1).

Metric	1990	2006
Diameter	8	5
Density	0.043	0.029
Modularity Number of Communities	20	30
Average Path length	2.16	3.45

Table 2.

The highest the values for density and diameter of the network, less are the connectivity, and consequently more fragmented the landscape are, i.e. more difficult for species to travel through patches (average path length). That reality is highlighted by the existence of more isolated communities.

The following figure (fig. 6) is also very helpful for us to understand network evolution through time.

As we can see, from 1990 to 2006 there is a lost of connection between nodes (habitat) on the NW part of the network, which leads to the creation of more sub-graphs and consequently lost of connectivity.

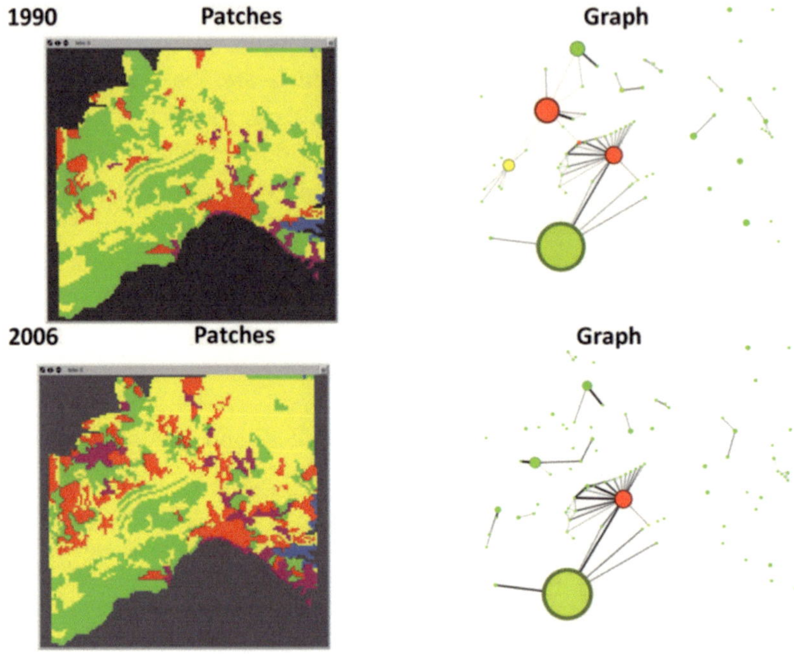

Figure 6. Temporal analysis connectivity

5. Conclusions and further research

We can conclude that our spatial-temporal bottom-up approach model suites the empirical knowledge on the habitat analysis so it can uncover some reality, as habitat fragmentation on space and species dangerous of isolation and extinction, accordingly to some business-as-usual kind of trend spatial policy.

Although, we consider that the model as to be more refine and robust in order to better fits reality and applies for well better spatial planning decisions. The fact that we needed to migrate data through different software's is a major minus. Therefore, our further research is either considering a creation of a friendly graphic user interface that could run as a plugin for some GIS software, or build our own GIS-Multi-agent-complex network based open source software.

Also, in a more short time table, we are already experiment and parameterize some individual metrics of nodes accessibility as an upgrade of connectivity metrics. We consider those metrics helpful for understand resilience and robustness of the network, i.e. what could happen to the network connectivity, if we lose this or theses particular nodes? Which nodes (habitats) lose their connectivity to the network? Which nodes is the core or the periphery of the network? Which nodes are from the cluster or the hub of the network? Those are only some of the answers that we could already get answers, both we believe at this research stage of our experimental project, that some other algorithms can be easily created and tested.

Author details

Paulo Morgado
Instituto de Geografia e Ordenamento do Território, Universidade de Lisboa, Lisboa, Portugal

Marina Toger
Technion - Israel Institute of Technology, Faculty of Architecture and Town Planning, Haifa, Israel

Patricia Abrantes
Instituto de Geografia e Ordenamento do Território, Universidade de Lisboa, Lisboa, Portugal

Jérémy Fiegel
Laboratoire d'Informatique, Institute de Système Complexes Paris-Île de France (ISC-PIF), Paris, France

6. References

[1] Wilson, E.O. ed., F.M.Peter, ass. Ed. (1988) Biodiversity, National Academy Press. http://www.nap.edu/catalog.php?record_id=989. Accessed 2012 March 23
[2] Benguigui, L., Czamanski, D. Z., & Marinov, M. (2001). City Growth as a Leap-frogging Process: An Application to the Tel-Aviv Metropolis. Urban Studies, 38(10), 1819-1839
[3] Batty, M. (2005) Cities and Complexity: Understanding Cities with Cellular Automata, Agent-Based Models, and Fractals, The MIT Press, Cambridge, MA
[4] EEA, Landscape fragmentation in Europe, Joint EEA-FOEN report No 2/2011 www.eea.europa.eu—landscape-fragmentation-in-europe. Accessed 2012 February 14.
[5] Sylwester, Alexandra. (2009) Green Infrastructure - supporting connectivity, maintaining sustainability. European Commission, DG Environment, discussion paper, http://green-infrastructure-europe.org/download/Discussion%20Paper%20Green%20Infrastructure%20Aleksandra%20Sylwester.pdf. Accessed 2011 November 21.
[6] Bennett, F. A. (1998, 2003). Linkages in the landscape. The Role of Corridors and Connectivity in Wildlife Conservation. In The World Conservation Union (ed.): IUCN Forest Conservation Programme, Conserving Forest Ecosystems Series No. 1. IUCN, Australia
[7] Longley, P.; Goodchild, M.; Maguire, D.; Rhind, D. (2001) - Geographic information systems and science. John Wiley & Sons, Ltd., West Sussex
[8] RAMAS http://www.ramas.com/software.htm. Accessed 2012 March 24.
[9] Alfasi N., Benenson I., Almagor J. (2012).The actual impact of comprehensive land-use plans: Insights from high resolution observations Land Use Policy, doi:10.1016/j.landusepol.2012.01.003
[10] Tscharntke, Teja. (2011). Landscape moderation of biodiversity patterns and processes. The Ninth Nekudat Hen Symposium, Ramat Hanadiv
[11] Ramalho, Cristina E. and Hobbs, Richard J. (2011). Time for a change: dynamic urban ecology Trends in Ecology and Evolution
[12] R.T.T. Forman & M. Godron (1986). "Landscape ecology" Jonhn Wiley & Sons,Inc

[13] MacArthur, R. H. & Wilson, E. O. (1967). The Theory of Island Biogeography. Princeton, N.J.: Princeton University Press

[14] Levins, R. (1969). Some demographic and genetic consequences of environmental heterogeneity for biological control. Bull. Entomol. Soc. Am., 15, 237–240.

[15] Hanski, I. (1998). "Metapopulation Dynamics." Nature 396: 41–49

[16] Saura Santiago (2009). "Measuring connectivity in landscape networks: towards meaningful metrics and operational decision support tools" Ecological Networks: Science and Practice 16th Annual UK-IALE Conference Edinburgh

[17] Marull, Joan, & Mallarach, J. M. (2005). A GIS methodology for assessing ecological connectivity: application to the Barcelona Metropolitan Area. Landscape and Urban Planning 71

[18] Fahrig L. Baudry J. *et al.* (2011). "Functional landscape heterogeneity and animal biodiversity in agricultural landscapes" Ecology Letters, 14: 101–112 doi: 10.1111/j.1461-0248.2010.01559

[19] Cantwell, Margot D. and Forman R.T.T. (1993). Landscape graphs: Ecological modeling with graph theory to detect configurations common to diverse landscapes. Landscape Ecology vol. 8 no. 4 pp 239-255 (1993) SPB Academic Publishing bv, The Hague

[20] Gonzalez A., Rayfield B., Lindo Z. (2011). "The Disentangled Bank: How Loss of Habitat Fragments and Disassembles Ecological Networks" American Journal of Botany 98(3): 503–516,; http://www.amjbot.org/ doi:10.3732/ajb.1000424

[21] S.Ferrier & M. Drielsma (2010). "Synthesis of pattern and process in biodiversity conservation assessment: a flexible whole-landscape modeling framework." Diversity and Distributions, 16, 386–402

[22] Foresman, T. (ed.) (1998). The History of Geographic Information Systems. Perpectives from the Pioneers. Prentice Hall, New Jersey.

[23] Watts, D. (2004) - *Six degrees. The science of a connected age*. Vintage, London.

[24] Cardillo, A. *et al* (2006). Structural properties of planar graphs of urban street patterns. *Physical Review E*, 73: 1-8.

[25] Chorley, R.; Hagget, P. (1969). *Network Analysis in Geography*. Edward Arnold, London.

[26] Barabási, A-L. (2003). *Linked. How everything is connected to everything else and what it means for business, science, and everyday life*. Plume, New York.

[27] Wilensky, U. (1999). NetLogo. http://ccl.northwestern.edu/netlogo/. Center for Connected Learning and Computer-Based Modeling, Northwestern University. Evanston, IL. Accessed 2012 March 25.

[28] Bastian M., Heymann S., Jacomy M. (2009). Gephi: an open source software for exploring and manipulating networks. International AAAI Conference on Weblogs and Social Media.

[29] Taaffe, E. and Gauthier, H. (1973). Geography of Transportation. In Foundations of Economic Geography Series. Prentice Hall, Englewood Cliffs

Chapter 7

Electronic Waste Management in Ghana – Issues and Practices

Martin Oteng-Ababio

Additional information is available at the end of the chapter

http://dx.doi.org/10.5772/45884

1. Introduction

Man has always been proficient producer of waste; however, towards the end of the 20th century saw the upsurge of a new, noxious clutter: the electronic detritus that has come to be known as e-waste. The consumption of electronics: televisions, computers, cell phones, video games, iPods, etc has increased over the last few years, making the electronic industry the world's largest and fastest growing enterprise [1]. The boom in the consumption of electronic products also come with a price to be paid –the management of the end-of-use products, or the e-waste. According to UN estimates, between 20 to 50 million tonnes of e-waste are generated worldwide annually, accounting for about 5% of all municipal solid waste. Not only is the figure representing the fastest growing municipal waste stream, it also has the potential of increasing further. In the case of mobile phones, for example, 98 million phones are said to be discarded in America annually [1].

The increased consumption and production of EEE have been facilitated by rapid economic growth, an increasing urbanization and globalization [2]. These have become major drivers of change, providing forceful leverage to socio - economic and technological growth in most developing societies, and contributing significantly to the digital revolution worldwide. Indeed, new electronic gadgets and appliances have infiltrated every facet of live today, providing society with more comfort, health, security and easy information acquisition and exchange [3]. Ironically, e-waste has become an emerging challenge as well as a business opportunity of tremendous significance. This is due to the volumes being generated and the content of both toxic and valuable materials in them. The fraction of iron, copper, aluminium, gold and other metals in e-waste is over 60% while plastics account for about 30%, with hazardous pollutants comprising about 2.7% [4].

Some recent studies [5,6] indicate that the society is 'unconsciously' creating its own toxic footprints. A research by Swerts argues that "the same hyper-technology that is hailed as a 'crucial vector' for future modern societal development has a not-so-modern downside to it: electronic waste" [7]. The fact is that the increasing 'market penetration' in the developing countries, 'replacement market' in the developed countries and 'high obsolescence rate' make e-waste one of the fastest growing waste streams. Currently, the average life span of a computer has shrunk from 6 years in 1997 to less than 2 years as at 2005, generating a flourishing export trade in used computers from developed to developing countries through up to 75% of such shipments are normally unusable [8].

The resultant waste is posing a serious challenge in disposal and recycling and creating ugly solid waste management (SWM) scenes in most developing societies. The fact is that, managing the normal waste from households in these countries already appears to be an insurmountable task [9]. It is therefore seen as more complicated if the so-called e-waste invasion from developed countries finds an easy entry into the developing countries all in the name of free trade [10]. Admittedly, the absence of proper mechanism, regulations and standards of disposal make these high-tech products often end their lives in the 'normal' waste stream meant either for recycling or landfilling [9]. The situation becomes worrying in situations where studies in China and India have shown that unregulated disposal of such wastes can contaminate soil, groundwater, and air, as well as affect all those involve in their processing, as well as the nearby communities [11,5].

Without doubt, most of the e-waste disposed of in developed countries eventually arrives in African countries through both legal and illegal means [9], where it is processed under risky conditions by poor and marginalised population. This condition of risk includes toxic health and environmental dangers. However, at the same time, access to livelihoods, access to technology, upgrading of technical skills and know how, the extension of useful life of electronics and material reuse also occur [12,13]. The e-waste processing sites in Ghana exemplifies the challenges Africa policy makers face with respect to e-waste and its impacts on health and the environment.

This paper looks at the magnitude e-waste trade in Ghana by analyzing the growing trade in electronic products ostensibly "to bridge the digital divide". It also examines its current management practices. The goal is to help raise awareness about the growing e-waste menace and encourage critical debate around the issues and hopefully, enable further action. The objectives of the study were achieved through years of research in the subject area [see 9,14,13,15]. The data were further updated by in-depth interviews with the key stakeholders in both public and private sectors, especially at Agblgbloshie which is the hub of e-waste activities in the country (see Figure 1).

The paper is structured as follows. The next section explores the meaning of e-waste and poses the question whether e-waste is a reality or myth. The third section examines the global dynamics of e-waste and presents an overview of the Ghanaian situation. This is followed by a discussion on the e-waste circuitry in Accra including the limitations that confront the current management system. The conclusion examines a way forward on how to make the e-waste recycling in the Ghanaian economy in particular and possibly, other developing countries in general more environmentally friendly without compromising its economic virtues.

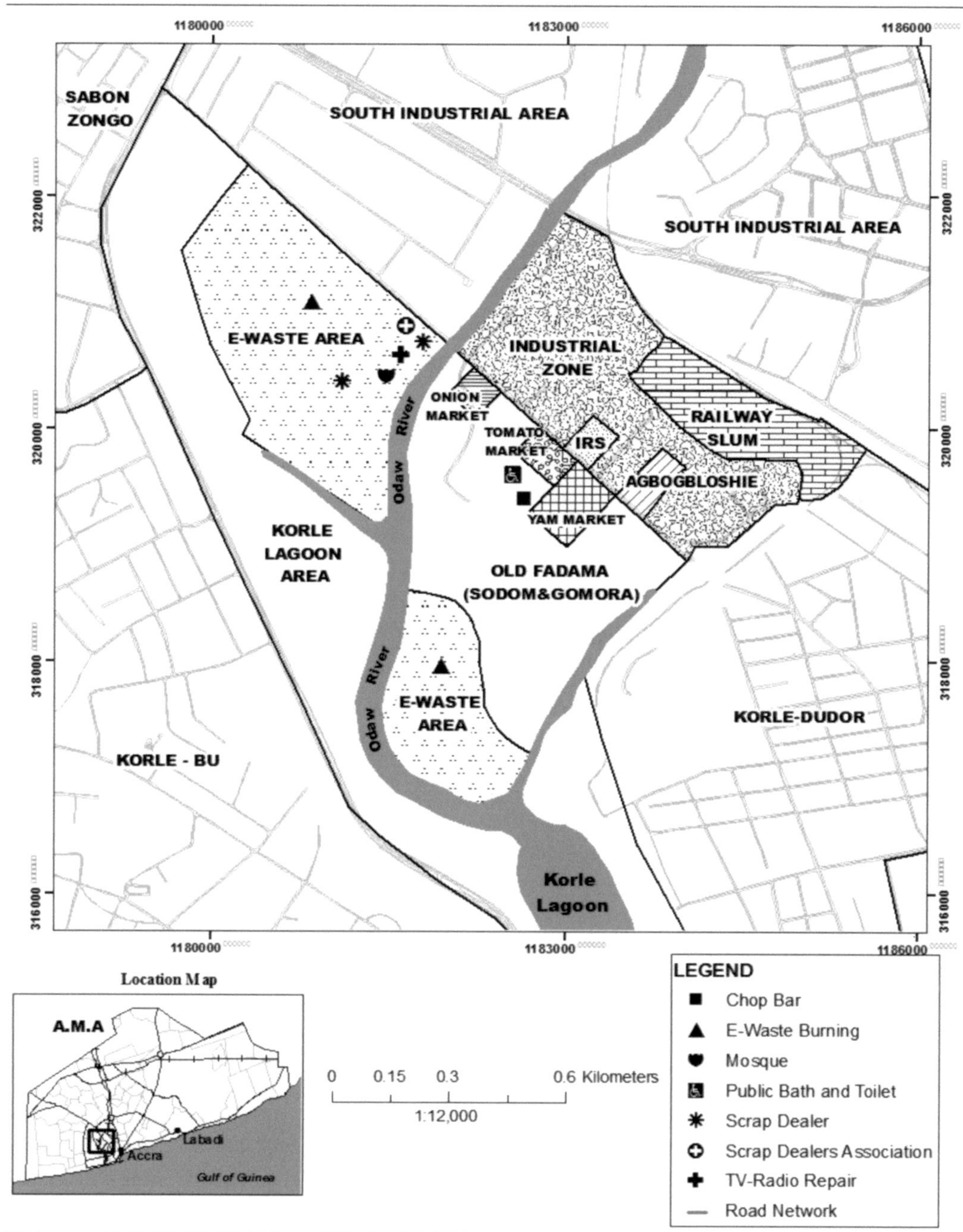

Source: Oteng-Ababio, 2012

Figure 1. Map showing the study area

2. What is e-waste?

In general, e-waste describes old, end-of-life electronic and electrical equipments (EEE) or waste generated from any equipment running on electricity or a battery including computers, laptops, TVs, DVD players, mobile phones, MP3 players, etc., which have been disposed by their original users. It has been categorized into three main groups, and these are; large household appliances like refrigerator and washing machine; information technology (IT) and telecom like a personal computer (PC), monitor and laptop; and consumer equipment like television sets. Each of these e-waste items has further been classified with respect to 26 common components which form their 'building blocks' and are therefore readily 'identifiable' and 'removable.' These include metals, compressors, plastics, glasses, wiring/electrical, transformer, circuit board, fluorescent lamp, brominated flamed retardant (BFR), etc.

E-waste also contains more than 1000 different substances, which make it either 'hazardous' or 'non-hazardous'. The presence of elements like lead, mercury, arsenic, cadmium and flame retardants beyond threshold quantities in e-waste classifies them as hazardous waste. Generally, EEEs are largely classified under three major heads, as: 'white goods,' like household appliances (air conditioners, dishwashers, refrigerators and washing machines); 'brown goods,' like TVs, camcorders, cameras, and 'grey goods,' including computers, printers, fax machines, scanners, etc. The grey goods are comparatively more complex to recycle due to their toxic (hazardous) composition.

2.1. The e-waste blues – A myth or reality?

The literature is replete with conflicting statements on whether e–waste is 'stunning whitewashed of reality' or otherwise. Greenpeace for example argues that e-waste is being exported often illegally to Ghana from Europe and the U.S [5]. In the e-waste yards, unprotected workers many of them children dismantle computers and T.Vs with little more than stones in search of metals that can be sold. The remaining plastics, cables and casing are either burnt or simply dumped. Brook as long ago as 1988, had also revealed that as safety laws in Europe and the USA push toxic waste disposal cost up to $2,500 a ton, waste brokers are turning their attention to the closest, poorest, most unprotected shores – West Africa [16]. Jim Puckett, a former Toxic Director of Greenpeace paints a glimmer picture of the main recycling site in Ghana. He writes:

> *It [Agbogbloshie] is a place where the developed world's old techno-crash waste has been tossed up by the hidden currents of today's consumerism and commerce, and has found a strange resting place..... In these global waysides, questions beg for answers; they cry out from the bone yards where these fallen icons of our proud information age lie as rotting fruit the progeny of centuries of technological advancement. Machines which months ago could process a billion instructions per second have found their end as metal and plastic skeletons in the world's most sorrowfully poor communities, to be subjected to hammer and fire, emitting deadly smoke and fume [6].*

However, Larry Summers, a former Economist of the World Bank in 1991, reportedly justifies the economic sense of the exportation of e-waste to developing countries. According to Summers:

> *the less developing countries especially those in Africa, are seriously under polluted and thus can stand to benefit from pollution trading schemes as they have air and water to spare; environmental protection for health and aesthetic reasons is essentially a luxury of the rich, as mortality is such a great problem in these developing countries that the relative minimal effects of increased pollution would pale in comparison to the problems these areas already face [cited in 4].*

Incidentally, a former Deputy Minister of Local Government in Ghana in 2008 subtly collaborated Summers' assertion, by emphasizing that, 'there is no dumping of e-waste in Ghana' [17]. Reacting to the *Daily Graphic* report, an Officer of the EPA in Ghana in 2008 ranted during a local radio discussions that 'there is no need for people to be concerned; ... the situation is not as scary as the media is making it look like'.

Notwithstanding the many negative commentary and contestation about e-waste, recent studies have demonstrated succinctly that e-waste contains valuable metals like copper, gold, and silver that are lost if not recovered properly, and which have to be compensated for by intensified mining activities, which ultimately lead to severe sustainability impacts. A conundrum is created as to whether e-waste recycling is an "economic boom or an environmental doom". The nexus becomes more complex particularly at Agbogbloshie, the hub of e-waste activities in Ghana, where there is nothing like "waste"; where every object, component, and material has "value". On the daily basis, computers and televisions are regularly bought and sold, assembled, disassembled, and reassembled. They disintegrate into their constituent materials-plastics, glass, and metals. Plastic printer cases are smashed with rudimentary tools including hammer, spanner, chisel and even the bare hands.

3. The global dynamics of e-waste

Millions of computers purchased around the world every year become obsolete and leave behind lead, cadmium, mercury and other hazardous wastes. Recent studies have shown that a workplace computer has a life span of about 2 -3 years, whilst that of a household is 3 – 5 years [18]. Additionally, Mundada et al, revealed that in 2004, 315 million computers became obsolete while 183 million new ones were sold [19]. They also noted that since 2005, for every new computer put on the market comes with an obsolete one. The US EPA estimates that in 2007, 29.9 million desktops and 12 million laptops were discarded in the USA; that is over 112,000 computers were discarded daily [20]. In the same year, a total of 205.5 million units of computer products were disposed of out of which only 48.2 million or 18% was recycled while the rest was trashed – in landfills or incinerators (see Table 1).

Similar studies concluded that only about 10% of the total waste generated are recycled while about 80% are exported into developing countries, most of which end up in landfills and incinerators [21,22,23]. In the European Union (EU), the volume of e-waste is expected to increase by 3 to 5% a year while developing countries are expected to triple their output of e-waste by 2010.

Products	Total disposed (millions on units)	Trashed (millions on units)	Recycled (millions on units)	Recycle ratio (by weights)
Television	**26.9**	**20.6**	**6.3**	**18%**
Computer products	**205.5**	**157.3**	**48.2**	**18%**
Cell phones	**140.3**	**126.3**	**14**	**10%**

Source: US Environmental Protection Agency, July 2008.

Table 1. E-waste generation and handling processes in the USA (2007).

Proportionately, e-waste is currently not a large part of the waste stream though it shows a potentially higher growth rate than any other category of municipal waste. In the USA for example, between 2005 and 2006, the total volumes of municipal waste increased by only 1.2%, compared to 8.6% for e-waste [20]. Some studies have also revealed that about 90% of e-wastes in some developed countries end up in landfills [23]. Such a tendency has increased the agitations by civil society groups over the negative environmental impact of improper handling of e-waste. Today, beyond doubts a large proportion e-waste from the developed world is being exported – in some cases illegally – to developing countries including Ghana [24], where the appropriate end-of-life management systems are non-existent [9].

4. Ghana and e-waste: an overview of the issues

Beyond doubt, there has been a phenomenal growth in the ICT sector in Ghana in the last decade due to its application in the national growth process (schools, internet cafes, etc) [25]. Probably the single most relevant and comprehensive document on ICT and development in Ghana is the ICT for Accelerated Development Policy, which hopes to "transform Ghana into an information-rich, knowledge-based and technology driven high income economy and society" [26]. It aims to introduce computers into all schools to allow children who might otherwise not have access to quality educational opportunities to use the laptops to access knowledge and provide them the opportunity to engage their own capacity for learning, regardless of their physical location or financial limitations. Between 2010 and 2011, the Ministry of education, in conjunction with rlg Communications Limited, a local private computer assembling company successfully distributed 60,000 laptops to school pupils throughout the country.

The introduction of the Global System for Mobile Communication (GSM) has also increased the preference for mobile telephony. This has reduced fixed telephone services from 206,300 lines in 2000 to 143,900 by 2008 and thus creating large quantities of obsolete telephone sets which are being thrown away or stored for perceived value. Figure 2 shows the trend in the number of mobile and fixed telephone lines in the country from 2000 to 2009. From the figure, mobile phone subscription increased from 90,000 in 2000 to more than one million in 2004 and to almost 15 million by the end of 2009. The teledensity (in respect of mobile phone alone) thus grew from about 5 phones to 100 people in 2000 to 49 phones in 2008, an increase of about 190%.

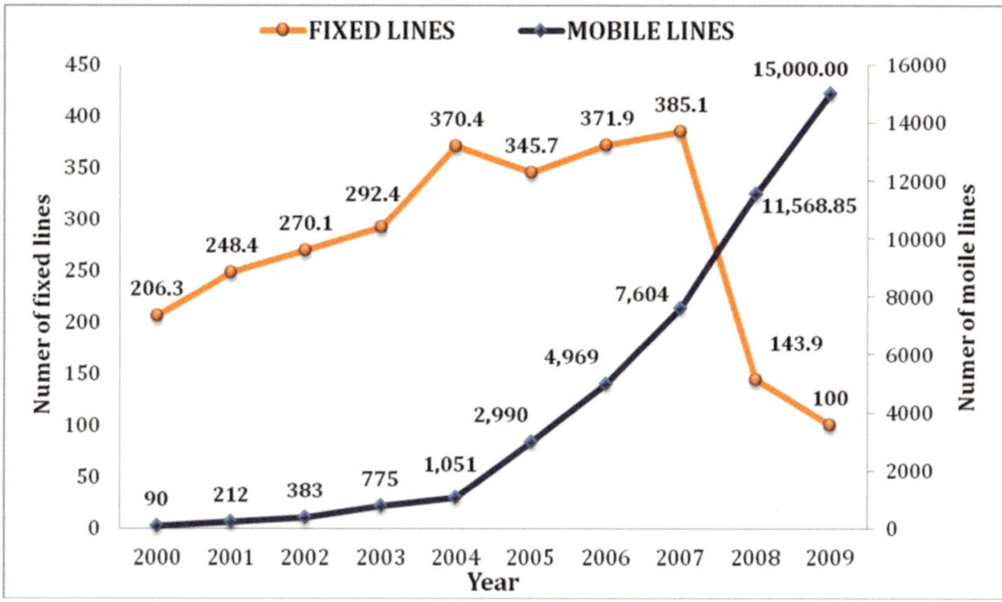

Source: Oteng-Ababio 2010; **2009 data from National Communication Authority (NCA) Records 2010

Figure 2. Number of Mobile and Fixed Telephone Lines in Ghana 2000 – 2009 (in thousands)

One important characteristic of e-waste scenario in Ghana is the fact that the rate at which electronic gadgets became obsolete is also not known. This is because of inappropriate data management practices and the fact that a number of them come in already old. Additionally, Ghana's land frontiers are porous, particularly along certain stretches of its three land borders (Cote d'Ivoire, Togo and Burkina Faso), making shipments through unmonitored routes possible. Figure 3 presents the trend of used computers imports into Ghana between 2004 and 2011, which until 2004, had been unattractive due to the then associated high import duty. Trade blossomed only when the government zero-rated (tax exempted) the importation of computers and computer accessories in 2004, primarily to make the product affordable and promote the use of ICT in the Ghanaian economy. From figure 3, a total of 1.3 million kg of used computers and accessories were imported to Ghana in 2004, reaching 10.3 million by 2011.

A major characteristic about the advancements in ICT in Ghana is the increased dependence on used or refurbished products, due mainly due to financial considerations. The reality is that although the poverty rate in the country fell from 51% in 1991/92 to 28.5% in 2005/06, yet an estimated 44.8% of the population still live on less than one US dollar per day [27]. It goes without saying that a large segment of the population could not genuinely afford a new computer if they were to join the global ICT revolution. During the studies, some used computers could be obtained for as low as 30% of the cost of a new product of similar brand in Accra. For example, while a new Toshiba A110 series laptop computer cost $1,200, a refurbished one of the same make was going for about $150 [9].

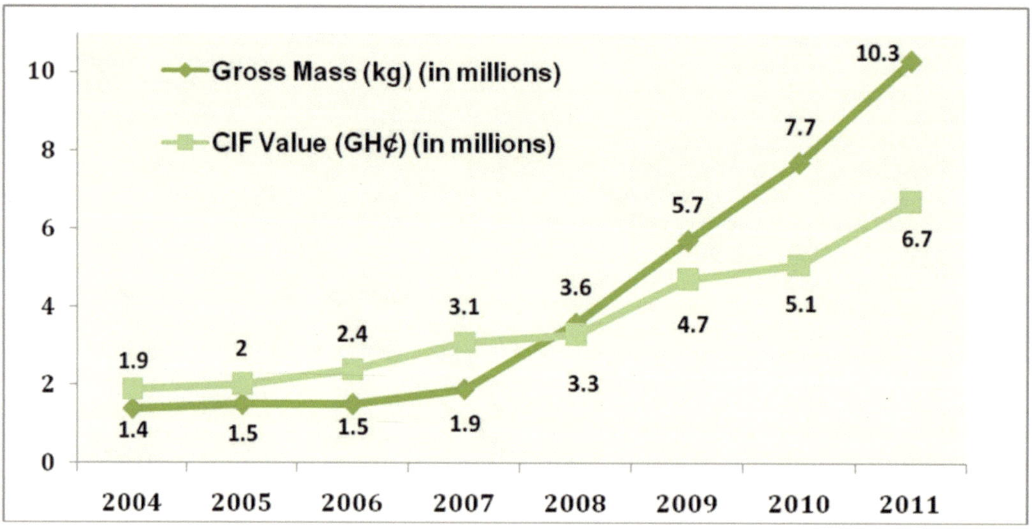

Source: Oteng-Ababio 2012

Figure 3. Trends of Used Computer Imports into Ghana: 2004-2011

Computers are playing a huge role in the Ghanaian socio-economic developmental agenda. Yet, the authorities lack the technology and infrastructural capacity to handle such waste. Accordingly, crude (open) dumping and recycling have become the lot of most local authorities who are constitutionally mandated, and responsible for waste management. This task is expected to be executed in a society where majority seem to lack appreciation of the threat pose by improper disposal practices. The need for a conscious public awareness (creation) campaign, which may culminate in the promulgation of the appropriate legislations regarding the proper handling of e-waste, has been long overdue. Ultimately, the quest to satisfy potential and actual human consumption demand should not necessarily lead to negligence of ecological and health concerns or to blatant dumping of junk products on the stakeholders as being claimed by some environmental NGOs and other media houses.

5. Think Globally, Act locally: Accra's e-waste circuitry

5.1. The legal framework for e-waste management

Despite a wide range of environmental legislation in Ghana, there are no specific laws for e-waste recycling [28,9]. In recent times, the Government in conjunction with its development partners has initiated processes through workshops and seminars, to enhance awareness creation about environmentally sound e-waste management. In 2005, a National Working Group was constituted by the EPA to help formulate a strategy for e-waste recycling but the outcome of their deliberations is yet to be made public. Ghana is however a signatory to the Basel convention which seeks to provide a framework for the international regulation for e-waste.

Research shows that in the 1970s and 1980s some developed countries used to export hazardous wastes to developing countries for final disposal which before long culminated in serious environmental pollution [29]. To remedy this problem, the Basel Convention came into effect in 1992. The Basel Convention requires that prior notice of any proposed export of certain hazardous items should be given to the government of an importing country and approved by it. But the Basel Convention does not regulate secondhand items and some e-waste scrap (including printed-circuit boards).

In principle, the Convention does not solve the new environmental problem caused by the recycling of e-waste. To solve the new problem, it is argued that the Basel Convention should be amended in such a way that hazardous wastes must not be exported from developed countries to developing countries for any purpose (even for recycling). In 1995 such a statement was presented as the Basel Total Ban, but it has yet to be agreed upon.

The convention also prohibits trade between Annex VII countries (OECD, EU and Liechtenstein) and non-Annex VII countries. The convention contains language that exempts prohibited trade in cases where an Annex VII country has signed a bilateral trade agreement with a non-Annex VII country so long as that agreement contains equivalent provisions for "environmentally sound" treatment of waste (the convention however fails to define "environmentally sound"). Despite the ratification, the convention becomes operational and applicable only when it has been properly "domesticated" which the government of Ghana has failed to do till date.

5.2. Informal e-waste recycling

In the main, the collection and re-cycling of e-wastes is by the informal sector [25,13]. The practice however exhibits a highly stratified system, comprising collection, recycling, refurbishment and reuse activities and eventually the disposal of the residuals. Generally, the EEEs are processed informally in small workshops using rudimentary methods such as manual disassembly and open burning. The appliances are stripped of their most valuable and easily extracted components which are processed to directly reusable components or secondary raw materials in a variety of refining and conditioning processes. There are also indications that some selected components like printed wiring board are selected for export probably to Asia for recycling [13]. The remaining parts are dumped or stockpiled directly. Figure 4 gives an overview of the current end-of-life management practices in Ghana.

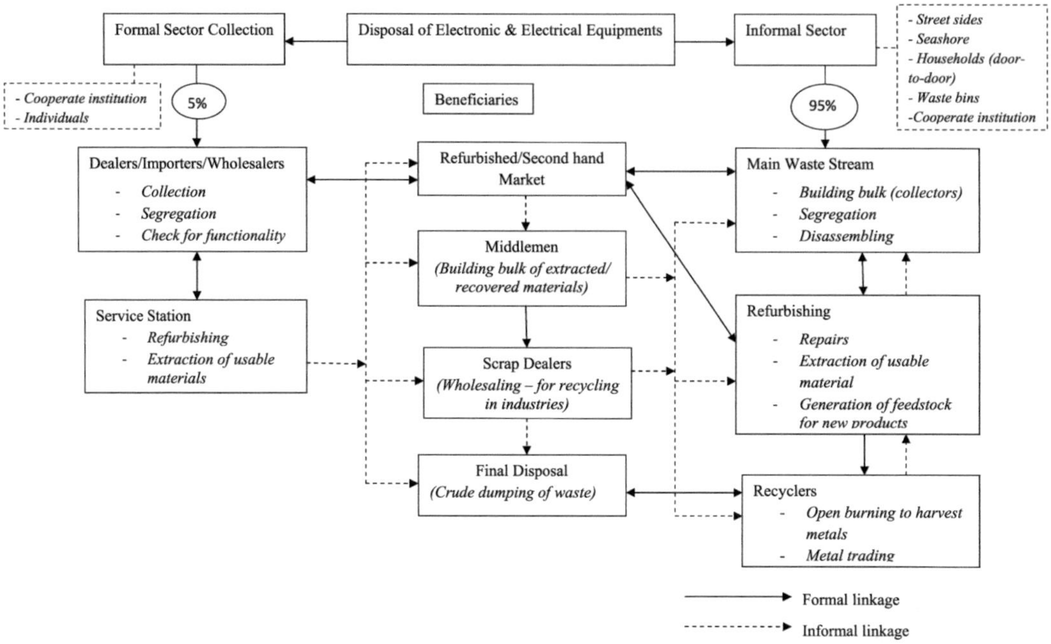

Source: Oteng-Ababio: 2012

Figure 4. The current recycling and disposal practices in the study area

5.2.1. Collection

The collection process is the first point of entry into the e-waste economy. The collectors are mostly youthful and constitute majority of the workforce in the e-waste trade. They are the actors who execute door-to-door collections of used electrical and electronic equipments (EEE) from private homes, institutions, dump sites and transfer stations. It can thus be concluded that waste collectors "make a living" by creating their own jobs as opposed to "earning a living" in regular formal employment. However, they do not operate in a separate economic realm since their operations depend on both the local and international formal economy. Moreover, there can be various loops from informal activities back to formal industry in terms of supplying recycled inputs. The challenge however is that like most informal activities, they operate in a subordinate position within global and local economies.

Initially, collectors did not have to pay anything for items dumped at street corners, neighbourhoods or dump sites. However, with increasing competition occasioned by increasing youth unemployment and the entrance of more prospective scavengers, the "waste" has begun to attract a competitive price. During the fieldwork, it was learnt that a collector has to pay $1–2.5 for an obsolete desktop computer. Some collectors also directly engage in the dismantling and recovery of metals including the burning of cables and wires to liberate copper, but there are a few who "sell their booty" to middlemen, who also serve as the intermediaries between the collectors/recyclers and scrap dealers.

An emerging dynamic in the collection of e-waste is the increasing spatial extent that collectors have to explore. Initially, many collectors operated within Accra and its environs and commuted daily between Agbogbloshie and their targeted mining neighborhood for the day. With increasing competition, the city appears fully mined and collectors have to increase their orbit for scavenging, spend days in targeted areas, build bulk before returning to base at Agbogbloshie with their booty. This has implication for the sustainability of the enterprise. For example, collectors now need huge "financial capacity" in order to spend days at a targeted destination and build bulk.

5.2.2. Refurbishment and reuse activities

Generally, reuse of older electronic products is a common practice in Ghana and the most environmentally preferable option in dealing with e-waste. It is also economically the means through which many people can access electronic products. It further conserves energy and raw materials needed to produce new once and reduces pollution associated with energy use and manufacturing. Unfortunately, since most used electronic imports are rarely tested for functionality, there is high level of refurbishment and repair, and this serves as a disincentive and time-consuming.

Additionally, repairing and refurbishing have emerged as important segments of e-waste (mis)management. Refurbishers transform old/nonfunctioning products by replacing defective components. They engage in cleaning and repairing activities in order to make the refurbished product more appealing and affordable to the populace. The findings reveal that the cost of a secondhand desktop computer at Agbogbloshie which hovered around $60 in 2010 had dropped considerably 2012. Table 2 presents the current price list of the value chain of some of the electronic items on offer at Agbogbloshie.

Prices (GHC)	Cell phone	CRT monitor	Desktop computer	Laptop	Fridge	Air conditioner
Consumers	1.5 - 5	2 - 5	2- 10	5 - 10	10-20	10 – 40
Refurbished	15 - 50	5 -10	10 - 50	10 - 50	20-50	30 – 90

Source: Field work, 2012

Table 2. Prices of some selected refurbished electronic equipments at Agbogbloshie

The area is also seen as having extensive inventories of accumulated parts to service the reuse cluster while the city refurbishers travel to the area to source parts. Local re-users capitalize on available stocks and so have "earned" reputations as the most rapid installers of reused components in the country. Accordingly, reuse traders, shopkeepers, and "individuals in the known" send devices from all over the country to be repaired in Agbogbloshie. It was also established that some of the refurbishers have also opened outlets outside Accra, especially in the Northern regions where they also coordinate the repairs and sale of refurbished electronics. Figure 5 shows some refurbishers of computer system units at the study area.

Source: Field work, 2011

Figure 5. Electrical refurbishing shops in Agbogbloshie

5.2.3. Crude recycling

Informal dismantling and recycling of e-waste aimed at material recovery is emerging as 'a lucrative business' in Ghana. At the Agblobgloshie e-waste management site, their primary activities include manual disassembly of obsolete computers, monitors, televisions, etc to isolate metals (copper and aluminium). There is also open burning of certain components to isolate copper from plastics in which they are encased, particularly from plastic coated wires and cables (see figure 6).

Source: Field work, 2011

Figure 6. An open burning of e-waste to harvest copper at Agbogbloshie

Much of the work is carried out by children, using only rudimentary tools and with no protective equipment. The recovered materials have ready market; copper is sold at 22 US cents (0.22 USD) per half kilo while plastic is sold at 1 US cent (0.01 USD) per kilo (see 13). Admittedly, this crude practice results in some loss of resources and environmental pollution yet it remains a major source of livelihood for many of the urban poor, especially the displaced youth from the North, who are compelled to choose between living in perpetual poverty or working in "perceived poison".

5.2.4. Disposal

Ironically, most obsolete electronic devices are usually stored for a while for a perceived value (physical or emotional) before disposal. Even in both public and private establishments, these items are usually stored until directives are issued for their disposal. Until recently when 'crude informal recycling' became prominent in areas like Abglogbloshie, most of these products end up in the landfills. This is not unique to developing countries as even in the US, 3.2 million tonnes of e-waste were sent to landfills in 1997 [23]. In recent years however, courtesy the informal recycling practices, only residues like ashes from other recycling operations [30] and casing are dumped. Some of these materials without doubt, contain toxic chemicals which have the potential to pollute the soil and groundwater through leaching.

5.3. Formal e-waste recyclers

Even though the informal e-waste activities are highly visible and have indeed established a circular flow within the Ghanaian economy, valuable resources are lost through the current process. Several causes including insufficient collection efforts and resources; inappropriate recycling technologies; and above all, illegal 'imports' (or movement) streams of e-waste into regions with inappropriate recycling infrastructures have been identified. With the burgeoning trade in e-waste and the growing public environmental consciousness, there have been clarion calls on the government to institute formal recycling system. It is envisaged that the formal sector would be able to manage e-waste in an environmentally friendly manner, using "state-of-the-act-technique" that will ensure better environmental management and enhanced resource recovery [31].

The Government is in the process of formulating a e-waste management bill and a draft has been sent to cabinet for consideration. It is however not clear what role the proposed legislation has carved for the informal sector; whether it is going to be at the expense of the well mesh-worked system or would complement it. From all indications, the latter might potentially be the case, even though majority of Ghanaians engage with second hand electronics. The planned policy is to be rolled out at a time most of the local authorities lack the financial resources to acquire the needed sophisticated capital-intensive recycling technologies or provide appropriate landfills. Furthermore, very few of the well established electronic companies operate official collection and take-back facility/services.

One of the very few companies operating a semblance of a take-back system is rlg Communications, a private Ghanaian company established under the companies code of 1963 (ACT179) on the 23rd day of February 2001. The company begun as a sale and service centre for mobile phones, but currently assembles mobile phones and laptops not only for the Ghanaian market but has offices in China, Nigeria and The Gambia. It has a monthly production output of 10,000 mobile phones and 8,000 laptops.

In terms of end-of-life operations, the company occasionally advertises for its customers to bring old (not necessarily non-functional) for an upgrade. The company has branch offices in all the ten regions in Ghana which undertake phone and computer repairs and sales. They also receive old products for upgrade or refurbishment. Thus, people do not return a product because it has reached end-of-life but their desire to upgrade it. Even that, there are instances where upgrading becomes problematic when the product was sourced not directly from rlg office or where the customer misplaces the purchase receipt and/or warranty card. The company has no recycling facility so the "unwanted electronics" are sent to China for processing and/or reuse. Averagely, the company exports 1, 300 pieces of mobile phones every 2 months and about 100 pieces of computers every four months.

Even though the operation of rlg is commendable, it is nonetheless limited in scope and geo-spatial extent. The company at best operates only from the 10 regional capitals even though its products are visible at all nooks and cranny in the country. Meanwhile, there is not enough incentive for people to voluntarily deposit the used electronics to the 'pseudo collection centers'. Above all, the company does not "take back" its own used product let alone those from other companies. It could therefore not be a proper yardstick to advocate for the complete disbandment of the informal sector. Suffice to state that at the present level of the country's development, the formal sector cannot solely be depended on in terms of managing e-waste in Ghana, at least not in the immediate future.

6. Limitations to proper e-waste management in Ghana

Although government seems to be demonstrating some readiness to improve on the status quo, the major obstacles to safe and effective management of e-waste remain. First, the absence of appropriate legislation dealing specifically with e-waste appears to be the main challenge. The current laws guiding the management of hazardous, solid and radioactive waste including local Government Act (1994), Act 462 and Environmental Sanitation Policy of Ghana (1999) were passed before the e-waste problem emerged. The EPA in 2005 announced it was developing guidelines to regulate the importation of used electronic gadgets but nothing has happened since. The government is a signatory to the Basel Convention but has failed to rectify it till date. The earlier a e-waste regulatory policy is enacted the better will be the drive towards sustainable e-waste management process [32].

Secondly, the increasing importation of second-hand computers, at times illegally, most of which are not tested for functionality, is equally posing a serious challenge. This has been facilitated by the government's unfettered open-door policy (duty free and lax regulations) coupled with the high incidence of poverty as well as chronic institutional corruption. Additionally, lack of reliable data (difficulty in inventorisation) poses a challenge to policy makers wishing to design an e-waste management strategy and to an industry wishing to make rational investment decisions. There is also a lack of safe e-waste recycling infrastructure in the formal sector and thus reliance on the capacities of the informal sector pose severe risks to the environment and human health. Additionally, the existing e-waste recycling systems are purely business-driven that have come about without any government intervention.

Finally, there also appears to be a high level of ignorance of the toxicity of e-waste not only among the general public but even within government circles. The public might not be wholly ignorant but for economic reasons, people are being challenged to choose between 'poverty and poison'; i.e. working in such recycling facilities and being exposed to health hazards but making a living or remaining unemployed. Educating the general public on the impact of improper management practices will go a long way in the fight for environmental and human-friendly practices.

7. Conclusion

Generally, this study has demonstrated how e-waste scavenging has emerged and become embedded in specific networked places within highly differentiated circuits that produce geographically uneven development. The findings show that e-waste has also emerged as a challenge for local authorities especially in terms of its end-of-life management which is currently driven by the informal sector. At the same time, it has turned out to be a 'saviour' for not only those who depend on the survival economy for livelihood, but also who are economically challenged to join the ICT revolution. That notwithstanding, the informal sector or the survival economy remain marginalized and excluded from the waste management system.

Indeed, participants in the sector currently represent the basis of Ghana's e-waste management system. They are the reason for its effectiveness, but occupy the weakest position in the waste management system in general, with minimal income and precarious working and living conditions. In the absence of a well developed formal sector, it is important that the local authorities and government machinery in general have knowledge about the role of informal economy in the e-waste recycling and processing system. The institution of friendly policies and regulations cannot only help abate the negative tendencies inherent in the practice but more importantly, improve productivity and working conditions without compromising the sector's flexibility traits.

The government also has the responsibility to enforce the tenets of the international treaties like the Basel convention which have not been rectified. The 'extended producer responsibility' (EPR) for example focuses on the responsibility that producers assume on their products at the end of its useful life. The government has to ensure that the obvious dereliction of duties by agencies like the Standards Board, Customs, Excise and Preventive Service as well as environmental regulators, etc are halted. The present lethargic attitude of handling of municipal wastes should encourage the promulgation and enforcement of apposite legislations. Ultimately, the government should focused on technical and policy-level interventions, implementation and capacity building, and increase in public awareness such that it can convert the challenges of e-waste into opportunities.

Author details

Martin Oteng-Ababio
University of Ghana, Department of Geography and Resource Development, Legon, Accra, Ghana

8. References

[1] BAN/SVTC (2002) Exporting Harm: the High Tech Trashing of Asia. The Basel Action Network and Silicon Valley Toxics Coalition. February 25 2002.

[2] Remesh B.B, Parande A.K, Ahmed B.C (2007) Electrical and Electronic Waste: A Global Environmental Problem. Waste Management Research. 25: 307-318.

[3] Sinha S (2007) Downside of the Digital Revolution. Published in Toxics Link. Available: http: //www.toxicslink.org/art-view.php. Accessed 2007 Dec 28.

[4] Widmer R, Oswald-Krapf H, Sinha-Khetriwal A, Scnellmann M, Boni H (2005) Global Perspectives on the E-waste. Environmental Impact Assessment Review. 25: 436–458.

[5] Brigden K, Labunska I, Santillo D, Johnston P (2008) Chemical Contamination at E-waste Recycling and Disposal Sites in Accra and Korforidua, Ghana. Greenpeace International Amsterdam.

[6] Puckett J (2011) A Place Called Away. In: Hugo P (2011) Permanent Error. Prestell Publishers. ISBN 978-3-7913-4520-8

[7] Swerts T (2006) Waste or Opportunity? the Importance of a Progressive India E-waste Policy. Published in Toxics Link. Available: http: //www.toxicslink.org/art-view.php. Accessed 2006 Nov 17.

[8] UNEP (2005) E-waste: the Hidden Side of IT Equipment's Manufacturing and Use. Early Warnings on Emerging Environmental Threats. No. 5. United Nations Environment Programme.

[9] Oteng-Ababio M (2010) E-waste: an Emerging Challenge to Solid Waste Management in Ghana". Int. Development Planning Review (IDPR). 32:2.

[10] Toxics Link (2004) Scrapping the High-tech Myth: Computer Waste in India. Toxic Links, Delhi, India.

[11] BAN (2005) The Digital Dump: Exporting Re-use and Abuse to Africa. Basel Action Network. October 24 2005. www.ban.org.

[12] Oteng-Ababio M (2012) When Necessity Begets Ingenuity: Scavenging for survival in Accra, Ghana. African Studies Quarterly (Forthcoming).

[13] Grant R, Oteng-Ababio M (2012) Mapping the Invisible and Real African Economy: Urban E-Waste Circuitry. Urban Geography. 33:1 1-21. http://dx.doi.org/10.2747/0272-3638.33.1.1

[14] Oteng-Ababio M (2012) The Legal and the Reasonable: Exploring the Dynamics of E-waste Disposal Strategies in Ghanaian Households. Journal of US-China Public Administration. 9:1 38-52. ISSN 1548-6591.

[15] Agyei-Mensah S, Oteng-Ababio M (2012) Perceptions of Health and Environmental Impacts of E-waste Management in Ghana. Int. Journal of Environmental Health Research. 1-18. http://dx.doi.org/10.1080/09603123.2012.667795

[16] Brook J (1988) Waste Dumpers turning to West Africa. The New York Times. Sunday July 17, 1988.

[17] Daily Graphic (2008) A Ghanaian Newspaper, Daily Graphic. Monday 18th August 2008 No. 16743. Accra Ghana. p. 42.

[18] Boon J.E, Isaacs J.A, Gupta S.M (2001) Economics of PC Recycling. In: Proc. SPIE, 2001. http://www.coe.neu.edu/~smgupta/4193-07-SPIE. PDF.
[19] Mundada M.N, Kumar S, Shekdar A.V (2004) E-Waste: A New Challenge for Waste Management in India. Int. Journal of of Environmental Studies. 61:3 265-279.
[20] US Environmental Protection Agency (2008) Electronic Waste Management in the United States. EPA530-R-08-009 US Environmental Protection Agency. July 2008. http://www.epa.gov/osw/conserve/materials/ecycling/docs/app-1.pdf
[21] Ravi V, Shanker R, Tiwari M.K (2005) Analyzing Alternatives in Reverse Logistics for end-of-life Computers: ANP and Balanced Scorecard Approach. Computer and Industrial Engineering. 48: 327–356.
[22] Hicks C, Dietmar R, Eugster M (2005) The Recycling and Disposal of Electronic Waste in China – Legislative and Market Response. Environmental Impact Assessment Review. 25: 459–471.
[23] Antrekowitsch H, Potesser M, Spruzina W, Prior F (2006) Metallurgica Recycling of Electronic Scrap. In: Howard S.M, et al EPD Congress 2006. pp. 889–908. The Minerals, Metals and Materials Society, TMS, Warrendale. PA, USA.
[24] Hageluken C (2006) Improving Metal Returns and Eco-efficiency in Electronic Recycling – a Holistic Approach to Interface Optimization between Pre-processing and Integrated Metal Smelting and Refining. In: Proceedings of the 2006 IEEE International Symposium on Electronics and the Environment. 8–11 May 2006. San Francisco CA. 218–23.
[25] Prakash S, Manhart A, Amoyaw-Osei Y, Agyekum O (2010) Socio-economic Assessment and Feasibility Study on Sustainable E-waste Management in Ghana. Accra.
[26] ICT4AD (2007) ICT for Accelerated Development - One Laptop Per Child: Mission. Available: http://www.laptop.org/vision/mission/ Accessed 2007 Aug 30.
[27] World Bank (2007) Ghana Country Environmental Analysis. Report No. 36985 – GH Nov. 2, 2007.
[28] Kwakye M (2009) EPA Lacks the Capacity to Control E-waste Disposal. The Ghanaian Journal Posted August 28, 2009 at 11am.
[29] Shinkuma T, Huong N.T.M (2008) The Flow of E-waste Material in the Asian Region and a Reconsideration of International Trade Policies on E-waste.
[30] Roman L.S, Puckett J (2002) E-scrap Exportation: Challenges and Considerations. In: Proc. International Symposium on Electronics and the Environment 2002 IEEE. May 6–9 2002. San Francisco CA USA. pp. 79–84.
[31] Oteng-Ababio M, Amankwaa E.F (2012) The E-waste Blues in Ghana: A Case of a New World, New Waste, New Challenge. Population Space Place (Forthcoming).
[32] Daily Graphic (2005) A Ghanaian Newspaper, Daily Graphic. Saturday 5th November 2005. No. 149593. Accra Ghana. pp. 21-28.

Chapter 8

Remote Sensing Based Crop Coefficients for Water Management in Agriculture

Jorge Rocha, António Perdigão, Raquel Melo and Cristina Henriques

Additional information is available at the end of the chapter

http://dx.doi.org/10.5772/48561

1. Introduction

The climatic factors and their variability, both spatial and temporal, linked to precipitation decreasing and irregular distribution, due to climatic changes, have been gathering a higher weight in the definition of water management policies. These policies have important implications on agriculture. Using new technologies that allow a better use of water requires institutional changes in major areas. The first point is the need for base information with an adequate spatial and temporal resolution. The work we have done includes itself in the water efficient and sustained use, allowing the improvement of irrigation systems and it's the result of a jointly effort of several teams based on an international project.

PLEIADES (Participatory multi-Level EO-assisted tools for Irrigation water management and Agricultural Decision-Support) is a research and technological development project co-funded by the European Commission's Sixth Framework Programme within its Sustainable Development, Global Change and Ecosystems Priority. The project responds to the identified need for targeted research in the area of Integrated management strategies and mitigation technologies, topic Water in Agriculture: new systems and technologies for irrigation and drainage. A set of pilot Case Studies represents a sample of the wide range of conditions found in the European and Southern Mediterranean and in the Americas, covering Portugal, Spain, Italy, Greece, Turkey, Morocco, Mexico, Peru, and Brazil.

The Portuguese working area was the Caia irrigation area, a subsystem of Guadiana basin, located in the southeast of Portugal, near the border with Spain. PLEIADES was expected to generate new knowledge on the functioning and performance of these pilot areas. This in turn aimed at providing the knowledge and information base for decision makers at all levels on agricultural water needs and consumption. It also set out to provide the basis for assessing the benefits and threats potentially brought about by new technologies to all actors in changing environments. The project was also expected to generate new tools for irrigation water management, combining innovative sensor technology with flexible easy-to-use Decision-Support Systems for adaptive management. These tools were designed to help farmers to control water more efficiently and improve the environmental and economic performance of their irrigation systems.

In our work, we address the efficient and sustainable use of water for food production in water-scarce environments. We consider the economic, environmental, technical, social, and political dimensions through a synergy of leading-edge technologies and participatory approaches. These technologies provide easy access to information for all stakeholders while active participation will be effected by spatial information and innovative networking tools. Our aim is to improve the performance of irrigation schemes by means of a range of measures. Major technical innovation is made possible by the comprehensive space-time coverage of *EO* data and the interactive networking/connecting capabilities of Information and Communication Technologies (*ICT*).

The project was designed to assess and demonstrate in an operational perspective how the integration of Earth observation (*EO*) techniques in routine Irrigation Advisory Services (*IAS*) can improve the efficiency in the use of water for irrigation. The use of leading-edge Information and Communication Technology (*ICT*) tools in the generation and distribution of information makes the *EO* easily available to *IAS* and the farmers. The project WebGIS (www.pleiades.es) was one of the central outcomes of our project. Its key feature is the operational generation of irrigation scheduling information products from a virtual constellation of *EO* satellites and their delivery to farmers in near-real-time using leading-edge on-line analysis and visualization tools. It is supported by a methodology package to derive crop coefficients and further advanced parameters from *EO* satellite images in an operational processing chain.

The overall goal is to improve and optimise irrigation and drainage systems by means of new technologies. In this context, New Technologies (*NT*) include Earth observation, Geographical Information Systems, Information and Communication Technologies, and Decision-Support systems. In order to achieve this overall goal 3 specific objectives have been set:

1. In accordance with the identified needs of stakeholders, multi-level NT-assisted tools will be adapted and developed for farmers and other water managers to optimise their water use at farm, irrigation scheme and at river-basin levels.
2. To conduct trial campaigns in pilot areas with the active participation of users at farm and irrigation scheme level.
3. To evaluate the performances of the NT-assisted tools using an extended evaluation system covering technical, economic, environmental, social and political dimensions and involving stakeholders at all levels over the whole lifetime of the project.

PLEIADES aims to demonstrate that New Technologies (*NT*) can effectively support the optimisation of irrigation schemes and in the long run foster sustainability by providing comprehensive and timely spatial information that supports decisions made at many levels: farms, irrigation schemes and river-basins.

We recognise that improvements will come not only from technical innovations but also from changes in social factors related to water governance, participation and social learning. Thus the NT-assisted tools will be deployed to facilitate technical and social learning enabling farmers to act responsibly by fine tuning their on-farm practices in accordance with the river-basin water status.

This work addresses the efficient and sustainable use of water for food production in water scarce environments. It aimed to improve the performance of irrigation schemes by means of a range of measures that consider the economic, environmental, technical, social, and political dimensions through a synergy of leading-edge technologies and participatory approaches. Major social and technical innovation was made possible by the comprehensive space-time coverage of Earth observation (*EO*) data and the interactive networking/connecting capabilities of Information and Communication Technologies. The system we developed stands mainly over *FAO* normative, about culture water needs and the calculation of cultural coefficient (K_c) in a simple way, directly from remote sensing data. For that we simply use radiometric parameters derived from visible and infrared bands. Crop evapotranspiration can be calculated using the crop coefficient (K_c) defined as the ratio of total evapotranspiration (*ET*) by reference evapotranspiration (ET_0).

Earth Observation (*EO*) provides an objective evaluation of crop water demand; this information can be used at different decision levels (from the farmers to the river basin authorities) to promote a more efficient use of water resources in agriculture. A rational management of water resources for irrigation requires information characterized by high temporal and spatial variability, which cannot be monitored with traditional field inspections. *EO* is a mature technology, ready for being transferred to operational applications in agricultural water management. Detailed data on crop development and irrigation needs are timely distributed to final users by means of modern Information and Communication Technologies.

Three main usages of *EO* based products have been conceived:

1. Distribution of personalized information to a range of stakeholders (i.e. landowners, irrigation farmers and their associations) concerning crop and water status;
2. Integration in *GIS* based river-basin water management tool, for distributed water balance calculations.
3. A portfolio of *EO* based products has been set-up, and the methodologies for their retrieval have been defined, starting from past experiences and scientific knowledge available among the partners in the Consortium. *FAO*. methodology has been adopted as the standard procedure for computing crop water requirements from *EO* based products.

Three different levels of *EO* based products are distinguished:

1. Land-use (irrigated vs. non irrigated crops; crop inventory maps);
2. Basic (vegetation cover, Leaf Area Index, Crop Coefficients, potential evapotranspiration, Crop Water Requirements among others);

3. Advanced (reference and actual evapotranspiration, biomass, yield).

The conceptual approach for the derivation of *EO* model of the service is split up into the following steps:

1. Acquisition and analysis of high resolution satellite images in the visible and infrared spectrum;
2. Local agro-meteorological data acquisition (e.g. temperature, humidity, wind speed, sun radiation, rainfall);
3. Field validation through measurements in selected areas;
4. Elaboration of *EO* based products;
5. Data quality check and integration in a dedicated Geographical Information Systems (*GIS*) for irrigation management from field to district and hydrological basin scale;
6. Real-time distribution of personalized irrigation advices on a weekly basis directly to farmers by means of different communication systems (Internet, text and graphical messages by using *GSM/UMTS*).

The validation of the different methodologies for the retrieval of *EO* based products has been an important part of the work carried out within all the pilot areas. Intensive field campaigns carried out simultaneously to satellite acquisitions have produced a large dataset for calibration and validation purposes. Micrometeorological instrumentations have been installed for comparison between field measurements of crop water use and estimates from *EO* processing. New methodologies have been set-up i.e. for improving the estimation of canopy parameters and for calculating reference evapotranspiration from geostationary satellites (of particular relevance in areas with very limited meteorological data).

2. Virtual constellation and space segment operationality

Farm management requires monitoring of agricultural crops at high spatial resolution and frequent temporal coverage during the entire growing season. The necessary spatial resolution can be provided by the current high-resolution sensors (20-30m pixel size), like TM, ETM+, SPOT, LISS, ASTER, ALI, or in the case of agricultural plots with special spatial requirements by very-high resolution sensors (like Quickbird, Ikonos). However, canopy architecture and biophysical parameters describing the canopy, like leaf area index, fractional ground cover, biomass, evapotranspiration, water stress, evolve continuously during the crop growing season.

A single satellite with a 16 day repeat time (like Landsat) would provide little useful information, considering also that cloud conditions may increase the time period between useable images. The critical requirement of frequency of coverage combined with high spatial resolution has not been satisfied after more than thirty years of Landsat mission launching. Satellite constellations have been proposed for this purpose by some studies, but a practical solution for the near future is not at hand. Our solution is a virtual constellation of EO satellites that corresponds to the needs and user requirements of irrigation scheduling and precision agriculture. Therefore, we have developed a unified procedure to obtain consistent time series of vegetation parameters of interest from this virtual constellation.

2.1. Definition and operational aspects

The virtual constellation (*VC*) is defined as a set of *EO* satellites, each of which provides the necessary data to derive *NDVI* and other vegetation parameters at the spatial resolution required for the given application. In our case, for irrigation scheduling a spatial resolution of at least 30 m is imperative to resolve the major part of agricultural fields (plots). An additional key selection criterion is near-nadir observation, such that bi-directional effects are minimized. Landsat is the backbone of this *VC*, because of its excellent operational availability and low cost.

Due to the technical failure in Landsat7 all images taken on/after 31 May 2003 have no useable data in a significant part of each image. The software correction offered by USGS involves spatial degradation, which moves the spatial resolution of ETM+ out of the interest margin of our *VC*, except for areas located directly in the centre of a scene (non-affected area). This sensor failure clearly demonstrates the vulnerability of the operational space segment for this and similar applications.

Currently available alternative platforms are less ideal for operations, since they are either much more expensive (IRS, ALI, Spot), more complicated (Spot, due to changing view angles), and/or not operational (Aster, no rush service). A number of emerging platforms may add more elements to the *VC*. Our experience in PLEIADES has shown that all can be used to complete the TM time series. Ikonos and Spot were successfully tested in the Italian pilot area. The experience of Spot programming for this area showed that on average 5-6 images per month can be obtained (fairly cloud free and with incidence angles less than 15º).

We want to stress here again that the space segment is the most vulnerable part of the entire operational system. For this reason, urgent actions are required to ensure the capability to obtain adequate *EO* images at the adequate coverage frequency and low cost. As a practical near-term solution for the case of cloudiness and/or satellite sensor failure, a contingency scenario was developed to base the PLEIADES operational system on a synergistic combination of *EO* data, field data, and an expert system of local crop coefficient (K_c) curves.

These curves have been developed from the synthesis of previous campaigns, specially tailored to the crops and climatology of a given area. In the case of an *EO* data failure (either missing image or clouds), the system would draw on a default list of K_c curves (per crop, crop cycle, sowing date) from (in order of priority) field data, the local expert system data base, and the look-up tables recommended by the Food and Agriculture Organization (FAO) [1]. Medium-resolution sensors (like AVHRR or MODIS) are also used to derive support data for this purpose.

2.2. Methodology to derive PLEIADES parameters from EO data

All biogeophysical parameter calculation from EO starts with a pre-processing of EO data, composed of three main steps: i) Geometric correction and image re-sampling; ii) Cross-satellite intercalibration; iii) Atmospheric correction.

Semi-automatic procedures have been developed in order to elaborate K_c maps from *EO* data in the minimum possible time. Pre-processing requires approximately half of the elaboration time of the entire process. Once georeferenced surface reflectance has been calculated in each pixel, the algorithms for determining K_c are quite straightforward.

The elaboration for step (i) is based on consolidated procedures available in each pilot zone. This step does not necessary requires a standardization, and it strictly linked to the topographical mapping standard adopted in each area. The principal recommendation coming from PLEIADES is to adopt procedures that do not alter substantially the radiometric content of data; as such, first-degree relationships should be preferentially used for coordinates transformation and nearest neighbour techniques for pixel resampling.

2.3. Cross-sensor intercalibration

When using different sensors from our virtual constellation to generate time series of maps of geobiophysical parameters, a reliable methodology is needed to intercalibrate the observations from different sensors at different observation scales in different platforms. Intercalibration between observations or cross-calibration of sensors aims at developing relationships that allow to translate reflectances and spectral vegetation indices from one sensor to another.

For this purpose, we have performed an observational study, comparing reflectances and *NDVI* from near-synchronous image pairs of ETM+ as the reference sensor and TM, LISS, Aster, Quickbird, and AVHRR. Linear relationships were found for the intercalibration of reflectances and *NDVI* from one sensor to another, for all sensors, provided that some spatial aggregation is performed.

The main source of data dispersion in our linear cross-sensor translation equations is the geolocalization uncertainty inherent in the process of geometric correction. Consequently, spatial aggregation needs always to be performed if (different or the same) sensors are to be used to derive time-series of biogeophysical parameters over heterogeneous areas.

The homogenous zone approach developed here is recommended as an excellent tool for deriving robust new cross-sensor relationships, provided that the selected homogeneous crops cover the full *NDVI* range. The linear cross-sensor relationships derived from one image pair are shown to be valid for the whole season and for all areas with similar vegetation and climate. We recommend repeating the procedure once or twice a year in order to check the temporal stability of the radiometric calibration coefficients.

Although the differences in most cases are small, we maintain different equations for different spatial aggregation sizes. Table 1 gives the coefficients of the resulting equations for the example of 100 m grid cell side length (except for 75 m for Aster), which is the minimum recommended grid size.

Sensor	Coefficients of linear equation $NDVI - ETM = a \cdot NDVI(sensor) + b$	
	a	**b**
TM	1.0336	+0.0128
LISS III 1D	1.1672	-0.0454
Quickbird	1.0443±0.008	+0.0191±0.005
Aster L1B	1.1304±0.022	-0.0002±0.019

Table 1. Summary of NDVI cross-sensor translation equations for elements in our virtual constellation. Measure of uncertainty is defined as deviation from the mean.

2.4. Operational atmospheric correction

An automatic procedure has been developed for atmospheric correction. This procedure uses parameters extracted directly from the image instead of recurring to external data on synchronous on-site vertical profiles of atmospheric data, which are usually difficult to obtain. It results in a substantial reduction of processing time as compared to approaches based on radiative transfer calculations. It was found that approach (i) for *NVDI* K_c calculation is less sensitive than approach of analytical K_c (ii) for the effects related to the atmospheric correction.

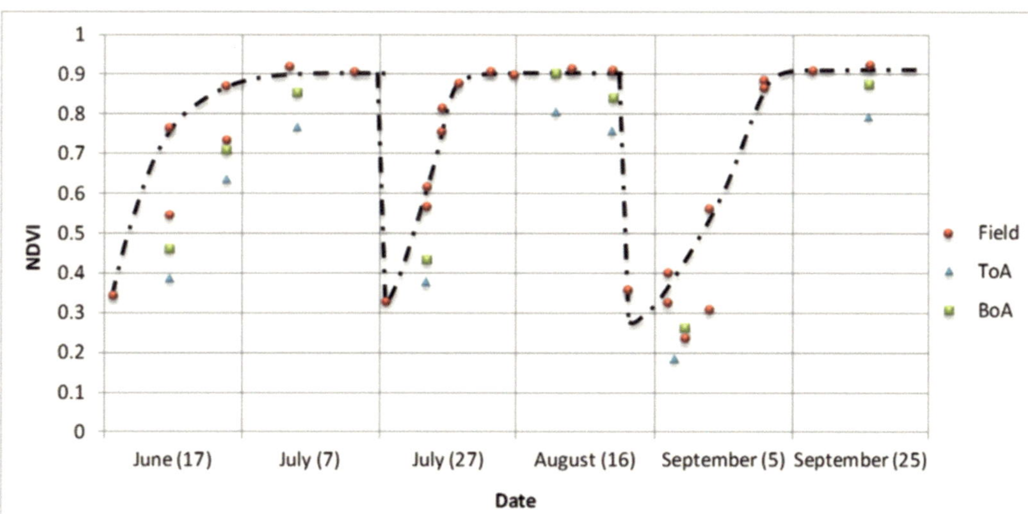

Figure 1. Example of the implementation of the atmospheric correction (*AC*) module on the satellite derived NDVI for an alfalfa plot. Top of atmosphere (*ToA*) *NDVI* refers to the *NDVI* without *AC*, whereas of atmosphere (*BoA*) *NDVI* corresponds to *NDVI* with *AC*. The reference field measurements of *NDVI* are shown as solid circles.

A validation of the performance of this module has been performed by using field measurements of spectral reflectance carried out with the spectroradiometer GER 3700 (331 - 2509 nm spectral coverage, 1.5 - 9.5 nm resolution) for two land uses: (i) an alfalfa plot and (ii) a bare soil. Figure 1 illustrates the outcome of the AC-module in terms of the *NDVI* values for the alfalfa plot. Field measurements of *NDVI* showed the greatest scattering due to both, heterogeneity of the canopy cover in the alfalfa field and, the small field of view of the spectroradiometer in comparison to Landsat pixel size. Taking into account this variability, we can conclude that atmospherically corrected *NDVI* values are in agreement with in field measured *NDVI*. Similar results are obtained for the bare soil plot.

3. Crop coefficient and reference evapotranspiration

Crop evapotranspiration can be calculated using the crop coefficient (K_c) (1) defined as the ratio of total evapotranspiration (*ET*) by reference evapotranspiration (ET_0). Combining K_c (from field measurements or from satellite images) with ET_0 from agrometeorological station observations allows us to calculate crop evapotranspiration. This coefficient integrates the effect of characteristics that distinguish a typical field crop from the grass reference, which has a constant appearance and a complete ground cover.

Factors that determine the crop coefficients are crop type, climate, soil evaporation and crop growth stages [1, 2]. For this purpose FAO has proposed tabulated average values distinguishing by crops that can be applied knowing its phenology.

In case of annual crops under standard conditions (disease-free, well fertilized, grown in large fields, under optimum soil water conditions and achieving full production under the given climatic conditions), the K_c curve for the whole growing season can be calculated considering the initial (K_{cINI}), medium (K_{cMID}) and end stage (K_{cEND}).

$$K_c = \frac{ET}{ET_0} \tag{1}$$

Mainly at the initial and end period, due to lower values of crop cover, soil evaporation has a large effect on K_{cINI} and K_{cEND} [3, 1, 4]. Therefore, vegetation indices (*VI*) are better related to transpiration of crop than to K_c [5] in those periods. This introduces a great variation in K_{cINI} and K_{cEND} daily values depending on soil water status, i.e. on frequency of wetting by irrigation and rainfall.

The dual crop coefficient approach proposed by [6] splits K_c into separate coefficients, one for crop transpiration K_{cb} (basal crop coefficient), and one for soil evaporation (K_e). The soil evaporation coefficient, K_e, describes the evaporation component of *ET*. When topsoil is wet, after irrigation or rainfall, K_e is maximal.

Estimation of K_e requires knowledge of soil water balance [1]. Wright [6] introduced the idea of a basal crop coefficient in which the soil evaporation component of *ET* was minimal due to a dry soil surface but adequate soil moisture in the crop root zone was available.

$$ET = \left(K_{cb} + K_e\right) \cdot ET_0 \tag{2}$$

In PLEIADES we use two approaches to obtain the crop coefficient from satellite imagery: one, directly from *NDVI*, named K_c - *NDVI*, based on the relationship between *NDVI* and the basal crop coefficient, and another, named analytical K_c, is based on the direct application of the Penman-Monteith equation. The *NDVI* is the main operational parameter to monitor vegetation status using Earth Observation.

4. Crop coefficient from NDVI and canopy biophysics parameters

Relevant canopy biophysics parameters are green fractional cover, fraction of absorbed photosynthetically active radiation, primary production, Leaf Area Index (*LAI*), basal crop coefficient. All they are involved in canopy evapotranspiration. The ability of *NDVI* to describe canopy biophysics parameter has been shown as follows:

1. *NDVI* is related linearly with green fractional cover [7, 8];
2. *NDVI* is related linearly with the fraction of absorbed photosynthetically active radiation (*fAPAR*) [9];
3. *NDVI* is related with primary production (dry biomass) by means of Light Use Efficiency (*LUE*) models [10, 11, 12]:

$$P = \int_0^t \left(a \cdot NDVI + b\right) \cdot PAR \cdot \varepsilon \cdot W \cdot dt \tag{3}$$

where P is primary production, PAR is Photosynthetically Active Radiation, ε is the efficiency of crop to transform PAR into dry mass, W is a water stress coefficient, and a, b are constants. Using these *LUE* models we can consider that, under non-water stress, *NDVI* on plateau stage can be seen as a good estimator of the dry matter accumulation rate, depending on crop and environmental variables [13].

It establishes a relationship between *NDVI* and crop growth rate (*CGR*) which agrees with the idea that considers *NDVI* as an estimator of the canopy photosynthetic power. This way, [14] consider that vegetation index can be legitimately used to provide an estimate of growth rate.

1. *NDVI* is related exponentially with Leaf Area Index (*LAI*) [15]. Is well known that *NDVI* begins to saturate for a value of *LAI* equal to 3 reaching a plateau for $LAI > 3$ (Figure 2 a);
2. *NDVI* is related linearly with the basal crop coefficient K_{cb} [16]. This relationship is a relevant basis for the K_c – *NDVI* approach.

The facts pointed out in (3) and (4) may appear contradictory (saturation of *NDVI* for $LAI > 3$ on one hand and the linear relation of *NDVI* with K_{cb} on the other). This seeming paradox is due to the usual reasoning that relates higher *LAI* with higher evapotranspiration. This reasoning arises from associating more leaf surface with more transpiration. However, already [17] stated that the evidence seems conclusive that transpiration in most mesophytic crop plants and other mesophytic vegetation well supplied with water increases with leaf area to *LAI* of about three (Figure 2 b).

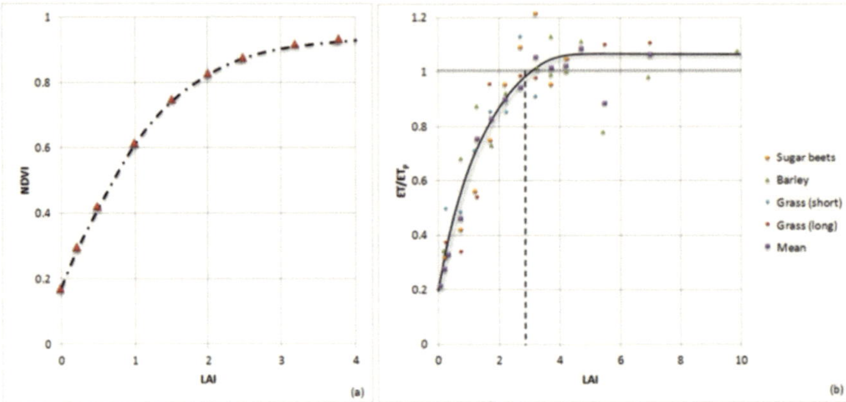

Figure 2. Typical *NDVI-LAI* curve (a), and ratio of actual to potential evapotranspiration (ET/ET_P) as a function of the *LAI* (b).

Accounting the *LAI* saturation in the relation with evapotranspiration has lead to the concept of active *LAI* (Allen et al., 1998). The active *LAI* is defined as the index of the leaf area that actively contributes to the surface heat and vapour transfer. It is generally the upper, sunlit portion of a dense canopy. For practical applications, however, the active *LAI* is an ambiguous concept due its dependence on canopy architecture and its interaction with sunlight.

The basal crop coefficient is clearly related with green fractional cover (f_c) (Figure 3). In fact, the procedure to estimate K_{cb} is based in the knowledge of f_c [1], despite of ambiguities of the green f_c concept, mainly in the maturation stage. The relationship between *NDVI* and *CGR* for well watered crops is based on the ability of *NDVI* to estimate *fAPAR,* introducing this fact in the *LUE* model. *CGR* is also related with the transpiration rate.

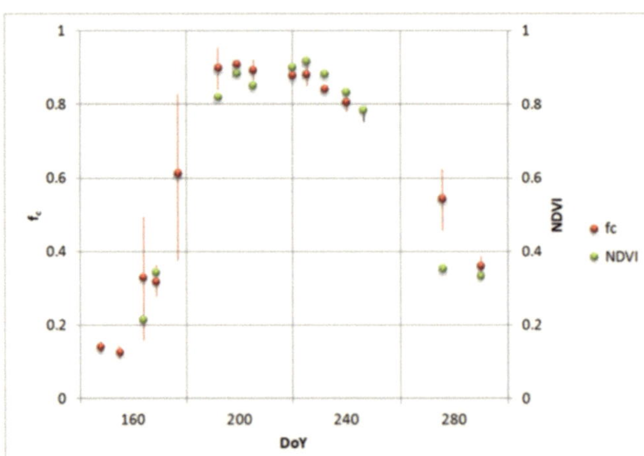

Figure 3. Green Fraction Coverand *NDVI* versus *DoY* at field scale for corn.

The relationship between *NDVI* and *CGR* exhibits a strong dependence on crop and environmental variables (solar radiation, temperature, etc.) [13]. This is due to the different nature of *NDVI* and *CGR*. *NDVI* depends only on canopy characteristics, while *CGR* and so the transpiration rate, are strongly also dependent on surface and environmental variables. The basal crop coefficient is the ratio of canopy transpiration rate over the reference canopy transpiration rate.

So, the empirical relationship between *NDVI* and K_{cb} shown by many authors, e.g. [18] and [19], (Figure 4) could be explained by considering *NDVI* as a measurement of relative *CGR*. Further research will be need in this subject. Despite limitations due to variability associated with canopy structure, background soil, and calibration uncertainties, *NDVI* can be used advantageously to estimate crop water requirements [20] in accounting its relationship with K_{cb}.

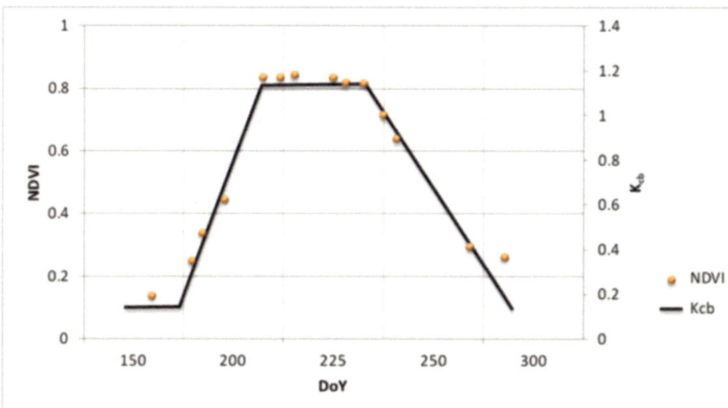

Figure 4. Temporal evolution of crop coefficient (K_{cb}) and *NDVI*, in maize. K_{cb} is estimated from green plant cover using FAO methodology [19].

Taking into account similarities between the crop coefficient curve and vegetation index, [21] established the potential for modelling crop coefficient as a function of vegetation index. This relation was derived from reflectance observations at field scale in the wavelengths ranges [0.63, 0.69 μm] and [0.76, 0.90 μm], measured at nadir and two meters above corn. A linear transformation of the *NDVI* was developed by equating the *NDVI* at effective cover and for dry, bare soil at the experimental site to the K_{cb} at effective cover and for dry soil evaporation, respectively. Similarly, [22] obtained (4) for two research sites in Colorado using alfalfa as reference evapotranspiration surface

$$K_{cb} = 1.181 \cdot NDVI - 0.026 \quad (4)$$

In order to minimize the presence of soil background, other vegetation indices (*VI*) have been used to compute K_{cb} [23]. This provides a particularly useful tool for satellite images where soil brightness and colour can vary. One of these *VIs* used is Soil Adjusted Vegetation Index (*SAVI*) [24] which for the same conditions as (4) gives the relationship [5]:

$$K_{cb} = 1.69 \cdot SAVI - 0.16 \quad (5)$$

Thus, crop coefficients derived from spectral measurements (K_{cs}) are independent of the time parameters, day of planting and effective cover, and represent a real-time crop coefficient. The use of spectral crop coefficients facilitates irrigation scheduling on a field-to-field basis over a large region if the fields can be observed spectrally, because planting and assumed effective cover dates are not required.

The spectral information would be sensitive to leaf loss due to hail, stress caused by disease and water deficit, cold or wet conditions that delay early growth, and warm temperatures and drought that speed senescence [16]. At field scale, further work was performed in order to improve scheduling irrigation events on corn compared to other traditional *DoY* based methods resulting in estimated crop water use reduced by 15% [25].

5. Crop coefficient and NDVI relation from field observations

Intensive experimental campaigns were conducted within pilot zones. The research field has a permanent lysimeter station and it is water controlled following FAO 56 specifications [26]. Coinciding in time with spectral acquisitions, biomass (kg.m^{-2}), Leaf Area Index (*LAI*), and Green Fraction Cover (f_c) were measured to describe the phenology of crops.

By the knowledge of crop stages, K_{cb} values have been estimated taking into account the effect of varying relative humidity and wind velocity from standard conditions ($RH = 40\%$, $v = 2$ m.s^{-1}) (Allen et al., 1998). Reflectance in red and near infrared to compute *NDVI* is obtained by integrating spectral reflectance in the range of ranges [0.63, 0.69 μm] and [0.76, 0.90 μm].

Evolution in time of *NDVI* and f_c for maize is represented in Figure 5. *NDVI* reaches its maximum value, when crop reaches also full effective green cover in coincidence to maximum of K_{cb}. As can be seen in Figure 3 for maize, ranges of maximum and minimum values for f_c and *NDVI* coincide in time obtaining comparable curves.

Variation in behaviour of f_c allows determining K_{cbINI} (0.15), and K_{cbMID} (1.15). To determine K_{cbEND} it is necessary to estimate water content of plant. The resulting 54% on $DoY = 277$ suggest a value of $K_{cbEND} = 0.5$. The average curve for K_{cb} adapted for crop height, minimum relative humidity and wind velocity is represented in Figure 5 along with *NDVI*.

From linear regression we obtain the equations for the reflected-based crop coefficients for corn (Figure 6) in case of *NDVI* and *SAVI*:

$$K_{cbNDVI} = 1.37 \cdot NDVI - 0.017 \qquad \left(R^2 = 0.99\right) \tag{6}$$

$$K_{cbSAVI} = 1.76 \cdot SAVI - 9.10^{-2} \qquad \left(R^2 = 0.99\right) \tag{7}$$

To perform the comparison between (6) and (7), which are grass based reference evapotranspiration, and (4), which is alfalfa based reference evapotranspiration, we have realized the following steps:

1. Derive from (4) a new *SAVI* based equation, using the relationship NDVI = 1.2735•*SAVI*+0.02106, obtained from the definition of *SAVI* with a value for L = 0.5 (where L is the adjusting factor to account for f_c in the *SAVI* definition equation), the same as used for (7);
2. Multiply alfalfa-based K_{cb} by a factor 1.15 to convert them in grass based K_{cb}, according the procedure described in [27]. Equations (8) and (9) show that the obtained results are very similar to (6) and (7).

$$K_{cb,grass} = 1.36 \cdot NDVI - 0.031 \quad (8)$$

$$K_{cb,grass} = 1.73 \cdot SAVI - 0.009 \quad (9)$$

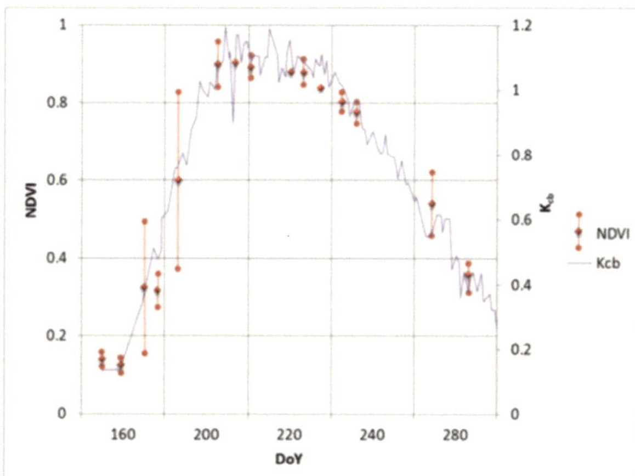

Figure 5. Field observations of K_{cb} and *NDVI* versus *DoY* for corn.

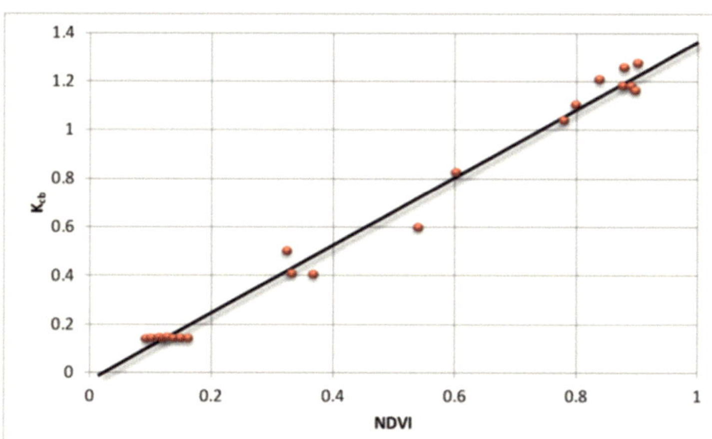

Figure 6. Linear regression between values of K_{cb} and measured *NDVI* for maize.

In the case of wheat, the evolution in time is not as representative as in maize. We observe in Figure 3 that on *DoY* = 95, f_c reaches a first local maximum before it continues growing in coincidence with emerging ears (with active photosynthesis), supposing a rapid increase in f_c, and thus in K_{cb} and spectral indices.

Applying (6) and (7) to data obtained for wheat, we see in Figure 7 that K_{cb} obtained from *NDVI* reproduces the evolution in time of f_c.

This relationship facilitates calculations of transpiration taking into account that only points over dry soil were considered, but without limiting crop transpiration. Evaporation of soil introduces an important contribution to K_c during days after irrigation or rainfall.

This means that water soil balance must be taken into account to get the contribution of evaporation in K_c. Over large areas, where variability of soil colour and brightness can influence the *NDVI*, *SAVI* and other *Vis* designed to normalize soil background effect should be used.

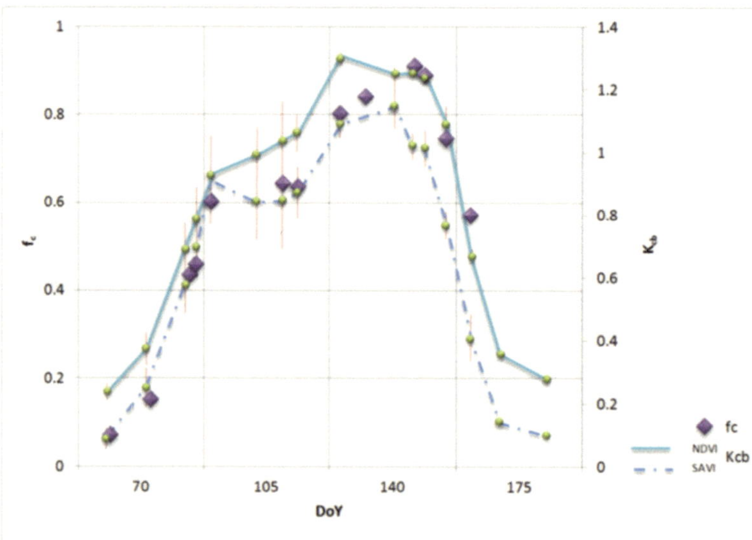

Figure 7. Evolution in time of observed f_c for wheat. The K_{cb} *NDVI* and K_{cb} *SAVI* values for wheat have been obtained from the linear relationships in (6) and (7).

6. Crop coefficient from NDVI: Operational point of view

Equations (6) and (7) provide the grass based basal crop coefficient from *NDVI* and *SAVI* data. These *VIs* are calculated for Landsat TM broadband from field radiometry data. Equation (4) provides alfalfa-based basal crop coefficient from *NDVI*.

Neale et al. [28] review the use of canopy reflectance observations to obtain crop coefficients over large areas. Similarities were found between the mean crop coefficient for small grain to the ratio of the perpendicular vegetation index (*PVI*) for wheat to *PVI* of wheat at full canopy cover. Heilman et al. [3] investigated the relationship between percent cover and reflectance-based perpendicular vegetation index (*PVI*) for alfalfa.

Neale et al. [22] related the crop canopy reflectance to basal crop coefficient for corn, developing an operational technique for estimating actual crop *ET*. The reflectance based crop coefficient (K_{cr}) was derived by nearly transforming the seasonal normalized difference vegetation index (*NDVI*) using the percent shading and leaf area measurements to establish the *EFC* and relate it to the basal crop coefficient by [6]. In several studies, *NDVI* has been directly used to predict K_c [5, 22, 23, 29, 30].

The operational procedure to estimate K_c from satellite imagery is based on the linear relationship between *NDVI* and basal crop coefficient described earlier. Landsat is the reference imagery to estimate *NDVI* (spectral broadband calibration). The attractiveness of Landsat is the high resolution (30 m in the visible and near infrared bands and 60 to 120 m in the thermal band) so that individual fields can be observed. The methodology that is described here has been checked against the preliminary results of all PLEIADES pilot areas for the following crops: Barley, wheat, maize, opium plant, sugar beet, alfalfa, pea, potato, onion and garlic. So we can establish the limits of applicability of this approach.

6.1. Dual crop coefficient NDVI approach

Wright [6] proposed a dual basal crop coefficient approach which splits the total crop coefficient into crop transpiration (K_{cb}) and soil evaporation (K_e) fractions, see Figure 8. The K_{cb} component represents the crop evaporative conditions from soil conditions whose surface is dry (direct evaporation from soil surface is minimum), and the crop growth is not limited by water, insect, climatological or physiological factors. The dual crop coefficient concept (see also equation (2)) expressed as

$$K_c = K_{cb} + K_e \tag{10}$$

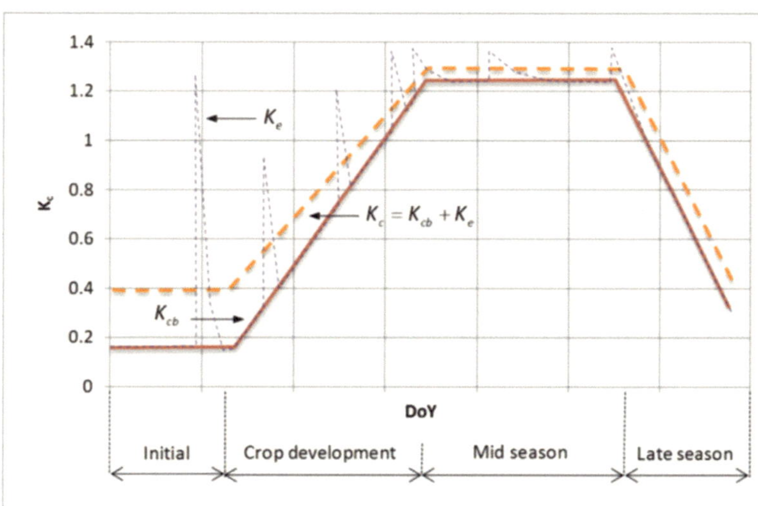

Figure 8. Crop coefficient curves showing the basal K_{cb}, soil evaporation K_e and the corresponding single $K_{cb} = K_{cb} + K_e$ curve [1].

We assume that there is a linear relationship between K_{cb} and *NDVI* as stated earlier. This time, however, the linear relationships are adjusted to values of $NDVI_{MAX}$ and $NDVI_{MIN}$ from satellite imagery rather to those from field radiometry. Table 2 gives the corresponding values.

	NDVI	K_{cb}	f_c	K_c
Minimum	0.16	0.15	0.0	0.4
Maximum	0.80	1.15	0.8	1.2

Table 2. Maximum and minimum values of NDVI and derived parameters observed in set of Landsat images.

The resulting linear relationship is:

$$K_{cb} = 1.5625 \cdot NDVI - 0.1 \quad (11)$$

The soil evaporation part in (10), K_e, is related with bare soil fraction, and is strongly dependent on wetting state of bare soil fraction, because the evaporative power of soil changes strongly if the soil is wetted or if the soil is dry. Irrigation system (gravity, sprinkler, drip, etc) and irrigation frequency, coupled with type and stage of crop, are the factors that determine the time of different bare soil wetting states. We propose a first approach to take into account these factors assuming *NDVI* as a good estimator of ground fractional cover, f_c, (and so, of bare soil fraction, 1 - f_c). The other factors are parameterized by means of a parameter β:

$$K_e = \left(1 - f_c\right) \cdot \beta \quad (12)$$

The parameter β is estimated empirically, from the values of K_{cINI} or K_{cMID} and can be modified on the basis of ancillary or local information. It is crop (and stage) dependent. Assuming a linear relationship between *NDVI* and f_c for all crops, and considering again the *NDVI* maximum and minimum values from satellite imagery and the corresponding f_c as given in Table 1, we obtain the relationship

$$f_c = 1.3514 \cdot NDVI - 0.2811 \quad (13)$$

6.2. Single crop coefficient NDVI approach

A common β parameter value is 0.25, obtained considering an f_c value of 0.8, K_c equal to 1.2, and K_{cb} equal to 1.15. Taking β as 0.25 and combining (10), (11), (12), and (13), we obtain a direct relationship K_c - *NDVI*

$$K_c = 1.2246 \cdot NDVI + 0.2203 \quad (14)$$

We also obtain a relationship K_c - *NDVI* directly from Table 1 in the same way as above, by considering a linear relationship between the maximum *NDVI* and the maximum K_c (at effective full cover) and the minimum (bare soil) *NDVI* and bare soil K_c, respectively. The resulting relationship is:

$$K_c = 1.25 \cdot NDVI + 0.2 \quad (15)$$

Equations (14) and (15) are very similar. By its simplicity we assume (15) as the operational formula to derive K_c from *NDVI*.

Table 3 shows the comparison between K_c values obtained from *NDVI* by means of (15), and the values for K_{cINI}, K_{cMID} from [1] for the crops studied in the field during the pilot campaign. We observe good agreement for crops with higher effective ground cover, although (15) seems to overestimate K_{cINI} slightly for spring crops.

Crop	NDVI min	NDVI max	K_{cINI} (15)	K_{cINI} FAO56	K_{cMID} (15)	K_{cMID} FAO56
Alfalfa	0.16	0.80	0.40	0.40	1.20	1.20
Barley	0.16	0.80	0.40	0.30	1.20	1.15
Garlic	0.16	0.44	0.40	0.70(0.40)*	0.75	1.00
Maize	0.16	0.78	0.40	0.30(0.40)*	1.17	1.20
Onion	0.16	0.53	0.40	0.70(0.50)*	0.86	1.00
Opium poppy	0.16	0.80	0.40	No reference	1.20	No reference
Pea	0.16	0.77	0.40	0.50(0.40)*	1.16	1.15
Potato	0.16	0.78	0.40	0.50(0.45)*	1.17	1.15
Sugar beet	0.16	0.78	0.40	0.35(0.45)*	1.17	1.20
Wheat	0.16	0.80	0.40	0.30	1.20	1.15

Table 3. Comparison between the averaged K_c values obtained from *NDVI* by means of (15), and the values for K_{cINI}, K_{cMID} from FAO56 [1] for the main crops in the pilot zone.

Significant deviations between *NDVI* based and FAO56 based K_{cMID} are found for crops like garlic and onion, which exhibit low ground cover in the stage of maximum development, in contrast with the rest of crops studied. The higher bare soil proportion of those crops can introduce and reinforce effects on K_c related with irrigation frequency, irrigation system utilized, environmental aspects and others. Further research is under way to study this behaviour.

It should be also noted that (14) and (15) are applicable for the initial crop development and mid-season phases only. The application for the late season phase, when the crop is maturing, requires a slight correction because ground cover (green and dry) remains nearly constant in that phase. Assuming a constant value of 0.8 for ground cover (Table 1) and combining (11) and (12), we obtain:

$$K_c = 1.5625 \cdot NDVI - 0.05 \quad (16)$$

Summarizing, the operational equations will be (15) for the initial, development and mid season and (16) for the late season.

7. Caia (Portugal)

The Caia irrigation scheme is located in the Caia watershed in east-central Portugal close to the Spanish border. The Caia river is a tributary to the transnational Guadiana river basin. It is a relatively flat area at a mean elevation of 200 m above sea level, with mean annual precipitation of 537 mm. The irrigation infrastructure was established in 1967, with a central dam of 203 hm^3 capacity and a metered canal network serving 7,237 ha. Except for a concentrated small-scale plot area of 600 ha, the fields are generally large, on average 35 ha. The main crops are corn, wheat, tomato, and sugarbeet, which are all grown for industrial commercialization.

The water resources from the dam are mainly used for irrigation (over 90%) and population supply (less than 10%). Pressure on water quantity is increasing and water quality is already under pressure, as ecological standards are very close to the regulatory limits.

The Associação de Beneficiários do Caia (ABCaia) is the local water user association (Irrigation District Board). It has the mandate for water management in the irrigation scheme and represents the local farmers in the River Basin Council. At the beginning of PLEIADES, there was an incipient traditional IAS (provided by the Centro Operativo de Técnicas de Rega, COTR) with a newly installed agrometeorological station and a small GIS facility at the Irrigation District Board. The goal was to build up an innovative irrigation advisory capacity, in order to cope with the rapidly increasing pressure on water quantity and quality.

7.1. Local crop expert database and field protocols

The first pilot campaign was a fruitful training and learning phase that consolidated the local team and established local field sampling protocols. It also laid the foundations for the local GIS-based expert database on crop phenology. Figure 9 shows the crop coefficient curves for the major crops. An important emphasis in the Caia pilot zone is on tomato, which has phonological cycles that can vary enormously between plots.

During the following pilot campaign, extensive field data were collected to extend the expert database for all major crops of the area. They were also used for validation and local calibration of EO-derived products.

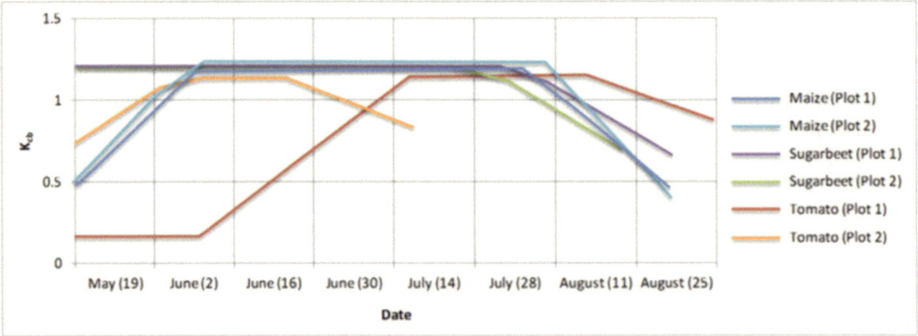

Figure 9. Crop coefficient curves for major crops in Caia area, from field campaign.

The local team developed and implemented a work strategy having as objective to be close to satellite overview conditions. In that sense nadir pictures of crop canopy have been taken as close as possible to overpass satellite period, to determine Green fraction cover, and the crop coefficient has been calculated using farming information concerning amount of used irrigation water, agrometeorological data and field phenological stage to correct and calibrate K_{cb} curve.

For this campaign a spatial extrapolation of the crop coefficients was performed, based on field work land use maps. Figure 10 shows the temporal and spatial evolution of crop coefficient in the pilot zone, for the three studied crops.

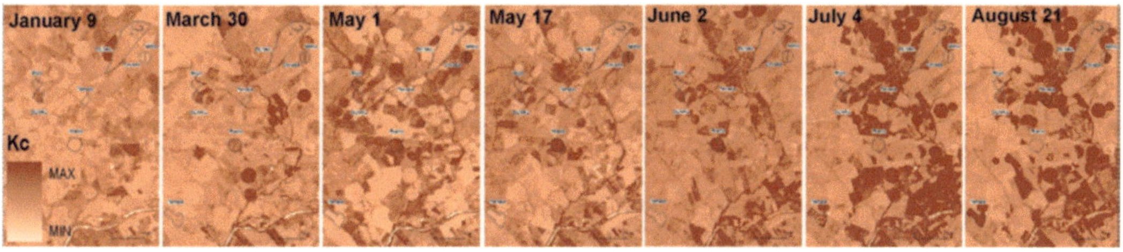

Figure 10. Spatial and temporal evolution of crop coefficient in Caia pilot Zone.

7.2. Calibration and validation of FAO and *EO* derived crop coefficients

The crop coefficient has been calculated in three different ways: the first (line in Figure 11) is based on the concepts of the FAO-56 (Allen et al., 1998) methodology for K_{cb} calculation, with its general tabulated values; the second is K_c *NDVI* calculated with *EO* PLEIADES methodology (circles and squares, from Landsat 7 and 5, respectively); third way is K_{cb} from field observations, adjusted with additional field data (triangles). Figure 11 shows maize monitored on "Melinho" and "Botafogo" test fields. The length of the "Initial stage" of phenological development is based on information concerning seeding dates, but FAO-56 standard methodology stipulates that crop green development occurs earlier than has been observed in the field. The EO-derived data are rather close to the field observations. The slight scatter indicates that the crop stage have several variations, depending on soil and crop water stress parameters.

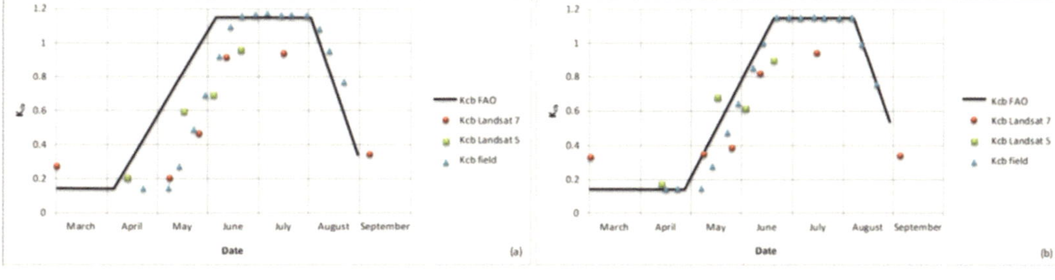

Figure 11. Maize field K_c evolution during campaign versus *EO* derived and FAO-56: Melinho plot area (a) and Botafogo plot area (b).

Tomato is planted in a way that full canopy is not reached in any growing stage. This means that soil evaporative fraction is always present, resulting in a better water management at plot level. Due to the fact that it is a multi-stage plant, i.e. one plant can be on multiple phenological stages at the same period, irrigation must be a compromise for the average stage of the field. Based on the experience of field technicians, the different K_{cb} curves (Figure 12) have been calculated with data obtained at plot level, FAO-56 methodology and *EO* derived data.

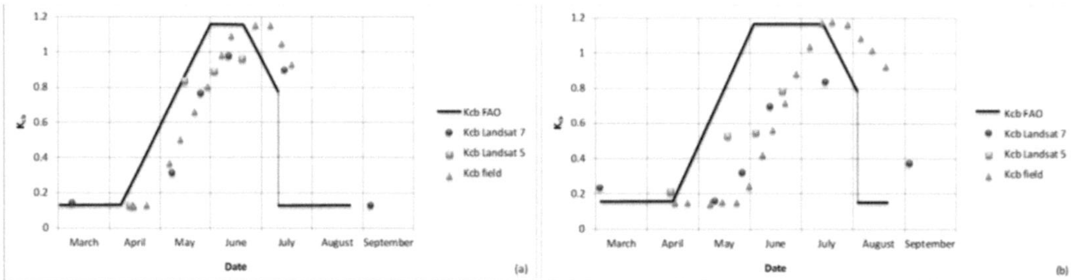

Figure 12. Tomato field K_c evolution during campaign versus *EO* derived and FAO-56: D. Joana plot area (a) and Sta. Isabel plot area (b).

The results demonstrate again a huge difference between standard FAO-56 values (line) and the K_c values obtained from field survey (triangles). As with maize, the *EO* derived data present closer values to what has been observed on the plots. Given the huge deviation of actual crop coefficients from FAO-56 values the local expert database on crop phenology plays a crucial role in the quality of an *IAS* under implementation.

7.3. Real-time demonstration and participatory evaluation with farmers

Field monitoring has been carried from seeding until harvesting dates, only during the months of July, August and September of 2005. A real-time on-place demonstration campaign of PLEIADES took place, surveying hot spots in each of the 6 pilot fields and delivering *IAS* information to farmers.

A "Farmer Report" was supplied to each individual land owner covered by the ground truth operation (Figure 13). In Figure 13 is clearly identified the difference in corn maturation in July, using drop-to-drop irrigation (a) and pivot irrigation (b).

Several meetings were organized to promote and explain the content of the information to the farmers and to get their feedback. The farmers found that the *EO* derived information, along with field survey data, can be very useful for them and that such an irrigate advisory service can help them with irrigation and farming strategies at their farm holding level.

Although, one of the monitored fields had a change for sugar beet with a winter variety, with impacts on data correlation, on the other two the same season of maize and tomato has been used.

The weekly procedure has been maintained in order to collect phenological stages data to define K_{cb} field curves, for evaluation of EO-derived data. For the demonstration campaign the Portuguese Meteorological Institute provided K_c *NDVI* data, obtained from Landsat 5, in almost real time, so that they could be compared with field data. The local team made their spatial integration and data extrapolation based on ABCaia GIS tools and related geographical information data available for this pilot zone.

Figure 13. Farmer report from D. Isabel (a) and Melo (b) fields.

Knowing, through data from the meteorological station at Caia (Figure 14), the relationship between potential evapotranspiration (ET_P) and evapotranspiration measured from *EO* (ET_{EO}), it is possible to establish the relationship expressed in (2).

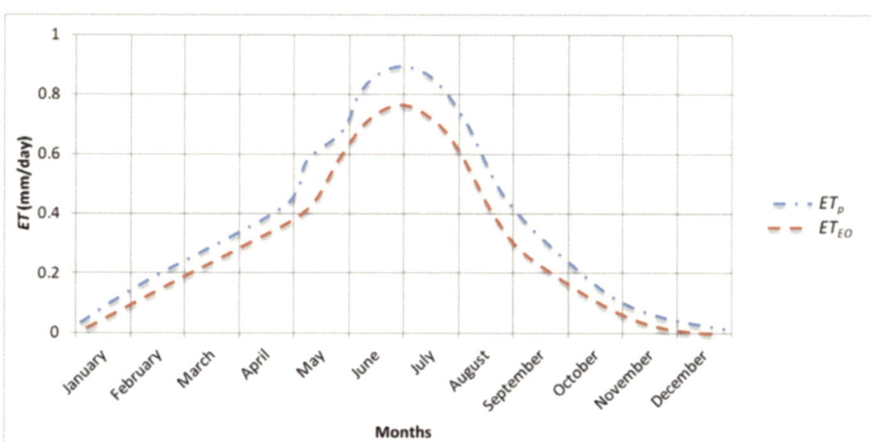

Figure 14. Evapotranspiration in Caia area.

Hence, having a value of, for example, ET_0 = 7.2 for the 2nd of June we can make the correspondence for the all area (Figure 15). This kind of data is very important to farmers because there is a well known relation where precipitation plus irrigation is equal to evapotranspiration plus terrain drainage.

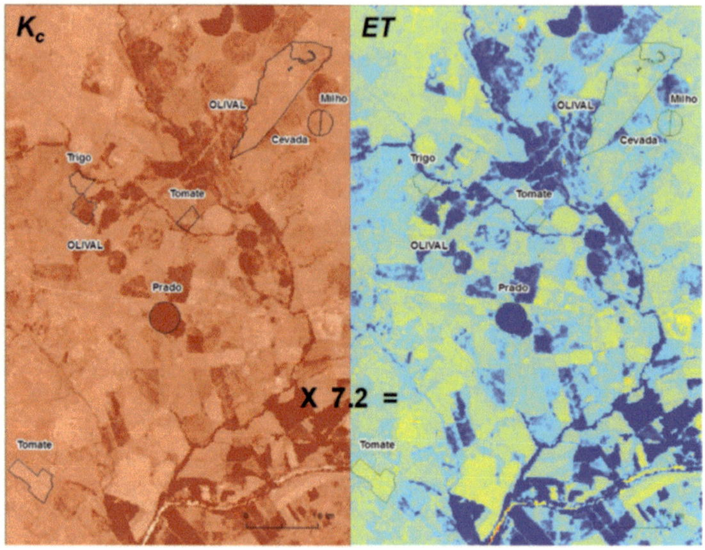

Figure 15. Crop coefficient and Evapotranspiration in Caia area at 2 of June.

7.4. Future perspectives

For the intensive and extensive campaign the local ground team has been organised and defined protocols and set up a strategy to collect data as close as possible to satellite overpass conditions. During the project period a local effort has been made to implement irrigation advisory services on the pilot zone area, resulting in ongoing use of agrometeorological station data provided by the Centro Operativo de Técnicas de Rega (COTR) and the Farmer Association (ABCaia) supplying information on irrigation parameters. Survey data have been available due to the cooperation of the Agronomy School of Elvas University (ESA) and also made available to them jointly with with ABCaia aiming to set up strategies for development of local technical skills to support irrigation development in accordance with environmental requirements.

The foundations have been laid for the PLEIADES prototype to start being operative in the Caia pilot zone. An efficient and dedicated local operating team has been consolidated, with collaborating entities. A local *GIS* based expert database on crop phenology and *EO* methodology calibration has been developed and the necessary *ICT* infrastructure has been established (Figure 15). The user participation is incipient but promising. A funding model still needs to be found for further sustainable implementation. Increasing pressure on water quantity and quality may provide an important motivation.

The Server is based on leading edge online GIS Technology. The system architecture is web based and composed by a modular group of components, which makes maintenance of the system easy. Those components are based on the XML language.

The web services have been programmed also to let the server be distributed in different *PCs* to share the load of the system. The server has been programmed for a Windows system, using open source libraries and toolkits like FOP, Xerces C++, GDAL, GD,

MapServer, SWFF, etc. GDAL has been used for obtaining the satellite images information, getting the pixel value and creating charts with GD showing the time evolution.

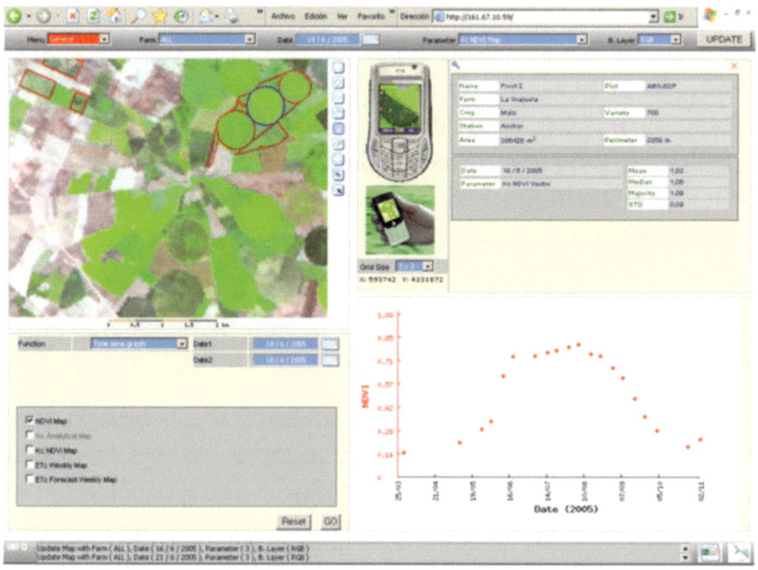

Figure 16. PLEIADES WebGIS and mobile applications example.

This toolkit, as well as the queries form, helps the user in taking decisions about water, crops, etc. MapServer lets the system offer a visual support for the user, supporting a wide variety of image formats, as well as vector format and databases. Xercess is the parser for the XML messages that implements the DOM interface providing quite an easy interface for the programmer for accessing the XML information.

The system receives information from images and stations, pre-processing some of them to provide formatted data. It also offers a Quality Control (QC) interface for the IAS manger to control the information for the farmer, so that the final information distributed to the user ("published") includes only correct and accurate information. The system is prepared for assimilating the satellite image as far as possible, so that the system gives a real time toolkit for analysing the satellite information with the spatial information.

The mobile phone client has been programmed in Flash. A simpler interface has been designed because of the limitations of the mobile technologies. This client lets the farmers access the information of the satellite wherever they are without needing a computer or an internet connection and with an easy interface that lets them zoom into the image or ask for parameters information.

8. Conclusions

PLEIADES was designed to assess and demonstrate in an operational perspective how the integration of Earth observation (*EO*) techniques in routine Irrigation Advisory Services

(*IAS*) can improve the efficiency in the use of water for irrigation. The use of leading-edge Information and Communication Technology (*ICT*) tools in the generation and distribution of information makes the *EO* easily available to *IAS* and the farmers.

The PLEIADES WebGIS (www.pleiades.es) is the central outcome of the project. Its key feature is the operational generation of irrigation scheduling information products from a virtual constellation of *EO* satellites and their delivery to farmers in near-real-time using leading-edge on-line analysis and visualization tools. It is supported by a methodology package to derive crop coefficients and further advanced parameters from *EO* satellite images in an operational processing chain.

PLEIADES basic products were generated and transmitted to a sample of farmers normally within 2 days from overpass, thus completely matching the weekly operational irrigation scheduling cycles.

Participatory evaluation with selected farmers shows that the farmers feedback is very positive, both on the information quality as on the added value of the spatial information (within-plot heterogeneity and between-plot variations). The reliability and accuracy of the information has been con-firmed by the comparison of different approaches to derive crop coefficients from EO and validation with field data in all pilot zones.

The major improvement achieved by the use of *EO* in the generation of basic IAS information products like crop coefficients is twofold. Firstly, the spatial coverage is enhanced significantly, both extending to larger areas and providing within-field heterogeneity information. Secondly, the spatially resolved *EO* data can easily be combined with cadastral information in a geographical in-formation system (*GIS*), which allows for personalization of the irrigation scheduling recommendation.

Conventional *IAS* provides average irrigation recommendations per crop type, while the new space-assisted *IAS* is able to provide specific recommendations for each individual plot, based on the actual state of that plot.

The fast image delivery and quality controlled operational processing make the EO-based crop coefficient maps available at the same speed and quality as ground-based data (point samples), while significantly extending the spatial coverage and reducing service cost. The uptake of users at *IAS* and farmer level is encouraging.

Advanced products have made a significant step towards operationality while maintaining satisfactory levels of accuracy. First exploitation steps including full operational implementation are indicators of the success of the prototype and the project.

The space segment is the most vulnerable part of the entire operational system. After the sensor failure of Landsat 7, the backbone of the actual system is Landsat 5, due to its excellent operationality and low cost (22 years old, with no replacement in sight). Urgent actions are required to ensure the capability to obtain adequate *EO* images at the adequate coverage frequency and low cost.

Author details

Jorge Rocha, Raquel Melo and Cristina Henriques
CEG, Institute of Geography and Territorial Planning, University of Lisbon, Portugal

António Perdigão
DGADR, Direcção Geral de Agricultura e Desenvolvimento Rural, Lisbon, Portugal

9. References

[1] Allen, R.G, Pereira, L.S., Raes, D. y Smith, M. (1998). Crop evapotranspiration. Guidelines for computing crop water requirements. *UN-FAO, Rome, Italy.*

[2] Bailey, J.O. (1990). The Potential Value of Remotely Sensed Data in the Assessment of Evapotranspiration and Evaporation. *Remote Sensing Reviews,* 4(2):349-377.

[3] Heilman, J.L., Heilman, W.E. and Moore, D.G. (1982). Evaluating the Crop Coefficient Using Spectral Reflectance. *Agronomy Journal,* 74:967-971.

[4] Allen, R.G., Morse, A., and Tasumi, M. (2003). Application of SEBAL for western US water rights regulation and planning. University of Idaho, Kimberly, Idaho, USA. *Proceedings of ICID-CIID International Workshop: Use of remote sensing of crop evapotranspiration for large regions*: *September 17, Montpellier, FRANCE.*

[5] Choudhury, B. J., Ahmed, N. U., Idso, S. B., Reginato, R.J., and Daughtry, C.S.T. (1994). Relations Between Evaporation Coefficients and Vegetation Indices Studied by Model simulations. *Remote Sensing of Environment,* 50:1-17.

[6] Wright, J.L. (1982). New Evapotranspiration Crop Coefficients, *Journal of Irrigation and Drainage,* 108(1):57-74.

[7] Calera, A., Martínez, C., Meliá, J. (2001). A procedure for Obtaining Green Plant Cover: Relation to NDVI in a Case of Study for Barley. *International Journal of Remote Sensing,* vol 22, no 17, 3357-3362.

[8] Carlson, T.N., Ripley, D. (1997). On the Relation between NDVI, Fractional Vegetation Cover, and Leaf Area Index. *Remote Sensing of Environment,* 62:241-252.

[9] Sellers, P.J., Dickinson, R.E., Randall, A., Betts, A.K., Hall, F.G., Berry, J.A., Collatz, G.J., Denning, A.S., Mooney, H.A., Nobre, C.A., Sato, N., Field, C.B., Henderson- Sellers, A. (1997). Modelling the Eschanges of Energy, Water, and Carbon between Continents and the Atmosphere. *Science,* Vol 275, 52-509.

[10] Monteith, J. L. (1977). Climate and Efficiency of Cropproduction in Britain. *Phil. Trans. Res. Soc., London, SeriesB.* 281, pp. 277 - 329.

[11] Asrar, G., Fuchs, M., Kanemasu, E.T., and Hatfield, J.L. (1984). Estimating absorbed photosynthetic radiation and leaf area index from spectral reflectance in wheat. *Agronomy Journal* 76:300-306.

[12] Gower, S. T., Kucharik, C. J., and Norman, J. M. (1999). Direct and indirect estimation of leaf area index, fapar, and net primary production of terrestrial ecosystems. *Remote Sensing of Environment,* 70: 29-51.

[13] Calera, A., González-Piqueras, J., Meliá, J. (2004). Monitoring barley and corn growth from remote sensing data field scale. *International Journal of Remote Sensing*, vol. 25, no 1, 97-109.

[14] Monteith J.L., Unsworth M.H. (1990). Principles of Environmental Physics., *E.Arnold Ed., London,* pp.290

[15] Gilabert, M.A., Gandía, S., Meliá, J. (1996) Analyses of Spectral-Byophysical Relationships for a Corn Canopy. *Remote Sensing of Environment,* 55: 11-20.

[16] Bausch, W.C. and Neale, C.M.U. (1987). Crop Coefficients Derived from Reflected Canopy. *Transactions of the ASAE,* 30(3):703-709.

[17] Rosenberg, N.J., Blad, B.L., and Verma, S.B. (1983). Microclimate; The biological environment 2nd ed. *New York; John Wiley & sons,* pp.13-15,170.

[18] Calera, A., A. Jochum, and A. Cuesta (2003): Earth observation techniques to improve Irrigation Advisory Services. Proc. Envirowater, Albacete, 4-6 September.

[19] Jochum, A.M., A. Calera, and A. Cuesta (2003): Irrigation management from space: Towards user-friendly products. Proc. ICID Workshop "Use of Remote Sensing of Crop Evapotranspiration fro Large Areas", 17 September, Montpellier. (submitted *Int'l J. Irrigation & Drainage*)

[20] Moran, M.S., Inoue, Y., Barnes, E.M. (1997). Opportunities and Limitations for Image-Based Remote Sensing in Precision Crop Management. *Remote Sensing of Environment,* 61, 319-346.

[21] Bausch, W.C. and Neale, C.M.U. (1987). Crop Coefficients Derived from Reflected Canopy. *Transactions of the ASAE,* 30(3):703-709.

[22] Neale, C.M. U., Bausch, W.C., and Heermann, D.F. (1989). Development of Reflectance-Based Crop Coefficients for Corn. *Trans. ASAE* 32(6):1891-1899.

[23] Bausch, W.C. (1993). Soil Background Effects on Reflectance-Based Crop Coefficients for Corn. *Remote Sensing of Environment,* 46:213-222. 46:213-222.

[24] Huete, A.R. (1988). A Soil Adjusted Vegetation Index (SAVI). Remote Sensing of Environment, 25:295-309.

[25] Bausch, W.C. (1994). Remote Sensing of Crop Coefficients for Improving the Irrigation Scheduling of Corn. *Agricultural Water Management,* 27:55-68.

[26] Gonzalez-Piqueras, J., Calera, A. and Gilabert, M.A. (2003). Estimation of Crop Coefficients by Means of Optimized Vegetation Indices for Corn, *Proceedings of the SPIE, Barcelona, September, 8-11th.*

[27] ASCE Hydrology Handbook 2nd Ed . (1996). Chapter 4: Evaporation and Transpiration. *ASCE Manuals and Reports on Engineering* Practice No.28.

[28] Neale, C.M.U., Jayanthi, H., and Wright, J.L. (2003). Crop and irrigation water management using high resolution airborne renote sensing. Utah State University, Logan, Utah, USA. *Proceedings of ICID-CIID International Workshop: Use of remote sensing of crop evapotranspiration for large regions: September 17, Montpellier, FRANCE.*

[29] Bausch, W.C. and Neale, C.M.U. (1989). Spectral inputs improve corn crop coefficients and irrigation scheduling. *Transactions of the ASAE,* 32(6):1901-1908.

[30] Bausch, W.C. (1995). Remote sensingof crop coefficients for improving the irrigation scheduling of corn. *Agricultural Water Management.* 27:55-68.

Chapter 9

Certification and Integration of Environment with Quality and Safety – A Path to Sustained Success

Gilberto Santos, Manuel Rebelo, Síria Barros and Martinha Pereira

Additional information is available at the end of the chapter

http://dx.doi.org/10.5772/48414

1. Introduction

According to Wright [1] certification of products and processes began during the 1960's in the manufacturing industry, as a tool to control and assure the quality/conformity of products and services provided by suppliers to customers/consumers. Thus, the series of ISO 9000 was published first time, in 1987 and it was been created with a flexible character, to be reviewed periodically. Later, were published others normative references, which highlight the ISO 14001 in 1996 and OHSAS 18001 in 1999. This was also, the natural sequence of the certification processes in the organizations, i.e., began with the certification of quality management systems (QMS) followed by the environmental management systems (EMS) and after for the Occupational Health and Safety Management System (OHSMS). Hence, a high percentage of organizations with an EMS, in accordance with the ISO 14001, had also implemented, a certified QMS, in accordance with ISO 9001. At first the implementation of a QMS was particularly relevant in high demanding activity sectors, like the automotive and aeronautical industries, but it has rapidly extended to every activity sector, becoming a common requisite of any company worldwide and a factor of competitiveness and survival. Due to the increasingly demanding environmental legislation in developed countries, companies nowadays are required to seriously take into consideration not only environmental aspects associated to the production chain itself, but also to the life cycle of their products.

They are forced to implement suitable EMS to manage the environmental issues as for example at the level of the prevention and reduction of wastes. Consequently to promote and protect environment in a sustainable way. This is a particularly important issue for small and medium-sized companies (SMEs), which are considered to make up the vast majority of business in Europe)[2]. They are quoted as contributing 70% of global environmental pollution, with the majority coming from the manufacturing sector [3]. Thus, following both the Rio Conference in 1992 and GATT (General Agreement on Tariffs and Trade) negotiations, international standards have become important for succeeding and for getting access to the markets; at the same time, there has been an increase in the interest of environmental management. Such regulatory and competitive pressures have caused firms to take into consideration the environmental issues within their own production and market plans. Many firms have attempted to seek an effective environmental management system. These have led to implementation and development of the ISO 14001 standard for assessing environmental management processes. Today, all over the world, many firms are seeking ISO 14001 certification [4]. Thus, improved environmental management in the industrial sector is required to protect the environment, protect human health and property, and to satisfy environmental requirements associated with international trade [5]. In a sense, the value of the ISO 14001 certificate, as a proof of environmental performance, is a combination of the supplier's environmental ambitions, the advancement of supply chain practices of the customer and the ambitions of the certification bodies [6] . On other hand, according to Casadesús [7] and others authors, the interest shown by organizations and other entities linked by the implementation of environmental management systems (EMS), especially the family of ISO 14000 standards and the EMAS regulation in Europe, has grown spectacularly all over the world in recent years, even though a certain saturation has been detected in some countries. Thus, the implementation and management of all economical, environmental and social aspects within a company is gradually becoming a crucial requirement for any business and has become a widespread phenomenon around the world [8]. Large companies are increasingly requiring this management policy from their suppliers, establishing specific requests and performances that small and medium enterprises (SMEs) often find extremely difficult to accomplish. On other hand, there are many organizations which, either because of the demands of the market itself or because of other internal motivations, have implemented different Management Systems alongside their EMS.

In fact, although no reliable references on this matter have been found, it is quite plausible to think that the great majority of ISO 14001 – registered companies are also certified in accordance with the ISO 9001 standard [9]. Thus, quality management philosophy and methods have been imported into ISO 14001 from ISO 9000. As a result, it is not surprising that measurement and evaluation are enshrined as important hallmarks of an effective EMS. According Curkovic [38], a companys ability to reframe learnings from total quality management (TQM) is crucial to the successful implementation and use of environmentally responsible manufacturing (ERM) -based systems and procedures.

Generally speaking, this component of the EMS is considered effective when the contents of review meetings are well communicated, the focus of meetings is on improving the system, findings (i.e. about noncompliance, from various statistical charts, and audit results) are reported honestly, and corrective actions follow [10]. In some cases, the similarities between QMS and EMS systems can facilitate the integration of the two related management systems

[11]. The people that work in environmental management and at the same time are members of quality teams, assure that quality management goes hand-in-hand with environmental management. The actions that are carried out to achieve quality are, in many situations, the same actions necessary, for example, to achieve effective environmental management. In line with this, ISO 14001 has become compatible with the ISO 9001. Consequently, the integration of environmental issues (including environmental protection and pollution prevention in the management of organizations through the implementation of an environmental management system) allow acquiring a deep insight of the most important environmental aspects associated with its activity, and identifying the processes that need to be improved through the implementation of effective environmental measures [12]. On other hand, according Curkovic [38] TQM focuses on waste as it applies to process inefficiencies, whereas ERM focuses more on pollution in the form of air emissions and solid and hazardous waste. Because the two concepts share a similar focus, it makes sense to use many of the TQM tools, methods, and practices in implementing an ERM-based system.

Moreover, human resources are the most valuable resource of any company or country, but not always the most valued. Thus, the greatest asset of any organization, any region or any country, are people and their know-how [13]. Therefore, among others, another system to be implemented in the organizations is the OHSMS. Thus, according Fernández-Muñiz [14], several fields are showing increasing interest in safety culture as a means of reducing accidents in the workplace. The literature shows that safety culture is a multidimensional concept. Hence, nowadays, companies that search greater profitability and better organization implement the quality systems, aiming at a reduction of defective products and lost time, searching for the loyalty of customers and searching for excellence. The progressive implementation of ideas and techniques related with the quality management is one of the clearest demonstrations of organizational innovation in the industry in the last decades. From the standpoint of the risk prevention literature, it has been argued that the use of advanced quality management systems help reduce accident rates because quality management methods are based on the principle of prevention rather than corrective actions. Hence, the concept of an OHSMS has become common over the past 20 years [15]. The people that work in safety management and at the same time, are members of quality teams, assure that quality management has a great relationship with risk management.

More than ever, today, is in question the business sustainability of the organizations and the focus should be placed far much more than financial results. These results will not verify if that focus does not prioritize also, the satisfaction, balanced, integrated and growing of the customers and others relevant Stakeholders, that are clearly and objectively the employees for example [16]. But according to the ISO – IMS publication [17], a common objective of management system standards is to assist organizations to manage the risks associated with providing products and services to customers and other stakeholders. On other hand the management system of the organizations is frequently split into a number of parts or sub-systems, which must be managed separately with relative independence. These parts or sub-systems of an organization's management system reflect the different needs and expectations of the stakeholders. Many organizations use standards, such as those related to quality, the environment and safety, among others, to manage certain aspects of their performance.

Within this framework, in order to have quality and excellence in products (or services), as well as in the management of the companies that manufacture and provide them, it has become imperative for the companies define and implement quality, environment and safety management system, according to ISO 9001, ISO 14001 and OHSAS 18001 standards, respectively . At the same time the companies must also improve and optimize, continually, these management systems to allow them to true added value for the companies and their stakeholders [16].

The following questions may arise: How can these three management systems be integrated? Can they be integrated? According to Santos [18] this is a problem that the most developed companies started to experience some time ago, and it has been discussed by various authors, that we highlight: McDonald [19], Arifin [20] and Bernardo [21] among others, who provides a summary of the degrees of integration according to some authors, and Labodová [22], who reported on the implementation of integrated management systems using a risk analyses based approach. Thus, according with the ISO 72:2001 Guide [23], the experience with management system standards issued by the ISO shows that there exist a number of common elements, which can be arranged under the following main subjects: policy; planning; implementation and operation; performance assessment; improvement and management review, as stated by Santos [13].

Related to the integration of management systems, Karapetrovic [9] states that there are many organizations that implement different management systems, either as a result of market demand either for internal reasons. During the combined audits in accordance with ISO 19011 [24] for Quality and/or Environmental Management Systems audits, it can be observed that the integration of the systems is implemented in a partial way by the integration of procedures by phases.

When it is observed that it exist similar procedures, those are adapted to the two systems, for exemple, quality and environment, and jointly audited. However, the manuals still individually separated. As stated by Santos [25], although an audit be integrated / joint, named as by "combined audit", the respective reports are often kept separate so that any nonconformities detected in one of the systems do not implicate to stop the other system that complies with the respective standard. This could implicate the loose of clients, which would be a backward step for the company, and is considered one of the obstacles to a quicker integration. Suditu [26], associates to the integration of management systems internal motivations and corresponding benefits, by dividing them into: 1 - Organisational – improvement of the quality of the management by downsizing of three departments into one and reducing barriers between individual systems; 2 - Financial – reduction in auditing costs; 3-Employees – increased motivation, awareness and competences; 4 - External motivations and corresponding benefits, by dividing them into: Commercial – competitive advantage, improved market position, gain of new clients and satisfying current ones;

Communication – improved image of the organisation, improved relationships with Stakeholders and evidence of legal compliance; In fact, significant differences in these areas have not been found, and it is fairly plausible that the vast majority of companies certified under ISO 14001 are also certified under the ISO 9001 standard, according to Santos [13].

Therefore, a new necessity has emerged in organisations, namely to integrate these systems into a single IMS - Integrated Management System and we cannot forget the fact that the environmental improvement of the product must be considered in relation to the impacts on Quality & Health and Safety, as mentioned by Jørgensen [27]. Therefore, the idea of an IMS - Integrated Management System consists of establishing correspondences and to combine two or more independent management systems, for example in accordance with ISO 9001, ISO 14001, and OHSAS 18001. Evidence of this can be seen in table A.1 - of the annex A - of OHSAS 18001:2007 [28]. Despite having their origins in different aspects of company performance, the Quality, Environment and Safety Management Systems have a lot in common, as mentioned by Fresner [12] and Block [29], among other authors. The integration of these management systems is a path that can be followed, and the ISO 19011 standard - Guidelines for auditing management systems [24] is a good example of the future. According to Santos [13] the future lies in the integration of these management systems, managed by only one multidisciplinary team with training and skills in several areas, thereby economizing both financial and human resources.

The Portuguese industry, mainly, consists of Small and Medium Enterprises (SMEs), which activity and performance are crucial factors for the country development. SMEs provide 75% of the total labour force employed in industry, trade and services. According to the website of the Ministry of Economy and Innovation [30] SMEs are responsible of 99.5% of national business, generating 74.7% of employment and held 59.8% of sales nationwide. They are the basis of Portuguese economy. A good example of Management Systems Certification and Integration in Portugal is represented in Figure 1.

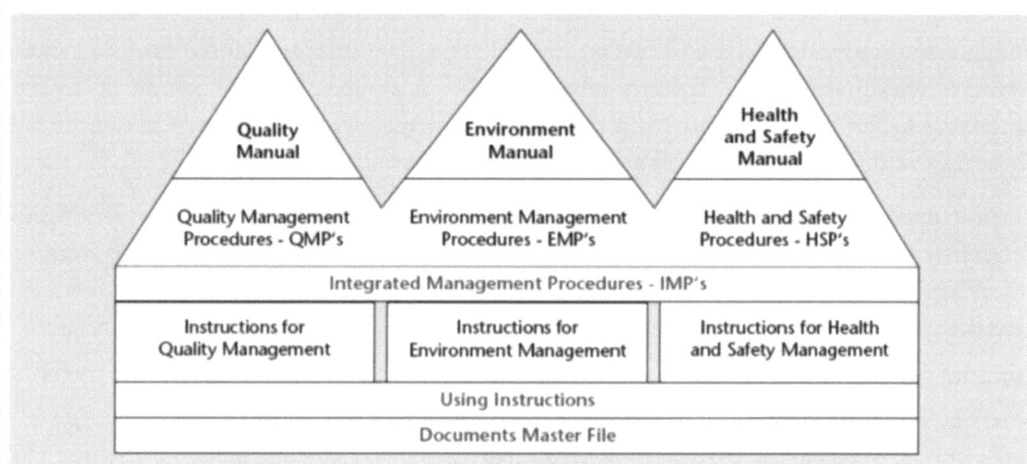

Figure 1. Documental structure of the Integrated Management Systems at Kupper & Schmidt [13]

For a long time, the only Foundry in the country to have its Quality, Environment and Health and Safety systems certified was Kupper and Schmidt, a SME that supplies exclusively the automotive industry and exports about 98% of its production to the European and American markets. Just a few years after starting its activity the company had its Quality System already certified by the main customers. In 1997 it was certified according to ISO 9002 standard, in 1998 by QS – 9000, in 1999 by VDA 6.1, and in 2001 by ISO/TS 16949 standards. In 2000 the company focused in Environment, and got the certification according to ISO 14001 standard. Since 2002 the company has its OHSMS certified according to OHSAS 18001 [31]. Presently, the company runs an Integrated Management System in what concerns to management procedures, operating instructions and documentation.

The objective of this work was to contribute to characterize the situation about the Integration of Environment with Quality and Safety in the Portuguese companies, as well, highlight the benefits obtained with the certification and integration and show that, it is a path to sustained success.

2. Methodology

In the last years there has been a significant increase in the number of Portuguese companies certified in Quality, Environment and Health and Safety. Some of them have a degree of integration of such systems which are not known. In order to make such characterization a survey based on a questionnaire was carried out in several Portuguese SMEs.

The questionnaire was sent via e-mail or delivered personally in hand, explaining and justifying its main objectives and it was organized in five sections, according to Table 1. Both qualitative and quantitative answers were asked, depending on the nature of the question and the available data to give an accurate response. The questionnaire was sent by e-mail together with a cover letter describing the objectives of the research and including some answering instructions to 300 SME with a certified quality management system. 162 companies were located in the centre region of Portugal (including Lisbon area), 114 in the northern region (including Oporto area) and 24 in regions located south of Lisbon. According to the data available from the "SMEs Portugal Association" this distribution is proportional to the location of SMEs within the Portuguese territory [32].

80 companies answered the inquiry, but only 46 have been validated, thus representing the sample size of the Portuguese territory. From those 46 companies, 20 (43%) were from the Trade/Services activity sector, 17 (37%) from the Industrial sector, 5 (11%) from the Electricity/Telecommunications sector and 4 (9%) from the Construction area (Figure 2).

About 80 questionnaires were received. The main criteria for validation were to be a SME, to have the ISO 9001 QMS certification and to have answered the main questions completely. Thus, only 46 were completed properly. Data was worked and some results presented [13 ; 33]. However, it is known that there are SMEs with other certified systems, among them the EMS. An additional criterion was to have the ISO 14001 EMS certification. Initially, 12 SMEs

Main Sections	Questionnaire main topics
General Description of the Company	• Number of employees; Volume of business; Branch of activity; Main products and markets; etc.
QMS	• Year of ISO 9001 certification; Main reasons for Quality certification. • Main difficulties; Main benefits that arose from certification (new costumers, image, competitiveness, business increase, quality improvement, customer satisfaction, products innovation, organization improvements, etc.); Main drawbacks; Quality tools that the company uses.
EMS	• Year of ISO 14001 certification; Main reasons for Environmental certification; Main difficulties; Main benefits that arose from certification (waste reduction, waste destination, environmental costs reduction, etc); Main drawbacks.
OHSMS	• Year of OHSAS 14001 certification; Main reasons for Health and Safety certification; Main difficulties; Main benefits that arose from certification (reduction of the number of accidents, productivity increase, absenteeism, etc);Main drawbacks.
IMS	• Running independently or integrated; Degree of integration (integrated systems, what is integrated, reasons for integration, management structure, etc); Main benefits of integration (cost reduction, management simplification, etc); Main difficulties; Main drawbacks.

Table 1. Main sections and question main topics of the questionnaire [13].

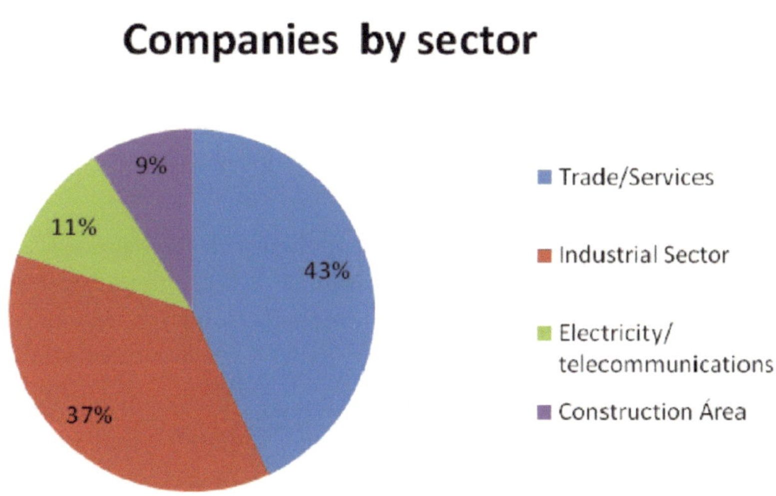

Figure 2. Distribution of participating companies by sector of activity

(26,1%) fulfilled such criterion but later, 5 questionnaires was completed what totalized 17 SMEs (36,9%). This was the sample for SMEs with EMS certified. Also initially only 7 SMEs (15,2%) had the OHSMS certified according OHSAS 18001. Later it was completed and validated more 5 questionnaires, which totalize 12 SMEs (26,1%) with the OHSMS certified according OHSAS 18001. This is the sample for SMEs with OHSMS certified.

Once the information has been collected, the next step consisted of the analysis and interpretation of data. An *Excel* file has been created with the collected data, and then exported to SPSS (Statistical Package for Social Sciences). SPSS is powerful software to support statistics, which provides complex statistical calculations. However, due to the small sample size, two obstacles came up: knowing what statistical test to use and to interpret correctly the results from calculations. Thus:

1. Statistical techniques - It was used primarily to inductive and descriptive statistics: average, frequency plots, the principal component analysis, cluster analysis and statistical inference to find the important conclusions about the population inferred from analysis of the sample, bearing in mind the limitation its reduced size.
2. KMO and sphericity test of Bartlett's - the KMO and Bartlett test are two statistical procedures that allow measuring the quality of the correlation between variables. The Kaiser-Meyer_Olkin (KMO) is a statistic test that varies between 0 and 1 and compares the zero-order correlations with the partial correlations observed between the variables.
3. Internal consistency Cronbach's Alpha - Cronbach's Alpha is commonly used to measure internal consistency of a group of variables (items). It can be defined as the correlation that is expected to get between the scale and other scales used the same hypothetical universe with an equal number of items that measure the same characteristic.
4. Cluster analysis - is also a technique of exploratory multivariate analysis that allows to group variables into homogeneous groups or to compact one or more common characteristics. Each observation belonging to a particular cluster is similar to all others belonging to that cluster, and is different from the observations belonging to other clusters. We chose to use the method of cluster analysis to detect groups in the original variables, the issues where the sample was small, and the principal component analysis had not reached a conclusion. The analysis of the survey was directed to the most relevant items of the questionnaire.

The principal component analysis to data was applied with the help of SPSS, which allows an investigation of the multivariate structure, and a better interpretation of data. The internal consistency of the subject is given by Cronbach's Alpha index which was applied to the 19 selected variables, considering 5 analysis components, which was 0.901. As this value was greater than 0.7, the responses were considered as trustworthy and free of errors. [33]. On this context, we used punctuation 4 for "Greater Impact", 3 for "Impact", 2 for "Little Impact" and 1 for "No impact". The resume of results of this work was published in the Journal of Cleaner Production [13].

Moreover, it was made a case study. The investigation was developed in business environment at an Portuguese Company - the Itron – Portugal, that over the years has been adopting, in whole or in part, gradually and individualized standards or specifications of different Management Systems, relevant to the ISO 14001 (Environmental Management Systems) ISO 9001 (Quality Management Systems), OHSAS 18001/NP 4397 (Safety and Health Management Systems). While it is imperative to assess the perception of employees of the Company on the structuring, implementation and evaluation of the integration model and its validation in a real work environment, it was developed an internal research supported in an inductive approach in the context of the investigator's experiences with the object and the environment of the study.

The technique used to evaluate the contributions to the structure of the model proposed of IMS_QES was the internal investigation by questionnaire to the Collaborators of the company. The total population was 160 employees which are the organization structure of the different branches and levels. The sample that was considered - 49 employees, represent around 30.62% of the total Collaborators. 42 Collaborators answered to the questionnaire which is an overall reply rate of 86%. This rate represents a strong adherence, at all hierarchy levels of the Organization.

In a first phase, the model was developed supported on the Lean philosophy , in order to the simplification and consequent resolution of problems and gains in efficiency of existing management systems in the company (referred as Sub-systems), which along the years were adopted gradually, but individualized and often isolated between to each other.

Were considered four main questions and for each of them several topics as described in the table 2.

Main Sections	Questionnaire main topics
1 - Importance of motivation factors for the implementation of the IMS_QES.	• Improving the image of the Company • Rationalizing and optimizing the management of the resources (financial, material and human) • Facilitating the management of the three components for the Sustainable Development of the Organization and of the Business • Increasing requirements of the Costumers • Competitive advantage and strategic vision of the Business • Natural evolution of the Management Systems in the Company • The continual improvement of Quality, Environment and Safety Management Systems involves their integration

2 - Stakeholder influences on the performance and evolution of an IMS-QES	• ITRON Group - Orientations and evolution of QES and Sustainability policies • Customers, competitors, insurance companies, suppliers and official QES entities • Financial institutions • Local Management group and Collaborators
3 - Main Internal difficulties for the development of the IMS-QES model and its implementation	• Deficit of human and material resources due to the strongly competitive environment and costs reduction; • Resistance to change • Diversity of products and services vis-a-vis Customer's requirements and legal and other requirements • Do not explicitly fit in the objectives and priorities of the ITRON - Portugal • Lack of internal competences in Integration Management Systems • Do not exist an international standard for the integration of Management Systems
4 - Potential benefits resulting from the implementation of the IMS-QES	• Elimination of conflicts between individual Systems, and consequent resource optimization, namely human resources; • Elimination of several organizational waste including at the level of bureaucracy associated to the sub systems Quality, Environment and Safety • Common management policy, objectives, goals and KPIs - Key Process Indicators related to QES performance • Improvement of the internal and external image of the company in the aim of the QES issues • Involvement and consolidation, by ALL employees, of a continual improvement culture, attitudes and values of global QES scope • Reduction of the number of internal and/or external audits and to Suppliers • Improvement at the level of coordinated and integrated management of the Risk associated to the Safety of People, Infrastructure, Environment, and Products • Greater employee valorization and motivation • Integrated management of several components of Sustainability in a Global Market; • Improvement of the partnership relationships with Suppliers of goods and services;

The resume of results of this work of investigation performed at Itron - Portugal, was presented at the International Symposium on Occupational Safety and Hygiene SHO 2012 [16].

Table 2. Main sections and question main topics of the questionnaire used on the investigation [34].

3. Survey results

The first system that was certified in majority of the companies was the QMS. When this system was consolidated, then the EMS was certified. Only when this system was consolidated, companies certified the OHSMS which has started quite recently. After the three certifications, they started in a staggered way to develop integrated procedures to integrate two systems (quality and environment or safety), and whenever possible, the three systems (quality, environment and safety) [13 ; 25] .

This has been, more or less, the general rule that Portuguese SMEs have adopted, fact that was confirmed by the number of certifications, where the quality stands out in the first place (QMS), followed by environmental certification (EMS) and finally the safety certification (OHSMS).

However, we know that some companies have adopted another rule in sporadic cases. After the certification of quality according to ISO 9001, common procedures have been designed for the Environmental Management System (EMS) and the Occupational Health and Safety Management System (OHSMS) at work certification, which worked as the embryo for the implementation of a set of Integrated Management procedures [13].

3.1. The main benefits that companies have gained from QMS certification

Certification is certainly a strategic option for developing organizations in the sense of a wiliness to improve and gain market share. With the reality of the growing number of certified companies, it is important to analyze and quantify the benefits that companies have gained from the certification of their quality system.

Considering the descriptive analysis of responses and the analysis of each item separately, it can be concluded that certification of QMS led companies to improve their internal organization, to make it easier the access to information, improved the internal evaluation of the management systems (namely through continuous audits in time) and had a beneficial effect on the company image, as depicted in Table 3.

	Major Impact	Impact	Little Impact	No Impact
Internal Organization of the company	**72%**	22%	6%	0%
Continuous assessment through internal audits	**54%**	44%	2%	0%
Company image	**52%**	46%	2%	0%
Ease of access to information	**44%**	39%	15%	2%

Table 3. Aspects of the company management where QMS had the greatest impact/benefit [13]

It is clear from Table 3 that these were, in fact, the major benefits of QMS certification. 72% of the companies referred that the major achievement was in their internal organization, and only 6% have referred that certification had little impact on this particular aspect. 52% of the companies stated that their image was improved as a consequence of certification. The basis for these statement was the increase in the number of new or potentially new customers, the drastic reduction of customers rejections and increase of customers satisfaction (evaluated not only by the decrease in the number of rejections, as well as by the increase in the sales volume for those same customers and the increased satisfaction with the overall quality of the products). Moreover, any company reported to have not felt some sort of impact in any aspect of their management structure, as inferred by the results presented in column "no impact" of Table 3.

3.2. The main benefits gained from the EMS certification

In what concerns to EMS certification, the main reasons referred for certification have been environment promotion and protection, improvement of the company's image on this field, improvement of life quality inside and in the surroundings of the enterprise and a marketing strategy, with particular relevance for the first three. The main benefits of EMS certification were classified as having had major, relevant and low or any impact in several fields. Those fields where a major impact of certification have been detected were environment protection (evaluated by the decrease in the amount of residues and disposal costs), better integration of the company in the community, rationalization of natural resources, implementation of recycling techniques (and consequent decrease of costs related to raw-materials acquisition), legislation compliance and consequent decrease of penalty costs. Those companies with non-certified EMS refer to lack of investment support, high implementation costs when compared with the foreseen benefits of certification, or irrelevant environmental risks associated to their activity as the main reasons for their non-certification option/decision.

3.3. The main benefits that companies have gained from the OHSMS certification

The main benefits that companies have gained from the OHSMS certification were: "Improvement of working conditions" clearly took the majority with 91.6% of the responses. Another important question with great impact was "Ensuring compliance with legislation" with 83.3% of responses. Legislation exists which must be fulfilled and most of Portuguese SMEs with certified OHSMS fulfill that legislation work; another strong question was "Notice to workers about the risks and dangers at work" and 75% of the SMEs responded affirmatively. Hence, it can be concluded that there is "better internal communication for workers about the risks and hazards" when an OHSMS is implemented and certified.

4. The Integration of Management Systems (quality, environment and safety)

More than ever, today is in question the sustainability of the organizations and the focus should be placed far beyond the financial results. These will no longer verify it is not prioritize also the continual satisfaction , balanced, integrated and growing of the customers and other stakeholders taking as benchmarks for orientation and acting the dimensions of sustainability - economic, environmental and social and too the organizational and operational factors of the competitiveness.

There exist a set of management system standards that apply to any type of organisation and activity. Others will certainly be created. These standards like as: NP EN ISO 9001:2008, NP EN ISO 9004:2011, NP EN ISO14001:2004, OHSAS 18001:2008/NP 43397:2008, NP EN ISO/IEC 17025:2005, SA 8000:2001 /NP 4469-1:2008, NP 4457:2007, ISO 31000:2009, ISO/IEC 27001:2005, NP 4427:2004, PAS 99:2006, cover a wide array of different disciplines, aims and activities of organisation and operation of the Enterprises including the interfaces and satisfaction of all theirs stakeholders. In itself, this fact leads to a greater perception and awareness of challenges and brings sustainability to businesses, with which Organisations face a fiercely competitive global market in which the only thing that does not change is change itself [34]. Several of these standards, having been published recently, require that a lot of supplementary effort first be put at the level of the learning process needed considering the route that must be taken by organisations towards full implementation, which should be carried out in a balanced manner with added value.

In this context, Quality, Environment and Safety Management Systems certification have a great impact in companies, at internal, external and operational levels. The certification of these Management Systems includes many common procedures that can be run in an integrated basis, as illustrated in Figure 3.

In fact and according to the ISO – IMS publication (The integrated use of management systems standards) [23], a common objective of management system standards is to assist organisations to manage the risks associated with providing products and services to customers and other stakeholders. On other hand the management system of the organisations is frequently split into a number of parts or sub-systems, which must be managed separately with relative independence. These parts or sub-systems of an organisation's management system reflect the different needs and expectations of the stakeholders. Many organisations use standards, such as those related to quality, the environment and safety, among others, to manage certain aspects of their performance.

The synergy that an Integrated Management System (IMS) can offer have driven organizations into higher levels of performance at a cost lower than that associated to independent certification management systems. The simple schematic Figure 3 represents the vision of an IMS, suggesting that they have common information and procedures and at same time others standards, used in the companies, are involved in the organization and functioning of them.

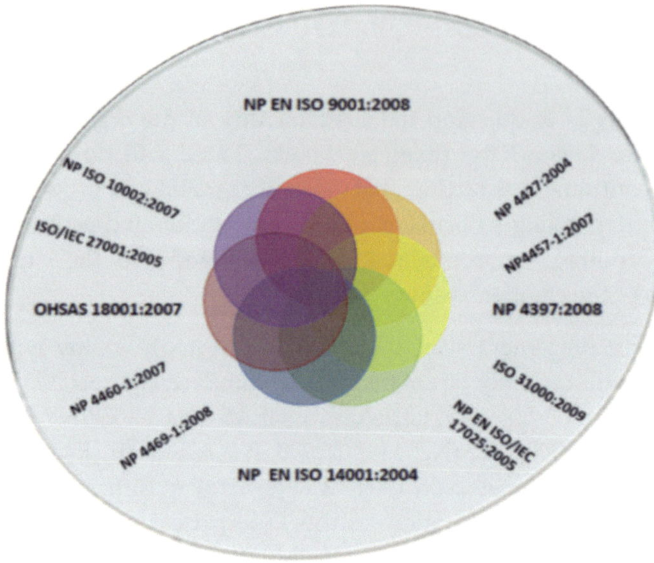

Figure 3. Integration - a route towards sustained success [34]

Integration can be achieved at different levels, leading to partially or fully integrated systems. A partial integrated system keeps their manuals separated using, as far as possible, integrated procedures. A fully integrated system is based in a single manual that integrates unified management systems requirements.

Integration of Quality, Environment and Safety Management Systems are known to have potential benefits over the companies' efficiency. Moreover, integrating IMS-QES enhances organizational performance. Thus, in a team, all employees contribute to the mission and to the defense of the values of a unique organization, bringing up the overall development of the company, thus improving the image of the organization. However, an unsuitable integration process can bring additional problems, like increased strictness, inefficiency and even professional and cultural conflicts. Thus, a careful reflexion about the way to start and run an integration process is crucial.

4.1. Similarities between the Management System standards which enhance the integration - Matrix of compatibility of the requirements and of support to the integration

One of the activities that forms part of the scope and objectives of the case study to which we have paid particular attention is the compatibility of the requirements of the standards, in context and framework of the characterization of the company's situation, backed up by an analysis of these standards. According to the Table 4 this compatibility represents, at our understanding, the starting point for consequents activities of integration, simplification and optimization, to achieve a level of the strictly necessary and consequently the three subsystems - QMS, EMS, and OHSMS are integrated to the maximum extent possible.

PHASE I - PLAN	ISO 9001	ISO 14001	OHSAS 18001
1 - INTEGRATED MANAGEMENT POLICY			
1.1 - Management commitment Continuous improvement	5.1 5.3 8.5.1	4.2	4.2
2 - PLANNING			
2.1- Identification of : product requirements, aspects, impacts, hazards and risks and their assessment	5.2 7.2.1 7.2.2	4.3.1	4.3.1
2.2 - Identification , access to and updating of legal requirements and other requirements of Stakeholders		4.3.2	4.3.2
2.3 - Definition of objectives, targets and Programmes of QES management and improvement	5.4.1 5.4.2 8.5.1	4.3.3	4.3.3
2.4 - Definition of the plans of response to emergency situations	8.3	4.4.7	4.4.7

PHASE II - DO	ISO 9001	ISO 14001	OHSAS 18001
3 - IMPLEMENTATION AND OPERATION			
3.1 - Resources, organizational structure, roles, responsibilities and authority	5.1 5.5.1 5.5.2 6.1 6.3	4.4.1	4.4.1
3.2 - Training, awareness, competence and qualifications	6.2.1 6.2.2	4.4.2	4.4.2
3.3 - Communication, participation and consultation of the Stakeholders	5.5.3 7.2.3	4.4.3	4.4.3
3.4 - QES Management System documentation	4.2.1	4.4.4	4.4.4
3.5 - Control of documents	4.2.3	4.4.5	4.4.5
3.6 - Product realization Operational control	7.1 to 7.5.5	4.4.6	4.4.6
3.7 - Operationalisation of contingency plans.	8.3	4.4.7	4.4.7

PHASE III - CHECK / PHASE IV - ACT	ISO 9001	ISO 14001	OHSAS 18001
4 - CHECKING AND CORRECTION			
4.1 - Performance monitoring and measurement of processes and products	7.6 8.1 8.2.3 8.2.4 8.4	4.5.1	4.5.1
4.2 - Evaluation of compliance	8.2.3 8.2.4	4.5.2	4.5.2
4.3 - Incident investigation	---	---	4.5.3.1
4.4 - Non-conformities; corrections; corrective and preventive actions Control of nonconforming products	8.3 8.4 8.5.2 8.5.3	4.5.3	4.5.3.2
4.5 - Records Control	4.2.4	4.5.4	4.5.4
4.6 - Combined Internal QES Audits	8.2.2	4.5.5	4.5.5
5 - MANAGEMENT REVIEW			
5.1 - Critical analysis and combined QES Management System review	5.1 5.6.1 5.6.2 5.6.3 8.5.1	4.6	4.6

CONTINUAL IMPROVEMENT OF THE IMS_QES

Table 4. Matrix of compatibility of the standards requirements and of support to the integration of the Sub-systems [16 ; 34]

The ISO defines as principles of Quality Management: Customer focus; leadership; involvement of people; process approach; system approach to management; continual improvement; factual approach to decision making and mutually beneficial supplier relationships. According to the ISO Guide 72:2001, anyone drafting these and other management standards must take into account that the corresponding management systems must consider the following phases of the PDCA cycle - *Plan, Do, Check, Act*: Policies and principles; planning; implementation and operation ; performance assessment. Within this framework, taking into account the structuring of standards - management responsibility/planning; resource management; product realization/operational control and measurement, analysis and improvement, on the matrix of the Table 4 we have shown the requirements of the ISO 9001, ISO 14001 and OHSAS 18001 (NP 4397:2008), as well as we have established correspondences, made them compatible with each other and associated with the phases of the PDCA methodology - "Plan-Do-Check-Act". With this matrix, we aim to orientate and align the organizational structure of the company in the same direction, while at the same time creating a structured and useful work referential to support an effective alignment and correspondence of the Sub-Management Systems of Quality, Environment and Safety with consequent compatibilities between each other, for consequent implementation of the IMS-QES. From this matrix we can also, at the same time, to made a correspondence with the Deming Cycle, in this circumstance for the Integrated Management System, as well as a set of stages (1.1; 2.1...2.4; 3.1...3.7; 4.1...4.6 and 5.1) associated with each other these phases of the PDCA cycle, according to Figure 4.

4.2. Model of development of the IMS-QES based on Deming`s Cycle (Plan-Do-Check-Act)

The continuous improvement of the global performance of a Company shall be an objective always present in the development of the IMS [34]. The Company should therefore potentiate for each stage: Plan, Do, Check, Act, a careful and methodical analysis of the differences that effectively can be observed in terms of standards requirements under clauses equivalent involved and for each phase and each stage of development of the IMS-QES, according to the model of Figure 4 to ensure its compliance and evidence of it, in full conformity.

First of all, the definition, approval and communication of the Integrated Management Policy, a common requirement to the different normative references, which must take into account and be consistent with the Mission and Vision of the Company, these supported on a strategy and specific objectives which in turn , support the implementation of that policy and its consequent effectiveness.

The planning of activities in the aim of the Integrated Management System - Phase I (Plan) - is perhaps the most important [34]. In fact, a neglected planning will lead to inefficiencies that can be translated into potential deviations to the objectives. It is therefore fundamental to invest resources and expertise at this stage, via a thorough and careful work, in order to respond effectively to all requirements arising from the involved standards and others applicable requirements in this phase of the planning of the IMS [34].

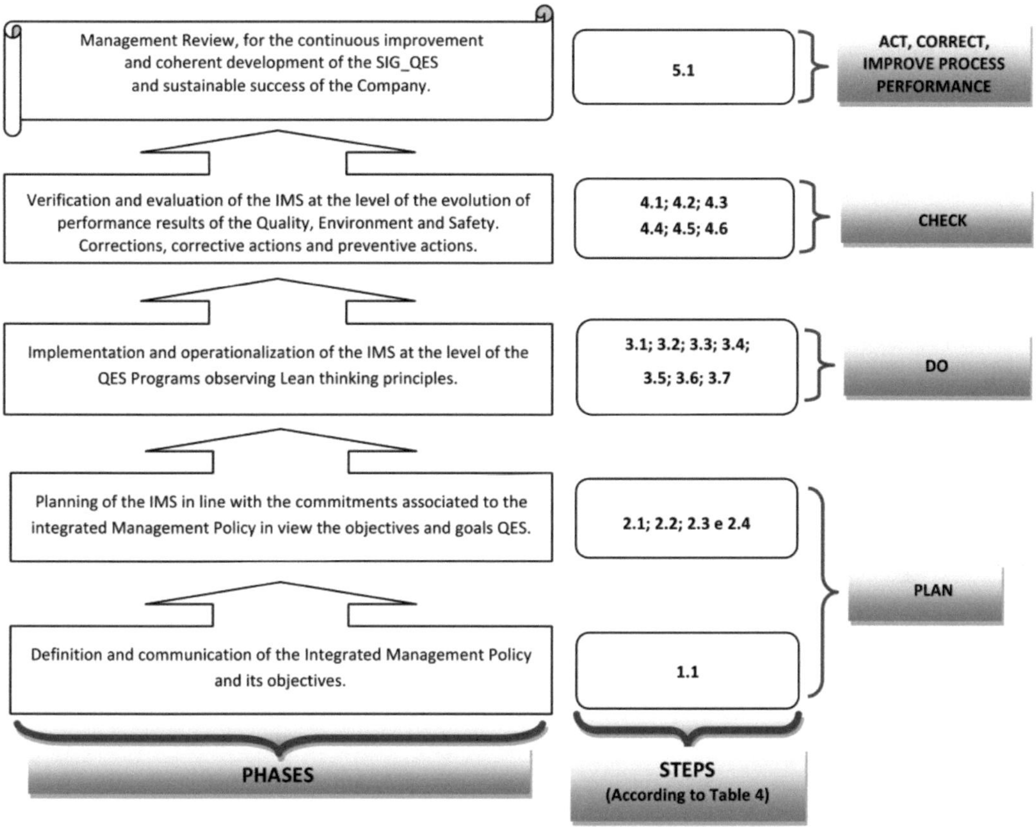

Figure 4. Model of development of the IMS-QES [16 ; 34]

Following is the Implementation and Operation - "Do", the Company should, in this Phase II - Do - promote, the "Make / Do" in coherence with the pre-planned. Corresponds mainly to clauses: 7 - Product Realization, of ISO 9001 - and 4.4 - Implementation and operation of the NP 4397 and ISO 14001 and in the case of ISO 9001 should be considered associated with the product realization, other complementary clauses, particularly in context of resource allocation (6.1, 6.2, 6.3, 6.4) and management commitment (5.1, 5.5.1).

In the Phase III - Check, we identified six steps (4.1 to 4.6) designed to meet the requirements of clauses: 8 - Measurement, analysis and improvement of ISO 9001, 4.5 - Checking of the ISO 14001 and OHSAS 18001/NP 4397. With the exception of step 4.3 - Investigation of incidents resulting from a specific sub-section, the 4.5.3.1 - Incident investigation, the OHSAS 18001/NP 4397 has no correspondence in the ISO 9001 and ISO 14001.

At the end, in the Phase IV - Act, we identified the step 5.1 - Critical analysis and review of the Management System, which refers to the requirements of clauses: 5.6 - Management review of the ISO 9001 and 4.6 - Management review of the ISO 14001 and OHSAS 18001/NP 4397. The Management review process can be done in an integrated manner [34]. It requires, in itself, a very careful preparation face, particularly, to the level of various information that supports the inputs, as is schematized in Figure 5.

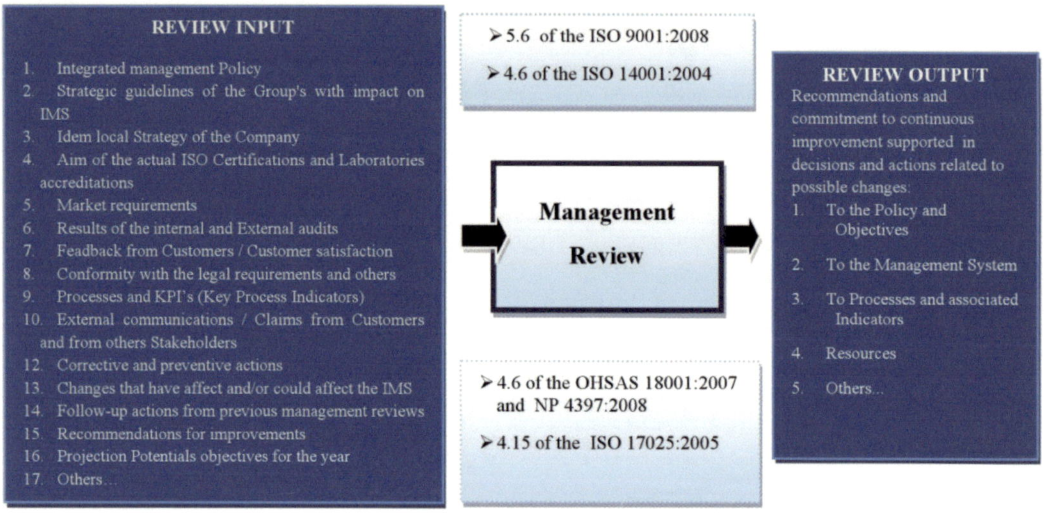

Figure 5. IMS-QES - Management Review - Standards, inputs and outputs [16 ; 34]

The continuous improvement of the global performance of organizations must be always a present goal in a perspective of sustainability [34]. The development Model of the integration of the Management System of Safety and Health at Work with the Quality Management Systems and Environmental Management at the Company should therefore to potentiate, for each phase: Plan, Do, Check, Act, a careful and methodical analysis of the differences that effectively are observed at the level of normative requirements under the equivalent clauses and for each step of their development as the advocated model of integration.

The compatibilization of the normative requirements supported by an analysis of similarities, of the normative referentials of Management Systems, which promotes the integration and it's formatting on a Matrix of compatibilization of the requirements and of support to the integration of the referential - NP EN ISO 9001:2008, OHSAS 18001:2007 / NP 4397:2008 and NP EN ISO14001:2004, establishing correspondences, matching them with each other and associate them, according to ISO Guide 72:2001, the following phases of the methodology *PDCA - Plan, Do, Check, Act*: Policy and principles; Planning, Implementation and Operation, Performance Evaluation, Improvement, Management Review, is one of the activities that in the aim and objectives of the integration model was given special attention in context of characterization and framework of the situation in the Company. That compatibilization constitutes, the starting point for subsequent activities of integration, simplification and optimization, to a level of the strictly necessary and consequent integration maximized as desired of the three sub-systems – the EMS, the QMS and the OHSMS in context of strong competitiveness [34].

As can be expected, there are several difficulties involved with implementing an Integrated Management System (IMS). However, Beckmerhgeni [35] points out that *"the management systems implemented separately in an incompatible way results in costs, an increased probability of faults and errors, duplicated efforts, the creation of unnecessary bureaucracy and a negative impact near the Stakeholders, particularly Employees and Costumers"*.

The integration of the Environmental System supported on the ISO 14001:2004 with the Management Systems of Quality - ISO 9001:2008 and Occupational, Health and Safety - OSAS 18001:2007 represent real added value both in the present and, fundamentally, for the future, not only for the Company, as well as for a whole range of Stakeholders.

Environmental management in organizations is therefore a fundamental process for an more and more integrated and responsible management of sustainability [34]. Aims essentially, an efficient resource management, consumption, wastes and effluents, the use of cleaner technologies and compliance with applicable environmental legal requirements for protecting and promoting environmental and pollution prevention.

An Environmental Management System (EMS) is translated in practice in a continuous cycle of planning, implementation and operation, checking, analyzing and improving of the actions that drives an organization to continuously improve their environmental performance, covering a wide range of issues, including those with strategic implications as well as of competitiveness. The ISO 14001:2004 [36] is integrated in a family of standards developed by the ISO in the fields of Environmental Management according to Table 5.

The ISO 14001:2004 describes the requirements that an EMS of an Organization must meet in order to help the Organization to manage the impact of their activities on the environment, bringing advantages that are considered of high relevance to the promotion of continuous improvement in environmental performance of the Organization and to the prevention of pollution, taking into account the socio-economic surroundings. It allows an organization to develop and implement a policy and objectives, taking into account legal requirements and other requirements that the organization subscribes, and information about significant environmental aspects which the organization can control and influence. While a reference for an EMS, the ISO 14001:2004 contains guidelines that enable their implementation and the requirements that can be objectively audited for certification.

Today, the success of an organization goes through Sustainability [34]. According to ISO 9004:2009 [37], the sustained success of an organization is considered the result of its ability to achieve and maintain their long-term objectives that pass unfulfilled, in a consistent way, the needs and expectations of its Stakeholders, in a balanced and long-term as well.

Furthermore the fact that, in turn, other normative references of Management Systems, namely the ISO 14001 and OHSAS 18001/4397 NP, are valuable auxiliaries in the evolutive management of Organizations and, hence, on the three strands of its Sustainability: the Economic, the Social and Environmental. In fact, today Quality, Environment, Safety and Occupational Health are fundamental pillars of the Sustainable Development and therefore should be considered at the level of the management activities and of strategic planning of Organizations that want to take a leadership role, at present, but fundamentally in the future.

SERIE	CONTENTS	STANDARDS
ISO 14000	Environmental Management Systems	- ISO 14001:2004 – Environmental management systems – Requirements with guidance for use - ISO 14004:2004 – Environmental management systems – General guidelines on principles, systems and supporting techniques - ISO 14005:2010 – Environmental management Systems - Guidelines for the phased implementation of an environmental management system, including the use of environmental performance evaluation. - ISO 14006:2011 – Environmental management Systems - Guidelines for incorporating ecodesign
ISO 19011	Audits to Management Systems	- ISO 19011:2011, Guidelines for auditing management systems
ISO 14020	- Environmental label	- ISO 14020 - Environmental Labelling: General Principles - ISO 14021 - Environmental Labels and Declarations: Self-Declaration Environmental Claims, Terms and Definitions - ISO 14022 - Environmental Labels and Declarations: Self-Declaration Environmental Claims, Symbols - ISO 14023 - Environmental Labels and Declarations: Self-Declaration Environmental Claims, Testing and Verification - ISO 14024 - Environmental Labels and Declarations: Environmental Labelling Type I, Guiding Principles and Procedures
ISO 14030	Environmental performance evaluation	- ISO 14031:1999 - Environmental management -- Environmental performance evaluation - Guidelines
ISO 14040	Life cycle assessment	- ISO 14040:1997 - Environmental management -- Life cycle assessment -- Principles and framework - ISO 14041:1998 - Environmental management -- Life cycle assessment -- Goal and scope definition and inventory analysis - ISO 14042:2000 - Environmental management -- Life cycle assessment -- Life cycle impact assessment - ISO 14043;2001 – Environmental management – Life cycle assessment – Life cycle interpretation
ISO 14050	Understanding of Terms and Definitions	- ISO 14050:2002 – Environmental management – Vocabulary.

Table 5. ISO Standards in the fields of Environmental Management [34]

Q M S	100% are certified by ISO 9001			
	Reasons for certification	**Very Important**	Empowering knowledge workers	Business benefits
			Better Quality of the products	Improvement of company image
		Important	Pressure from customers	Community relations
	Benefits	**Major Impact**	Improvement of the company image	Ease of access to information
			Internal organization of the company	Continuous assessment through internal audits
		Impact	Better productivity and profitability	Deadlines
			Cost reduction	Number of defects / complaints
			Competitive position	Satisfaction of internal and external customers.
		Little Impact	Absenteeism	Product innovation
	Tools	**Most commonly used**	Surveys	Charts (graphics)
			Sheets of data collection	
		Less used	Scatter diagram	Cause and effect diagram
	Difficulties		High certification costs	Difficulties to change company's culture
			Difficulties to motivate personal	
E M S	36.9 % are certified by ISO 14001			
	Reasons for non-certification	**Main reason**	Lack of investment support	A form of marketing
		Secondary reason	Benefits do not outweigh the costs	Investments are high
		Less Relevant	Environmental risks are low	
	Reasons for certification	**Very Important**	Progress but always protect the environment	Raising public
			Improved quality of life inside and outside the company	
		Important	Obtaining an asset in terms of marketing	
	Benefits	**Major Impact**	Environmental protection	Rational use of natural resources
			Business image in the community	Recycling
			Ensuring compliance	Prevention of environmental risks
		Impact	Clear, organized work environment	Waste generation
			Awareness of employees in environmental issues	
		No impact	Insurance costs	Costs of raw materials and equipment
			Ability to obtain financing at low interest rates	
	Drawbacks		Overall costs increased on the early stages of certification.	
	Main Difficulties		High certification costs	Difficulties to change Company's culture
			Difficulties to motivate personal	
O H S M S	26.1 % are certified by OHSAS 18001			
	Reasons for non-certification	**Main reason**	Investments are high	Certification is seen as a cost and marketing
		Secondary reason	Benefits do not outweigh the costs	

		Less Relevant	Risk of accidents at work are low	
	Reasons for certification	**Very important**	Eliminate or minimize risks to workers	
		Important	Improving the organization with the reduction of accidents at work	
		Less Important	Rate of absenteeism due to occupational diseases	
	Benefits	**Major Impact**	Improvement of working conditions	Better internal communication
			Improvement company image.	Ensuring compliance with legislation
		Impact	Less number of accidents	Reduction in cost of accidents and occupational diseases
		No impact	Dissemination in the media of data on safety and health of the company	
	Drawbacks		No drawbacks have been referred	
	Main Difficulties		High certification costs	Difficulties to change Company's culture
			Difficulties to motivate personal	Increase bureaucracy
			Management difficulties in the early stages of certification	
IMS	**Main Benefits**		Management costs reduction	Unification of internal audits
			Reduction of bureaucracy	Better definition of responsibilities and authority
			Easier compliance of legislation	Increased performance and efficiency
			Improved external image of the company	Better and easier communication system.
			Improved organization	Simplified management systems resulting in less confusion, redundancy and conflicts in documentation
			Increased employee training	Optimized resources(financial and humans)
	Drawbacks		One problem on a single system affects the overall management system	Initial costs increase associated to an increase in non-conformities;
			Initial higher organizational problems	Continuous update of every documentation with negative impact in the management activity itself
			Complex organizational system	Incompatible concepts between systems
	Main Difficulties		Long time to implement the integration process	Insufficient integrability of the standards
			Deep changes in the management system due to operational changes	Difficulties to implement systems integration
			High difficulty associated to training and changes in the organization methods and culture	Higher difficulty and cost to implement all systems simultaneously when compared with individual implementation

Table 6. Summary of the benefits and impact grade characterization that companies obtained with certification of the management systems (Adapted of [13])

Also in this context of Sustainability, the integration of the EMS whith others management systems is, nowadays, a strategic decision of the organizations, influential of its competitiveness and continuity in the Market, fostering a better and increasing satisfaction of all Stakeholders, increasing the image and brand reputation near them [34] . Are too examples of adavantages of the integration, also highlighted by the survey respondents: the elimination of conflicts between individual systems with optimization of Resources; the improvement at the level of the Coordinated and Integrated Management of the Risk associated to the Safety of the Persons and Company Assets, Environment and Quality of the Products; the reduction of the number of internal and/or external audits and audits to Suppliers and spent time versus associated costs; the creation of added value for the business through the prevention and/or elimination of several types of operational and organizational wastes.

4.3. Summary of the benefits and impact grade characterization that companies obtained with certification and integration of the management systems

The three standards ISO 9001, ISO 14001 and OHSAS 18001 have a common underlying principle: continuous improvement based on Deming`s Cycle (Plan-do-Check-Act) [16 ; 34]. According to Labodová [22], basically one PDCA approach governing the aspects of quality, environment and safety, consistent with clear priorities, shows hot spots in companies, makes it easy to focus procedures and responsibilities on important areas.

According to the survey, 100% of the companies of the sample had their QMS certified. 36.9 % of these companies had also their EMS certified, and only 26.% had OHSMS certification. In Table 6, a summary of the main benefits, classified according to their impact relevance, as well as the main reasons for the three certification systems is presented. Moreover, the main factors that justify the option for non-certification of EMS and OHSMS of the remaining companies are also presented. The main factors that justify the option for non-certification of QMS are not presented, since every company participating in the survey had their QMS certified. Are also presented the main benefits, drawbacks and difficulties what concern Integration of Management Systems.

5. Conclusions

The theoretical fundamentation was supported on the concepts of Quality Management Systems, Environment and Safety, Integrated Management Systems, orientations, guidelines and ISO standards as well as a group of other sources of knowledge, such as Integrated Management Systems QES.

Were identified the main benefits of implementing an EMS and it´s certification after receiving the QMS certification on Portuguese SMEs performance. In fact the main benefits that the Portuguese SMEs have gained from the referred certifications have been, among others, prevention of environmental risks, environment protection, improvement business image in the community, ensuring compliance with legislation and rational use of natural resources. In addition, the main reasons for non-certification were analyzed.

Those companies with non-certified EMS mentioned "lack of investment support" and "consider the certification as a form of marketing". Drawbacks that have been concluded were "overall costs increased on the early stages of certification", because there is a better control. The main difficulties uncovered in relation to EMS certification can be pinpointed to difficulties in changing company culture, high certification costs, and difficulties to motivate personal. With respect to environment, there is still a lot to do in Portugal. Several Portuguese SMEs have already implemented the EMS and others will follow, mainly, because money can't pay for an environmental disaster.

In terms of the potential benefits of the integration of the Environmental Management System (EMS) with the Quality Management System (QMS) and Occupational Health and Safety Management System OH&SMS and it´s certification, will be mostly a whole range of opportunities for performance improvements of the organization as a whole, and throughout it's supply chain such as: the elimination of conflicts between individual systems with resource optimization , creation of added value to the business by eliminating several types of wastes, reducing the number of internal and external audits, common policy management, greater valuation and motivation of Collaborators, among others.

The proposed model of IMS_QES has led to the simplification and consequent resolution of problems and gains in efficiency of existing management systems in the company (referred as Sub-systems), which along the years were adopted gradually, but individualized and often isolated between to each other.

All these benefits will enhance the competitiveness of the Companies and its differentiation by the positive in the face of competition, national and/or international.

The developed model of an Integrated Management Systems (IMS), generic, flexible, integrator, evolutionary and lean namely what concerns to Quality (ISO 9001), Environment (ISO 14001) and Health and Safety at work (OHSAS 18001/ NP 43397:2008), but too to others standards, such as, Risk Management (ISO 31000:2009), Social Accountability (SA 8000) and Information Security, among others, could be progressively more integrated and allow the Companies to even more capable, enhancing in this way too the development of the businesses and consequently their sustainable successes .

Author details

Gilberto Santos*, Manuel Rebelo, Síria Barros and Martinha Pereira
College of Technology, Polytechnic Institute Cávado e Ave, Imasys Research Centre, Campus do IPCA 4750-810 Barcelos, Portugal

6. References

[1] Wright T (2000) IMS-Three into One Will Go!: The Advantages of a Single Integrated Quality, Health and Safety, and Environmental Management System. The Quality Assurance Journal 4: 137–42.

* Corresponding Author

[2] Zorpas A (2010). Environmental management systems as sustainable tools in the way of life for the SMEs and VSMEs. Bioresource Technology 101, 1544 -1557.

[3] Burke S, Gaughran WF (2007). Developing a framework for sustainability management in engineering SMEs. Robotics and Computer-Integrated Manufacturing 23, 696 -703

[4] Turk A M (2009). The benefits associated with ISO 14001 certification for construction firms: Turkish case. Journal of Cleaner Production 17 (2009) 559–569.

[5] Kwon D M, Seo M S, Seo Y C (2002). A study of compliance with environmental regulations of ISO 14001 certified companies in Korea Journal of Environmental Management 65, 347-353.

[6] Nawrocka D, Brorson T, Lindhqvist T (2009). ISO 14001 in environmental supply chain practices Journal of Cleaner Production 17 1435–1443.

[7] Casadesus M, Marimon F, Inaki Heras H (2008). ISO 14001 diffusion after the success of the ISO 9001 model. Journal of Cleaner Production 16. 1741 – 1754

[8] Zeng SX, Jonathan JS, Lou GX (2007). A synergetic model for implementing an integrated management system: an empirical study in China. Journal of Cleaner Production 15, 1760 -1767.

[9] Karapetrovic S, Casadesús M (2009). Implementing environmental with other standardized management systems: Scope, sequence, time and integration. Journal of Cleaner Production 17. 533–540.

[10] Fryxell G E, Szeto A (2002). The influence of motivations for seeking ISO 14001 certification: an empirical study of ISO 14001 certified facilities in Hong KongJournal of Environmental Management 65, 223–238

[11] Tarí J J, Molina-Azorín J F (2010). Integration of quality management and environmental management systems Similarities and the role of the EFQM model. The TQM Journal Vol. 22 No. 6, pp. 687-701

[12] Fresner J, Engelhardt G (2004). Experiences with integrated management systems for two small companies in Austria. Journal of Cleaner Production, Vol. 12, pp. 623-631.

[13] Santos G, Mendes F, Barbosa J (2011). Certification and integration of management systems: the experience of Portuguese small and medium enterprises Journal of Cleaner Production 19 1965-1974

[14] Fernández-Muñiz, B, Montes-Peón J M, Vázquez-Ordás C J (2009). Relation between occupational safety management and firm performance. Safety Science 47, 980–991.

[15] Robson L S, Clarke J A, Cullen K, Bielecky A, Severin C, Bigelow P L, Irvin E, Culyer A, Mahood Q (2007). The effectiveness of occupational health and safety management system interventions: A systematic review. Safety Science 45, 329–353.

[16] Rebelo MF, Santos G (2012) - Integration of the Occupational Health and Safety Management System with the Quality Management System and Environmental Management System - from the Theory to the Action. International Symposium on Occupational Safety and Hygiene SHO - 2012. Minho University 9, 10 March 2012. ISBN 978-972-99504-9-0

[17] ISO (2008)The integrated use of management system standards – Switzerland.

[18] Mendes F, Santos G (2009). Impacto de la certificacion de los sistemas integrados de gestion en las PMEs portuguesas. Forum Calidad nº 198, 46-51

[19] McDonald M, Mors T A, Phillips A (2003). Management system integration: can it be done? Quality Progress; 67-74.

[20] Arifin K, Aiyub K, Awang A, Jahi J M, Iten R. (2009). Implementation of Integrated Management System in Malaysia: The Level of Organization's Understanding and Awareness European Journal of Scientific Research ISSN 1450-216X Vol.31 No.2:188-195.

[21] Bernardo M, Casadesus M, Karapetrovic S, Heras I (2009). How integrated are environmental, quality and other standardized management systems? An empirical study. Journal of Cleaner Production 17: 742–50.

[22] Labodová A (2004). Implementing integrated management systems using a risk analysis based approach. Journal of Cleaner Production; 12: 571–80.

[23] Guide ISO 72:2001, Guidelines for justification and development of management System standards.

[24] ISO 19011:2011, Guidelines for auditing management systems

[25] Santos G et al. (2008). Implementação de Sistemas Integrados de Gestão (Qualidade, Ambiente e Segurança) – ISBN 978-972-8953-26-3; Publindústria.

[26] Suditu C (2007). Positive and negative aspects regarding the implementation of an integrated Quality, Environment, health and Safety Management System. Annals of the Oradea University, - Fascicle of Management and Technological Engineering, Volume VI (XVI),

[27] Jørgensen T H (2008). Towards more sustainable management systems: through life cycle management and integration, Journal of Cleaner Production 16, 1071-1080.

[28] OHSAS 18001:2007, Occupational Health and Safety Management Systems – Requirements - BSI, first published July.

[29] Block M R, Marash IR (2002). Integrating ISO 14001 into a Quality Management System. ASQ – Second edition. Milwaukee.

[30] http://www.min-economia.pt/ 2007.

[31] Santos M G, Barbosa J, Pedro C.(2004). Integration of "QES" systems on the small and medium size foundry companies in Portugal. Communication presented at the 66th Would Foundry Congress. Istanbul, Turkey.

[32] htpp://www.pme.online.pt/ 2007.

[33] Mendes, F. (2007). O Impacto dos sistemas QAS nas PMEs portuguesas – Master Thesis – Univ. Minho.

[34] Rebelo M F (2011). Contribuição para a estruturação de um modelo de sistema integrado de gestão QAS. Master Thesis. Polyt Inst Cavado Ave. Portugal.

[35] Beckmerhageni A; Berg H. P.; Karapetrovic S. V.; W. O. Willborn (2003). Integration of Standardized Management Systems: focus on safety in the nuclear industry. International Journal of Quality & Reliability Management, Vol 20 N º 2, pp. 210-228.

[36] ISO 14001:2004, Environmental management systems. Requirements with guidance for use.

[37] ISO 9004:2009 – Managing for the sustained success of an organization – A quality management approach.

[38] Curkovic S, Melnyk S A, Handfield R B, Calantone R (2000). Investigating the Linkage Between Total Quality Management and Environmentally Responsible Manufacturing. IEEE Transactions on Engineering Management, Vol. 47 Nº4, pp.444-464.

Chapter 10

Uncovering the Relation Between Environmental Damage and the Rate of Rainfall Received Through a Life Cycle Assessment (LCA) Study on Potable Water Production in Malaysia

Amir Hamzah Sharaai, Noor Zalina Mahmood and Abdul Halim Sulaiman

Additional information is available at the end of the chapter

http://dx.doi.org/10.5772/45851

1. Introduction

Global climate change is a threat that people are not aware of or do not give enough attention to. In Malaysia, many people take this problem very lightly. Global climate change is usually linked to an increase in world temperature that involves complex processes and would take a long time to take effect. However, the impact from global climate change is getting more apparent and can be felt by people of the world resulting from the increase in world temperature. The increase in world temperature, better known as global warming, is a new threat to the safety of people and nations, and could be even worse than weapons of destruction. This is due to the fact that global warming does not discriminate between skin colour and national boundaries. In Malaysia, signs of climate changes due to global warming can be seen through various disasters and symptoms affecting several areas throughout the country such as increasing sea level, haze, flood and water shortages. Climate change not only involves Malaysia, but is a global problem that must be dealt with cooperatively by every nation.

1.1. What is global climate change?

Global climate change, or global warming, is usually associated with the increase of world temperature. Global warming is an indication that there is an increase in surface temperature on land, in the ocean or a combination of both on a vast scale. Global warming is caused by the burning of fossil fuels such as coal, oil and gas, resulting in the increase of green house gases such as carbon dioxide, methane and chlorofluorocarbon (CFC) in the atmosphere. The accumulated green house gases trapped in the atmosphere causes the Earth to become warmer.

The actual causes and factors contributing to global warming are still being debated, but the fact is that evidence shows that global warming is caused by human activities and lifestyle. Development, deforestation, industries, factories and transportation are among the activities that contribute to worldwide climate change. Global climate change occurs because of the increase in green house gas in the atmosphere generated by the rapid development in production industries, timber processing, agriculture, transportation and other industries, tipping the balance against nature. Increased carbon dioxide gas in the atmosphere is generated from the burning of fossil fuel such as petrol, coal and natural gas.

According to the 4th Assessment Report (2007 Climate Change) Inter-Governmental Panel of Climate Change (IPCC), 98% of the temperature increase on Earth is caused by the release of carbon dioxide gas. Carbon dioxide concentration in the atmosphere increases from 280ppm recorded in the 18th century (prior to the industrial revolution era) to 379 ppm in 2005. In 2099, if humans maintain the current momentum of fossil fuel use as it is right now, the concentration of carbon dioxide in the atmosphere is predicted to increase to 700ppm. This situation would render the environment of the Earth no longer suitable for human habitation. Green house gas emission by Malaysia is still small compared to other countries in this world (see **Table 1**), nevertheless, a control system is needed to reduce the emission of green house gas into the atmosphere.

Country	Year	Green house emission (million tan)
Malaysia	1990	138.0
	1994	144.0
Thailand	1990	225.0
Australia	1990	572.0
Japan	1990	1215.9
	1994	1276.1
USA	1990	5895.9
	1994	6130.0

Table 1. Green house emission by country (Minister of Science, 2000)

1.2. The impact of world climate change on the country

Climate change is believed to have sped up the weather process causing sudden changes within a period of time. It is threatening the safety of both the population and the nation, from the physical and psychological aspects. If not managed properly, the combination of both would create a crisis in peoples' lives, public safety/order, political stability and economic integrity. The physical effects of climate change to the country are as follows:

1.2.1. Natural disaster

A change of monsoon wind direction, rising sea level and irregular rainfall are among the results of climate change. Increasing temperature of Earth will create a huge threat to human life resulting in an increased frequency of natural disasters over time. Typhoons, floods, drought and plagues of diseases related to the environment are among the natural disasters that would follow the unpredictable climate change. With 9% of land area in Malaysia (29,000 sq.km) exposed to flooding, 2.7 million people in this country might be the victims. The ever-changing weather would cause the country to face the possibility of a natural disaster. As an example, in 2006 and 2007, the state of Johor experienced flooding, resulting in 109, 831 victims to be evacuated (Bakar et al., 2007). Other states experiencing a similar situation at that time were Melaka, Pahang, Kelantan and Negeri Sembilan. A total of 344 flood evacuation centres were set up. It was reported that 17 lives were lost in this extreme flooding disaster (BBC News, 2007) linked to Hurricane Utor (The Star, 2006). In October 2008, a flash flood happened in the northern peninsular states of Kedah and Perak due to the hotter sea surface in the region (Bernama, 2008).

1.2.2. Water source

Rapid development in the urbanization process and industries increases the demand for water especially in the industrial and agricultural sectors, and also for domestic use. Climate change could cause Malaysia to experience increased or decreased rainfall. Increased rainfall would mean that Malaysia would be exposed to flooding, causing destruction to the infrastructure and property, and loss of life. Meanwhile, a decrease in rainfall would cause a reduction in clean water sources for domestic use, and the agricultural and industrial sectors.

During the 1977 - 78 period, drought devastated paddy production in most of the irrigation schemes in northwest peninsular of Malaysia. In 1982 and 1991, drought was responsible for the critical drop of water levels in the state of Kedah, resulting in cancellation of the off-season crop production. In 1991, the state of Melaka faced critical water problems with water levels falling below critical levels, thereby forcing severe water rationing for months in the state (Climate Ark, 2002). In 1998, the El Nino related drought caused severe water stress in the states of Kedah, Penang and Selangor. The state of Selangor was forced to impose severe water rationing in Kuala Lumpur and Petaling Jaya for many months (Angela, 2002). According to the Seventh Malaysia Plan reports, several countries experienced water shortages after 2000, including Kedah, Pulau Pinang and Selangor (Rahman, 2007).

1.3. World climate change and the need for sustainable development

Development can increase the quality of human life. This can be gauged by the increasing quality of products, services, health, society and culture. Development is endless and will keep expanding to fulfil the changing human vision and mission, but the development process usually has negative effects on the environment, thus demanding a more environmentally sound plan. Unplanned and unsustainable development has placed great pressure in every dimension of the environment (air, water, soil, health, etc.) that ultimately causes problems that not only involve local issues, but also global such as global warming. Sustainable development is the need of the hour; it can only be achieved through effective environmental management (Khan et al., 2002). Effective environmental management can be achieved through the various environmental assessment tools including life cycle assessment (LCA).

2. Methodology of Life Cycle Assessment (LCA)

There are four main phases in LCA as suggested in ISO 14040 series:

2.1. Goal and scope definition (ISO 14040)
2.2. Life cycle inventory (LCI) (ISO 14041)
2.3. Life cycle impact assessment (LCIA) (ISO 14042)
2.4. Life cycle assessment and interpretation (LCAI) (ISO 14043)

2.1. Goal and scope definition

In goal definition and scoping, the use of the results is identified, the scope of the study is stated, the functional unit is defined and a strategy and the procedures for data collection and the data quality assurance are established.

2.1.1. Objectives

The objective of this research is to get a clear picture of the impact potential produced from potable water production when two phases are involved, namely, production stage and construction stage using an LCIA method (in this case the Ecopoints method). This research will identify which impact is greater by comparing them using normalization and weighting procedures so that suggestions to reduce the impact can be recommended.

2.1.2. Functional unit

Functional unit is quantified performance of a product system as a reference unit in a life cycle assessment study (ISO14000, 2000). A constant value must be created to make the comparison (Miettinen & Hamalainen, 1997). Functional unit for this study is the production of $1m^3$ of treated water a day that fits the standard quality set by the Ministry of Health, Malaysia.

2.1.3. Description of the system under study

There are two stages which Became the basis of comparison in this study, namely, the production and construction stages.

2.1.3.1. Production stage

Raw water extracted from rivers will go through the following processes in the water treatment plant (Sastry, 1996):

- **Screening.** Removing large, floating rubbish on the surface of the water.
- **Coagulation and flocculation**. The coagulation process is the process of forming particles called floc. A coagulant needs to be added to form floc. The coagulants that are normally used include aluminium sulphate, ferric sulphate and ferric chloride. Tiny flocs will in turn attract each other while at the same time pulling the dissolved organic material and particulate to combine, forming a big flocculant particle. This process is called flocculation.
- **Settling**. Aggregated flocs settle on the base of the settler. The accumulation of floc settlement is called settling sludge.
- **Filtration**. Part of the suspended matter that does not settle goes through filtration. The water passes through a filtration system consisting of sand layers and activated carbon or anthracite coal.
- **Disinfection.** This process is needed to eliminate the pathogen organisms that remain after filtration. Among the chemicals used for disinfection are chlorine, chloramines, chlorine dioxide, ozone and UV radiation.

2.1.3.2. Construction stage

The main building Materials used when building water treatment plants are concrete and steel. Concrete is a type of composite material which is usually used in construction. It is a combination of the following:

a. Cement
b. Fine aggregate/sand
c. Coarse aggregate
d. Water

The quality of the concrete which is produced depends on the quality of the raw materials that are being used such as cement, coarse aggregate and water, rate of mixing, the method of mixing, transportation and compression methods. If the raw materials used are not good quality, the concrete produced will be low quality and this causes the concrete to be weak and unable to fulfil the fixed specifications. So, concrete technology warrants that all the materials used should first be tested and certified through fixed standardizations before being used in construction work.

Steel increases the tensile strength of the concrete structure. Reinforcement steel functions to increase the tensility strength of the concrete structure. Types of reinforcement steel that are used are as follows:

i. Mild steel reinforcement /*mild steel*
ii. Reinforcement steel with high tensility
iii. Fabric steel (*fabric*)

The steels that are provided are 12m long, with diameters of 6mm, 8mm, 10mm, 12mm, 16mm, 20mm, 22mm, 25mm and 32mm. The reinforcement steel will be cut and moulded according to the concrete structure design. Reinforcement steel with high tensility is used as the backbone of the concrete structure because of its high strength. Mild steel reinforcement is usually used as reinforcement fixation where high tensility is not needed. Fabric steel (*fabric*) is used in a wide concrete surface area such as floors and comes in sizes of 2.4m x 1.8m with steel diameter 4mm to 12mm. The distance between each steel rod is different based on the types of fabric. Reinforcement steel that is used should be free from any dirt and rust, so it has to be protected from water and humidity.

2.2. Life Cycle Inventory (LCI)

The inventory of the studied LCA system includes information on the input and output (environmental exchanges) for all the processes within the boundaries of the product system (see **Figure 1**). The inventory is a long list of material and energy requirements, products and co-products, as well as wastes. This list is referred to as a material and energy balance, the inventory table, or the eco-balance of the product (Guinée, 2002). This LCA study is a streamlined LCA with background data for electricity, chemicals and transport using a database contained in the Jemaipro and Simapro 7 software. Foreground data collected from the treatment plant are: (see **Table 2**)

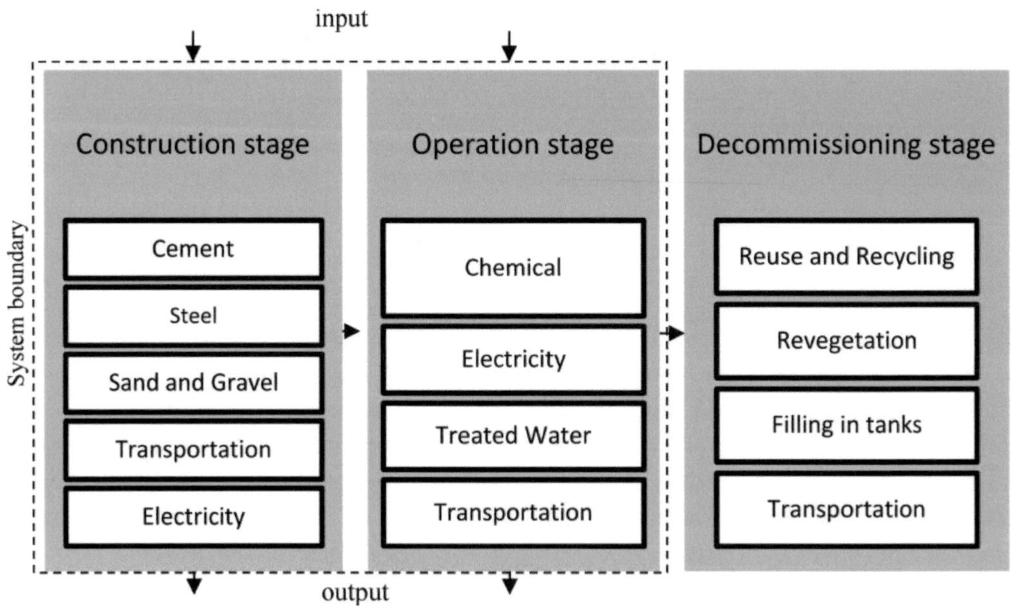

Figure 1. System Boundary of Potable Water Treatment Plant

- Electricity usage, and
- Chemicals for water treatment such as aluminium sulphate (alum), polyaluminium chloride (PAC), chlorine and calcium hydroxide (lime)
- Building materials such as steel, gravel, sand and cement

Filtration material (activated carbon and anthracite) and coagulant (ferrochloride) are not included in this study because no water treatment plants in Malaysia use these materials.

Background data for all building materials and chemicals is obtained from Japan Environmental Management Association for Industry (JEMAI) - PAC, BUWAL 250 - chlorine, alum and electricity, ETH-ESU 98 - lime, LCA Food DK - tap water, and IDEMAT 2001 - cement, steel, sand and gravel.

Construction Stage		Production Stage	
Steel (kg)	8.78	Alum (kg)	22.55
Cement (kg)	30.72	Chlorine (kg)	3.65
Gravel (kg)	70.72	PAC (kg)	16.85
Sand (kg)	47.15	Lime (kg)	11.12
Electricity (kwh)	0.09	Electricity (kwh)	397.28
Tap water (litre)	477.26		

Table 2. Foreground data for construction stage and production stage.

2.3. Life Cycle Impact Assessment (LCIA)

The impact assessment is used to identify the significant potential environmental effect by using the results of life cycle impact analysis (LCI). LCIA is very different from other techniques, such as environment impact assessment (EIA) and risk assessment, because the approach uses functional units. LCIA comprises four elements, namely, the classification, characterization, normalization and weighting, but normalization and weighting are optional elements (Koroneos, Dompros, Roumbas & Moussiopoulos, 2005). According to Jolliet, Brent et al., (2003), the classification of LCI due to the impact categories is through the impact pathway which begins from LCI results until the end-point. The explanation on impact pathway is also touched upon in ISO (Jolliet, Brent et al., 2003) where:

> 'LCI results are classified into the impact categories and category indicators that can be stated in any LCI results (mid) with the end-point category.'

In accordance with the above explanation, two approaches have been developed to explain the inter-connection of the LCI results with the environmental impacts via mid-point or end-point approaches (Heijungs et al., 2003; Jolliet, Margni et al., 2003; Jolliet et al., 2004; Ortiz, Francesc & Sonnemann, 2009; Sleeswijk, van Oersc, Guinée, Struijsd & Huijbregtsb, 2008; Soares, Toffoletto & Deschenes, 2006). According to Bare et al. (2000), the main difference between both models is the methodology used and how category indicators are presented to translate the achieved impact categories.

Figure 2 illustrates that the impact pathway begins from LCI results until the end-point. The emission of ozone depletion gases is used as an example for the characterization of ozone depletion gases that can be conducted either until mid-point or end-point. Impact at mid-point is the ozone layer depletion and impact in the end-point is the protected area involving human health, natural biotic environment and manmade environment.

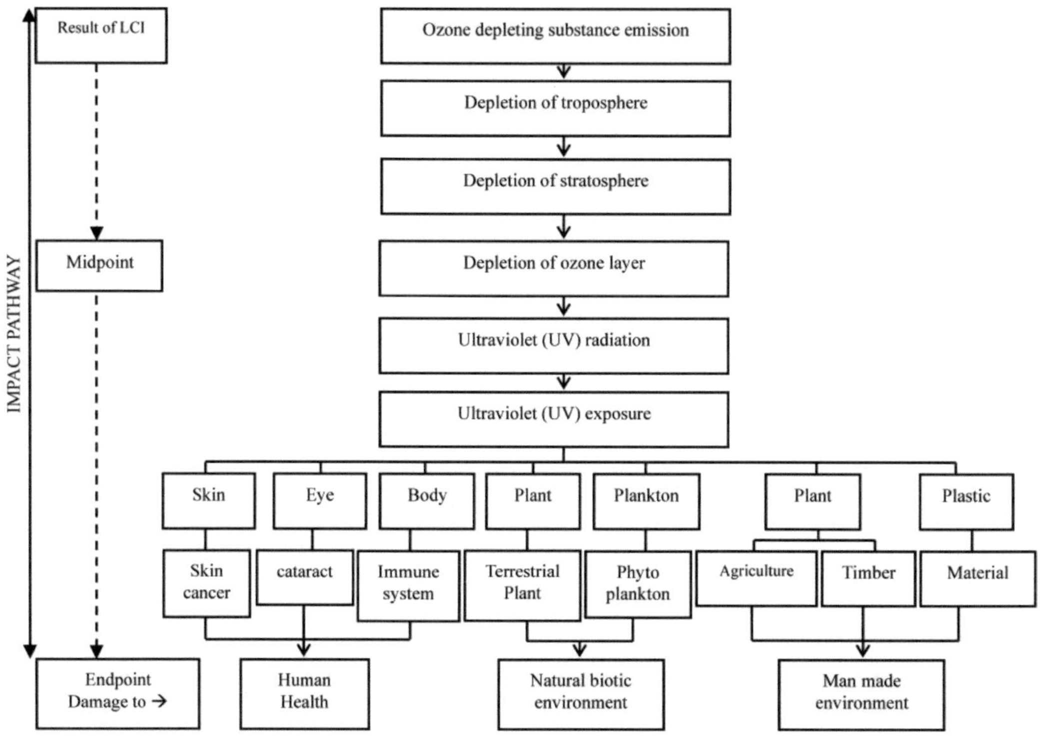

Figure 2. Impact pathway connecting the emission to several deterioration categories.

2.3.1. Mid-point approach

The LCIA mid-point approach is also known as the problem-oriented approach (Dreyer, Niemann & Hauschild, 2003; Ortiz et al., 2009) or the classical impact assessment method (Jolliet, Brent et al., 2003; Jolliet et al., 2004). The term mid-point refers to the category indicator for each impact category which is expressed in the mid pathway of impact between LCI results and end-point (Josa, Aguado, Cardim & Byars, 2007). Mid-point translates the category impact into real phenomenon such as climate change, acidification and aquatic toxicity (Sleeswijk et al., 2008). An example of the methodology that was developed using the mid-point approach is CML 2001 (Dreyer et al., 2003; Heijungs et al., 2003), EDIP 97 and TRACI (Jolliet et al., 2004).

2.3.2. End-point approach

The end-point LCIA methodology is also known as the damage-oriented approach (Dreyer et al., 2003). According to Heijungs et al., (2003) this approach looks at the elements inside the impact pathway that consist of independent values for society. The term 'end-point' refers to the category indicator for each impact category located at the end of the impact pathway as in Figure 2. End-point indicator translates the category impact based on the area of protection such as human health, natural environmental quality, natural resources and the manmade environment (Bare & Gloria, 2008). Examples of end-point methodology are Eco-indicator 95 and 99, EPS 92, 96 and 2000, and LIME 2003 (Pennington et al., 2004). According to Reap et al. (2008), there are several factors affecting the level of confidence and suitability of LCA research result which include the options of LCIA methodology either using the mid-point or end-point approach. Reap et al. (2008) state that the end-point impact category is less comprehensive and possesses higher levels of uncertainty compared to the mid-point impact category. Nevertheless, the mid-point impact category is difficult to interpret, especially in the process of decision making because the mid-point impact category is not directly correlated with the area of protection (i.e., damage to human health, ecosystem quality and resource depletion) which is practiced by the end-point.

2.3.3. Eco-Indicator 99 evaluation method

LCIA for this study uses the Eco-Indicator 99 method (Eco-indicator 99 is an end-point method as mentioned in previous sub-topic) where 11 impacts classified into three damage assessment areas (refer **Table 3**) as listed below:

Damage Assessment	Unit	Impact
Human Health	DALY	Carcinogen, radiation, respiratory organic and inorganic
Ecosystem Quality	PDF*m^2yr PAF*m^2yr	Climate change, ozone layer and acidification Ecotoxicity
Resources	MJ surplus	Land use, minerals and fossil fuels

DALY: Disability Adjusted Life Years (years of disabled living or years of life lost due to the impacts)
PAF: Potentially Affected Fraction (animals affected by the impacts)
PDF: Potentially Disappeared Fraction (plant specie disappeared as result of the impacts)
MJ: Surplus Energy (MJ) (extra energy that future generations must use to excavate scarce resources)

Table 3. Damage Assessment and Impact According to Eco-Indicator 99.

2.3.4. Steps in Life Cycle Impact Assessment (LCIA)

Generally there are three steps in LCIA:

2.3.4.1. Classification and characterization
2.3.4.2. Normalization, and
2.3.4.3. Weighting

2.3.4.1. Classification and characterization

Classification is the step in which the data from the inventory analysis (the substance emissions) are grouped together into a number of impact categories (Bovea and Gallardo, 2003). Grouping to the impact categories is done according to their ability to contribute to different environmental problems. While characterization is the effect of each item on each impact category. A typical way is to use equivalency factors, in some instances also called potentials is, for example, global warming potential for a substance indicates its two relative potentials to increase the global warming effect compared to CO2, whose GWP is set to one. In ISO 14040 series classification and characterization are two basic mandatory elements. Below are the results of the analysis done in this study (see **Figure 3**).

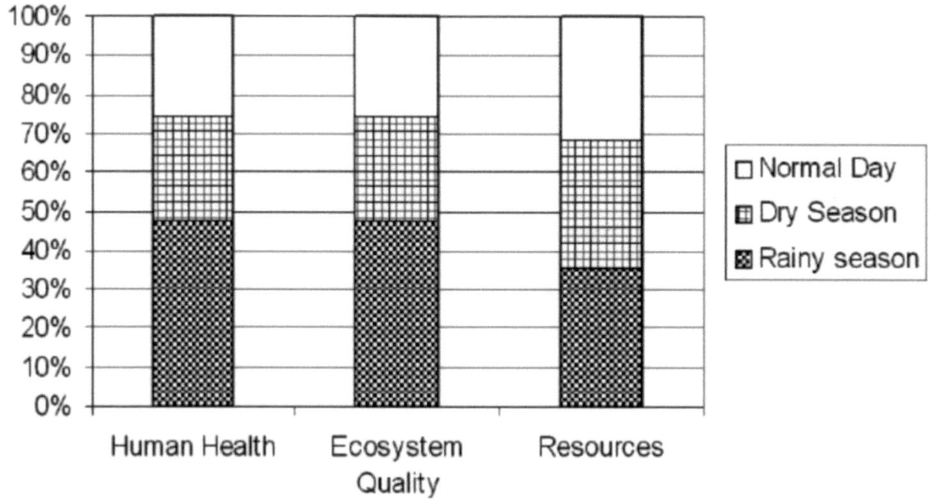

Figure 3. Characterization According to Three Protection Areas

Analysis is focused on three damage categories, namely, damage to human health, damage to ecosystem quality and damage to resource depletion. Analysis found that during the rainy season, the situation has the potential to cause higher damage to human health (0.0407 DALY) and ecosystem quality (1.92E3 PDF*m2yr) at about 46% each compared to the dry season and a normal day. However, during the dry season, even though the value is much lower than during the rainy season, it is still higher than a normal day, but the difference is not too large (the difference in value of 0.0014 DALY in Human Health damage and 0.07E3 PDF*m2yr in ecosystem quality damage). In damage to resources, the rainy season is still higher compared to the dry season and a normal day, but the contribution does not exceed 40%. Meanwhile, the dry season still contributes higher compared to a normal day. Values for the rainy season, dry season and a normal day are 578, 536 and 521 MJ surplus, respectively. Analysis shows that two main substances contribute to these three damage categories. The substances are polyaluminium chloride (PAC) and natural gas. The PAC chemical production process releases sulphur oxides and nitrogen oxides that contribute to human health and ecosystem quality damage, while electricity generation contributes to natural resources depletion damage.

Impact Category	Unit	Rainy Season	Dry Season	Normal Day
Carcinogens	DALY	7.16E-06	6.62E-06	6.4E-06
Resp. Organics	DALY	1.17E-07	1.06E-07	1.01E-07
Resp. Inorganics	DALY	0.040633	0.023199	0.02182
Climate Change	DALY	8.06E-05	7.5E-05	7.2E-05
Radiation	DALY	1.71E-08	1.85E-08	1.17E-08
Ozone Layer	DALY	4.79E-09	4.27E-09	4.2E-09
Ecotoxicity	PDF*m2yr	2.036848	1.871666	1.851406
Acidification/Eutrophication	PDF*m2yr	1915.167	1093.544	1028.461
Land Use	PDF*m2yr	0.053107	0.057419	0.036363
Minerals	MJ Surplus	0.011961	0.01291	0.008192
Fossil Fuels	MJ Surplus	577.824	535.6501	521.4128

Table 4. Characterization to Impact Category for Different Situations.

From impact analysis (refer **Table 4**), the rainy season surpassed other categories in all impact categories (carcinogens, respiratory organics, respiratory inorganics, climate change and ozone layer) in human health damage, except for the radiation impact category. For the radiation impact category, the rainy season falls into second place followed by a normal day. In the rest of the categories included in the damage to environmental quality category and natural resources depletion, the highest impact is contributed most by the rainy season (ecotoxicity, acidification/eutrophication, land use, minerals and fossil fuels.

2.3.4.2. Normalization

Normalization expresses the magnitude of the impact scores on a scale which is common to all the categories of impact. Impact scores and resource consumptions from characterization are related to a common reference in order to facilitate comparisons across impact categories (Huijbregts et al., 2003). The impact scores are usually expressed in person equivalents, PE. The PE represents the annual impact from an average person and is useful for bringing together the rather diverse environmental impacts on a common scale. Normalization is an optional element in the ISO 14040 series. The following are the results of the normalization analysis for the impact category obtained (refer **Table 5**)

Normalization shows the damage to human health quality as the main item. Impact to respiratory inorganic is pointed out as the main cause. The value for this impact category in the rainy season, the dry season and a normal day is 2.65, 1.51 and 6.6E-06 DALY, respectively. Meanwhile in the damage to ecosystem category, acidification/eutrophication is valued at 0.37, 0.21 and 0.20 for the rainy season, the dry season and a normal day, respectively. However, damage to ecosystem quality is in second place after damage to human health. The resource category is last in ranking as the impact contribution is the lowest among all the impact categories. For this damage, the fossil fuels impact category is higher compared to minerals. The contribution value for the rainy season, the dry season and a normal day is 0.07, 0.063 and 0.062, respectively. The rainy season remains the highest contributor for this impact category.

Impact Category	Rainy Season	Dry Season	Normal Day
Carcinogens	0.000466	0.000431	0.000417
Resp. Organics	7.64E-06	6.92E-06	6.6E-06
Resp. Inorganics	2.64518	1.510249	1.420503
Climate Change	0.005246	0.004881	0.004689
Radiation	1.11E-06	1.2E-06	7.61E-07
Ozone Layer	3.12E-07	2.78E-07	2.73E-07
Ecotoxicity	0.000397	0.000365	0.000361
Acidification/Eutrophication	0.373457	0.213241	0.20055
Land Use	1.04E-05	1.12E-05	7.09E-06
Minerals	1.42E-06	1.54E-06	9.75E-07
Fossil Fuels	0.068761	0.063742	0.062048

Table 5. Normalization to Impact Category for Different Situation.

2.3.4.3. Weighting

Weighting is also known as valuation. Weighting is the last step in LCIA where a ranking is undertaken of the different environmental impact categories and resource consumption, reflecting the relative importance they are assigned in the study (Soares et al., 2006; Pennington et al., 2004). The aim of this step is to arrive at a further interpretation and aggregation of the data of the impact assessment. The importance of the impact categories in relation to each other is a value-bound procedure based on an assessment of the relative environmental harm. This assessment will therefore reflect social values and preferences (Consoli et al., 1993). Weighting is another optional element.

Weighting analysis shows a result similar to normalization analysis where the first ranking remains with damage to human health followed by damage to ecosystem quality and damage to resources (refer **Figure 4**). Values for damage to human health are 795, 455 and 428 Pt for the rainy season, the dry season and a normal day, respectively. Damage to ecosystem quality is at second place with a difference between 340 Pt (rainy season), 369.4Pt (dry season) and 408 Pt (normal day) compared to the damage to human health category. Damage to resources category is in last place, indicating that this category is seen as contributing the least impact compared to damage to human health and ecosystem quality. Values for the three situations are 20.6 Pt (rainy season), 19.1Pt (dry season) and 18.6 Pt (normal day).

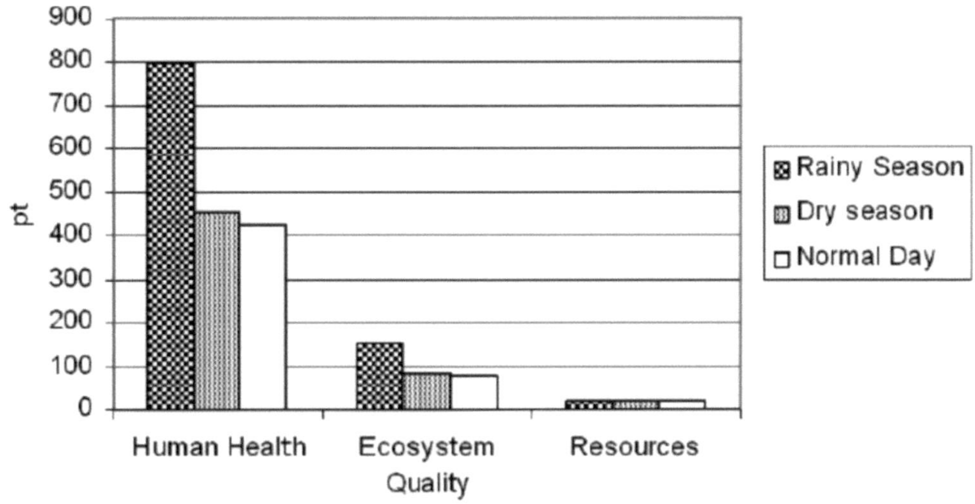

Figure 4. Weighting in Damage Assessment for Three Different Situations.

2.3.5. Life Cycle Assessment Interpretation (LCAI)

Interpretation is the phase of the LCA where the results of the other phase are interpreted according to the goal of the study using sensitivity and uncertainty analysis. The outcome of the interpretation may be a conclusion serving as a recommendation to the decision makers, who will normally consider the environmental and resource impacts together with other decision criteria (such as economic and social aspects) (Hauschild et al., 2005).

2.3.5.1. Improvement assessment

From the analysis conducted, There are two weaknesses identified:

1. Production of PAC: contributed to damage to human health quality and damage to ecosystem quality. The process in producing this chemical releases two other dangerous chemicals; nitrogen oxides and sulphur oxides.
2. Depletion of fossil fuel, namely, natural gas: natural gas is used in electricity generation.

These weaknesses could be overcome using more environmentally friendly alternatives such as:

a. Replacing PAC with alum. PAC is a coagulant that could be replaced with other chemical substance such as alum. In this case, the water treatment plant uses both coagulants in similar quantities. Thus, the suggestion is for the complete PAC replacement with alum to depict the impact of the replacement.
b. Natural resource depletion - natural gas: the existing advantages that water treatment plants have must be given attention and complete dependence in natural gas used as fuel for electricity generation is to be avoided. The main advantage is that the constant flow of water in water treatment plants could be used to generate electricity. Other than that, the location of water treatment plants usually exposed to solar radiation is also an

advantage that should not be taken lightly. The use of solar panels could assist in getting alternative electricity sources from this existing advantage. Thus, a suggestion made to reduce the impact of electricity generation is the use of solar panels (25%), hydroelectric generators (25%) and natural gas (50%).

Background data for solar panels and hydroelectrics in Simapro 7 software is used to compare the actual results (natural gas usage) with the suggested corrective measure using the combination electricity generation process. The results of the weighting analysis for the corrective suggestions are shown in **Figure 5**.

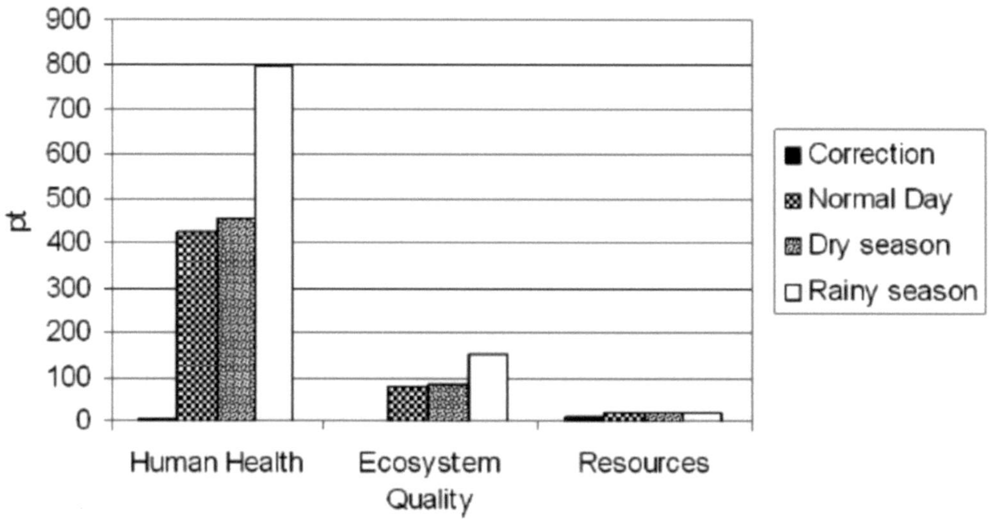

Figure 5. Weighting Analysis with Corrective Measures to Overcome Damage to Human Health, Damage to Ecosystem Quality and Damage to Resources

From the analysis conducted, both damage to human health and ecosystem quality can be reduced by more than 90%. For example, in the damage to human health category, the original value for the rainy season (795 Pt), the dry season (455 Pt) and a normal day (428 Pt) can be reduced to 2.65 Pt in total (the total of each three different situations). This is the same with damage to ecosystem quality, the original value for the rainy season (150 Pt), the dry season (85.4 Pt) and a normal day (80.4Pt) can be reduced to a mere 0.48 Pt in total (the total of each three different situations). However, the use of hybrid technology in electricity generation can reduce 50% of the damage to resources. The analysis also found that the use of hybrid technology could also potentially have an impact on human health and ecosystem quality.

3. Conclusion

The goal of this study is to analyse the damage that would happen from the use of chemical use and electricity generation based on two extreme climate conditions in Malaysia: the rainy season and the dry season. Both these climate conditions are also linked to world climate change which is a hot topic nowadays. Based on the analysis of environment impact using the Ecoindicator 99 evalution method, it is found that the rainy season and the dry season contribute higher to environmental damage, namely, damage to human health, damage to ecosystem quality and damage to resources if compared to normal days. The rainy season is found to contribute much higher impact compared to the dry season and a normal day. The dry season still contributes impact to the environment, but insignificantly compared to normal days. The consumption of alum, PAC and chlorine increases during the rainy season to ensure that treated water fits the standard set. The increase in these chemicals has the potential to cause environmental damage. The chemical that was identified to contribute to this damage is PAC. The production of this chemical releases two hazardous by-product chemicals, namely, nitrogen oxides and sulphur oxides. Nevertheless, the damage could be reduced by completely replacing the coagulant with a more environmentally friendly coagulant such as alum. After the replacement of the chemical, the burden on ecosystem quality and human health could be reduced by up to 80% -90%. Damage to natural resources depletion is caused by electricity generation using natural gas. Advantages at the water treatment plants could be fully utilized to avoid dependence on natural gas. This includes the running water in the water treatment system and water treatment plant locations that are exposed to solar radiation which could be harnessed to generate electricity. Hybrid electricity generation could reduce the complete dependence on fossil fuel natural gas in water treatment plants.

Even though the replacement of PAC with the more ecosystem friendly alum could reduce the damage to ecosystem quality and human health quality, the disadvantage of alum use is that it generates a high quantity of sludge. Though there are claims that sludge produced from water treatment plants is not dangerous compared to sludge produced by wastewater treatment plants, it can have negative effects on the environment, especially if this sludge is released directly into rivers, as is still currently practised by some water treatment plants in Malaysia. The proportion - 25% (solar panels): 25% (hydroelectric): 50% (natural gas) - suggested to reduce the dependence on natural gas might be improved further, as it is currently only able to reduce it by about 50%. However, if this suggestion is put into action, it would at least reduce the use of fossil fuel natural gas.

World climate change demands sustainable development practices. Without sustainable development in environmental management, damage to the environment is inevitable. Global warming is an example of this caused by anthropogenic activities. The results are disasters, such as frequent droughts and floods, that occur in Malaysia. This situation applies to physical disasters and also indirectly causes environmental damage when river water needs to be treated with higher chemical dosages during the rainy and dry seasons. By using LCA as the environmental assessment tool, environmental damage could be foreseen. Among the potential damage from these two extreme climates are damage to human health (especially respiratory inorganic impact), damage to ecosystem quality (especially acidification/eutrophication impact) and damage to resources (especially impact to fossil fuels).

Author details

Amir Hamzah Sharaai
Department of Environmental Management, Faculty of Environmental Studies, Universiti Putra Malaysia, Serdang, Malaysia

Noor Zalina Mahmood and Abdul Halim Sulaiman
Institute of Biological Sciences, Faculty of Science, University of Malaya, Kuala Lumpur, Malaysia

4. References

Angela, M., No Water Rationing in Klang Valley, in *The Sun*. 2002: Shah Alam.

Bakar, S.A., M.F. Yusof and W.S. Amly, *Johor hampir lumpuh*, in Utusan Online. 2007: Johor Baharu.

Bare, J. C. & Gloria, T. P. (2008). Environmental impact assessment taxonomy providing comprehensive coverage of midpoints, endpoints, damages, and areas of protection. *Journal of Cleaner Production*(16), 1021-1035.

Bare, J. C., Hofstetter, P., Pennington, D. W. & Udo de Haes, H. A. (2000). Life Cycle Impact Assessment Workshop Summary. Midpoints versus Endpoints: The Sacrifices and Benefits. *The International Journal of Life Cycle Assessment*, 5(6), 319 - 326.

Bovea, M. D. & Gallardo, A. (2006). The influence of impact assessment methods on materials selection for eco-design. *Materials and Design*, 27, 209-215.

BBC News, Malaysia floods test aid efforts, http://www.bbc.com/news/ in BBC News. 2007.

Bernama, Residents Advised To Be Cautious Of Inter-Monsoon Period, in *Bernama*. 2008: Kuala Lumpur.

Bovea, M.D. and A. Gallardo, 2006. The influence of impact assessment methods on materials selection for eco-design. *Materials and Design*, 27: 209-215.

Climate Ark. Malaysian state begins water rationing amid dry spell 2002 [cited 23 July 2008]; Available from: http://www.climateark.org/.

Clarke, J. (1998). Concrete Reinforced with Fibre Reinforced Plastic. *Materials World*, 6(2), 78-80.

Consoli, F., et al., 1993. Guideline for life cycle assessment: A code of practice. *Society of Environmental Toxicology and Chemistry workshop report*: Sesimbra, Portugal.

Dreyer, L. C., Niemann, A. L. & Hauschild, M. Z. (2003). Comparison of three different LCIA methods: EDIP97, CML2001 and Eco-indicator 99. Does it matter which one you choose? *The International Journal of Life Cycle Assessment*, 8(4), 191-200.

Goedkoop, M., Schryver, A. N. & Oele, M. (2007). Introduction to LCA with Simapro 7. Amersfoort: PRé Consultants.

Goedkoop, M. and R. Spriensma, 2001. *The Eco-indicator 99 - A damage oriented method for Life Cycle Assessment Methodology Report*. 3rd ed. BB Amersfoort: Pre Consultants. 132.

Guinée, J. B. (2002). *Handbook on Life Cycle Assessment: Operational Guide to the ISO Standards:* Springer, Netherlands

Hauschild, M., *Why LCA?* 2007, Department of Manufacturing Engineering and Management, Technical University of Denmark.

Huijbregts, M.A.J., et al., 2003. Normalisation figures for environmental life-cycle assessment The

Netherlands (1997/1998), Western Europe (1995) and the world (1990 and 1995). *Journal of Cleaner Production*, 11: 737-748.

Heijungs, R., Goedkoop, M., Struijs, J., Effting, S., Sevenster, M. & Huppes, G. (2003). *Towards a life cycle impact assessment method which comprises category indicators at the midpoint and the endpoint level. Report of the first project phase Design of the new method* [Electronic Version]. Retrieved 7 Mac 2007, from http://www.leidenuniv.nl/cml/ssp/publications/recipe_phase1.pdf

Hauschild, M., J. Jeswiet and L. Alting, 2005. *From Life Cycle Assessment to Sustainable Production: Status and Perspective. Annals of the CIRP*, 54/2/2005: 535-555.

ISO14000, *Malaysian standards handbook on environmental management*: MS ISO 14000 Series - 2nd Ed. 2000, Shah Alam, Malaysia: SIRIM.

Jolliet, O., Brent, A., Goedkoop, M., Itsubo, N., Mueller-Wenk, R., Peña, C., et al. (2003). *Life Cycle Impact Assessment Programme of the Life Cycle Initiative. Final report of the LCIA Definition study* [Electronic Version]. Retrieved 17 September 2007, from http://lcinitiative.unep.fr/includes/file.asp?site=lcinit&file=F7BF1ABF-8B98-4A95-9FDE-3E32EB7C4EC4

Jolliet, O., Margni, M., Charles, R., Humbert, S., Payet, J., Rebitzer, G., et al. (2003). IMPACT 2002+: A New Life Cycle Impact Assessment Methodology. *The International Journal of Life Cycle Assessment*, 8(6), 324-330.

Jolliet, O., Müller-Wenk, R., Bare, J., Brent, A., Goedkoop, M., Heijungs, R., et al. (2004). The LCIA midpoint-damage framework of the UNEP/SETAC life cycle initiative. *The International Journal of Life Cycle Assessment*, 9(6), 394-404.

Khan, F.I., V. Raveender and T. Husain, 2002. Effective Environmental Management Through Life Cycle Assessment. *Journal of Loss Prevention in the Process Industries*, 15: 455-466.

Mangena, S. J. & Brent, A. C. (2006). Application of a Life Cycle Impact Assessment framework to evaluate and compare environmental performances with economic values of supplied coal products. *Journal of Cleaner Production*(14), 1071-1084.

Miettinen, P. & Hamalainen, R. P. (1997). How to benefit from decision analysis in environmental life cycle assessment (LCA). *European Journal of Operational Research*, 102, 279-294.

Minister of Science Technology, 2000. and the Environment Malaysia, Malaysia Initial National Communication (INC). Minister of Science Technology and the Environment Malaysia: Putra Jaya,. p. 131.

Moberg, Å., Finnveden, G., Johansson, J. & Lind, P. (2005). Life cycle assessment of energy from solid waste-part 2: landfilling compared to other treatment methods. *Journal of Cleaner Production*, 13, 231-240.

Ntiamoah, A. & Afrane, G. (2008). Environmental impacts of cocoa production and processing in Ghana: life cycle assessment approach. *Journal of Cleaner Production*, 16, 1735-1740.

Pennington, D. W., Potting, J., Finnveden, G., Lindeijer, E., Jolliet, O., Rydberg, T., et al. (2004). Life cycle assessment part 2: Current impact assessment practice. *Environment International*, 30, 721-739.

Rahman, H.A. *A survey on a river pollution in Malaysia* in Geographic Conference 2007. 2007. UPSI

Sastry, C. A. (1996). *Water Treatment Plants*. New Delhi: Narosa Publishing House.

Sleeswijk, A. W., van Oersc, L. F. C. M., Guinée, J. B., Struijsd, J. & Huijbregtsb, M. A. J. (2008).

Normalisation in product life cycle assessment: An LCA of the global and European economic systems in the year 2000. *Science of the Total Environment*, 390, 227-240.

Soares, S. R., Toffoletto, L. & Deschenes, L. (2006). Development of weighting factors in the context of LCIA. *Journal of Cleaner Production*, 14 649-660.

Typhoon Utor to Blame, in *The Star*. 2006: Cyber Jaya.

Chapter 11

Integrated Water Management Around a Large Chemical Platform

Cristian Teodorescu, Margareta Nicolau, Adriana Cuciureanu,
Marinela Petrescu, Doina Guta, Ciprian Tetiu and Rodica Baciu

Additional information is available at the end of the chapter

http://dx.doi.org/10.5772/47436

1. Introduction

According to the EU Water Framework Directive (subsequently WFD) and its set of guidance documents [1,2], the main deadlines in implementing a coherent policy for water resources in the EU included:

a. By 2004 at the latest, each Member State had to produce an analysis (to be revised in 2013 and every 6 years afterwards) that must include:
   - an analysis of the characteristics of each river basin district;
   - a review of the impact of human activity on water and an economic analysis of water use;
   - a register of areas requiring special protection;
   - a survey of all bodies of water used for abstracting water for human consumption and producing more than 10 m^3/day or serving more than 50 persons.

b. By 2009, all Member States had to develop management plans for each river basin district, taking account of the results of the analyses and studies carried out in the previously mentioned analyses. These plans will cover the period until 2015, will be thoroughly revised in 2015 and then every 6 years thereafter.

c. The management plans must be implemented in 2012. These management plans aim specifically to:
   - prevent deterioration, enhance and restore bodies of surface water,
   - achieve good chemical status of water bodies by 2015 at the latest;
   - achieve good ecological status of water bodies, again, by 2015 at the latest, by eliminating / reducing pollution from discharges and emissions of hazardous substances;
   - protect, enhance and restore the status of all bodies of groundwater, prevent its pollution, and ensure a balance between groundwater abstraction and replenishment;

- preserve protected areas.

d. A specific mention is dedicated to how the management plans will be devised and implemented. The WFD asks Member States to encourage participation by all stakeholders in the implementation of this Framework-Directive, specifically with regard to the management plans for river basin districts.
e. Starting from 2010, Member States must establish water pricing policies that provide adequate incentives for users to efficiently use water resources and that the various economic sectors contribute to the recovery of the costs of water services, including those relating to the environment and resources. Member States must introduce arrangements to establish an effective, proportionate and dissuasive system of fines and penalties for the event of breaches of the provisions of WFD.

The WFD is a robust response that could be replicated in areas were irrational water consumption led to ecological disasters (e.g., the tragic fate of the Aral Sea [3]). But having a coherent legislation is only one part of the problem. Implementing it needs a dedicated work and collaboration among all stakeholders [4-8]. In taking decisions that affects the environment, the river basins ask for a high degree of responsibility and for a scientific, conservative approach [9], knowing that failing to carry out correct actions will lead to irreversible consequences at global level.

As an EU Member, Romania is committed to fully implement the WFD, and observe strictly all the deadlines in it. Adopting and inserting the WFD in the national legal framework was, by far, the easiest step in this process. As for implementing it, the number and dimensions of obstacles are important. Some of them are detailed below:

a. the Romanian industry inherited a very low effectiveness and efficiency because the majority of the industrial platforms, big water consumers, were built before 1990, when the concern about environment, resource conservation, water quality, etc., was limited. As a first consequence, industrial platforms could extract all the water they needed, sacrificing in many cases the habitat, the quality and quantity of groundwater, the needs of communities;
b. Very little attention was paid to polluting water bodies so that at present, in many areas, the rivers and the underground water content in fertilizers, insecticides, nitrates/nitrites and other chemicals, etc., is unacceptably high;
c. Though one may say that Romania has enough water resources for its industry and habitants, recent drought years revealed the dangers hidden in this assertion and asked for swift and coherent measures and plans in order to efficiently manage the existing water resources, without compromising the environment;
d. There is little or no co-ordination and communication among all stakeholders using water from the same river basin or from the same underground water body. Establishing an industrial facility, a hotel or touristic resort in a given location does not take into consideration the detailed, long-term impact impact upon the water resources of the area and the consequences for all other stakeholders already consuming water from the same sources;

e. At enterprise level, the water management is in almost all cases reactive, i.e., the management responds to penalties, fines or other enforcing measures taken by the Romanian Environmental Guard or similar enforcing bodies. Pro-active policies at enterprise level are scarce, uncoordinated with other stakeholders, inefficient since they include only the interest of a given company;
f. There is a lack of practical tools (mathematical models, data mining, benchmarking procedures), at enterprise level that can manage the water resources in the same way in which fuel or raw materials are managed. Though available, detailed and comprehensive, reference documents [10] are not, in many cases, used in day-by-day work on industrial platforms;

The paper presents the outcomes of a Project carried out in the period 2009-2011, having as general objective the implementation of the WFD at a chemical platform (including the surrounding areas and stakeholders).

Specific objectives were:

- Set up a task force, at local level that can take care of the WFD specifications and deadlines even after the Project will end;
- Know-how transfer to the mentioned task force (management tools, mathematical tools, specific indicators connected to all aspects of water management);
- Establish a mode of cooperation among all stakeholders interested in the sustainable use of the same water sources.

The Project was coordinated by the Romanian National R&D Institute for Industrial Ecology (INCD-ECOIND), Bucharest.

Its role in the Project was to:

- establish the focal area of the project;
- set up a Project Advisory Board that will guide the Project team and assess its results;
- carry out an interactive SWOT analyses, with the help of local experts and managers;
- devise, together with the local experts, material balances for water, in the focal area;
- set up a mathematical model for optimal use of water in the focal area;
- run and analyze various scenarios in order to test the model usefulness and validity.

Though some of the processes in the focal area were analyzed against best available techniques and some options for reducing water consumptions were generated, the main focus of the Project was to articulate a management system that includes all the important water users in the area and to optimize the water use at the level of this system. The optimization of each inner component of the structure of the system was not the objective of the present Project (only the VIROMET platform subsystems have been considered). The main reason for this limitation is that a mathematical model that could carry out the optimization of a system including many imbricate subsystems (companies and other water users) needs the knowledge of many technological parameters of each consumer. Many of these parameters are confidential. Nevertheless, by providing ranges of consumptions by each system component, companies that constitute the system studied could infer relevant conclusions about how big their share in the system is and how they could contribute to its sustainability.

2. The VIROMET Water Management System

The first task of the Project Team was to clearly establish the borders of the system to be scrutinized. This led to the Water Management System centred on the VIROMET platform (subsequently VIROMET WMS), illustrated in Figure 1.

Mainly, it includes:

1. multiple water intakes from adjacent rivers. The intakes are owned and managed by VIROMET;
2. a pre-treatment station that produces water for technological purposes (industrial water streams in Figure 1). The station is owned and operated by VIROMET;
3. s system of distributing the industrial water to all industrial stakeholders, namely:
    a. a ion-exchange resin factory (Platform B in Figure 1);
    b. several SMEs in the vicinity of the VIROMET platform;
    c. a station for producing drinking water, subsequently used internally by VIROMET or distributed to Platform B and other users, including small communities. The Station is owned and operated by VIROMET;
    d. multiple industrial installations, components of the VIROMET platform;
4. a wastewater treatment station that collects wastewater from a number of surrounding water users. The station is owned and operated by VIROMET. Wastewater from the SMEs that take industrial or drinking water from VIROMET WMS is not collected and processed by this station.

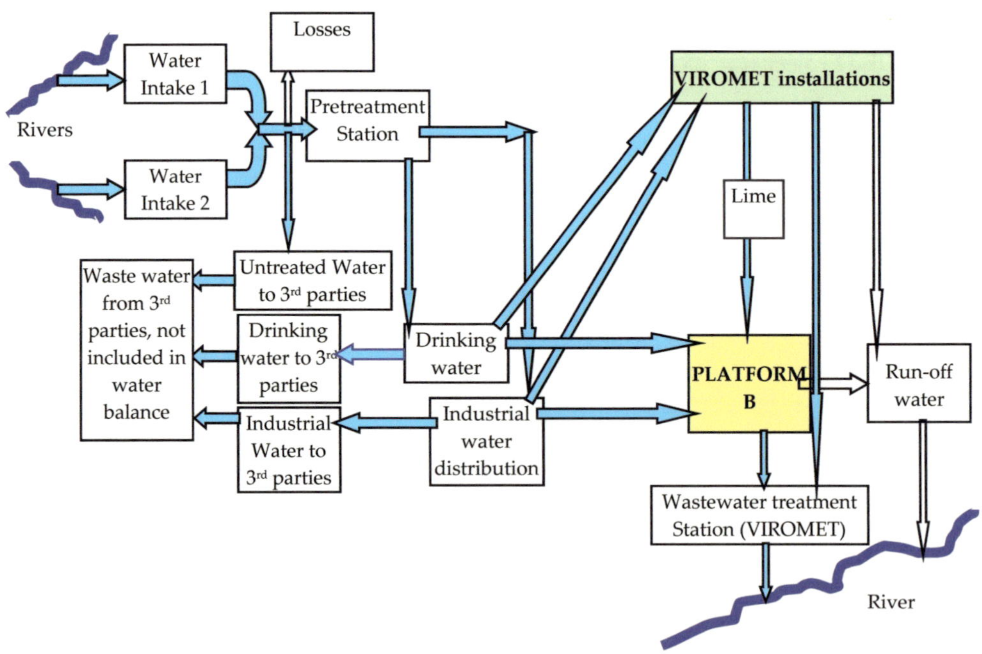

Figure 1. The VIROMET Water Management System structure. Dark arrows denotes measured or estimated flows

As a first remark, VIROMET is the main provider, the main consumer and the single wastewater operator in the area considered.

The water intake facilities, the pre-treatment and drinking water stations owned and operated by VIROMET provide 3 different qualities of water for end-users:

- water taken from the rivers, delivered as is to consumers;
- industrial water (for technological uses). This includes also the advanced purified water (deionised) used by local boilers.
- Drinking water distributed to local industrial consumers and to adjacent communities.

This being the situation, the main focal point of the Project was the VIROMET platform.

This is a very large chemical platform, founded by Czechs in 1937, then operated by Germans and used during WWII. After 1990, the portfolio of installations and products was constantly shrinking, mainly because the capacity of the platform was correlated to other industrial consumers that ceased to exist. At present, VIROMET is a private company, listed on the Bucharest Stock Exchange. It has some 850 employees (in 2009) and a turnover of 33 MEuro (2008). For the needs of the presdent project, the following main sections of the platform have been included in the study:

1. The Methanol installation;
2. The resin production facility;
3. The plastic ware installation (canisters, barrels, other similar products);
4. The R&D, small scale production installation;

3. The project task force

Following the WFD specifications, the Project aiming at establishing a performing water management system in the VIROMET area had to identify and include all relevant stakeholders and establish a procedure of consultation, cooperation and coordination among them, a rather new approach for the Romanian business environment, but a must, in the lines of the WFD.

The identified stakeholders were:

- The VIROMET Company, the main focal point of the Project;
- Another, adjacent chemical platform, producing adsorbents and ion-exchange resins – the second major water user (Platform B in Figure 1);
- Several SMEs interested in constant and good quality water feed from the VIROMET water intake and pre-treatment system;
- Adjacent small communities. It must be underlined that the location of the VIROMET platform is in the vicinity of the Fagaras Mountains, the tallest in Romania (+2500m), an area still covered by large virgin forests, with many protected zones for plants and animals. The beautiful landscape provides large resources for developing touristic activities in the area and, in fact, especially after the year 2000, many touristic facilities

have been commissioned (hotels, winter sports services), especially by private investors. This new and abrupt development of the zone, without accurately, objectively evaluating the short and long-term resource potential of the area, led to additional stresses upon the water resources (rivers, underground resources). All water management systems must therefore integrate these end-users, in trying to evaluate their contribution to the water footprint and limit, minimize and repair the damage done to water resources.

In order to accurately evaluate the situation of the water resources in the focal zone and to take into account the interests of all stakeholders, it is recommended that a Project Advisory Board (subsequently, PAB) be set up at the beginning of the Project.

This PAB has the following characteristics:

- It consists of recognized, uncontested personalities in the area (managers of large companies, reputed specialists, local mayors, retired experts). The PAB must enjoy an intrinsic legitimacy, unanimously accepted by the local people;
- Their role is to guide and censor the Project team members during the Project implementation by analysing the pertinence of their findings, their suggestions, their initiatives. In order to be accepted and implemented, these initiatives must not contradict the history, traditions, customs of the local people and communities, must preserve the specificity of the region; members of the PAB are believed to be in the best position to know exactly what is best for the local people, from the point of view of preserving water bodies and the beautiful surrounding environment;
- Members of the PAB are in no way remunerated by the Project. Their participation is fully voluntary but will impact upon the local people in the focal zone.

4. The existing situation. SWOT and DPSIR

A detailed SWOT analysis was carried out by the Project team. Main findings are included in Table 1. Issues in Table 1 are, in majority, available for all chemical platforms in Romania.

Nr	Internal factors		External factors	
	Strengths	**Weaknesses**	**Opportunities**	**Threats**
1.	The platform is in the vicinity of important river courses, easing the water intake operations	The pre-treatment technologies and equipment need upgrade. Consumptions in some modules of the system are not measured but only roughly estimated.	Restructuring the water management system will be made in line with last minute norms and most advanced technologies [10-12]	Climate change induces major events at an accelerated pace (drought, flooding). Retrofitting should take into account new conservative capabilities for water management.

Nr	Internal factors		External factors	
	Strengths	Weaknesses	Opportunities	Threats
2.	Water distribution and wastewater collection system have extra capacity, due to the shrinking of the activity of the platform. Existing extra capacity assures that there are enough places for expanding the industrial activity in the focal area.	Maintaining and operating a water distribution and wastewater collecting systems not adjusted to the exact needs of the users is very costly.	Any new investment in updating the water distribution and collecting network must be aligned to foreseeable trends in UE, about water resources so it will fit the long term requirements for the water networks.	There is little communication, coordination, no integrated management system of water resources in the focal zone. Uncorrelated use of water can lead to major disfunctionalities, especially in a drought year.
3.	The personnel operating the water and wastewater system is very experienced and skilful.	Existing water network cannot recycle water on the platform (no pumping facilities, no dedicated recycle piping). Practically, the water flows in a single direction, without recycling loops that could increase the efficiency of the system and reduce associated costs.	Opportunities to access EU funds for retrofitting the water system	Adjacent communities are exposed to risks caused by poor operation of the wastewater treatment facilities in the focal zone. Water and soil polluted with chemicals may affect large areas, reducing the touristic potential and gravely affecting the environment and public health.
4.		Many technologies (especially small scale ones, used in SMEs) in the focal zone are not aligned to BAT, mainly because the production level does not legally impose such an alignment. Consumption is therefore inefficient for many installations. This adds to stress upon the water resources and also increase the amount of wastewater to be treated.	Investment needed for updating the water and wastewater networks in the focal zone could be supported, partially, by other stakeholders, via some mutually beneficial arrangements with VIROMET (public-private ventures)	Environmental norms in Romania are aligned to the EU ones. But this reduces the profit of Romanian companies, by comparison with industrial facilities not observing these norms, situated in other geographical zones (outside EU, CIS, Asia, etc.)

Nr	Internal factors		External factors	
	Strengths	**Weaknesses**	**Opportunities**	**Threats**
5.		The parameters assured by the wastewater treatment station are difficult to monitor or modify, in case the legal limits for pollution will become stricter (instrumentation of the stations is in many instances inadequate).	The platform can offer technical assistance, maintenance and expertise to adjacent water users in the field of water distribution and wastewater collecting networks	Not having a sound water resource management system will add to the financial burden of the Company, since Banks become more interested in the level of environmental concern of their clients
6.		There are no identified and implemented options for water reuse outside the platform (irrigation, sport facilities, etc.). Neither the residues of the wastewater treatment station are used in any way.		There is a gap between enforcing agencies and enterprises. Instead of acting together, coherently, joining forces, learning one from another, they act as enemies. This does not add to the know-how and awareness of the companies.
7.		There is neither sound and comprehensive strategy nor specific managerial, operational tools that can lead to more effective and efficient water use (DPSIR evaluation [13], eco-efficiency indicators, environmental management accounting [14], etc.).		

Table 1. Main findings of the SWOT Analysis

The most coherent way to evaluate critically and in full detail the existing situation in the focal area is the DPSIR approach [13].

A multitude of DPSIR indicators were collected or derived (population, density, geographic area, climate, household area per person, unemployment, criminality, cars / 1000 inhabitants, sanitation, water supply characteristics, emissions to air, water, soil, etc.), but they will not be described here

The objective of using DPSIR was threefold:

- to test the potential of an analytical framework, recommended at EU level, in a practical case;
- to get the specialists of the focal zone used to a powerful tool, to exercise such an analysis whenever need be, not necessarily in the case of water management but in any environmental studies;
- to generate a basis for future DPSIR assessment that will evaluate the trend of the processes and pressures in the area.

For the structure and elements of the DPSIR approach, see Reference [13].

Essentially, the DPSIR tries to find objective responses to the following issues:

- *What is the state of waters in the focal zone?*
  - Quality? (Nutrients, pesticides, heavy metals, ecological quality)
  - Quantity? (Runoff, availability, demands, water stress)
- *Dynamics of water bodies*
  - Is the situation improving or getting worse, year by year?
  - Are the trends observed leading outside the legal limits?
- *What are the causes of the problems?*
  - Industrial
  - Agricultural
  - Human Communities
  - Pressures on the environment
- *Policies*
  - Are they the right ones, effectively and efficiently working towards targets?

The DPSIR evaluation carried out in the focal zone pointed out the aspects included in Table 2.

DPSIR element	Structure	Findings
Drivers	Industry	Large chemical platforms and diverse SMEs in the area operate technologies difficult to control (e.g. bakers using polluting additives, car service stations improperly disposing of paints, batteries, wastewater) The City of Victoria depends in great measure on the success and sustainability of the VIROMET platform, because a large share of the active population works for VIROMET. There are few other opportunities for employment in the area.
	Energy	VIROMET Power plant operates on gas
	Agriculture	Large areas of maize, potato, vegetables, wheat cultures, orchards
	Animal breeding	The area includes large pastures for livestock and is particularly renowned for the quality of dairy products from cows, sheep.
	Aquaculture	Trout breeding facilities commissioned, especially during the last decade

DPSIR element	Structure	Findings
	Households	The general trend is for population to move from agglomerated cities to a country side that is very far of being prepared to host them. Little or no sanitation or sewerage systems available, no environmentally sound landfill facility.
	Tourism	The area has a large touristic potential
Pressures	Climate change	Last 20 years included several drought years and flooding events as well in Romania but the focal zone was little affected
	Point source pollution	Accidental
	Diffuse source pollution	Mainly from households that irrationally use fertilizers and insecticides. Increasing awareness, education, trading must contribute to sort out this complex problem.
	Water abstraction	Not subject to any regulation or voluntary agreement by individual households
State	Water quantity	Could support the existing level of human activity, but not in case of extreme climatic events. All future development of the area must align to a coherent Action Plan
	Groundwater status	A large number of analyses carried out by ECOIND showed that the groundwater is polluted in many areas, especially with oil products, chemicals, fertilizers and the like. Nitrites-nitrates level too high.
	Ecological status	Acceptable, improving but needs permanent monitoring
	Chemical, physical, biological	Large industrial platforms and service companies (e.g., gasoline stations) have induced soil and underground water pollution in many areas
Impacts	Loss of habitats/species	Irrational deforestation led to reduced habitat areas
	Ill health	No reported illnesses specifically connected to the state of water
	Droughts/floods	Potential for drought or floods that would have grave consequences upon the communities
	Loss of amenity	Potential loss of amenity very high, due to incoherent development of the focal zone, especially of household buildings and touristic facilities
	Eutrophication	Present in some small rivers, near pollution sources
	Acidification	Present in some small rivers, near pollution sources
Responses	Water use restrictions	Little cooperation among stakeholders
	Alternative supplies	Need large investment. The capacity of the VIROMET water providing system seems enough for the period to come
	Subsidized water prices	Water intake from underground, by individual households not measure or paid for

DPSIR element	Structure	Findings
	Improved information	Needed by all stakeholders
	Demand side management	Not present
	Voluntary agreements	Not in operation
	Regional conflicts	No
	Waste water treatment	Yes, but only for major industrial water users. Small communities still wait for sanitation, in-expensive wastewater treatment facilities

Table 2. DPSIR findings.

The conclusions of the preliminary evaluation of the water resources in the focal zones are:

- The City of Vitoria and other small communities depend on the fate of the VIROMET platform
- Though the water sources seem to cover the present needs of the industrial activities and of the communities in the VIROMET WMS, there are obvious pressures upon this system, due to rapid development of the small communities and touristic facilities that must accommodate more and more people coming from large cities. Future industrial investment in the area will add more stress upon the water bodies under scrutiny;
- There is a need of coordination of action of all stakeholders, in order to implement a rational framework for water use that preserves the environment in a particularly sensitive area;
- "Historical" pollution (caused by industrial platforms operated before 1990, without any environmental concern) must be confined by taking special measures (barriers installed in soil, extracting and purifying the underground polluted water).
- There is plenty of room for improving the water management in the focal zone by reducing consumptions, by improving the industrial technologies operated in the area (especially in SMEs), by increasing the awareness and limiting the irrational use of fertilizers, pesticides, and the like;
- By applying the DPSIR framework of analysis, managers realize that they have at hand a procedure that indicates the causes, impacts and places of actions in order to improve the water system performances. The approach is straightforward and accessible, yet thorough and comprehensive.

The following section will show how the Project team tried to improve the water management in the focal area.

5. Internal benchmarking and alignment to BAT

Benchmarking is a management technique that consists of permanently searching and implementing of significantly better practices than those currently in use in a given area. It leads to better performances by investigating the performance and practices of other organisations (benchmark partners) [15].

Benchmarking is key to creating crisis that facilitate the change process in an enterprise.

Applying benchmarking techniques has the following advantages:

- Employees are able to perceive the improvement, in the short term;
- Increase responsibility of the personnel and challenges operational complacency;
- Points to weaknesses and suggest what actions are required to get better;
- Creates a motivating atmosphere leading to continuous progress;
- Generates a challenge to attain and go beyond the performances of competitors;
- It is a permanent source of realistic and achievable targets.

As can be seen, benchmarking uses comparison and reference to other benchmarking partners, but in a zone where the state of equipment and the level of technologies is not quite state-of-the-art, putting side by side the performances of benchmarking partners and the results of the company that is searching for improvement may be discouraging. It could become obvious that the improvement one is searching for is not possible without huge investment and would be achieved only after large time delays.

The Project team chose, instead, the internal benchmarking approach. As a first step, before comparing to others, the managers would have to identify the best performances recorded by the company itself. As those performances were recorded, they have been attained without any investment or new technologies, only by carefully operating the existing equipment, paying attention to all details, fully respecting the operating procedures. It is the task of company and installation managers to identify the best results and performances (benchmarks) achieved by the company and analyze how they were possible. By simply putting order in the operational stage, by motivating personnel to responsibly carry out their work, the company could align its performance to its own best level, already recorded.

To illustrate the procedure, in the case of VIROMET WMS, Figure 2 shows the production of the platform and the water consumption to achieve that production, all data expressed in % from average yearly values.

The conclusions of Figure 2 are obvious:

- the water consumption follows the water production profile but there are marked deviations from average;
- The specific consumption varies reaches a lowest 87% of the average value, instead of being practically constant, all the year long;
- The benchmarking emerging from this figure is the month of April, where the performances of the platform were the best;
- Analyzing the consumption data in connection with the production parameters and with all operational data recorded, managers will have to identify why the specific consumption could be as low as 87% of the average value. By simply identifying and

observing the conditions that led to the lower specific consumption, the platform will see its water specific consumptions decrease by 13%.

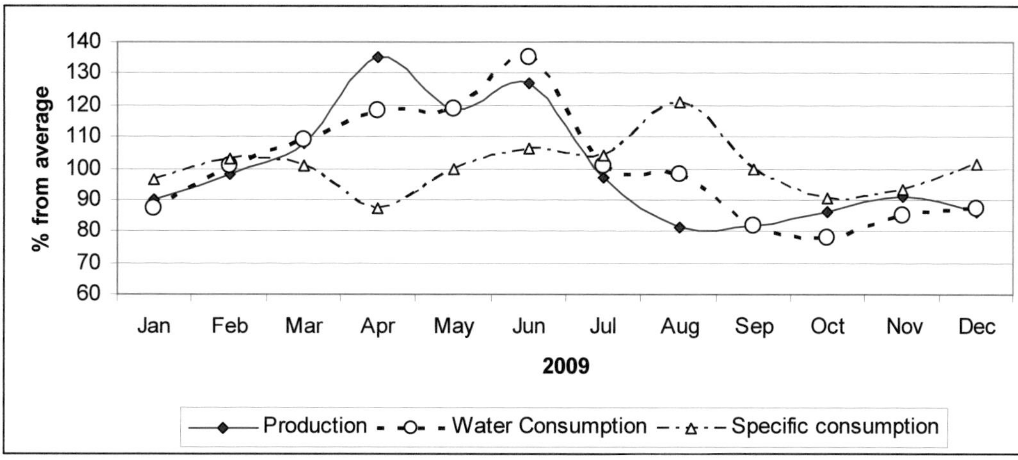

Figure 2. Monthly production and water specific consumptions.

Cleaner Production options generated during the comparison of the existing technologies with BAT produced a list of actions to be taken by the platform managers in order to improve the installation performances.

The list included:

- Tangible elements:
  - Proactive maintenance;
  - Use of piping and instrumentation from non-operational units to increase the performance of the operational ones;
  - Installing additional measuring devices in order to monitor the water consumed in places where there is poor instrumentation and control;
  - Replacing the ceramic bed in the absorption towers with more efficient packings;
  - Water flows that may be recycled on the platform, instead of being directed to the wastewater treatment facility. Though this will call for investment (piping, pumps, some heat-exchange equipment), it is estimated that this will lead to no more than 4-5 years return on investment;
  - Establish points of intake for monitoring the water quality in the VIROMET WMS;
  - Identifications of water flows that could be send to adjacent consumers (for irrigation purposes) instead to be sent directly to the wastewater treatment station or directly to the river.
- Intangible elements
  - Training the personnel, especially the operators of the water system;
  - Motivating the operators;
  - Direct R&D efforts toward reduction of water intake;
  - Encouraging innovative solutions from operators;

- Increase the know-how and expertise: ask personnel to assist operators in other installations, in order that everyone becomes aware of the problems confronting the entire platform.

6. Optimization of the Water Management System

In order to optimize the performances of the water management system at the focal point, the Project Team chose to develop a mathematical model comprehensive enough, so that its results are useful for the local managers but simple enough in order for the same managers to understand it, apply it and contribute to its improvement.

There are many software products that can design, model, simulate, and optimize water and wastewater systems at different scales [16-18]. Instead of using a software that comes as a black box that no one can access and understand how it works, the Project team chose to use the facilities and tools offered by the MICROSOFT EXCEL software, already existing in many companies and much more accessible and intuitive.

The MICROSOFT EXCEL software includes a mathematical package that can solve linear and non-linear programming problems, given a minimal set of data that must include:

- the optimization criterion (to be minimized, maximized or set equal to a prescribed value);
- the design variables (that must be adjusted in order to get the optimal value of the optimization criterion. The design variables are, e.g., water flowrates among different parts of the VIROMET WMS, productions of the VIROMET WMS components, etc.:
- the set of model parameters (e.g., cost of unit of water, cost of treatment of 1 m^3 wastewater, specific consumptions per physical unit of different products of the VIROMET WMS, etc.);
- a set of restrictions that take into account that production capacities are limited, piping transporting capacities are limited, water amounts needed by different users must observe some minimal and/or maximal levels, all water flows must be non-negative, etc.

The structure of the mathematical model is detailed in the subsequent paragraphs, for the case of minimizing the water consumed by the VIROMET WMS by linear programming [19].

The model is based upon the following assumptions:

- it considers and includes 4 sections of VIROMET (Methanol, R&D, Resin, Plastic ware)
- because of confidentiality reasons, the contribution of other water consumers is included in the mathematical model in the set of restrictions (e.g., a consumer will not accept a water flowrate under a given level and is not interested in buying water over a given level. These minimal and maximal levels are included in business contracts signed by VIROMET as this is the water provider of other stakeholders. In principle, the model could be easily expanded to include and optimize processes at any water

consumer in the focal zone, by simply adding extra terms (design variables) to the optimization criterion and by expanding the set of restrictions. Yet such a process needs a total transparency and access to data that are confidential at every partner in the water management system studied;

- cost levels are those available for the 2010 year but the MICROSOFT EXCEL application can be readily include new, adapted cost values without problem.

The application is set to minimize the following criterion (written in condensed format):

$$\min_{P_i} \sum_{i=1}^{4} P_i \times w_i$$

Here, P_i is the production of a given component (i) of the VIROMET WMS and w_i is the water consumed per 1 unit of product.

The set of restrictions are:

- Productions vary between two limits: a minimal threshold (generally = 0 but the managers may consider to adopt different minimal threshold, if, e.g., based upon past years, they are sure that they could sell a minimal production) and a maximal level (generally the production capacity of a given installation but, again, managers are free to choose any limits they consider suitable)

$$P_{i,\min} \le P_i \le P_{i,\max};$$

Here the index *min* and *max* refers to the already mentioned minimal, maximal thresholds for the productions

- In the case of other optimization criteria, different restrictions may be used. For example, the industrial water consumed will vary between a minimal value (e.g., 0) and the maximal capacity of the water transporting network or of the water treatment installation. The upper value could also include limits imposed by climatic events (the water amount will then be, for example, an acceptable fraction of the available river flowrate, fraction that will preserve the environment and ensure enough drinking water for the communities);
- Freshwater consumed on the platform will also obey similar restrictions:
- All variables are non-negative

$$P_i \ge 0.$$

As can be seen, there is no fixed structure for the optimization criteria and for the set of restrictions. In this way, a simple MICROSOFT EXCEL application is flexible enough to suit many cases of interest for the local managers / stakeholders.

Table 3 illustrates the simplest way to use the model: the managers wish to know the exact amount of water needed to produce a given quantity of a product, by one of the members of the water system. Such a facility of the application is necessary when there is an opportunity to sign an additional contract with a client, situation in which the managers should know what that contract will add to the existing structure of the water consumption.

Calculated parameter:	***Water consumed, m³***	***894380***
Productions, tons	**Value**	
Methanol	75000.00	
Water consumption	**Value, m³**	**Max available**
Industrial water	892757	1200000
Freshwater	1623	20000
Total	894380	
Water available	**Value, m³**	
Industrial water	307243	
Fresh water	18377	

Table 3. Scenario 1: amount of water needed for a given production

As seen from Table 2, the platform still have 307243 m^3 of industrial water and 18377m^3 of fresh water available for other fabrications so the additional methanol production will be possible without any problems.

Table 4 is the result of the application put to minimize the water consumed by the VIROMET WMS.

Optimized Criterion:	***Water consumption, m³***	***932435***
Productions, tons	**Value**	
Methanol	35000.00	Contracted
R&D	0.00	
Resins	40.00	Contracted
Plastic ware	8.00	Contracted
Water consumption	**Value, m³**	
Water sold as is	16000	Contracted
Freshwater to Platform B	75000	Contracted
Freshwater to 3rd parties	17000	Includes communities
Industrial water to Platform B	2900000	Minimal value
Industrial water to 3rd parties	750000	Minimal value
Water for VIROMET	**1678400**	Includes losses
Industrial Water to VIROMET	922016.368	Max = 1661616
Fresh water to VIROMET	10418.338	Max = 16616

Optimized Criterion:	***Water consumption, m³***	***932435***
Total Water to VIROMET	932434.706	
Water available	**Value, m³**	
Industrial water	739599.632	
Fresh water	6197.662	

Table 4. Minimized consumption of water.

The structure of the water consumption still takes account that signed contracts should be fulfilled. Other consumptions are at the minimal level provided by the restrictions.

Table 5 illustrates the maximization of the VIROMET production, given a limited amount of water available.

Optimized Criterion:	***Maximal production, Euro/yr***	***29260434***
Productions, tons	**Value**	**Remark**
Methanol	71480	Potential
R&D	30	Contracted
Resins	97745	Potential
Plastic ware	50	Contracted
Water consumption	**Value, m³**	**Remark**
Water sold as is	160000	Contracted
Freshwater to Platform B	5000	Contracted
Freshwater to 3rd parties	17000	Includes communities
Industrial water to Platform B	2900000	Minimal value
Industrial water to 3rd parties	750000	Minimal value
Water for VIROMET	1998400	Includes losses
Industrial Water to VIROMET	1978416	Max = 1978416
Fresh water to VIROMET	19784	Max = 19784
Total Water to VIROMET	1998200	
Water still available	**Value, m³**	
Industrial water	0.00	
Fresh water	0.00	

Table 5. Scenario 3: Maximizing production, given the water amount available.

In this case, the scenario includes two levels for already contracted products (for the R&D department and for the Plastic ware department). The production of all other components of the system is arranged so that highest profit possible is attained, all the available water is consumed and all the contracts with 3rd parties, observed. As can be seen, in the end, there is no industrial water or freshwater left, when the production is maximized.

This zero-level availability refers to the fact that the system cannot consume more water than the above amounts, though the adjacent rivers could provide for some extra quantities. But the restrictions were imposed in order to preserve the habitat. The environment could not sustain, in this case, larger water intakes by the water system studied.

Optimized Criterion:	***Maximal production, Euro/yr***	*12850587*
Productions, tons	**Value**	**Remarks**
Methanol	44682	Potential
R&D	30	Contracted
Resins	55471	Potential
Plastic ware	48	Potential
Water consumption	**Value, m^3**	**Remarks**
Water sold as is	100000	Contracted
Freshwater to Platform B	90000	Contracted
Freshwater to 3rd parties	20000	Includes communities
Industrial water to Platform B	2000000	A 50000o m^3 reduction to provide enough water to communities
Industrial water to 3rd parties	800000	Agreed level, for agricultural uses in the area
Water for VIROMET	1192000	Includes losses
Industrial Water to VIROMET	1180080	Max = 1180080
Fresh water to VIROMET	7499	Max = 7499
Total Water to VIROMET	1187570	
Water still available	**Value, m^3**	
Industrial water	0.00	
Fresh water	0.00	

Table 6. Drought year. Optimal production structure.

The scenario includes a drastic reduction of water (20%) for PLATFORM B and a major increase in the amount of water given to communities for agricultural purposes. In this case, the model finds the optimal production profile at VIROMET in order to reach the maximal possible production value, given the water restrictions. The zero-level availability refers, again, to the fact that the system cannot extract more water from the available resources than the above amounts without affecting the environment.

In this case, the set of restrictions is modified to take into account the limited water availability.

Though this is only a potential situation, it becomes clear that the new restrictions may gravely affect, in this scenario, the productions and benefits of one or more of the companies in the VIROMET WMS. Though the above scenario will probably not materialize in the years to come in the focal area, provisions must be made that such situations do not generate conflicts.

The simplest way in allocating scarce resources is that managers (and PAB) convene and add to each flow in the system a priority. The mathematical model should be modified such priority coefficients or weights and this can be easily done. But the use of priorities and allocation keys, even convened and accepted by all water consumers of the system does not solve the problem. Any allocation generates animosity and can lead to conflicts. Some of the consumers could leave the system, making all the management and optimization work pointless.

The solution should be mathematical as well as ethical. If some of the water consumers cannot use (or can use only a part of) this valuable resource for their processes, they cannot generate profit and cannot survive. Adopting a consensus policy and a co-operative approach to sort out the potential conflicts leads to a procedure to share the profits and benefits generated by those companies that consumed water, among all members of the water system. Though the procedure could be thought as unfeasible because it implies total and permanent transparency and frankness from all stakeholders, it works in the Israeli kibbutzim [20]. In order to build trust among all stakeholders, they must be able to verify the accounting registers of all those companies that have used water.

The last scenario presented illustrates the derivation of the optimal value of an eco-efficiency indicator, a new metrics in the focal area.

The criterion gives the minimal value of the m^3 of water consumed per monetary unit of production of the VIROMET WMS. The optimal value of the criterion and the production structure is presented in the Table 7.

Optimized Criterion:	***Minimal water consumed m^3 / 1000 euro production value***	***38.729***
Productions, tons	**Value**	**Remarks**
Methanol	32805	
R&D	268	
Resins	54723	
Plastic ware	34	
Water consumption	**Value, m^3**	**Remarks**

Water sold as is	250000	Contracted
Freshwater to Platform B	90000	Contracted
Freshwater to 3rd parties	20000	Includes communities
Industrial water to Platform B	2900000	Contracted
Industrial water to 3rd parties	30000	Contracted
Water for VIROMET	1872000	Includes losses
Industrial Water to VIROMET	930831	Max = 1853280
Fresh water to VIROMET	18532	Max = 18532
Total Water to VIROMET	949363	
Water still available	**Value, m^3**	
Industrial water	922449	
Fresh water	0.00	

Table 7. Computing an eco-efficiency indicator.

As can be seen, there is still some industrial water available but it cannot be used because the restriction imposed to the drinking water was reached. The model points to a limitation of the system that could cause problems. In order to avoid such circumstances, managers should look to expand the capacity of the fresh water production facility.

By calculating regularly such eco-indicators, the managers can appreciate if the performance of the platform is getting better and make the necessary adjustments.

A large number of similar scenarios and optimization criteria were tested, showing the capabilities of the model that still remains simple and versatile.

7. Forecasting

Knowing what will be the trends in a water system in the near future (3-6 months) is essential for the managers of the companies and consumers included in the VIROMET WMS. They can buy the necessary amounts of reagents, set up plans for maintenance, negotiate future contracts, evaluate the future risks that could hamper their production, etc.

The mathematical model of the system included also a forecasting procedure, also implemented in MICROSOFT EXCEL.

To illustrate the approach, Figure 3 shows the monthly amount of water entering the VIROMET wastewater treatment facility from all consumers connected to it.

First of all, moving average graphs are derived, using 2 values and 4 values in averaging, respectively (both situations are illustrated in Figure 3). It can be seen that going from 2 to 4 value moving averages does not sensibly improve the situation and that still no pattern seem to emerge.

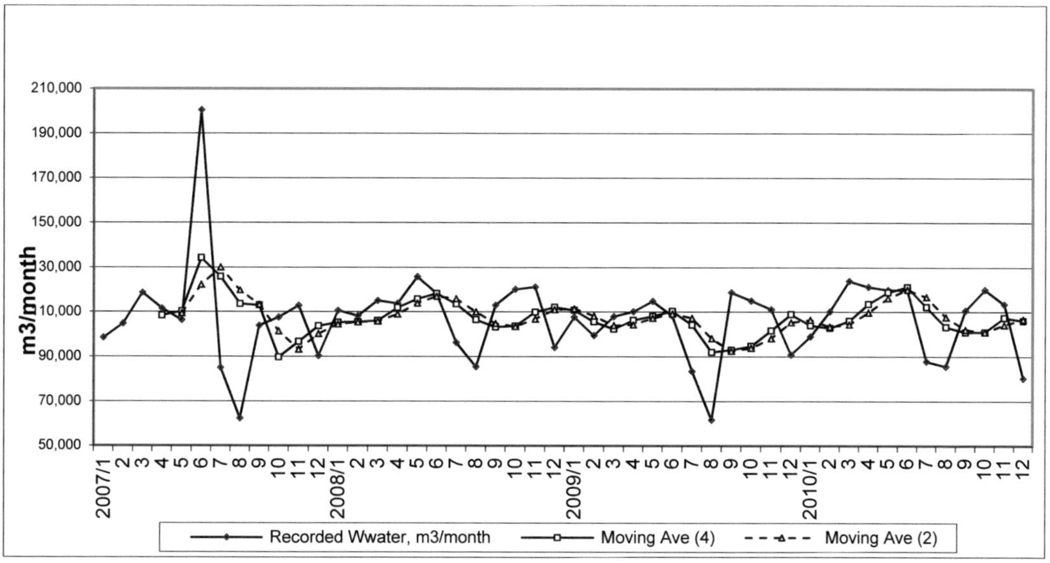

Figure 3. Actual and smoothed variation of wastewater treated in the VIROMET facility.

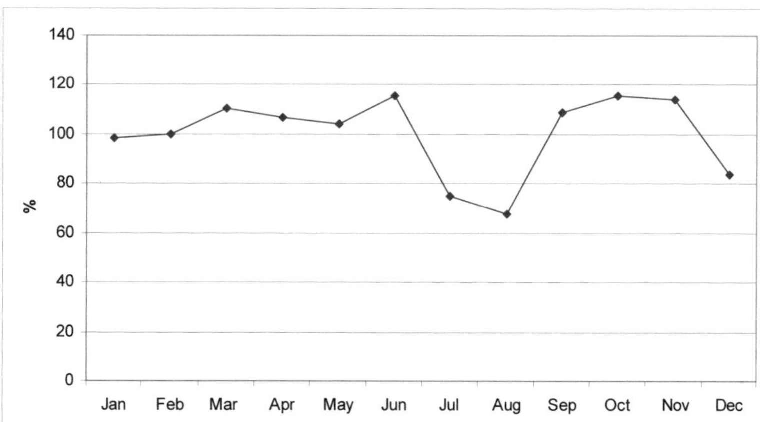

Figure 4. Seasonal Index for the wastewater reaching the treatment facility.

Apparently, there is no pattern governing the graph in Figure 3.

Applying a procedure for calculating a seasonal index for the amount of wastewater [21] leads, nevertheless to a conclusion, illustrated in Figure 4. Data used to infer such an index are taken from the years 2007-2010.

It seems that there is a gap in July and August as well as at the end of the year.

If the pattern stands, July –August or December would be the right periods to carry out revisions, technical repairs, and maintenance of the wastewater treatment plant.

March and October-November are the months when the wastewater facility works well over its annual average.

In order to test the seasonal index, it is used to predict the consumptions in the first half of 2011. Results are presented in Table 8.

Month	Predicted value	Actual value	Absolute deviation	Relative deviation, %
Jan 2011	105,672	114,091.00	-8,419	-7.38
Feb	107,696	113,637.00	-5,941	-5.23
Mar	118,487	124,225.00	-5,738	-4.62
Apr	114,587	121,675.00	-7,088	-5.83
May	112,175	122,339.00	-10,164	-8.31
Jun	124,466	121,953.00	2,513	2.06
Total	683,082	717,920	-34,838	-4.85

Table 8. Predicted and actual values for the wastewater reaching the treatment facility.

Deviations of actual recorded values from predicted values are, in all cases, less than 8.5%. For the entire first half of 2011, the relative error is only 4.85%. The seasonal index calculations are self-learning (additional data can be added at any time, once available, in order to make the index more accurate).

A large number of variables of the VIROMET WMS were subjected to forecasting in order to provide the local managers with useful decision support tools.

8. Conclusions

The study presented a methodology that assess and optimize an expanded system, including multiple water users and offers to local managers simple yet versatile tools to analyze various scenarios for their businesses.

A synthetic outline of the methodology follows:

1. Observe the legal framework governing a river basin but adopt a flexible approach, tailored to the area studied – legislation cannot include all the particular aspects of a given case. Stakeholders' actions should therefore be complementary and implement the general norms in their letter and in their spirit.
2. Organize the task force that will carry out the Project, identify stakeholders, set up a Project Advisory Board; it is important that all major consumers of water in the area are attracted in the Project;
3. Establish communication procedures, encourage transparency, dialogue, identify potential sources of conflict and be prepared to solve such conflicts. An external facilitator could prove essential for such instances since it is presumed that the facilitator has no interests in the area and his judgement will remain fair and objective;

4. Evaluate the existing situations: interactively carry a SWOT and a DPSIR analysis; refer to PAB and local managers to validate the results of these analyses;
5. Identify sources of reducing consumptions (cleaner production options, comparison to BAT, recycling possibilities, etc.);
6. Carry out internal benchmarking and establish the best achieved performance of the water system. Ask managers to identify and replicate the situations that led to that best performance; as a first step in a more elaborated data-mining process, this procedure will lead, in the short time and with practical no investment to sensible improvements;
7. Establish the exact structure and characteristics of the water system to be optimized (components, material balances, water quality analyses);
8. Build up a mathematical model for optimization using MICROSOFT EXCEL software package: interactively establish various relevant criteria, identify, in each case design variables, model parameters, set of restrictions. This operation should be carried out together with local managers and experts, they being the end-users of the model;
9. Test, together with local experts, the validity of the model by analysing as many scenarios as possible, simulate past known situations, examine all the possibilities of the model;
10. Update the key performance indicator system of the system by adding new metrics, e.g., eco-indicators that will assess the trend of the water system toward sustainability; include the new metrics in the strategy metrics operating at the water system level;
11. Help local people convene upon allocation keys for the case the water available is under the threshold needed by the system; establish a procedure for conflict resolution, in case some of the water users will not be able to access all the water they need for their processes; suggested approach: establish a way to distribute the profit accumulated by some of the stakeholders to all the members of the water system, in a transparent and fair manner;
12. Help local managers with other decision support tools, e.g., forecasting. Derive the seasonal index for as many variables in the system as possible, keeping the forecasting model open to include future available data;
13. Test procedures for expanding the mathematical model to include additional consumers.

The main advantages and disadvantages of the methodology and its are summarized below:

- The approach calls for support and participation of all stakeholders, in a given area, using the same water sources; if major water consumers are excluded from the system, the results of the optimization program will be grossly incorrect and generate major risks (e.g., in the situation when the water supplying system is under stress by climatic events);
- The characteristics of a given water system are better known and can be monitored in the long term by local experts so their involvement in the Project is essential. Setting up a Project Advisory Board could be a solution;
- This Project Advisory Board is of great help in assessing the accuracy of the assumptions made by the mathematical analysis, the precision of the set of restrictions

as well as the water allocation keys in cases this resource is scarce (drought, long and heavy winters);

- In case the set of restrictions imposed to some consumers affects their performance and profit, procedures should be in place to give them a fair compensation;
- The optimization model uses the MICROSOFT EXCEL capabilities, therefore it can be easily implemented on most computers;
- The forecasting procedures should be used for short-term planning not only in the case of water consumption but should be extended to all main operations and issues affecting the system and its components (power and fuel consumption, product demand, planning maintenance and revisions, etc.);
- Adding new components to the water system should be done only after evaluating (with the help of the mathematical model presented) the capacity of this water system to sustain new consumers. As an option, all members of the water system could agree to some new allocation rules;
- The parameters of the mathematical model (specific consumptions, restrictions) should be revised via a detailed work of implementing cleaner production and best available techniques options that will lead to reduced water consumptions by each consumer in the water system;
- In order to increase the relevance of the set of results obtained with the mathematical model, consumers should agree to disclose the performances of their internal processes so that they can be included in the optimization criteria. This calls for confidentiality arrangements among all the users of the mathematical model.

Author details

Cristian Teodorescu, Margareta Nicolau,
Adriana Cuciureanu, Marinela Petrescu and Doina Guta
INCD-ECOIND, Sector 6, Bucharest, Romania

Ciprian Tetiu and Rodica Baciu
VIROMET S.A., Victoria, Brasov County, Romania

9. References

[1] Directive 2000/60/EC (The Water Framework Directive) (2000), available at http://eur-lex.europa.eu/LexUriServ/LexUriServ.do?uri=CONSLEG:2000L0060:20011216:EN:PDF Accessed Feb 3, 2012.

[2] WFD Portal (2012), Guidance Documents, http://circa.europa.eu/Public/irc/env/wfd/library?l=/framework_directive/guidance_documents&vm=detailed&sb=Title Accessed April 15, 2012.

[3] Heaven, S. et al. (2002), "Water Resources Management in the Aral Basin: A River Basin Management Model for the Syr Darya", Irrigation and Drainage, 51(2): 109-118.

[4] Blackstock, K.L. et al (2012), "Linking Process to Outcomes – Internal and External Criteria for A Stakeolder Involvement In River Basin Management Planning", Ecological Economics (in press), available at http://www.sciencedirect.com/science/article/pii/S0921800912000791 Accessed April 15, 2012.

[5] Asian Develppment Bank Project #40687 Report (2008), "River Basin Water Resource Allocation and Management Policy".

[6] Watson, N. (2004), "Integrated River Basin management – A Case for Collaboration", Intl J. River Basin Management 2(4): 243-257.

[7] Biswas, A.K. (2004). "Integrated Water Resources Management: A Reassessment." *Water International* 29 (2): 248–56.

[8] Gallego-Ayala, J. and Juízo, D. (2011), "Strategic Implementation of Integrated Water Resources Management in Mozambique", Physics and Chemistry of the Earth, Parts A/B/C, 36(14-15): 1103-1111.

[9] Yuqiong Liu et al (2008), "Linking Science with Environmental Decision Making: Experiences from an Integrated Modeling Approach to Supporting Sustainable Water Resources Management", Environmental Modelling & Software, 23(7): 846-858.

[10] BAT/BREF, (2012), "Common Waste Water and Waste Gas Treatment/Management Systems in the Chemical Sector", European Commission, Joint Research Centre, Institute for Prospective Technological Studies. Seville, Spain, www.eippcb.jrc.es/reference/cww.html , Accessed Apr 15, 2012.

[11] Klemes, J.J.(2012), "Industrial Water Recycle-Reuse", Chemical Engineering (in press), available at http://ac.els-cdn.com/S2211339812000214/1-s2.0-S2211339812000214-main.pdf?_tid=11a573affe374f04f4e2d7c67eea5dc2&acdnat=1335276236_f52a24741b3899 5e754988d3f4bbab22 Accessed Apr 15, 2012.

[12] DIRECTIVE 2008/1/EC (The IPPC Directive), available at http://eur-lex.europa.eu/LexUriServ/LexUriServ.do?uri=OJ:L:2008:024:0008:0029:en:PDF Accessed Feb 3, 2012.

[13] Kristensen, P. (2012), The DPSIR Framework, available at http://enviro.lclark.edu:8002/rid=1145949501662_742777852_522/DPSIR%20Overview.p df Accessed Feb 1, 2012.

[14] Environmental Management Accounting (2012), available at http://www.unido.org/fileadmin/import/26164_EMApartIcropped.5.pdf accessed Feb 3, 2012.

[15] An Introduction to Benchmarking (2012), available at http://www.smthacker.co.uk/introduction_to_benchmarking.htm Accessed Feb 3, 2012.

[16] WEAP (2012) - Water Evaluation and Planning System Software. Details available at http://www.weap21.org/ Accessed Feb 3, 2012.

[17] Hydromantis Software (2012). Information available at http://www.hydromantis.com/products.html , Accessed Feb 3, 2012.

[18] Triana, E. et al (2010), "Neural Networks Approach to Stream Aquifer Modeling for Improved River Basin Management", Journal of Hydrology, 391(3-4): 235-247.

[19] Thie, P R and Keough G E (2010), "An Introduction to Linear Programming and Game Theory", 3rd Edition, Wiley & Sons, Chichester, New-York, USA, 460pp.
[20] Lipchin C. et al (2009), Integrated water resources management and security in the Middle East, Google Book, available at http://books.google.com/books?id=7rgHmpppZ-wC&pg=PA261&lpg=PA261&dq=kibbutz+water+management&source=bl&ots=E8vbgpiDrl&sig=zQEy_23nhs9PpP7xNCiGFFYthRA&hl=en&ei=-bPBSqe1KYue4QbGsqmLCA&sa=X&oi=book_result&ct=result&resnum=4#v=onepage&q=kibbutz%20water%20management&f=false Accessed March 2, 2012.
[21] Hall O P, Adelman, H E, (1991), "Business Statistics", Irwin, Boston, USA, 797pp.

Chapter 12

Knowledge-Based Development in Small Communities – Efficient Management Based on Local Expertize

Cristian Teodorescu, Lucian Constantin,
Margareta Nicolau, Aurelia Ballo and Cristiana Cosma

Additional information is available at the end of the chapter

http://dx.doi.org/10.5772/47437

1. Introduction

The "Lisbon Strategy" was adopted in 2000 by EU countries as a response to the challenges of globalisation and ageing. It set the strategic goal for EU "to become the most dynamic and competitive knowledge-based economy in the world by 2010, capable of sustainable economic growth with more and better jobs and greater social cohesion and respect for the environment" [1]. As the deadline approached, it became clearer that the above strategic goal will not be attained and the Strategy as a whole will be a failure. The main findings of a thorough analysis of this failure, carried out by EU officials were, among others [2]:

- The Lisbon Strategy has helped build broad consensus on the reforms that the EU needs and generated mutually acceptable solutions, still waiting to be implemented;
- It is not always possible to demonstrate a causal link between Lisbon reforms and EU economic growth and jobs outcomes, in the period 2000-2010. These results would be probably the same, without having a "Lisbon Strategy";
- The strategy should have been organised better to focus more on critical elements which played a key role in the origin of the economic crisis (risk in financial markets, speculative housing markets, credit-driven consumerism, wage increases outpacing productivity gains);
- The delivery gap between strategy commitments and actions has not been closed;
- Communication, awareness and public support for the objectives of the Strategy remained weak at EU and at national level;
- The non-binding character of the Lisbon Strategy contributed to its failure, and this lesson needed to be taken into account by the new Europe 2020 strategy.

The main idea emerging from the above remarks is that the Lisbon Strategy has failed not because of its generous, munificent objectives but because of inadequate operational implementation, lack of organizational measures and reforms, hesitation to generate and accept new solutions to the new challenges of globalization and economic crisis.

EU is now on the way to implement a new 10-year strategy, the "Europe 2020" [3].

There are three main priorities in the new EU targets for 2020:

- Smart growth: developing an economy based on knowledge and innovation. Again, the role of knowledge is the most important;
- Sustainable growth: promoting a more resource-efficient, greener and more competitive economy;
- Inclusive growth: fostering a high-employment economy delivering social and territorial cohesion.

In trying to give a substantial answer to the simple question: "where does EU want to be by 2020?" the European Commission proposes the following headline targets for 2020:

- 75 % of the population aged 20-64 should be employed;
- 3% of the EU's GDP should be invested in R&D;
- The "20/20/20" climate/energy targets should be met;
- Early school leavers should represent under 10% ;
- At least 40% of the younger generation should have a tertiary degree;
- 20 million less people should be at risk of poverty.
- The Commission proposes that EU goals are translated into national targets and trajectories that should become binding, one way or another for the Member Countries.
- There are seven flagship initiatives to catalyze progress under each priority theme:
- "Innovation Union" to improve framework conditions for research and innovation;
- "Youth on the move" to facilitate the entry of young people to the labour market;
- "A digital agenda for Europe" to speed up the roll-out of high-speed internet;
- "Resource efficient Europe" to help decouple economic growth from the use of resources;
- "An industrial policy for the globalization era" to improve the business environment, notably for SMEs;
- "An agenda for new skills and jobs" to modernise labour markets and empower people;
- "European platform against poverty".

Though these intentions show that the future instruments meant to implement the Europe 2020 Strategy may be better structured than those accompanying the Lisbon Strategy, they are mainly regarding the actions to be carried out at EU central level and by Member States. The approach for implementing the new strategy does not differ much from the Lisbon Strategy operational implementation: framing the general action plans by the Commission and asking Member States to take measures they believe appropriate, in line with the actions set up by the Commission. This is essentially a top-down, *ex-cathedra*, approach that may work when it is about labelling products, eliminating food additives, limiting the use of some pesticides, setting up new standards for the TV broadcasting but it certainly will not work when it comes to real people, to apply unavoidable austerity measures, to changing mentalities and modify habits in a 10-year interval, etc. Such an approach will probably encourage bureaucracy and corruption, asking for larger spending (from the money provided by the EU tax-payers) to prevent such consequences.

The present study recognizes and accepts the high-value of the general objectives included in the Europe 2020 Strategy but presents a case of a possible *bottom-up* approach in building of a knowledge-based EU, starting from the lower level, of rural communities. It presents the results of a Pilot Project carried out in the period 2009-2011 in a small rural area of Romania, in the Suceava County. The Project was led by the Romanian National R&D Institute for Industrial Ecology (INCD-ECOIND, Bucharest) that provided the facilitators, experts, laboratory infrastructure, background information.

The main lines of the present work are:

1. Identifying the driving forces for the Project and their particularities for Romania;
2. Defining essential concepts (development vs. growth);
3. A discussion of the sources of sustainable knowledge-based development;
4. Detailing the approach for implementing elements of knowledge-based development in the selected focal area: how to evaluate, mobilize and articulate the local expertize and energies to contribute to the Project;
5. Description of main results of the Project in 2 Projects addressing energy and environmental issues in the focal area.

The literature in the field is huge but it deals mainly with recognized centres of knowledge (urban environment where the intellectual capital is concentrated) and with knowledge-based development metrics [4-9]. Both aspects are important: the first, because it may illustrate more rapidly the benefits and the limits of knowledge-based development there where the intellectual capital is more active and reacts more swiftly, the second because knowledge-based development management does not have its recognized metrics, which is an absolutely essential tool to assess the correctness of the approach. Measuring, comparing knowledge and making decisions in these circumstances is difficult because the intangible, weightless character of knowledge [10, 12].

Though not as spectacular as new inventions or breathtaking discoveries, the Project demonstrates that there is an important amount of latent knowledge and expertise in small communities that could contribute, even in the short term, to the well-being of local people. It is important to create a friendly environment for dialogue and communication, to involve local people to the decision making process, to build on their ideas and perception, to generate communities of practice. This knowledge is not necessarily about advanced physics or nanotechnologies but can generate satisfaction and well-being, can contribute to a better life of communities.

2. Driving forces for the project

Aligning the Romanian society to the quality of life standards of other EU members requires much more than a bunch of legal norms, much more than Agencies and Commissions for

Development, Sustainability, Human Rights, Environment, Social Inclusions, etc., even if these institutions have a role to play and are aligned to the EU legal system, Agencies and Commissions.

Europe 2020 will succeed if and only if its advantages will reach, in the short term, every level of the society, if and only if every small community and its members will be convinced to act along the Europe 2020 objectives because they will be the first and most important beneficiaries.

What the Lisbon Strategy and the Europe 2020 lack is to stress that implementation calls for dedicated specialists not staying at a desk, setting up questionnaires, action plans, producing tons of deliverables and intensively using the internet capabilities but going down, innovating and generating new approaches for communicating, convincing, campaigning, working with people, understanding their needs, building on what they already have, respecting them, fighting bureaucracy, inertia, poverty. The Europe 2020 will be a success only if it will be more than a Strategy but a Crusade against poverty and exclusion.

In general for the Eastern Europe and in particular for Romania, now included in the EU, the last 50 year history presents a heavy legacy.

In its earliest stage, the Communist Regime destroyed the traditional social structures that proved their value and sustainability for centuries; it levelled the society and replaced old structures by central planning and governance. Small communities had not had to think – taking initiatives was considered insidious. They simply had to wait from the Central Government and from the Communist Party structures (the only legitimate depositaries of what is good and necessary) what they have to do. Before coming with Europe 2020 action plans in a Society that has lost its habits and capacity for self-governance and its appetite for change and innovation, it is important to adapt such action plans and take the right measures to re-build the necessary social bonds that will accept and implement the action plans.

Along the recent developments in anthropology research (e.g., the concept of "community of practice", coined by Wenger [13]), one can say that there is a lot of work to do in the field of re-structuring Romanian communities, in transforming them in live, dynamic, full-fledged organisms, aware of their capabilities and potential, ready to take action, to aggregate in communities of practice rather than of interest. This takes time and dedication.

3. Knowledge-based development of small communities

Development or growth?

While the Europe 2020 document includes in its title the term "growth", the term "development" would probably be more appropriate. The Strategy itself uses both terms: "sustainable growth" (starting with its title) and "sustainable development" [3, page 20] without making any difference and this can generate confusion.

Attaining a level of personal satisfaction and well-being does not necessarily reduce to growth and in many cases it does not imply growth at all. *Development* is a systemic policy intervention aiming at the economic, cultural, health, security and social well-being of people. *Growth* is connected mainly to market productivity and rise in GDP, being only one aspect of the process of economic development [14].

The findings of recent EU documents on sustainable consumption [15, 25] are relevant for the matter:

- An average European citizen uses about four times more resources than one in Africa and three times more than one in Asia (but half of one in USA, Canada or Australia).
- Resource use per person increased by 9.1% in the EU-27 between 2000 and 2007, reaching some 17 tonnes per person annually. Of the 8.2 billion tonnes of materials used in the EU in 2007, minerals and metals accounted for more than 50%, while fossil fuels and biomass were approximately 25% each.
- 87% of EU citizens agree that Europe could use its natural resources more efficiently, and 41% think that their household produces too much waste.
- The average floor area of dwellings increased from 81 to 87 m^2 since 1990, while the number of people per household decreased from 2.8 to 2.4.
- Europeans travel more kilometres by car. Although cars on average become more fuel-efficient, overall fuel consumption for private cars does barely go down, mainly because more kilometres are driven (rebound effects).
- An estimated 89 million tonnes of food ends up as waste each year in the EU (180 kg per citizen).
- In 2008, every citizen on average threw out 444 kg of household waste, and indirectly generated 5.2 tonnes of waste in the European economy.

It is hard to believe that the economic *growth* could be the solution to these facts. Using fewer resources, generating less waste, travelling less, building smaller houses, throwing away less food will not prompt up the growth of production in the corresponding industrial sectors. Even increasing the amount of recycled waste is not the right solution because the main problem is not to generate the same amount of waste and recycle as much as possible of it but generating less and less waste in the first place.

What EU should look for is a new state of mind that accepts that well-being does not mean growth in production and associated consumerism but longer-lasting appliances, repair services, less owning, more sharing options. Some of these options are unacceptable connotations of well-being, nowadays. Abundance, maybe – waste, no thanks!

The conclusion is that the term "growth" should be better defined or replaced by a more appropriate term (development).

The driving force for sustainable progress being the present distance of the Romanian Society to the correct and inspiring objectives included in the Lisbon and Europe 2020 Strategies (conveniently updated, as already discussed), the present study will detail a practitioner view, a *modus operandi* that will try to translate them in practice in a way that could inspire other practitioners and, perhaps, will suggest some modification of the Europe 2020 Strategy that refers explicitly to its operational implementation. Most documents and studies addressing KBD refers to where knowledge is supposed to be concentrated, (academic media, renown universities, large IT companies, laboratories for nanotechnologies, nuclear physics, space technology, large urban areas, etc.). It seems there is a gap between these entities and the communities, especially small communities. Knowledge seems to flow from the large knowledge generating structures mentioned above toward the rest of the society that has only to wait and enjoy the results of scientific studies and experiments carried out in R&D entities.

The authors believe that:

- There is a huge heritage and dowry of knowledge accumulated at small communities level. Human kind has progressed based on the findings of humble inventors lost in small communities;
- Pretending that small communities have to wait and apply what comes from R&D Institutes denies the potential of generating valuable knowledge by everyone;
- People believe in their own ideas and experience. Encouraging the generation of such ideas, collecting and disseminating such experience will add to the well-being of small communities even if those new ideas have little or no connections with space technology, nanoparticles or advanced ultrapure materials;
- Neglecting the know-how accumulated by small communities in their historical development in trying to implement the Europe 2020 Strategy would be a recipe for failure. Small communities know better what is good for them, are already adapted and react swiftly to emergencies, know how to build good, inexpensive houses, how to reuse materials, how to spare energy, how to manage local resources to enjoy them the next year too, etc.;
- The simple fact that a commune or a village exists since the 12th or 15th Century constitutes the most severe performance indicator and certificate for sustainability and defies any other 21st century metrics. All other levels of society, up to national government and EU structures have a lot to learn from the knowledge and life experience collected in small communities about how they managed to persist against all odds;
- Sustainable development must not deny, destroy what already exists and replace it with expedients and substitutes generated in some R&D facility or in EU *ad-hoc* Agencies but must build on what proved to be right and durable in every community;
- Inspiring solutions for the current economic crisis are certainly to be found by studying small communities;
- What Europe 2020 could add to what communities already know and are familiar too is the dimension of swifter change and introduction of modern trends and tools of science and technology to their existing way of life.

An adapted definition of sustainability, on what small communities have lived for centuries may be derived from the well-known Brundtland Report, could be: "Sustainable development is development that meets the community needs of the present without compromising the ability of future generations and of neighbouring communities to meet their own needs".

History and the present time is full of examples of small, powerful communities across Europe and in the Mediterranean space that sustain the remarks above, not only for entire entities (villages, communes, cities) but also for segments of larger cities [16]. Large EU-financed Projects are also directed to evaluate and build upon the know-how of small communities in managing a valuable resource like water, in Northern Africa [17].

A constructive study dedicated to what is the level of preparation of Croatia to align to the knowledge-based society [18], as illustrated in the new EU documents gives a very comprehensive definition of knowledge-based economy:

"A knowledge-based economy is one in which the generation and exploitation of knowledge play the predominant part in the creation of wealth. A knowledge economy is not an economy of scarcity, but rather of abundance because information and knowledge can be shared, and actually grow through application. A key component in a knowledge-based economy is human capital, or, more accurately, its competencies. In traditional industries most jobs require employees to learn how to perform routine functions, which, for the most part, remain constant over time. In the knowledge-based economy, rapid changes force workers constantly to acquire new skills and to update their skills throughout their lifetimes".

Knowledge translated in: applications, information, human capital, competencies – wherever these keywords characterize a community, there are good premises for well-being.

An interesting experience in how to develop a sustainable strategy for a local community [19], as a primary tool for common action, provides a number of questions discussed by local people, questions that give substance to the concept of knowledge-based approach:

- What is good about living in this area (Chichester, UK)?
- What would you like to improve in the area?
- Taking away the limitations that exist. What would be your ideal vision of your area?
- Bringing limitations back into play, what priorities would you see for your area?

Evaluating the experience of the mentioned communities in EU as well as from other part of the world [5, 6, 20] led to a handful of results expected by local people from knowledge-based development of their small communities:

1. Integrated communal services (water, sanitation, IT, cable TV, good education, health assistance, transportation, etc.), dependable and of high quality, acting proactively toward prevention of risk materialization by using communication, by protecting vulnerable people;
2. Sustainability, environmentally sound development, job creation inside the community;
3. Increase in the quality of life should maintain, preserve and develop the local specific that differentiates a given community from its neighbours;
4. An increased decision power and more resources allocated to communities. People want to take their fate in their own hands.

In Romania, a recent study addresses the possibilities of knowledge-based development at the strategic level [21]:

1. *Top-down strategies* that neglect the feelings and specific wishes of the subjects;
2. *Expert strategies,* generating purely technical solutions, identified by experts for given areas and communities;
3. *Bargaining strategies*, based upon negotiations, compromise among all stakeholders. This category includes the *"bottom-up approach"* by which local decision makers identify problems and barriers, call for consultations among all community members, collect ideas and solutions and decide which are the best ones by the same mechanism of bargaining;
4. *Heuristic strategies,* having a high degree of rational and affective content. They come out from dialogue, participation, in the presence of experts, with community members. Community members will feel that they themselves have discovered the right solutions to their own problems;
5. *Participatory strategies,* based essentially on voting. The obvious disadvantages are that in small communities, this voting process can be easily manipulated and that not all the members of the same community have the same education and can discern and choose the right option.

A mix of these approaches, adapted to a given community and a given business and cultural environment will probably be the best solution for a given case. The present study will use mainly the bargaining approach, with some help from the experts' and heuristic strategies.

4. Fundamental principles

In putting local know-how and expertise to work for the benefits of the community itself, the following principles should be observed:

1. Equity in distributing profits and benefits generated by the knowledge-based development. A special code of good practices should be in operation at EU level to encourage and motivate innovative producers in small communities;
2. The depositary of traditional know-how may be individuals or groups. They should be fairly acknowledged and rewarded;
3. The knowledge heritage of a community is dynamic, upgrading, adapting. This will help implementing rapid changes required by the Europe 2020 strategy. Yet changes should be understood and accepted by the community, before being implemented. This approach would be applied to tangible matters (local products, handicrafts) and to intangibles (habits, healthy life, institutions, attitude toward deviants, etc.);
4. Whenever possible, know-how and expertise of local people should be protected by patents or trade-marks;
5. The mechanism for transmitting knowledge and expertise to future generations should be improved and institutionalized. Good practices generated in a community should be disseminated. Elderly people, known for their skills, should be able to transfer their know-how to younger generations, e.g., in schools;

6. Feedback: all applications, improvements, adaptations of techniques and products inspired from the know-how of a community should come back and be presented to the same community;
7. Implementing knowledge-based development in small communities is a multidisciplinary endeavour; all the work needed will be carried out by a team of engineers, economists, ecologists, analysts with many years of experience in as many as diverse projects as possible.

5. Sources of sustainable knowledge based development

The study examined five sources of sustainable, knowledge-based development of small communities (subsequently detailed din Table 1):

- Material and energy resources of the community;
- The human factor;
- Intangibles active in the communities;
- The environment;
- An articulated, sustainable strategy.

Table 1 details also what could be the contribution of supplementary R&D work in order to assess and certify that traditional materials, products, techniques are in line with current environmental, health and safety or other similar regulations.

Category	Details	Contribution of R&D work	Remarks
Material and Energy Resources			
Crafts, adding value to local materials and products	New life to traditional skills (glass, leather, wood processing, painting, engraving, sculpture, dwelling, etc.) and use of local ceramic ware, traditional tableware, textiles and clothing, etc. Reviving the production and processing of flax, hemp, silk. Traditional dyes, detergents, chemical mixtures, drugs from plants, etc.	Characterization of materials and processes against present EU health and safety standards. Material, energy balances for processes, comparison to BAT. Suggestions for using renewables. Evaluate the waste generation processes and methods for waste use and disposal.	Traditional products still constitute curiosities at national or international fairs. Communication and dissemination will help their spreading, export. RISK: traditional industry could use newer, environmentally aggressive but more profitable techniques (electroplating, synthetic dyes, etc.)
Traditional farming and animal breeding	Identification of traditional methods in agriculture and animal farming Identify plants that do not need fertilizers and pesticides. Respecting the traditional calendar of agricultural works. Traditional methods for plant and animal protection.	Comparing the traditional approach to the principles of ecology, accepted by EU. Help and train local people and business to access EU funds	Codes of practices needed at EU level to protect and promote traditional agriculture and animal breeding, targeting SMEs and small communities, to develop bio-economy methods.

Category	Details	Contribution of R&D work	Remarks
Diet	Traditional diets, food and drinks	Analysis of products and scientific evaluation of diet. Comparison to quality standards required in EU. Encourage and support local community to acquire ISO, HACCP certifications and register their trade-marks.	Traditionally, food and drinks are produced by bio-techniques with no chemicals or additives. Promoting and disseminating activities needed
Tourism, commerce, other services	Pensions, hotels conserving the local specific. Gaming and fishing. Traditional occupations could serve to develop new sport activities (river rafting supervised by local people that usually take the timber down the rivers)	Services at EU levels of quality standards. Encourage local owners to apply for certification, quality auditing, etc.	Encourage associations. Local, foreign language speaking guides for tourists needed.
Traditional buildings and households	Techniques for increasing the energy efficiency. Arranging households along traditional experience and habits. Respecting the specific local architecture.	Evaluating energy efficiency and environmental footprints (Life Cycle Assessment, [26]).	Preserving what differentiates a community from its neighbours
Water management	Identification of traditional water sources, water management techniques leading to the protection of rivers, lakes, underground water.	Water quality analyses. Solutions for wastewater treatment and sanitation. Evaluating the potential of recycling wastewater.	The Water Framework Directive [22] and lessons learned from the EU Zer0 Project [17] should be carefully observed.
Energy management	Small communities could become self-sufficient energetically.	Identification of solutions for producing and saving energy	Larger projects (e.g., wind turbines or micro-hydropower stations) could ask for more than one community to be implemented
The Human Factor			
Health	Traditional medicine, practices, drugs	Scientifically assess the efficiency and risk of practices and drugs	Intellectual protection of products
Education	Complementary to schools, communities should transfer know-how and experience through traditional methods (apprenticeship, social events, fairs, Sunday schools in churches, etc.)	Assessing curriculum.	Mobilizing local learned people in the educational process. Kindergartens managed with the aid of elderly, educated people.

Category	Details	Contribution of R&D work	Remarks
INTANGIBLES			
Cultural Heritage	Habits, customs, religious and other traditional holidays. Conservation of institutions, ethnographic particularities.	Assessing the content of such heritage elements and keeping them clean from influences that would compromise their existence and value.	Reviving the authentic traditions and culture.
Participation, communication, social inclusion and cohesion	Traditionally, an Elderly Council, enjoying an intrinsic legitimacy overviewed and solved many conflict inside the community. Community Gatherings should be revived and given decision power.	Educated people from the community (teachers, doctors, priests, technicians, other specialists) should promote permanent dialogue inside the community.	Sanctions issued by communities should be complementary and not contradicting the legal conviction. Search for pro-active, preventing rather than coercive initiatives and actions
IT, GSM, Cable TV	Should not replace traditional direct contact of people, social events, gatherings, etc.	Identifying and using local skilled computer specialists to devise tools for e-governance.	IT, GSM, Cable TV should remain a valuable tool to promote and develop local specific
Environment			
Environmental Protection, biodiversity, climatic change	Traditional methods for sustainable management of forests, pastures, rivers. Evaluate traditional responses to disasters.	Help training local people. Assess emergency plans for environmental accidents.	Encouraging local ideas and solutions for zero-waste communities
Symbiosis	Re-build traditional connections for exchange of materials, services, products, expertise among all interested stakeholders	Solutions for turning waste from local or adjacent sources to valuable resources	Co-operation with other communities to be institutionalized
Strategy and tools			
Strategy	Establish objectives, priorities in accordance with the local specific	Transferring expertise for strategic management.	Strategy should be the result of local people initiatives.
Institutions	Guilds, religious gatherings, celebrations of some agricultural events, etc., should be complementary to existing institutions	Include them in the strategy; make them contributors to the social bond.	Old institutions are intrinsic sources of local legitimacy and should be used to govern local communities
Management	Traditional methods for consultation, option generation, decision making.	Assess their efficiency.	New managerial tools are complementary to older ones, familiar to local people.

Table 1. Sources of sustainable, knowledge based development of small communities

6. Project vision and objectives

The Project Vision was: "the identification of small communities needs for adding value to their human, material, scientific and cultural capital, in order to sustainably increase their quality of life, in harmony with the environment".

The operational objectives of the Project were:

1. Selecting the focal area, establish a Project Advisory Board (subsequently PAB) and carrying out a preliminary diagnostic, by using tools like SWOT, in the focal area;
2. Interactive generation of options for development;
3. Selecting a number of agreed projects that will demonstrate the advantages of the approach and the potential of knowledge-based development of the focal area;
4. Know-how transfer to the local people, complementary to what expertise is already present in the focal area. This led to the application of some modern tools for strategic management of small communities;
5. Evaluating, where possible, the success of the Project by comparison to national or EU practices, etc.

7. The focal area

The Focal Area of the Pilot Project covered the territory of several communes in the Suceava County, in the Northern part of Romania. Initially, the communes of Ilisesti and Balaceana were envisaged but, during the Project, it attracted a number of other communities in the same County that took part, more or less actively: Scheia, Ciprian Porumbescu, Veresti, and Stroiesti. Some of these communes have more than one village. The total number of inhabitants is estimated at 20000.

Once some local projects started, their immediate success acted like attractors, like a critical mass for some more communes. So, at the end of the project, there were 11 communes involved and the number is increasing.

A Project Advisory Board was set up that included mayors of the above mentioned communes but also specialists and even a priest. The Project co-ordinator (INCD-ECOIND-Bucharest) underlined that the Project should and shall stay absolutely free of any political involvement or connotation. The role of the PAB was essentially a honorific one – members of the PAB were in no way remunerated for their participation but their contribution was essential because:

- PAB members know best what is good for the local communities;
- They are aware of the traditions, ethnography, particularities of the area, how to preserve them, how to take the most out of them;
- They know who are the best local specialists, skilled persons that could contribute to a given part of the project;
- They have the authority and legitimacy to guide and censor the Project team on its way.
- They constitute the element of continuity, after the Project ends.

The stakeholders identified during the Project were:

- The Commune halls (mayors, top clerks);
- The Church;
- Teachers from the local schools and doctors from local hospitals;
- Bank subsidiaries present in the area that are in the first line of crediting local business for starting new projects;
- Successful businessmen
- Educated retired people.

The role of the Project coordinator was:

- to identify the focal area and to obtain the commitment of local policy makers;
- to carry out an in-depth diagnostic of the focal area;
- to process, together with local specialists the SWOT findings and to generate a structured list of development options aligned to what communities knows, needs, can do;
- to facilitate dialogue and contact among all stakeholders, to smoothen communication and solve conflicts, if need be;
- To develop some R&D work (chemical analyses, balances, cost-benefit assessment, test techniques against BAT, evaluate potential use of waste, etc.).

8. The SWOT analysis

Contrary to what someone could expect from a local community, lost in a beautiful landscape in the rural area of Bucovina (the historical name of the region), there was no need for instructing local people in order to carry out a meticulous and painstaking SWOT analysis. Young educated people took part with great interest in the action and produced high value documents for the initial diagnostic of the area.

Table 2 illustrates the SWOT analysis of the Ilisesti commune as it came from the local specialists. Little intervention has been made (elimination of some repetition, reformulation of some findings). Most of the suggestions of this Table can be found in SWOT analyses conducted in other communities so Table 2 may be considered representative for what happens in all the focal area.

Section 1: Agriculture, forests, rural development	
Strengths	**Weaknesses**
- Majority of population work in agriculture - Large available areas of the commune enable cultivation of an important number of plants (wheat, maize, rye, potatoes, vegetables, pastures, orchards); - Animal breeding has a long tradition;	- Poorly equipped farms; - No collecting and processing capacities for the local products; - No irrigation systems; - Insufficient development of service sector; - Farmers own very small pieces of land – cultures are fragmented; - No centres for artificial animal breeding;

Opportunities	**Threats**
- Room for association of local farmers in larger organizations; - EU funds available, with the support of the local Agricultural Chamber; - Potential for wind energy production to reduce the energy bill of farmers; - Large potential for bio-products.	- Increased competition makes difficult the way of local products to EU markets; - Unpredictable, constantly changing legal system; - No protection / encouraging measures for domestic agriculture, SMEs; - Low awareness about EU norms and potential cooperation;
Section 2: Infrastructure and Environment	
Strengths	**Weaknesses**
- Access to an important national road, DN17 - Nearest Airport at only 30 km distance; - Fuel distribution capacities available for domestic fuel and for cars; - Existence of a human dispensary, a veterinary clinic and a pharmacy; - Good quality timber and wooden products; - The largest (600000m3/yr) wooden waste-processing facilities in the Northern part of Romania are only 40 km away; - There is a wheat mill and a large bakery in the commune; - A state-of-the-art meat processing unit operates in the commune; - Relatively good quality, reliable INTERNET, GSM and Cable TV networks.	- No sewerage system and no wastewater treatment facility; - Natural gas network is not present in the area; - Road network need maintenance and repair; - River pollution; - Little interest of local business for environment protection; - Environmental education - inadequate; - There is no eco-landfill available - Health infrastructure does not cover the local needs; - Forests affected by irrational exploitation - led to landslides, floods; - Large amounts of wooden waste pollutes valleys, rivers, obstruct roads, cause floods; - No domestic waste recycled.
Opportunities	**Threats**
- Installing a water supply system and investing in sanitation and wastewater treatment; - High potential for waste recycling.	- Little experience in elaborating and management of EU financed Projects; - The zone is exposed to transboundary pollution (Chernobyl).
Section 3: Economic	
Strengths	**Weaknesses**
- Area renown for good, diverse food, good traditional cuisine - Ecological products developing rapidly - Animal breeding offers large quantities of animal products; - Highly skilled workforce.	- Inadequate infrastructure of food and drink industry; - No relevant foreign investments; - Degradation of potential industrial sites, built before 1990; - Little marketing activities
Opportunities	**Threats**
- The area is known as the land of potato. Could become a source of ethanol, as car fuel - Workshops and other infrastructure available for industrial development.	- The area is little known abroad, to foreign investors; - Little flexibility of local people to market needs; - Increased share of undeclared work.

Section 4: Tourism	
Strengths	**Weaknesses**
- Large number of small pensions, hotels - Good access to important monuments; - Many protected areas; - People renown for their hospitality - Probably the only private ZOO in Romania	- No funds for investment in touristic infrastructure - No personnel for touristic activities - Little promotion of touristic potential - No sport facilities for adults and children.
Opportunities	**Threats**
- New forms of relaxation (rafting, paragliding); - Cultural tourism potential.	- Investors prefer other areas (sea-side, higher mountains); - Limited marketing and promotion.
Section 5: Education and Culture	
Strengths	**Weaknesses**
- The commune has a large school and a kindergarten; - A large public library; - Many traditional cultural, religious events still active;	- The IT infrastructure in the school is inadequate - Many pupils leave school before completing the 8 grade stage.
Opportunities	**Threats**
- School rehabilitation with EU funds; - Motivating teachers to work for the commune, not only in schools;	- Large number of families disrupted (father and / or mother working abroad, leaves children to grandparents' care);

Table 2. SWOT Analysis findings.

The conclusions of the first stage in the SWOT analysis are:

- There is a marked interest for the Project, illustrated by the celerity and level of detail of the SWOT submitted by local specialists to the Project Team;
- There are good traditions that must be revived so that they will contribute to social cohesion (farming, traditional food and drink, habits, fairs, etc.);
- Great potential for profiting from local resources;
- There is no coherent strategy for sustainable development in the focal area;
- Communities could become self-sufficient (energy, resources) and provide other regions with ethanol from potatoes, wind energy, traditional products;
- Little has been made to protect and promote local products;
- Great potential for improving communication, dialogue, decision making processes;
- The environment constitutes a big problem;
- There are social aspects that need special attention (early-leave of the educational system, separation of families and parents going to work abroad).

9. Generating options for community development

All the SWOT tables from the focal area were consolidated with the contribution of local specialists and PAB. In the process, some issues were dropped, some others were moved from one category to another, some others were rephrased.

A single table resulted after intense interactive work and discussions. It was an excellent opportunity for the local specialists to meet and learn to work together for the benefit of the communities. This was one of the most important intangible benefits of the Project.

The SWOT Table was further re-organized in the following manner. First, local specialists were asked to give each SWOT finding 2 scores, in the interval 1 – 10:

- One that answers the question: "is the issue important for the local community?" (1 = least important; 10 = very important);
- The other, answering the question: "Have the local communities the capability and means to carry out the task of the SWOT issue?" (1 = no means available; 10 = all required means are available).

Knowledge that came with the Project co-ordinator was included in this score.

An aggregate score has been subsequently derived, based on the two scores given to each SWOT finding, using the formula:

$$Aggregate\ Score = \frac{2}{\dfrac{1}{Score\ 1} + \dfrac{1}{Score\ 2}}$$

The resulted aggregate scores are in the same interval 1 (worse) – 10 (best). The formula, inspired from chemical engineering (series of resistances to heat or mass transfer) ensures that a SWOT finding will result highly opportune and feasible only if both *Score 1* and *Score 2* are high, because the aggregate score calculated with this formula is less than the lowest values of *Score 1* and *Score 2*.

Using these aggregate scores, the SWOT findings, reformulated and detailed as options for local development during interactive analysis, were classified in 4 categories, in the same way the options for cleaner production are usually classified:

- *A-type* options that could be easily implemented in the short term and are relevant for the local communities
- *B-type* options, requiring some investment but being relevant for the community;
- *C-type* options, requiring large funding and large time horizon for implementation;
- *D-type* options, left aside for the time being.

These options are presented in Table 3. They are organized in 4 main chapters (infrastructure, profit generation, social cohesion, and institutional framework)

Table 3 includes some particularities:

- Local people know what they need and what could be the solutions for their problems;
- The concern of local specialists for energy and environment (core themes of any EU strategy) is obvious and well structured;
- Large number of business ideas shows that local people are at current with community potential, to sustainable products (e.g., ethanol for cars, collagen from bones, etc.).

	Infrastructure	Profit generating activities	Social Cohesion	Institutional framework
A options	Landscaping Keep public spaces clean and neat River Banks maintenance against floods	Marketing and promotion of local products, services Trade marks for local products Symbiotic connections for adding value to local materials and services Adding value to local resources	Local companies should hire local people first Creating jobs for people with disabilities, elderly Increase environmental awareness Collecting facts and objects illustrating commune history and specific	Devising a Sustainable Strategy Cooperation with all stakeholders, neighbouring communes Assisting families having members working abroad Church to become involved in solving social problems, fight criminal behaviour
B options	Water supply network Acquiring a special bulldozer for cleaning the snow on the roads Upgrading the IT infrastructure in schools, kindergartens Reforestation	Promoting the use of renewables as energy sources New processing facilities for agricultural and animal products (traditional products) Waste recycling	Expanding schools Building a retirement home and a facility for people with disabilities Educational programmes for adult and young people	Defining and registering the BRAND of the Commune Lobby activities at local and central level for promoting interests of communities. Commune Meetings to be revived and become legitimate critics of the local people
C options	Modernize the road network Large work along rivers for protection against floods Wastewater treatment station Ecological landfill	Add value to local resources: Fruit processing units, brewery, and trout breeding. Produce / distribute stoves, small scale boilers operating on wooden-chips. Valuables from waste (pet-food and collagen from animal waste) Incineration of waste with energy recovery and use in a greenhouse	Building a larger kindergarten Financial support for young families to build their houses.	Asking for the status of town Church and monasteries to create housing services for retired people to live and work.

	Infrastructure	Profit generating activities	Social Cohesion	Institutional framework
D options	Wind energy park Micro hydropower installations	Ethanol from potatoes Building a sport and entertainment park	Building a new Commune centre, a Museum of ethnography	Local e-governance

Table 3. Classification of options for community development

- The concern of the same specialists about how to add value to local resources is also a noticeable. There are all kind of options (A, B, C, and D) for generating benefits from natural resources. This shows that the limiting step to the economic, sustainable development of the communities is not the lack of ideas or expertise but the funding and an encouraging business environment and an articulated marketing policy;
- There are some domains not covered by local specialists: marketing, economists familiar with EU funding, ecologists, etc. This should be a signal of concern for high-schools and universities, to pay attention to real needs of communities and adapt their curricula accordingly;
- The social aspects are also important. Local communities understand the risks generated by parents leaving children and look abroad for work, the early leaving of schools or the increased number of criminal acts. Unfortunately, this constitutes a national concern as hundred of thousands of Romanians work abroad. There is no coherent approach to support families having 1-2 members far from home. The only good part of this situation is that these workers come home with a life experience, knowledge and some savings that could be of great help for the community;
- The institutional dimension is very well represented (Commune Meetings, Church to get more involved in the social life).

Based upon the mentioned aggregate scores, a short list of options emerged (Table 4).

No.	Option	Score	Remarks
1.	Devising the sustainable strategy for the communes	8.7	Helps local decision factors, specialists, SMEs to act coherently for the good of all the community.
2.	Promoting the use of renewables as energy sources	8.4	In line with the 20/20/20 targets of EU. Tries to find knowledge-based, efficient solutions to the energy sector, at the community scale
3.	Recycling domestic waste	8.1	Adds value to waste and implements the requirements of development decoupled from the use of resources
4.	Local companies should hire local people first	7.3	Important social consequences
5.	Increase environmental awareness	7.4	The SWOT has shown deficit in understanding and preventing environmental aggression.
6.	Assisting families having members working abroad	6.9	A social problems for hundred of thousands of families in Romania
7.	Trade marks for local products	6.8	Needs expertise and extended work for setting up the application for the trade mark

Table 4. The shortlist of development options for local communities

10. Implementation of the selected options

A-options have been adopted by local authorities and institutions and will be implemented in the near future. Their implementation does not need the help of the Project Coordinator.

The many business ideas (especially C-options) will constitute the priority for future business development in the area because they are generated and endorsed by local specialists and managers and, as the analyses carried out during the project, they are sustainable solutions for the problems in the communities. Working together for their implementation will test the value of the approach used to generate such options and confirm the importance of Wenger's "communities of practice" [13].

Option 5 needs a special training programme that will be devised by local specialists, teachers, retired experts, in order to identify and centralize all the environmental problems that confront each community, to analyze their consequences and to increase the awareness of local people. It is important that discussions should take place in the months to come with the County Environmental Protection Agency and the Local Environmental Guard.

Option 6 represents a very tough issue though its solution could be simple. Co-operation with County Authorities and with Child Protection Institutions is needed. Local families without children or single people have expressed their availability to take care of the children left alone by parents working in Spain, Italy or elsewhere but, though the problem is pressing (at national level several cases of suicides were recorded) all arrangements need a detailed case-by-case auditing and a formal, legal approval of child protection authorities.

Option 7 is the task of local experts that know best what differentiate local products from similar products of other areas.

In the subsequent paragraphs, the implementation of Options 1, 2, and 3 will be detailed.

11. Sustainable community strategy

The paragraph details how the sustainable strategy of local communities was set up during the Project.

In interacting with local experts and decision-making authorities, all the elements needed to set up a sustainable strategy for the community were detailed.

The building of the strategy started with adopting the Vision / Mission / Fundamental Values.

The Vision identifies what local authorities and stakeholders will value most about the community. Example of Vision phrases suggested to local authorities were:

- "our commune – history and tradition aligned to the 21st Century"
- "experience Bucovina as it once was"
- "our commune lives by the legitimacy of history, the energy of its people and the beauty of the surrounding landscape"

A Mission Statement defines what should be the community primary objective. Its prime function is internal and its prime audience, the community leaders and community representative people. The Mission Statement communicates what the local community represents and how would it contribute to its welfare.

Example: "We promote innovative and responsible initiatives that:

1. Will generate new economic opportunities based upon our heritage;
2. Will protect the environment for future generations;
3. Will encourage co-operation with our neighbours;
4. Will respect and value the dignity of elderly and helpless people.

The Fundamental Values of the Strategy (a "Constitution" of the commune) must come from what elderly people and PAB believe are the most important values (old, cherished, respected, shared by all)

The strategic objectives must come out from the SWOT and subsequent analyses and must be accepted by the majority of the local people.

Contribution and critics are more than welcome at this stage, in order to set up a list of objectives understood by all community representatives in the Project.

Strategic objectives must address the three pillars of sustainable development (economic, environmental, and social) and also be aligned to the 4 perspectives that constitute the horizon of local communities:

1. The Budget perspective;
2. The Stakeholders perspective;
3. The internal processes and capabilities perspective;
4. The learning and growth perspective.

These four perspectives have been taken from the celebrated *Balanced Scorecard* developed by Kaplan and Norton [24], completed to take into account the three pillars of sustainable development. Such an approach leads to a specific structure of the associated strategy map, illustrated in Table 5.

At the intersection of lines with columns one will find in each cell one or more strategic objectives for the sustainable development of the community. Table 5 already includes some examples of strategic objectives but the final ones will be devised by local communities, after intense consultation and interaction with the local people.

Arrows may be added to the map in Table 5, interconnecting objectives and showing how one issue determines the fulfilment of another.

Also, a system of classification can be adopted (e.g., 1 to 5 stars) and mentioned in the strategic map, denoting how important a strategic objective is.

It was stressed that the number of strategic objectives should be kept at a minimum possible (15-25). Of course, a community could have more than 25 objectives, addressing more specific matters but the majority of them could probably be added to the list of operational objectives.

		Economic	Social	Environmental
		Vision-Mission, Destination Statement		
Perspectives	***Budget***	Adding value to local resources Absorbing EU and similar funds	Add value to traditional habits, products, services, landscape, historical heritage	Benefits from waste Green community
Perspectives	***Stakeholders***	Public-private partnerships	Business community to invest, apply corporate social responsibility	Cooperation and common, focused action for a better environment
Perspectives	***Internal processes***	Identify and develop local expertise Motivate local skilled people to stay and work in the community	Conserving, reconstructing the social bond The Church to contribute to reduce criminal deviances	Decoupling development from resources.
Perspectives	***Learning & growth***	Continual education. e-Governance	Support elderly, helpless, people in need, children left alone by parents gone to work abroad	Awareness
		Economic	***Social***	***Environmental***
		3 pillars of sustainable development		

Table 5. Strategic map with examples of objectives for sustainable development.

Such a structure of the strategic map commits the experts and the policy makers at the community level to address all the essential aspects of their community, its structure, its connections, its capabilities and its future, but also the 3 factors that should be considered in any sustainable development process.

Once they chose to organize their strategic objectives in this way, these objectives must fill in every cell, in order to show the dedication of local decision makers to set up a comprehensive strategy that covers all the issues. Also, they must be relevant and make sense for the local people, answer to their expectations.

Filling in just a page, the map is a powerful vector for communicating the strategy, the intentions of local community representatives to all stakeholders, to all interested persons or organizations. It constitutes the most visible part of the strategy, submitted to public scrutiny.

Implementing the strategy calls for hard work directed to:

- The setting up of clear-cut policies at community level;
- Detailed actions plans meant to identify and mobilize all resources needed in the accomplishment of a given strategic objective;
- Targets and deadlines for each objectives;
- Responsibilities for people that carry out the strategy implementation.

If need be, some of these elements could be confidential or made available to a limited number of people (e.g., stipulations of some public-private contracts).

Each strategic objective should be accompanied by one or more key performance indicators (KPIs).

Devising a list of KPIs constitutes the most difficult and delicate operation in the implementation of the strategy. They form the metrics of the strategy, the essential tools for evaluating how progress is made.

For a sound system of KPIs:

- They must completely characterize the strategic process and its evolution;
- There must be a balance between leading and lagging indicators [24];
- Some of the KPIs must be agreed with neighbouring communities (e.g., those referring to the management and exploitation of rivers, pastures, forests, other natural resources, waste);
- KPIs must be simple to derive and must be based, if possible, on existing metrics available at the community level;
- They must be easily understood by most people;

KPIs must be accompanied by transparent targets and deadlines in order to assess the progress or the flaws in the strategy.

Examples of KPIs:

- Income from tourism;
- Number of pupils that have left the educational system;
- Area of polluted environment reclaimed;
- Income from waste.

A final form of the community strategy will be produced by local authorities after consulting and interacting with all stakeholders.

12. Promoting the use of renewables as energy sources

The paragraph illustrates the implementation of one sustainable development option, based on local resources and expertise and addressing the energy domain. Forest people and timber producing companies in the Suceava County generate huge amounts of wooden waste (trunks with no economic value, branches, bark, and sawdust). The local Forest Authority asked for help from local companies to solve the problem of wooden waste left in forests, along the rivers, roads, etc. This waste currently alters the state of the local environment, obstructs streams and cause floods or landslides. The sawdust modifies the Carbon/Nitrogen balance in waters and soil and induces modifications in the microorganism population, alters the quality of surface and ground waters.

During the Project, an opportunity for funding was identified, from a Norwegian Fund. With Norwegian help, a local company acquired a second-hand truck with a special crane that collects wooden waste from remote places. Local specialists managed to repair the equipment and put it back in operation, in excellent conditions. Thus, local competences helped local company to expand and add value to waste, a rather new business in the area. In addition, it solves an important environmental problem and reinserts in the economic chain a valuable resource (firewood), saving important quantities of virgin resources (wood is the main source of energy for local communities).

Appreciations came from HE the Norwegian Ambassador in Romania who visited the area.

Table 6 centralizes the volume of wooden waste collected in 12 months.

Jul 2010	451
Aug	315
Sep	544
Oct	1004
Nov	519
Dec	437
Jan 2011	184
Feb	381
Mar	569
Apr	290
May	694
Jun	271
Total Jul 2010-Jun 2011	5659

Table 6. Wooden waste collected, m^3 (July 2010 – June 2011).

The benefits generated by using wooden waste as fire wood or for producing briquettes from sawdust, at a local manufacturing unit are shown in Table 7.

Biomass as a fuel is a sustainable solution for the energy balance of local communities.

Characteristics	Value	Units
Total biomass collected	4527.2	Tons
Virgin resources saved (forests)	26	ha
Main briquette characteristics: Higher Calorific Value	4443	kcal/kg
VOC content	80.3	g/kg
Sulphur	0.02	g/kg
Ash (may be used as fertilizer)	0.43	g/kg
Fossil fuel replaced (spared): Methane	536	Tons

Characteristics	Value	Units
Lignite (1.5%S; 25% Ash)	1184	Tons
Fuel oil	688	Tons
Benign CO_2 generated by burning 4527.2 tons biomass replaces the CO_2 generated by the following amounts of fossil fuels: Methane	1524.31	Tons
coal (lignite)	3352.39	Tons
fuel oil	1946.7	Tons
SOx from 4527.2 tons biomass	0.064	Tons
SOx from equivalent coal (lignite)	35.52	Tons
SOx from equivalent fuel oil (0.5% S)	6.88	Tons
Ash from equivalent lignite (to landfill)	296	Tons
Ash from equivalent fuel oil (to landfill)	13.76	Tons
Social benefits: Jobs created	26	
Cost of 1 Gcal produced by burning biomass	50	Euro
Cost of 1 Gcal produced in power plants and delivered in the heating system	60-150	Euro

Table 7. Benefits of wooden waste reinsertion in the economic cycle.

13. Recycling domestic waste

In Romania, only some 1% of the collected domestic waste is recycled (compare to EU level: 25-28%). The paragraph illustrates how a sensitive issue was sorted out during the Project, based on local ideas, resources and expertize. During the implementation phase of the Project, a second-hand waste sorting station (manufactured in 1980) and a baler were identified and bought by a local company. The equipment needed capital repair and maintenance to become operational again but this was done by using the skills and ability of local specialists. It is worth noting that the mentioned equipment was the first of its kind in the area but repairing and maintaining has been carried out smoothly by local skilled technicians. Using the sorting station and the baler, domestic waste collected from the focal area (6 communes in the initial phase, 11 communes at the end of the Project) was sorted and prepared to be taken by recyclers.

Detailed discussions with local managers led to a modern solution of "upcycling" some of the waste (PET bottles) to fibres, instead of "downcycling" (incineration or conversion to lower quality goods).

Table 8 presents the benefits of recycling the domestic waste, in a public-private partnership, a novel approach for the focal area but a sound option in the view of the Europe 2020 strategy.

Table 9 shows the amounts of waste sent to recyclers in the first half of 2011.

The efforts of identifying funds, equipment, retrofitting it, starting a new business are rewarding, as Table 9 illustrates.

In addition, a simple benchmarking operation pointed out that the recycling rate of domestic waste in the focal area was not 15% but increased steadily and attained a 35%

figure in August 2012, 35 times more than the national recycling rate (1%). Currently, monthly recycling rates are 30-34%.

6 communes (13 villages) in the focal area	They produce approx 2000 m^3/month domestic waste. At 150kg/m^3 density, this means 3600 tons/yr
If 15% of the collected domestic waste is not sorted and segregated, communities have to pay 1 Euro/m^3 for the 15% share of the amount of waste sent to landfill	
If minimum 15% of waste is sorted, 36000 Euro taxes are not paid and remain in the Community budget	Sorted waste can be sold, leading to an income of 54000 Euro/yr. Communities save 36000 Euro taxes and earn 54000 Euro from selling sorted waste = 90000Euro/yr
	The local company that sorts and bales the waste earns extra 104000 Euro/yr (300 Euro/ton of baled waste).
Environmental benefits	At least 540 tons waste diverted from landfill and reinserted in the value chain
Social benefits	5 new jobs, healthier environment

Table 8. Benefits from recycling domestic waste.

Month (2011)	Cardboard and paper	PET	Plastic sheet	Metal	TOTAL	Remarks
Feb	7360	16580			23940	
Mar	4673	29290		1420	35383	Arrangements made with a metal recycler
Apr	6099	50680	2525	467	59771	A plastic sheet recycler was identified. More added value to collected plastic
May	2420	55020	920	1470	59830	Starting with May 2011 PET are segregated in "white" and "coloured", adding supplementary value to PET waste collected
Jun	42760	57340	22462	4540	127102	

Table 9. Waste sent to recyclers (kg)

Table 9 illustrates continual efforts by local specialists and managers to diversify, to identify new smart and simple sources for adding value to the collected waste. In some cases efforts are still on the way (e.g., glass, textiles) because either the recycling infrastructure in Romania cannot process some kind of waste, either the recycler is too far from the focal area (transportation costs are too high). This leaves the door open for local managers to start or expand their businesses, to value locally the recycled waste available.

14. Conclusions

The study showed the great potential of small community development based upon ideas, skills and efforts of the same communities, without waiting for miracle panacea from Central Authorities.

Knowledge-based development means a lot more than looking for and implementing last minute cutting-edge technologies. It means working together with local people, evaluating their problems, looking for local, up-to-date, smart and efficient solutions that contribute to the well-being of communities. The role of external facilitators (the Project coordinator, in this study) is to provoke a substantial dialogue, to suggest possible solutions, to assess the efficiency of the solutions generated, to mobilize all local specialists, to help finding financial support, to smoothen communication and co-operation among stakeholders.

Apart of some material successes (35% recycle rate of local domestic waste, 31 new jobs, more than 4500 tons of biomass reinserted in the value chain, saving the equivalent amounts of fossil fuels, etc.), the project generated the lines for future development of the communities:

- A structured sustainable strategy;
- The associated strategic management tools;
- A shortlist of business ideas and development options believed to the most relevant and most suitable for the resources and know-how existing in the area;
- A framework of cooperation and dialogue, essential for future replication.

The project underlined the need of meticulous, in-depth work and co-operation of all stakeholders, of all those called to implement the generous objectives like those in the Europe 2020 Strategy and proves that innovation should not be limited to high-speed, efficient electric cars, IT stuff or high-yield solar panels. The approach of implementing the Europe 2020 Strategy should also be innovative and original in what regards commitment, organization, resources used, and human touch.

Author details

Cristian Teodorescu, Lucian Constantin,
Margareta Nicolau, Aurelia Ballo and Cristiana Cosma
INCD-ECOIND, Sector 6, Bucharest, Romania

15. References

[1] The Lisbon Strategy (2000), available at http://europa.eu/legislation_summaries/education_training_youth/general_framework/c10241_en.htm accessed March 5, 2012.

[2] Lisbon Strategy Evaluation Document Commission Staff Working Document (2010), Available at http://ec.europa.eu/europe2020/pdf/lisbon_strategy_evaluation_en.pdf , accessed March 5, 2012.

[3] Europe 2020 (2010), A European Strategy for Smart, Sustainable and Inclusive Growth, Communication from the Commission. Available at http://eur-lex.europa.eu/LexUriServ/LexUriServ.do?uri=COM:2010:2020:FIN:EN:PDF , accessed March 5, 2012.
[4] Carillo, F.J., Batre S., (2012), "Understanding and Measurement: Perspectives on the Evolution of Knowledge-Based Development", International Journal of Knowledge-Based Development, 3(1): 1-16.
[5] Tan Yigitcanlar (2011), "Position Paper: Redefining Knowledge-Based Urban Development", International Journal of Knowledge-Based Development, 2(4): 340-356.
[6] Shrestha, K.K. and Z. Mahjabeen (2011), "Civic Science, Community Participation and Planning for Knowledge-Based Development: Analysis of Sydney Metropolitan Strategy", International Journal of Knowledge-Based Development, 2(4): 412-432.
[7] Badenhorst, Anne and Charlotte Scarf (2011), "Local action to support knowledge-based development", International Journal of Knowledge-Based Development, 2(3): 316-333.
[8] Mansell, R. et al (1999), "Opportunities for Knowledge-Based Development: Capabilities, Infrastructure, Investment and Policy", Science and Public Policy, 26(2): 91–100.
[9] Metaxiotis, K et al., "Knowledge-Based Development for Cities and Societies: Integrated Multi-Level Approaches", IGI Global (2010). Online book chapters available at http://www.igi-global.com/book/knowledge-based-development-cities-societies/37271&f=e-book, accessed April 15, 2012.
[10] Bandyopadhyay, S.(2009), "Knowledge-Based Economic Development: Mass Media and the Weightless Economy", The Toyota Centre, Suntory and Toyota International Centres for Economics and Related Disciplines, London School of Economics, available at http://sticerd.lse.ac.uk/dps/darp/DARP74.pdf accessed April 15, 2012.
[11] Gudes, O. et al.(2011), "Using a Knowledge-Based Approach: the Way Healthy Communities Make Decisions", Queensland University of Technology, Brisbane, available at http://eprints.qut.edu.au/46539/1/KCWS_2011_Proceeding_Gudess.pdf , accessed April 15, 2012.
[12] Becca, A. et al (2006), "The New Fundamentals in Planning for Local Economic Development: Knowledge-Based Communities", Georgia Institute of Technology, available at http://www.pled.gatech.edu/pages/PLED_Report/2006/PLED_Knowledge-Based_Communities_-_Fall_2006.pdf , accessed April 15, 2012.
[13] Wenger, E., "Community of Practice – A Brief Introduction", available at http://www.ewenger.com/theory/index.htm , accessed April 15, 2012.
[14] Sen, A (1983), "Which Way Now?", Economic Journal, Vol. 93(372): 745-762.
[15] "Our Common Future" (1987), available at http://www.un-documents.net/wced-ocf.htm, accessed March 5, 2012.
[16] Harrow Core Strategy Preferred Options (2006), available at http://www2.harrow.gov.uk/mgConvert2PDF.aspx?ID=63118&J=1 accessed March 5, 2012.

[17] Zer0 (2009), EU Zero Outflow Community Project: http://www.zer0-m.org/ accessed March 5, 2012.
[18] Bejakovic, P (2005), "How Prepared is Croatia for a Knowledge Based Society" Zagreb.
[19] Chichester Sustainable Community Strategy, available at http://www.chichesterinpartnership.org.uk/index.cfm?articleid=9287 , accessed March 5, 2012
[20] Bhargava, P.M., (2009), "How to Make India a Knowledge Based Society?" Furqan College, Hyderabad.
[21] Moldoveanu, G. and Dobrin C.(2008), Theoretical and Empirical Researches in Urban Management, 3(7):1-10.
[22] The Water Framework Directive (2000), available at http://ec.europa.eu/environment/water/water-framework/index_en.html, accessed March 5, 2012
[23] Zer0 (2009), EU Zero Outflow Community Project: http://www.zer0-m.org/ accessed March 5, 2012.
[24] The Balanced Scorecard Institute (2012), http://www.balancedscorecard.org/ accessed March 5, 2012.
[25] Sustainable Consumption in a Time of Crisis (2012), http://www.clickgreen.org.uk/analysis/business-analysis/123320-europe-looks-to-break-cycle-of-unsustainable-consumption.html , Accessed March 17, 2012.
[26] Life Cycle Thinking and Assessment (2012), http://lct.jrc.ec.europa.eu/ accessed April 15, 2012.

Chapter 13

Industrial Clusters and Environmental Management

Massimo Battaglia, Tiberio Daddi and Francesco Testa

Additional information is available at the end of the chapter

http://dx.doi.org/10.5772/45886

1. Introduction

The "territorial approach" to Environmental Management began in the European context at the end of the Nineties [1, 2, 3]. These experimentations showed the possibility of a new application of the EMAS (Reg. EC/761/2001, modified today in the EC/1221/2009) requirements to territorial contexts in which many similar SMEs are "aggregated", and not only to a single organization or productive site. This interpretation lead to applications in industrial areas (a group of industrial companies located in a limited and constraintable area), but was not applicable as such to a wider territorial cluster [3, 4].

This wider applicability of EMAS gave the possibility of considering the Regulation a new available policy tool able to integrate the environmental managerial needs of enterprises (and in particular SMEs, suffering of a lack of human, technical and economic resources for the application of an Environmental Management System) and the interest of a Local Government to improve the environmental performances of a governed area [1, 5, 6, 7, 8, 9]. The adoption of EMSs represents a relevant opportunity for SMEs facing with challenges of globalization processes and the increased (and increasing) social attention and sensitiveness towards environmental protection [10, 11, 12]; on the other side, a new approach to territorial management is needed to overcome the limits showed by the traditional policy tools available for Local Governments, such as the Agenda 21 Local processes [13].

In this perspective, a "territorial" approach based on EMAS can be considered as a new opportunity to integrate industrial, territorial and environmental policies; in particular, this approach gains a great importance if we consider specific territorial and productive systems, known as *industrial clusters*[1] *and districts*. These contexts are characterized by the presence of

[1] Clusters have been officially defined by the Final Report of European Commission Export Group on enterprise clusters and network as follows: *"groups of independent companies and associated institutions that are collaborating and*

a large number of SMEs operating in the same productive sector and characterized by specific technical and social relationships among private and public actors. The cluster "organizational structure" and entrepreneurial culture is characterized by the presence of systematic business and non-business relations among the local actors, sharing technical solutions for integrating the processes and technologies in a common value vision [14, 15, 16, 17, 18, 19]. Also in an "environmental management" perspective, within these territorial agglomerations the high concentration of similar pollutant production processes of industrial activities causes similar and increasing environmental problems: firms in an industrial cluster, because of operating in the same sector, affect in a cumulative way the same local ecosystem, get environmental pressures on the same targets, and interact with the same local communities and authorities. In order to limit and efficiently manage their impacts, enterprises (mostly SMEs) operating in industrial clusters tend to cooperate strictly amongst themselves and with the other local stakeholders (local authorities, local trade associations, and other local intermediary institutions) [1, 3, 20, 21, 22].

On the basis of these premises, a new model of diffusion of voluntary management tools has been developed (named cluster approach), involving also (but not only) the environmental dimension [23, 24]. Specifically, in this contribution the proposed methodology fosters a co-operative and integrated approach for environmental management at the cluster level, based on the relationships existing between territorial environmental performances and proximity between firms and other local actors and stakeholders. This approach encompasses the implementation of the different steps foreseen by the EMAS Regulation at the cluster level [3, 4]; in particular the foreseen phases are the following:

a. set up of a *Promotion Committee*, coordinating environmental management initiatives of the different local actors and driving all planning actions at the cluster level; this must be representative of all the relevant public and sectorial interests of the cluster
b. carrying out a *Cluster Environmental Review,* identifying and assessing the main environmental criticalities of the cluster, by a territorial, sectorial and perceptive points of view
c. definition by Committee, of a *Cluster Environmental Policy*, identifying the commitment towards the continual improvement of the environmental performances within the cluster; elaboration of a *Cluster Environmental Programme*, based on the results of the Cluster Environmental Review, containing the concrete and measurable commitments for carrying out strategic and high-priority intervention for the cluster
d. promotion of collective initiatives addressed to local actors (SMEs, Local Authorities, suppliers, etc.) aimed at satisfying the commitments of the shared Programme.

The first step concerns the *creation of a Promotion Committee*, able to co-ordinate and integrate environmental issues within the cluster. This Committee must favor the synergies and

competing; geographically concentrated in one or several regions, even though the cluster may have global extensions, specialized in a particular field, linked by common technologies and skills, either science-based or traditional; clusters can be either institutionalized (they have a proper cluster manager) or non-institutionalized. The cluster has a positive influence on: innovation and competitiveness, skill formation and information and growth and long-term business dynamics".

coordination among the different local actors. The Promotion Committee identifies the priorities, the main actions for the improvement and the possible ways of their implementation; it should express the main public and private interests existing in the cluster. The members of the Committee should be the representatives of all the relevant public bodies involved in the environmental regulation and control (such as the Local Government carrying out territorial planning and local Agencies for the Environmental Protection) and the representatives of SMEs operating in the local productive sector (such as trade associations or local consortia of enterprises), legitimated to act on their behalf as a whole. The Committee should fix specific rules concerning, at least, the appointment of its members, the decision making process and the organisational rules guaranteeing the active role of the Committee. These rules should be formalized in a voluntary agreement signed by the members of the Committee.

The second step consists in carrying out a *Cluster Environmental Review* by the Promotion Committee. It should represent a useful tool for SMEs operating in the cluster, to support them in the identification of the environmental aspects that are important for them. represent a useful tool for SMEs operating in the cluster, to support them in the identification of the environmental aspects that are important for them. An effective and satisfying Cluster Environmental Review should provide a thorough and in-depth description of the cluster territorial area, both from a geo-morphological point of view and from a socio – economic one, identify the main environmental problems of the area, analyse the local production processes, evaluate all the environmental aspects directly and indirectly linked with each phase of these processes, and, finally, provide an overall picture of the projects and initiatives carried out up to that time. The evaluation process aims at identifying the main critical environmental aspects characterizing the cluster area and measuring the contribution of the characterizing productive sector in terms of pressure exerted on the environment. From this point of view it's important to focus on the role of the Promotion Committee in fixing the criteria for evaluating the local environmental aspects and establishing the weight of each one in the final algorithm. Because of this, the evaluation process is the result of a multicriteria analysis [25, 26], in which the opinions of the participants of the Promotion Committee can influence significantly the final result and the following planning phase.

The following step in the "cluster approach" to environmental management concerns the drafting of a *Cluster Environmental Policy* that includes the commitment of all the main local actors towards the continuous improvement of the environmental performances within the cluster. This Policy is developed by the Promotion Committee and has to be strictly consistent with the territorial and sectorial contexts and based on the results emerged by the previous evaluation process. The Cluster Environmental Policy is the milestone for all the environmental actions to be carried out in the cluster, and the point of reference for the policies to be drafted by every company operating in the cluster interested to contribute to these actions. On the basis of the Environmental Policy, a *Cluster Environmental Programme* is subsequently elaborated by the Promotion Committee. The Programme details the commitments for carrying out strategic actions for environmental improvement in the cluster. Such a Programme establishes these actions at the operational level and contains the detailed planning for their implementation. The Programme should be continuously monitored, revised and updated.

The final step consists in the implementation of initiatives for a *Cluster Environmental Management*; in order to achieve the Programme targets and objectives, the Promotion Committee can promote and implement a number of managerial, technical and organizational initiatives addressed to local actors. These cluster initiatives should consist both in direct implementation of tools addressed to local SMEs aiming at facilitating their single adhesion to EMAS requirements, and in developing and promoting activities aiming at monitor the compliance to targets and actions detailed in the Cluster Environmental Programme. This process should be implemented such as an Environmental Management System adopted by a single organization, with an orientation to a cyclic PCDA process based on checking the planned results and on monitoring, year after year, the real improvement of the local environmental performances.

2. Description of the case study

The described methodological approach was experimented in a pilot project co-financed by the European Commission (Environment GD), length 28 months and ended on February 2006, involving the Lucca paper Cluster. This project, named PIONEER project – "Paper Industry Operating in Network: an Experiment for EMAS Revision" (www.life-pioneer.info), was financed within the LIFE Action Programme. The methodology of the PIONEER project encompassed the implementation of the different steps foreseen by the EMAS Regulation at the cluster level, so to create a common basis for facing the local environmental problems and supporting all the individual organisations operating in the cluster that intended to use collective resources to achieve an individual EMAS Registration [3, 4]. The involved territorial area was the paper industrial cluster of Lucca, in the Tuscany region. This area is extended on a geographical surface of 750 km2, including the territories governed by 12 Municipalities. More than 100 companies, located in this area, operate in the local sectorial supply-chain (most of which SMEs), with a high level of aggregation, a considerable density per km2 and with an occupational capability of more than 5.000 employees. In this area, concentrating more than 80% of the Italian production of tissue paper, the industrial activities are deeply rooted in the social and institutional local context, and the production sites are mixed and integrated with many other civil, commercial, logistic, administrative and services activities.

The first step was the set up of a Promotion Committee for EMAS, aiming at designing and implementing an Environmental Management System for the paper cluster. The members of the Lucca Cluster Committee are:

- *Lucca Province – Environmental Department*, expression of the public interests and representative of citizens; the Lucca Province was the applicant of the project, too.
- *Association of the Industrials in the Lucca Province*, expression of the private interests and representative of local paper producers

- *Coordination organism of the paper Lucca Cluster*, an NGO representative of local paper industry and expression of the Lucca cluster development policies.

Powers, responsibilities and functioning rules of the Committee have been fixed in an official statute signed by all its members.

After constituting the Promotion Committee and detailing the powers, its first action was the Cluster Environmental Review; it focused on three levels of analysis:

- *first*: identification and assessment of the environmental problems and territorial criticalities of the area (by use of a specific set of environmental territorial Performance Indicators)
- *second*: focused on the industrial activities characterizing the cluster and the environmental impacts connected with their productive processes (by use of a specific set of environmental sectorial Performance Indicators); in particular by means of specific questionnaires addressed to the paper firms located in the cluster, all their environmental performances were measured
- *third*: survey about the environmental perception of local communities (carried out by means of questionnaires and directed towards a representative sample of the adult population of the investigated area); the survey aimed at giving an analytic framework and an evaluation of the environmental problems perceived by the local communities, of the way these problems are perceived, of the expectations the local communities show towards the economic, social and institutional actors operating in the territory.

After the previous data and information collection, a methodology of evaluation was prepared. The criteria of evaluation were the following ones:

- level of legislative compliance in respect to national and EU standards and limits
- inter-temporal trend of selected environmental indicators
- territorial and sectorial benchmarking
- results of the survey about the environmental perception of local communities

These criteria were applied to the different levels of analysis, and by the integration of the results a list of significant cluster environmental aspects was drafted.

In the following paragraph a focus concerning the indicators used in the "Cluster Environmental Review" and the evaluation methodology will be proposed.

On the basis of the Analysis results, a Cluster Environmental Policy was defined by the Promotion Committee; this Policy has been shared by the Agenda 21 Local Forum and formally approved by all members of the Committee on the 24th of September 2004 .

On the basis of the principles fixed in the Policy, a Lucca Cluster Environmental Programme was defined by the Promotion Committee; this Programme included 17 objectives and 50 detailed targets and was formally approved on March 2005.

The Lucca Cluster Environmental Programme specifies the following items:

- *Objectives*: medium–term targets referred to the cluster territory, aiming at improving the Cluster most significant environmental aspects
- *Targets*: short–term targets, consistent with the objectives, that are quantified and measurable whenever possible, and clarifies the implementation responsibilities for the actors operating in the cluster
- *Actions and sub – actions*: the Programme design in detail the activities to be carried out in order to achieve the foreseen objectives and targets
- *Resources*: definition of the human, economic and technical resources that are necessary to carry out each action
- *Timetable*: specification of the deadlines for each objective and target
- *Actors involved*: definition of the local actors (institutional or private ones, single companies or their consortia, etc.) that should be involved in carrying out the described actions for each target

The Lucca Cluster Environmental Programme was formalized by one voluntary agreement among all the most representative actors of the cluster who will be in charge of the fulfilment of the targets and objectives.

The following step of the project regarded the implementation of initiatives for the Lucca Environmental Cluster Management System. The main actions developed during the project can be summarised as follows:

- definition, documentation and communication of the organisational structure and of responsibilities, in order to improve the environmental management of the Lucca industrial cluster and make it more effective
- identification of the possibilities to improve the collective and co-operative management of the industrial operations, commercial, logistic activities and services (environmental pressures) linked with the significant environmental aspects for the territorial cluster
- definition and adoption of managerial and organisational procedures that can be shared and diffused to the different local actors (such as: emergency plan for the whole cluster, system of qualification of the paper industry suppliers on the basis of environmentally sound criteria, definition and updating of a register of all the relevant environmental laws for the local production, etc.)
- training and information initiatives within the Cluster EMS addressed to private and public actors;
- definition of procedures for favouring stable communication flows and exchange of information among the local actors
- planning and carrying out of auditing initiatives at the cluster level addressed to the whole *Cluster Environmental Management System* and to sample organizations, with the aim of evaluating their efficiency and, above all, effectiveness (compliance to legislation, to environmental standard of management and to planned objectives and targets)

During all the previously described process, a very relevant role was represented by the definition of a basket of Cluster Key Performance Indicators; these indicators were, at first, defined in the first phase of the process (during the *Cluster Environmental Review*) by the Promotion Committee for evaluating the local environmental performances, and then they were inserted in the local environmental performance monitoring procedure. In the next paragraphs a particular attention will be focused on the process adopted in the PIONEER project by the Promotion Committee for selecting the indicators in the Cluster Environmental Review and for the identification of the most critical local environmental aspects (multi criteria methodology for the evaluation process); moreover a framework of environmental performances at local level measured by the selected indicators will be detailed.

3. Focus on cluster environmental indicators

The first phase of the process of adhesion to EMAS for a single organization is the occasion for evaluating the level of its impacts on environment and identifying the priorities of action in terms of improvement of environmental performances. In accordance with the EMAS Regulation requirements, in the Environmental Review each single organization has to analyze its territorial context and the environmental issues characterizing the area; moreover the single organization, for each phase of its productive process, has to identify the connected environmental aspects and to evaluate their significance on the basis of specific defined evaluation criteria. These criteria have to be objective, repeatable and they have to consider the views of interested parties.

These requirements, in the cluster approach to EMAS, can be satisfied by the application of a specific objective methodology, shared by the members of the Promotion Committee, able to interpret the environmental significant aspects from territorial and sectorial points of view (evaluated also on the basis of the points of view of the stakeholders) [27, 28].

The first step of the PIONEER approach was a detailed environmental analysis of the local territorial context and subsequent selection of State indicators representative of the whole cluster area; the considered territorial aspects were: quality of air, water pollution, water availability, levels of wastes production, level of energy consumption, biodiversity, traffic and mobility, soil pollution, electromagnetism, noise pollution. For each aspect a specific set of performance indicators was selected; this selection was carried out by the Promotion Committee on the basis of the existing methodologies of environmental reporting and analysis (OCSE, United Nations, European Commission, Global Reporting Initiatives) [29].

Table 1 shows the territorial indicators used in the PIONEER project (with indication of the environmental issue and its unit of measurement).

These indicators were used by the Promotion Committee for drafting the Cluster Environmental Review; later on, in the phase of implementation of the cluster managerial activities, they were inserted in the Cluster Monitoring System (with a triennial term for monitoring them).

At the same time, the second level analysis started; a data collection process was implemented by the Promotion Committee with the aim of selecting a basket of sectorial performance indicators.

Issue	Selected indicators	Unit of measurement
Quality of air	- Annual average concentrations : CO, NO_2, PM10, O_3 (last three years)	CO: mg/mc ; NO_2: µg/mc PM10: µg/mc ; O_3: µg/mc
	- N° of events exceeding the levels of standard quality (for each year): CO, NO_2, PM10, O_3	Number per year
	- Biomonitoring process results: *Index of Air Purity*	% of territory with IAP < 25,5
	- Emission on air (CO, PTS, NOx)	T/kmq
Noise	- n°of municipalities with approved acoustic classification of the territory	% of classified territory
	- levels of noise monitored in the last three years	diurnal and nocturnal levels of dB(A) monitored
Electroma-gnetism	- Levels of electic field monitored in the last three years	V/m
	- Levels of electromagnetic induction monitored in the last three years	µT
	- Number of new radio stations installed in each municipality	N° for each year
Energy	- Pro – capite consumption of electric energy	kWh/inhabitant
	- Pro – capite consumption of LPG	l/ inhabitant
	- Pro – capite consumption of green petrol	l/ inhabitant
	- Pro – capite consumption of methane	mc/ inhabitant
	- Pro – capite consumption of diesel gas	l/ inhabitant
	- Pro – capite consumption of combustible oil	kg/ inhabitant
Water availability	- Levels of water stratum	m/year
	- Pro – capite water pickings and consumptions	mc/ inhabitant
Quality of subterranean water	- N° of events exceeding the levels of standard quality (for each year): chlorides, ammonium ion, nitrates	N°/year
	- N° of events exceeding the levels of standard quality (for each year): trichloroethylene.	N°/year
	- Annual average concentrations : chlorides, ammonium ion, nitrates, trichloroethylene	µg/l e mg/l

Issue	Selected indicators	Unit of measurement
	- Environmental State of Subterranean Water	ESSW Index
Use of soil	- Ratio between used agricultural surface and total agricultural one	%
	- Built surface	%
	- Wooded surface in respect to the total one	%
Quality of surface water	- Chemical monitoring of the main cluster rivers	LIM Italian Index of quality
	- Biological monitoring of the main cluster rivers	IBE Italian Index of quality
Wastes	- Pro – capite urban wastes	Kg/ inhabitant
	- Differentiated level of wastes	%
	- Per employed special wastes	Kg/employed
	- Level of hazardous produced wastes	%
	- Level of recovered paper	%
Quality of soil and subsurface	- N° of industrial sites forced to reclamation (in respect to the total existing ones)	%
	- Surface of industrial sites forced to reclamation (in respect to the total one)	%
Biodiversity	- Protected areas	mq
	- Agroforestal biodiversity	Shannon index
Traffic and transportation	- Density of streets in the cluster	Km / Kmq
	- Motorization rate	N° cars*100/ inhabitants.
	- Territorial density of vehicles	N° vehicles/km of streets
	- Level of registered industrial vehicles	%

Source: Cluster Environmental Review, PIONEER project – *www.life-pioneer.info*

Table 1. Territorial indicators

The sector of reference was the paper one; a questionnaire concerning the environmental aspects and performances was sent to all paper firms operating in the Lucca cluster (with information concerning their emission to air, quality and quantity of their water discharges, typologies and quantities of produced wastes, water and raw materials consumptions, and so on). On the basis of the obtained answers (about 70% of sent questionnaires were compiled, representing more than 80% of the whole cluster paper production), the Promotion Committee, for each environmental aspect characterizing the paper productive process, defined a basket of sectorial environmental performance indicators (many of these couched in terms of *coefficient of emission*). These indicators (as the territorial ones) were inserted in the monitoring plan defined by the Promotion Committee. In the following table environmental sectorial aspects and the selected indicators.

Environmental aspect	Selected indicators
Emission on air	• PM10 [kg/ton of produced paper] • CO [kg/ton of produced paper] • NO_x [kg/ton of produced paper]
Noise	• Levels of max dB[A] monitored on the firm perimeter
Energy	• Annual consumptions of methane [mc/ton of produced paper] • Annual consumptions of electric energy [kWh/ton of produced paper]
Water consumption	• Specific water consumption [mc/ton of produced paper]
Effluents on water	• Specific water emission [mc/ton of produced paper] • BOD [kg/ton of produced paper] • COD [kg/ton of produced paper] • SST [kg/ton of produced paper] • Emission of N [kg/ton of produced paper] • Emission of P [kg/ton of produced paper] • Chloride [kg/ton of produced paper]
Hazardous materials	• mc of asbestos/number of employed • mc of asbestos/number of firms
Wastes	• Specific production of wastes [kg of wastes/ton of produced paper]
Transportation	• Number of IN-OUT vehicles/ton of produced paper
Raw materials consumption	• Chemical substances specific consumption [kg of chemical substances /ton of produced paper] • Virgin fiber specific consumption [ton/ton of produced paper] • Package specific consumption [kg /ton of produced paper]

Source: Cluster Environmental Review, PIONEER project – *www.life-pioneer.info*

Table 2. Selected sectorial indicators

The third phase, transversal to the previous ones, was characterized by a survey about environmental perception of local communities, with direct interviews to 700 cluster inhabitants. The survey focused on ways of perception of both environmental territorial issues and the entity of pressures produced on environment from paper enterprises. As academic literature shows [30, 31], for many environmental aspects a big difference exists between real risk and perceived one: in a territorial perspective, the community perception of environmental issues represents an interesting way for learning the "point of view of the interested parties". Two questions addressed to citizens were used as criteria in the significance evaluation process of cluster environmental aspects and issues:

- Which are, in your opinion, the two most serious issues characterizing this territorial area?
- Which are, in your opinion, the two most significant environmental issues inducted by the large presence of enterprises operating in the paper industry?

Criterion of evaluation	Description of the criterion
Territorial section of the analysis	
Intertemporal trend of performances	Last three years trend of selected performance indicators
Comparison with overterritorial contexts	The comparison was carried out in respect to Regional and National performance
Comparison with quality standards and legislative limits	The comparison was carried in respect to standard legislation fixed by EU Directives and national laws
Results of survey about environmental perception	The question was: Which are, in your opinion, the two most serious issues characterizing this territorial area?
Indictments of citizens	The indictments of citizens to Public Authorities in the last three years represented another indirect indication of environmental perception of local communities.
Sectorial section of the analysis	
Intertemporal trend of performances	Last three years trend of selected performance indicators
Comparison with objective sectorial benchmark	The standards were: - Bref IPPC for paper industry - Ecolabel requirements for paper industry - Performance indicators concerning the paper industry (Environmental Assocarta Annual Report, SCA Group Environmental Report, data published by other paper clusters in Italy and Europe]
Presence of environmental laws concerning that specific environmental aspect	The legislative references were national and European ones
Results of survey about environmental perception	The question was: Which are, in your opinion, the two most significant environmental issues inducted by the large presence of enterprises operating in the paper industry?

Source: Cluster Environmental Review, PIONEER project – *www.life-pioneer.info*

Table 3. Criteria of evaluation

The first question was used as evaluation criterion in the territorial section of analysis; the second one was used as measure of perception of local communities of pressures produced by paper industry on environment. After selecting the performance indicators and summarizing the results of the survey about environmental perception of local communities, the Promotion Committee fixed a list of evaluation criteria for the identified environmental issues and aspects. In the Table 3 the list of evaluation criteria adopted in the PIONEER project.

The Promotion Committee, after approving the previous criteria, defined the weights to assign to each criterion in the evaluation algorithms of environmental significance. The application of the alghoritms to the baskets of territorial and sectorial indicators gave as result the list of significant environmental aspects and issues of the cluster. The shared algorithms were the result of opinions and interests expressed by the members of the Promotion Committee (in its turn representative of public, private and sectorial local interests).

The last step of the assessment phase was the integration of results of significance concerning the territorial section of the cluster analysis with the results of the sectorial section, with the aim of obtaining a measure of the pressure produced by the paper industry on environment in the cluster. The scheme of integration is represented in Table 4.

Environmental aspect	**Issue**
Emission on air	Quality of air
Effluents on water	Quality of surface water
Energy consumptions	Energy
Water consumptions	Availability of water resources
Wastes	Wastes
Transportations	Traffic and transportation
Noise	Noise
Odours	Odours
Visual impact	Use of soil
Hazardous materials	Quality of soil and subsurface

Source: Cluster Environmental Review, PIONEER project – *www.life-pioneer.info*

Table 4. Correspondence table

On the basis of the emerged results by integration of environmental issues and aspects (also in this case the weight assigned to each dimension was the result of a shared process among members of the Promotion Committee), the Cluster Environmental Programme was drafted.

4. The monitoring system and trend of performances

After the end of the project (February 2006), at cluster level the activities continued and the local Promotion Committee pursued to carry out its ordinary activities. In the environmental monitoring procedure the Promotion Committee established to monitor the cluster environmental performances at least biennially. This monitoring system involves both issue indicators (representing local environmental conditions), and indicators related to sectorial environmental aspects (representing the entity of environmental pressures from local paper enterprises). On the basis of the collected evidences, the Promotion Committee established initiatives to be acted at cluster level and the incidental new targets of improvement to be integrated within the Programme. Nevertheless, what relevant in this paper is not the framework of developed initiatives after the end of the project, but the trend of environmental performances characterizing the cluster from 2006 to today. This framework of performances has to be interpreted with respect both to local environmental conditions and to impacts of paper enterprises.

In the tables 5 and 6 we have selected in a synthetic way the trend of the monitored indicators from 2006 to 2011. The monitored indicators can be found in the "Updating charts of Lucca paper industrial cluster" available at www.life-pioneer.it. Last monitoring updating was November 2011; not all indicators are collected annually, and the trends refer to different periods.

Trends evaluation has been carried out by 5 different categories of data:

++ → relevant improvement of the performance
+ → slight improvement of the performance
= → steady condition
- → slight decrease of the performance
-- → relevant decrease of the performance

Issue	Selected indicators	Period	Trend of performance	Comments
Quality of air	Annual average concentrations	2006 - 2010	+	General improvement of performance
	N° of events exceeding the levels of standard quality	2006 - 2010	+	
	Biomonitoring process results	Not monitored	/	
	Emission on air	2005 - 2007	=	
Noise	N° of municipalities with approved acoustic classification	2006 - 2011	++	Relevant improvement
	Levels of noise monitored	2007 - 2010	+	
Electromagnetism	Levels of electric field monitored	2006 -2010	=	Initial situation already in compliance with regulation
	Levels of electromagnetic induction monitored	2006 -2010	=	
	Number of new radio stations	2006 - 2010	-	
Energy	Pro – capite consumption of electric energy	2003 - 2008	-	Steady situation (not critical)
	Pro – capite consumption of fuels	2003 - 2008	=	

Issue	Selected indicators	Period	Trend of performance	Comments
Water availability	Levels of water stratum	2006 - 2008	=	Steady situation (critical)
	Pro – capite water pickings	Not monitored		
Quality of subterranean water	N° of events exceeding the levels of standard quality	2006 - 2010	++	General improvement of performance
	Annual average concentrations	2006 - 2010	+	
	Environmental State of Subterranean Water	2007-2010	=	
Use of soil	Ratio between used agricultural surface and total one	Not monitored	/	Decrease of performance (but not critical)
	Built surface	Not monitored	/	
	Wooded surface in respect to the total one	2004 - 2007	-	
Quality of surface water	Chemical monitoring of the main cluster rivers	2006- 2008	=	General improvement of performance
	Biological monitoring of the main cluster rivers	2006 - 2011	+	
Wastes	Pro – capite urban wastes	2007 - 2010	+	General improvement of performance
	Differentiated level of wastes	2007 - 2010	++	
	Per employed special wastes	2007-2009	++	
	Level of hazardous produced wastes	2007-2009	=	
	Level of recovered special wastes	Not monitored	/	
Quality of soil and subsurface	N° of sites forced to reclamation	2006 - 2011	- -	Relevant decrease of performance
	Surface of industrial sites forced to reclamation	2006-2011	- -	
Biodiversity	Protected areas	2006-2011	=	Steady situation
	Agroforestal biodiversity	Not monitored	/	
Traffic and transportation	Density of streets in the cluster	Not monitored	/	Decrease of performance (critical aspect)
	Motorization rate	2006 – 2010	-	
	Territorial density of vehicles	Not monitored	/	
	Level of registered industrial vehicles	2006 – 2010	-	

Source: Updating charts of Lucca paper industrial cluster

Table 5. Trend of territorial performance indicators

Environmental aspect	Selected indicators	Period	Type of product		Comments
			Tissue	Packaging	
Emission on air	NO_x [kg/ton of produced paper]	2006 - 2010	=	-	Steady situation, influenced by a decrease of production in the period
	CO [kg/ton of produced paper]		=	+	
	PM10 [kg/ton of produced paper]		-	/	
Noise	Levels of max dB[A] monitored on the firm perimeter	Not monitored	/	/	
Energy	Annual consumptions of methane [mc/ton of produced paper]	2006-2010	-	+	General steady situation
	Annual consumptions of electric energy [kWh/ton of produced paper]		+	-	
Water consumption	Specific water consumption [mc/ton of produced paper]	2006-2010	- -	=	Worse performance, influenced by a decrease of production
Effluents on water	BOD [kg/ton of produced paper]	2006-2010	-	+	General steady situation
	COD [kg/ton of produced paper]		=	=	
	SST [kg/ton of produced paper]		=	+	
	Emission of N [kg/ton of produced paper]		=	-	
	Emission of P [kg/ton of produced paper]		=	=	
	Chloride [kg/ton of produced paper]		=	-	

Environmental aspect	Selected indicators	Period	Type of product		Comments
			Tissue	Packaging	
Hazardous materials	- mc of asbestos/number of employed - mc of asbestos/number of firms	Not monitored	/	/	
Wastes	Specific production of wastes [kg of wastes/ton of produced paper]	2006-2010	+	=	General steady situation
Transportation	Number of IN-OUT vehicles/ton of produced paper	Not monitored	/	/	
Raw materials consumption	Chemical substances specific consumption [kg of chemical substances /ton of produced paper]	2006 – 2010	-	-	Light decrease of performance
	Virgin fiber specific consumption [ton/ton of produced paper]		+	/	
	Package specific consumption [kg /ton of produced paper]		/	=	

Source: Updating charts of Lucca paper industrial cluster

Table 6. Trend of sectorial performance indicatorS

Data show an orientation to improvement of environmental conditions at cluster level, with respect to local cluster conditions (the only one decreasing datum refers to quality of soils and subsurface). On the contrary, with respect to data related to paper industry performances, we have a general steady of performances in the period 2006-2010, by the way in a period characterized by recession and decrease of productions.

5. Conclusions

From the PIONEER experience we can draw two different levels of conclusions; the first one (technical) regards the specific results obtained by the PIONEER project, while the second one regards the fallouts of the proposed methodological approach on territorial clusters and its reproducibility in other territorial and sectorial contexts.

From the first point of view, the PIONEER project produced results in terms of a high participation in EMAS by a relevant number of organisations; in fact, within the end of the project, 18 organizations obtained the individual EMAS Registration and, since 2007, the Promotion Committee obtained the award promoted by the Italian Competent Body for industrial clusters with a well implemented process for simplification of EMAS registration of local SMEs. Moreover, as described in the previous paragraph, local environmental conditions improved and pressures of enterprises (although in a recession phase) didn't increase.

From the second point of view, the comparison with other EU territorial contexts showed an high potential of transferability of the methodological approach; in fact, the PIONEER experience showed that there is a high number of synergies that can be obtained at the managerial and technological level to promote the inclusion and diffusion of innovative elements based on the partnership between the different firms operating within the same area. It is a question of exploiting the "co-opetition" attitude (co-operation between firms which also compete) and the collaboration between the enterprises and the other economic and institutional actors. These issues are the same for every industrial cluster, making the project methodology reproducible in each of these homogenous areas, characterizing the productive systems of many EU countries, in Italy, Spain, Portugal, Denmark, France, Finland and the UK. Moreover the project showed the operational feasibility of the methodological approach previously described, fostering an innovative approach to decision-making, based on multistakeholder involvement. According to this approach, EMAS has become a real territorial policy instrument, able to integrate industrial development targets and environmental quality ones, with a key–role played by the Promotion Committee.

Author details

Massimo Battaglia
Scuola Superiore Sant'Anna, Institute of Management, Pisa, Italy

Tiberio Daddi* and Francesco Testa
Sant'Anna School of Advanced Studies, Pisa, Italy

6. References

[1] Iraldo F. (2002). Ambiente, impresa e distretti industriali. Franco Angeli, Milan

---

* Corresponding Author

[2] Battaglia, M. (2008) Approccio metodologico per i sistemi produttivi territoriali, in: M. Frey & F. Iraldo (Eds) Il management dell'ambiente e della sostenibilita` oltre i confini aziendali, pp. 244–277 (Milano: FrancoAngeli).

[3] Battaglia M., Daddi T., Ridolfi R. (2008), Environmental Territorial Management: A New Approach for Industrial Clusters, in "Environmental Management" Editors: Robert H. Theobald, Nova Publisher - USA

[4] Frey, M. & Iraldo, F. (2008) Il Management dell'Ambiente e della sostenibilita` oltre i confini aziendali (Milano: FrancoAngeli).

[5] Amadei P., Croci E. e G. Pesaro (1998), Nuovi strumenti di politica ambientale – Gli accordi volontari. Franco Angeli, Milano

[6] Hillary, R. (1999). Evaluation of Study Reports on the Barriers, Opportunities and Drivers for Small and Medium Sized Enterprises in the Adoption of Environmental Management Systems. Routledge, London.

[7] Biondi V., Frey M., Iraldo F. (2000). Environmental management systems and SMEs: barriers, opportunities and constraints. Greener Management International, n. 29

[8] Del Brío JA, Junquera B (2003). A review of the literature on environmental innovation management in SMEs: implications for public policies. Technovation, 23: 939-948.

[9] Lepoutre, J., Heene, A., 2006. Investigating the impact of firm size on Small Business Social Responsibility: A critical review. Journal of Business Ethics, 67(3), pp. 257-273

[10] Steger, U. (2000). Environmental Management Systems: Empirical Evidence and Further Perspectives. European Management Journal, Vol. 18 No. 1, 23-37.

[11] Rennings, K., Ziegler, A., Ankele, K., Hoffmann, E., 2005. The Influence of Different Characteristics of the EU Environmental Management and Auditing Scheme on Technical Environmental Innovations and Economic Performance. Ecological Economics 57 (1), pp. 45- 59

[12] Iraldo, F., Testa, F., Frey, M., 2009. Is an environmental management system able to influence environmental and competitive performance? The case of the eco-management and audit scheme (EMAS) in the European union. Journal of Cleaner Production, 17, pp.1444–1452.

[13] Baldizzone G. (2000), L'Agenda 21 come strumento cardine delle politiche di sviluppo sostenibile, Ambiente e Sviluppo n. 5

[14] Becattini G. (1979) Dal settore industriale al distretto industriale. Alcune considerazioni sull'unità di indagine dell'economia industriale. Rivista di Economia e Politica Industriale 1

[15] Putnam RD, Leonardi R, Nanetti R. 1993. Making democracy work: Civic traditions in Modern Italy. NJ Princeton University Press: Princeton, NJ

[16] Becattini G. (1990). The Marshallian District as a Socio – Economic notion in Pyke F., Becattini G., Sengenberger W. "Industrial Sectors and Inter-firm Cooperation in Italy". ILO; Geneva
[17] Ferrucci L, Varaldo R. 1996. The evolutionary nature of the fi rm within industrial districts. European Planning Studies February, 4(1): 27–34.
[18] Ferrucci L., Varaldo R. (1997) a cura di. Il distretto industriale tra logiche di impresa e logiche di sistema. Franco Angeli, Milan
[19] Prusak, L., Cohen, D., 2001. How to Invest in Social Capital. Harvard Business Review. 79 (June), pp.86–93
[20] Arrighetti A, Seravalli G. (a cura di) (1999), Istituzioni intermedie e sviluppo locale. Donzelli, Roma
[21] Montini A., Zoboli R. (2003). Environmental impact and innovation in industrial districts in Cainelli G, Zoboli R. (Eds) "The evolution of Industrial Districts" Physica – Verlag
[22] Istituto AmbienteItalia (2003). Rapporto Ecodistretti, Milan
[23] Battaglia, M., Bianchi, L., Frey, M., Iraldo, F., 2010. An innovative model to promote CSR among SMEs operating in industrial clusters: evidence from an EU project. Corporate Social Responsibility and Environmental Management (CSR&EM), 17(3), pp.133-141
[24] Von Weltzien Høivik, H., Shankar, D., 2011. How Can SMEs in a Cluster Respond to Global Demands for Corporate Responsibility? Journal of Business Ethics, 101 (6), pp. 175–195
[25] Petri F. (1990). Who is afraid of choices? A proposal for multi – criteria analysis as a tool for decision-making support in development planning. Journal of international development
[26] Roy B. (1996). Multicriteria methodology for decision analysis. Kluwer Academic Publishers
[27] Daddi, T., Rizzi, F. (2008) Strumenti conoscitivi e sistemi informativi, in: M. Frey & F. Iraldo (Eds) Il Management dell'ambiente e della sostenibilita` oltre i confini aziendali, pp. 244–277 (Milano: FrancoAngeli)
[28] Battaglia, M., Daddi, T., Rizzi, F. (2012): Sustainable Tourism Planning and Consultation: Evidence from the Project INTER.ECO.TUR, European Planning Studies, 20:2, 193-211
[29] Niemeijer, D., Groot, R.D., Framing environmental indicators: moving from casual chain to casual networks in environment, Development and Sustainability 10 (2008) 89–106
[30] Sunstein R. (2004), La sicurezza ambientale tra percezione e approccio razionale, Edizioni Ambiente, Milano
[31] Maiello, A., Battaglia, M., Daddi, T., Frey, M. (2011), Urban Sustainability and knowledge: theoretical heterogeneity and need of a transdisciplinary framework. A tale of four towns, in Futures Vol. 43, Issue 10 ISSN 0016-3287

[32] http://www.life-pioneer.info - PIONEER project reports and documentation, by Province of Lucca, IEFE Bocconi and Sant'Anna School of Advanced Studies

Chapter 14

Overview of Past and Ongoing Experiences Dealing with the Environmental Management at Cluster Level

Francesco Testa, Tiberio Daddi, Fabio Iraldo and Marco Frey

Additional information is available at the end of the chapter

http://dx.doi.org/10.5772/47172

1. Introduction

Small and medium-sized enterprises make up a large part of Europe's economy, representing some 99% of all enterprises and 57% of the economy's added value (COM (2005) 551 fin). Therefore they also play a primary role in shifting the European economy to more sustainable production and consumption patterns. SMEs can have considerable impact on the environment, not necessarily through individual pressure, but through their combined total impact across sectors. It is widely accepted that it would be too complex and burdensome for companies and public authorities to determine the detailed contribution made by SMEs to pollution (e.g. air pollution), in terms of the environmental load from different types of pollutants (e.g. CO2, SOx, NOx, etc.) in each Member State. Indeed, in many cases the data does not exist. Nevertheless, the often quoted rough figure of a contribution of 70% of industrial pollution in Europe seems reliable, and a number of studies attempt to provide 'insights' into particular environmental problems deriving from SMEs for specific countries. For example, a British report estimated that SMEs accounted for 60% of total carbon dioxide emissions from businesses in the UK and concluded that there was substantial room for improvement in energy efficiency and emissions reductions among SMEs. Again, estimates from the Netherlands and United Kingdom suggest that the commercial and industrial waste from SMEs represents on average 50% of the total. These studies further support the claim that SMEs can exert considerable pressures on the environment (SEC (2007) 907). Numerous regional and national studies show that the majority of SMEs have low awareness of their environmental impacts and how to manage them. Most SMEs are 'vulnerably compliant', since they do not always know enough about legislation to ensure that they are compliant. This is mostly due to lack of awareness of the environmental impacts of their own activities, ignorance of environmental legislation, inability to tackle their environmental impacts, and sometimes the excessive administrative and financial burden of compliance. Compliance is further hindered by the perception that environmental protection is costly and has little benefit for the business (SEC (2007) 907).

This scenario shows:

- the potential threat to the environment and to the effectiveness of many Community environmental protection measures due to lack of knowledge and awareness;
- the potential risk for the health and safety of SMEs employees
- the potential advantages that an enhanced environmental management could bring to SMEs in terms of economic and/or financial benefits (of which SMEs may be unaware). The complexity of the issues involving the SMEs' environmental compliance and their environmental performance, other than their capacity to fully respond in time to the new challenges placed by environmental issues (which allows them to catch the benefits in terms of competitiveness and innovation) needs a multiple approach, capable of putting into action a set of complementary measures. These would include education and awareness raising, on-the-spot assessment and identification of problems, targeted consultancy, exchange of information and best practices, removal of administrative burdens, availability of shared tools.

Generally, the issues are related to the environmental impact of the SMEs production activities and services, involving all the environmental aspects (air, water, soil and sub-soil, biodiversity, noise, land, etc.) normally/usually disciplined by the EU and the national and local legislation. They also include crucial new challenges that the EU is facing, regarding global warming, energy efficiency, renewable energy sources, sustainable use of resources, waste reduction, re-use and recycling.

All these considerations have stimulated the development of the so-called "cluster approach" to manage the environmental issues of a large number of SMEs located in limited territorial areas. In recent years, the "cluster approach" has been mentioned in some important official documents of the European Union, such as:

- EC COM (2007) 379 "Small, clean and competitive - A programme to help small and medium-sized enterprises comply with environmental legislation" (ECAP Programme) which encourages the use/implementation of environmental management in industrial clusters or districts of SMEs, using specific cluster or supply chain approaches;
- the new Regulation (EC) no. 1221/2009 (EMAS III), which in article 37 mentions SMEs clusters applying for EMAS registration.

Networking and cooperation among organisations emerge from several studies and empirical evidence as some of the most important factors fostering the dissemination of formal EMS (such as EMAS). Many authors (inter alia:Biondi et al. 2000, Hillary 2004) emphasise that working with groups of companies is an useful and efficient way of adopting EMAS, particularly for SMEs. Moreover, the European Commission has recently confirmed the key role of networking for overcoming the constraints and barriers for EMS adoption among SMEs (European Commission 2007). In fact, the Commission has highlighted its commitment to promote and encourage the use of EMAS in industrial clusters or districts of SMEs, using specific cluster- or supply chain- oriented approaches, since they can reduce consultancy and audit/verification costs for SMEs, and facilitate additional sharing of knowledge and exchange of experience among participants.

The effectiveness of the networking approach is clear among organisations operating in the same sector (such as the industrial sector, but also service sectors such as tourism or public institutions operating at different levels) and among organisations operating in the same region (or territorial area).

In the first case, enterprises can co-operate by identifying and assessing similar environmental aspects and by finding technological and operational solutions that can be applied to similar production processes and products, as well as by defining organisational structures suitable for the same kind of production cycles. In the second case, co-operation is facilitated by the 'physical contiguousness' and there are synergies both in improving the environmental impact on the same local eco-system, and in interacting and communicating with the same stakeholders (local population, authorities, etc..).

In some experiences, a network has been created among SMEs within a 'cluster', in order to foster information exchange and experience diffusion and to define and apply common solutions to similar environmental, technical and/or organisational problems, or to share environmental management resources (Iraldo & Frey, 2007). A specific kind of co-operation within a cluster of organisations takes place in the supply-chain: when a large customer is willing to support small suppliers in the EMS implementation process, then all the smaller organisations involved in the supply chain can benefit greatly from networking. This approach proved effective in some Member States such as Germany ("Konvoi" approach), Spain (co-operation in the tourism supply chain), Nordic Countries (Denmark and Sweden). In Italy it has shown a real effectiveness in promoting the environmental compliance of SMEs by means of the so-called APO "Ambiti Produttivi Omogenei".

This study aims at collecting and describing the most important international experiences concerning the cluster approach, with the following objectives:

- To analyse different types of inter-company dynamics in the cluster;
- To give information on the diffusion of the cluster approach at international level;
- To understand possible correlations between cluster approach and environmental issues management;
- To analyze some excellent experiences on shared and collective management of environmental aspects.

The structure of this chapter is organised in sections according to the different types of clusters considered. After a brief introduction on the methodological approach, each section analyses the various forms of cluster with the following structure.

The first section focuses on the identification and analysis of the definition of cluster to which we refer, outlining its key features and describing the cooperative dynamics that consolidate within it. The second section explains the presence of the abovementioned type of cluster in Italy, describing its prevalence and relevance for the national economy. The third section aims at outlining the European framework and offers a comparison with the information applied to the national context. The last section identifies the elements that characterise the management of environmental issues (in a cooperative logic) in the type of cluster analysed,

This study is based on the results emerged from a research carried out within the ECCELSA (Environmental Compliance based on Cluster Experiences and Local Sme-oriented Approaches) Life project. For this reason the authors acknowledge all project partners1 who contributed to the results presented in this chapter.

2. The methodological approach

The first references to the "Cluster organizational model", as an approach that can develop synergies resulting in more efficient production than would occur within a single large plant, are found in Marshall at the end of the XIX century. In the first half of the XX century the benefits of agglomeration of economic activities were also confirmed by Austrian economist Shumpeter, who stressed the importance of the cluster system in terms of business competitiveness. In 1991, Michael Porter in his *Competitive Advantage of Nations* (1991), stated the "cluster theory" in which he identified the most potential for growth and development for industrial clusters as opposed to the single enterprises, thanks to the presence of vertical relations *[customer/supplier]* and horizontal relations *[common customers, technology, channels]*.

At European level, clusters have been formally recognized and defined in the Final Report of the 'European Commission Expert Group on Enterprise Clusters and Networks which offers a first "census" of the phenomenon, and in communication No. 652, October 2008[2].

In those documents, clusters are defined as "geographic concentrations of specialized companies that have workforce with advanced abilities and skills, and "support" institutions that make possible the spreading of knowledge and indirect positive effects as a result of their proximity".

According to this definition, the elements that characterize the cluster concept can be identified in *geographic proximity, specialization in production* and *interaction among different actors* in the cluster. Therefore, the definition of cluster is not restrictive; it covers a wide variety of approaches in which those elements are more or less relevant. In fact, the concept includes the classical configuration of the industrial district and the geographically confined

---

[1] The authors specifically acknowledge Sara Tessitore and Valentina Toschi (SSSUP); Besides Sant'Anna School of Advanced Studies (coordinator of the project) the other partners are: Ambiente Italia – Istituto di Ricerca, ERVET - Emilia Romagna Valorizzazione Economica del Territorio, IEFE Bocconi - Istituto di Economia e Politica dell'Energia e dell'Ambiente, SIGE - Servizi Industriali Genova, Gemini - Innovazione Sviluppo e Trasferimento Tecnologico.

[2] Communication from the commission to the Council, the European Parliament, the European Economic and Social Committee and the committees of the region 'Towards world-class clusters in the European Union: Implementing the broad-based innovation strategy ".

industrial areas such as the environmentally equipped production areas "APEA", as well as industrial parks up to the interactions along the supply chain. While being characterized by the presence of elements linked to the cluster, these network systems show major differences that allow to distinguish three main approaches within which to investigate.

For example, the Industrial District is a local system with the presence of a prevalent production carried out by a group of small independent firms highly specialized in different stages of the same production process. Production Areas instead represent an organizational model characterized by the territorial element but, unlike the District, they are concentrated in areas more easily defined and circumscribed geographically, and do not necessarily show the presence of one or more specialised production sectors.

Another organizational model that meets the definition of cluster is the supply chain: a network involving all stakeholders in the production chain, from the company that produces raw materials to the processing company, from carriers to distributors, from wholesale to retail. The feature of the production chain is the interaction between SMEs involved in a process of production, among which the exchange of information can flourish and the development of projects and a relationships based on trust may be encouraged. Compared to Districts and Production Areas the less important aspect in this context is the territorial delimitation and the concentration of enterprises.

The distinction between different aspects of the concept of clusters led to the identification of three inter-organizational interaction models on which to focus the analysis, particularly emphasizing the organizational arrangements implemented to manage common environmental problems. In detail, the types of clusters in this study are:

- Industrial districts, closely related to production;
- Ecologically equipped productive areas APEA and Industrial Parks;
- Supply chain management which aims to produce stable benefits for the companies that are part of the supply chain (buyers and suppliers) through process integration and long term relationships.

3. Industrial district cluster

3.1. "Industrial districts" cluster: Definition and boundaries

In literature, the industrial district is represented by a local system characterized by a main production activity performed by a group of small independent firms, highly specialized in different stages of the same production process. This peculiar entrepreneurial organizational model can develop synergies that result in a more efficient production than would occur within a single large plant. At the end of the XIX century Marshall (1890) had already highlighted the benefits coming from the agglomeration of economic activities in terms of availability of skilled labour and high level of specialization. Similarly, the Austrian economist Schumpeter in the first half of last century stated the existence of competitive advantages deriving from a business cluster. A significant contribution to the study of industrial districts and of internal relationship mechanisms able to generate competitive advantages for the cluster firms comes from G. Becattini who, in his article "From industrial sectors to industrial districts" introduced the concept of industrial district as a tool to support regional policies for territorial development.

Without any attempt to reorder the taxonomies in the field of industrial districts (which can be found in the literature according to the various configurations that such systems engage in), we mention some definitions and acronyms that have slightly different interpretation and, as a result, partially conflicting definitions of local systems with a high concentration of businesses: from industrial districts to system-sectors, from milieu to TPS (territorial production systems), to the RESS (regional economic and social systems). Connecting all the approaches underlying these concepts is the identification as the common element of the analysis of a system of usually small and medium enterprises operating in a homogeneous sector (or in sectors known as "auxiliary") and located in a limited socio-territorial area in which they have deep-seated social and economic relationships. The role which an industrial district can have in a competitive development of local production has stimulated in some national contexts the interest of policy makers. In Italy, Law 140, 1990, enacted to simplify and facilitate the set up of district areas, also fostered an institutional definition of the concept of Local Production System (LPS), which is an area characterized by:

- Homogeneous production contexts;
- High concentration of enterprises;
- Specific internal organization.

Based on that definition, the industrial district can be considered a specific LPS featuring:

- High concentration of industrial enterprises;
- Highly specialized production of business systems.

3.2. The "industrial districts" cluster in Italy

Law No 317 of 1991 "Action for innovation and development of small enterprises" introduced in Italy the concept of industrial district (art. 36) taken up and extended by Law No 140, 1999 with reference to the Local Production System. National legislation has given regional administrations the task to define criteria and procedures for the recognition of clusters and the legal form they should have once approved.

The Region of Lombardy under Regional Law (L.R.) 7 of 1993 and Regional Law 1 of 2000 regulates the procedures for the geographical boundaries of Industrial Districts, sets up the provision of development programmes in individual districts and the creation of innovative projects for the enterprises that belong to them. In 2000 Tuscany adopted Resolution No. 69 which defines districts as "monosectorial production systems with a high presence of small medium industrial manufacturers having a strong supply chain, social and institutional relationships also present in interprovincial areas". Italian Regions did not issue any rules to formally and rigidly acknowledge districts, so it remains a very flexible approach which is closely related to the characteristics of the area and primarily addressed to contexts consisting of SMEs.

The district approach has spread as a result of economic support programmes promoted at first at regional level (eg. I-C@AST programm - a textile project being reorganized - Lombardy 2006), and later by the Ministry of Economic Development (2003-2005 -2008)3. Art.3 of Law 266/1997 (so-called Bersani Law) for the first time establishes state funding for the Regions to be used in Industrial Districts identified on the basis of Law No. 317/1991, to fund programmes that improve service networks, in particular in the ITC sector.

Another important intervention of industrial policy in favour of Districts was carried out under the Decree for the liberalization of the electricity market, which envisaged the opportunity for those entities defined as "eligible customers" in Article 2 to sign supply contracts with any producer, distributor or wholesaler in Italy and abroad. Among those considered for this option are groups formed by companies whose total consumption reaches a value greater than 30 GWh, located exclusively in the same municipality or contiguous municipalities, or in areas identified with specific acts of regional planning. New policies to support technological development in the Districts were approved in 2001 and in 2005. The last supportive action took place under the 2008 Budget which included a call for bids to grant funding to districts also engaged in improving environmental performances at level of production area. Surveys by research bodies such as the "Districts Club" and the "IPI-Institute for Industrial Promotion" reported data showing a significant presence of the district system in Italy. Here is some information resulting from the last industry census carried out by ISTAT in 2000. In Italy there are 155 districts mostly located in the Northern regions. This model of production organization involves 4,929,721 employees and 1,180,042 businesses of the manufacturing, services and trade sectors[4]. The areas where this approach is prevailing are textiles with 45 districts, mechanics (38), household goods (35) and the tanning industry (20). Districts are less common in the production of paper and paperboard, in the food industry, chemicals and plastics.

The regions with greater presence of industrial districts (2001 data) are Lombardy (27), Marche (27), Veneto (22), Tuscany (15) and Emilia Romagna (13)[5]. Among the 27 districts of Lombardy the most important sectors are textiles and engineering, while in the Veneto and Marche regions districts are more present in the production of household goods. In Tuscany, along with textiles the most important districts are tanning and paper, while in Emilia Romagna the presence of districts in the mechanical industry is accompanied by the food industry and the production of household goods.

3.3. The "Industrial districts" in Europe

When comparing the characteristics of Industrial Districts as previously mentioned with the concept of cluster at European level, three important common characteristics emerge. Firstly, clusters are seen as *geographic concentrations* of specialized firms, of highly skilled

---

[3] Europe INNOVA Cluster Mapping Project: Report Italy Available in:
http://www.clusterobservatory.eu/upload/Policy_Report_Italy_20080116.pdf>

[4] Club Distretti, Map of Italian districts, 2005.

[5] ISTAT, Industry Census 2000.

Region	No Districts	Workers
Piedmont	12	297,034
Lombardy	27	1,745,042
Trentino-Alto Adige	4	46,814
Veneto	22	861,546
Friuli-Venezia Giulia	3	123,244
Emilia-Romagna	13	574,432
Tuscany	15	466,494
Umbria	5	61,823
MarcheMarche	27	435,063
Lazio	2	31,542
Abruzzo	6	96,859
Molise	2	4307
Campania	6	26,177
Apulia	8	144,096
Basilicata	1	9,927
Sicily	2	3,236
Sardinia	1	2,085
ITALY	**156**	**4,929,721**

Source: ISTAT 2001

Table 1. Presence of manufacturing districts in Italy

Industry	Industrial Districts	Local manufacturing units	Workers
Textile and clothing	45	63,954	537,435
Mechanics	38	56,816	587,320
Household goods	32	42,287	382,332
tanning and footwear	20	23,441	186,680
Food	7	3,781	33,304
Jewelry and musical instruments	6	13,010	116,950
Paperrmaking and printing	4	4,342	35,996
Rubber and plastic	4	4,779	48,585
TOTAL	**156**	**212,410**	**1,928,602**

Source: ISTAT 2001

Table 2. Types of industrial districts in Italy

and capable workforce, and of supportive institutions that improve the flow and the spillover of knowledge. Secondly, the cluster is useful to reach the functional objective to *provide a range of specialized and customized services* to a specific group of firms. Finally, clusters are characterized by some social and organizational elements, called "*institutional social cohesion tools"*, which link the different and interconnected actors, thus facilitating a closer cooperation and interaction between them.

At European level, two different approaches are used to identify clusters. The best known is based on *"case studies"*, the gathering of *qualitative information* through interviews with local experts or research on documents and publications. The second approach regards the various *quantitative techniques* that rely on the most sophisticated economic models and are based on statistical methods which encourage to identify clusters indirectly, by measuring the effects that are supposed to be detectable in the presence of a cluster.

There are hundreds of case studies that document the history, activities and impact of clusters on regional development, on employment and innovation. The European Cluster Observatory[6] has published 25 case studies of European clusters, related to the areas and the sectors indicated in the picture below.

Through the collection of "case studies" each cluster tells "its own story" and sometimes it is difficult to compare different results. Furthermore, due to the rapid changes occurring in clusters, the results may arise from old data. Therefore, by describing relationships, processes and interactions among actors this methodology becomes an excellent tool that can be used to complement statistical analysis. With regard to *quantitative techniques,* the approach used by the European Cluster Observatory is based on indirect measurement of the effects revealed by coordinated localization of those elements that are assumed to be detectable in the presence of a cluster, such as the concentration of workers or high productivity. There are other techniques for quantitative mapping of clusters but, unlike the operational methodology[7] of the European Cluster Observatory, they are not constantly updated according to changes which take place in the countries analyzed.

The first results came in June 2007 with the establishment of a framework of regional clusters in 31 countries, divided into 38 areas. For the first time, the quantitative analysis performed is based on a fully comparable and consistent methodology in all European countries.This method identifies clusters based on regional employment data collected by EUROSTAT and national and regional statistical sources. The approach used is deliberately based on the measurement of the effects that relationships and spillover have on the companies' choice of location, and not on the direct measurement of the dynamic interactions among the forces driving the cluster.

The quality and quantity of the knowledge that circulate and the *spillover* among firms located in a cluster depend on the size of the cluster, on its degree of specialization and on how well the areas are equipped and focused on the production in the main industries that make up the cluster. Therefore, the three factors *size, specialization* and *focus* can be chosen to assess whether the cluster has reached a "specialized critical mass" likely to *spillover* and develop positive relationships. Statistical mapping of Clusters by the European Cluster Observatory identifies over 2,000 regional clusters in Europe, among which clusters

[6] The European Cluster Observatory was founded in September 2006 by Europe INNOVA.

[7] Cluster Mapping methodology developed by the Institute for Strategy and Competitiveness, Harvard Business School.

classified as *"industrial districts"* are 1380[8]. The following table shows the geographical distribution of "industrial districts" clusters.

European States	No Industrial Districts	Percentage of total
Austria	34	2.5%
Belgium	19	1.4%
Bulgaria	33	2.4%
Cyprus	2	0.1%
Denmark	18	1.3%
Estonia	6	0.4%
Finland	16	1.2%
France	103	7.5%
Germany	269	19.5%
Greece	18	1.3%
Ireland	7	0.5%
Iceland	2	0.1%
Italy	158	11.4%
Latvia	3	0.2%
Lithuania	9	0.7%
Luxembourg	1	0.1%
Malta	4	0.3%
Norway	12	0.9%
Netherlands	29	2.1%
Poland	97	7%
Portugal	27	2%
United Kingdom	58	4.2%
Czech Republic	61	4.4%
Romania	75	5.4%
Slovakia	23	1.7%
Slovenia	8	0.6%
Spain	104	7.5%
Sweden	27	2%
Switzerland	34	2.5%
Turkey	83	6%
Hungary	40	2.9%

Source: European Cluster Observatory

Table 3. Geographic Distribution of "Industrial Districts" clusters

[8] Clusters that do not comply with the definition of Industrial Districts were not included in the total number of clusters considered by the European Cluster Observatory. Specifically, they are: Agricultural, Business services, Distribution, Education, Entertainement, Finance, Fishing, Hospitality, Sporting, Transportation.

Even taking into account the different spatial dimensions of the European countries analyzed, the table and the graph above show a greater presence of clusters in Germany, Italy, Spain and France. Within the above mentioned countries, there is a predominance of the Construction and Food sectors with 15.5% and 10.7% respectively on the total number of districts. The Construction sector is more prevalent in Germany (27 out of 269 districts), in Italy (21 out of 158), and Spain (17 out of 104). In France, however, the greatest number of districts is in the Food sector, with 19 out of 103 districts.

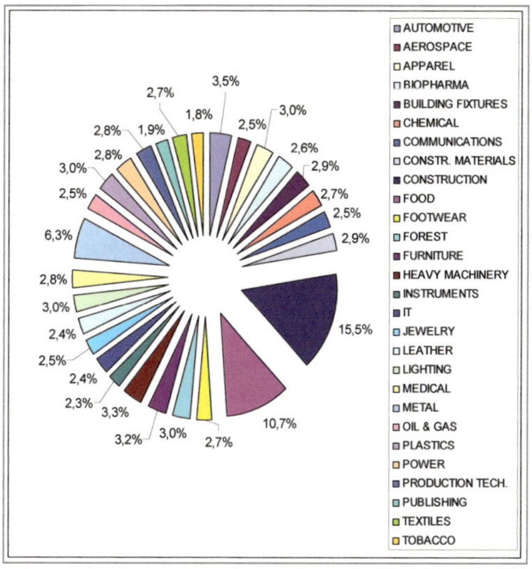

Source: European Cluster Observatory

Figure 1. Number of Districts by Industry

In recent years, many initiatives have been implemented in Europe in order to create favourable conditions for the establishment of new clusters and strengthen existing ones. To date, more than 130 specific national measures in support of clusters were identified in 31 European countries and registered by the INNO-Policy Trend Chart[9]. Nowadays, almost all European countries have specific measures for clusters or programmes developed at national and/or regional level, suggesting that they are a key element of the national and regional strategies in support of innovation

3.4. Cluster approach in the management of environmental issues

The size of clustering in a local context has critical relevance in the analysis of the environmental impact of industrial activities. When assessing the impacting factors related to a particular type of production, the characteristics of different local contexts in which that type of production produces its environmental effects have to be taken into account. Italy clearly shows how the environmental impact of some industrial sectors (textiles, tanning, ceramic) is localized around some areas where there is a high concentration of industries

[9] More detailed information in <http://www.proinno-europe.eu> and <http://cordis.europa.eu/erawatch>

from those sectors. In these cases local dimension becomes a key determinant of the significance of environmental issues for the entire industry sector and, at the same time, a key variable in coordinating an effective response by the companies.

There is no doubt that in terms of impacts on the environment, companies that operate in an industrial district have many elements in common.

First of all settlement, production and sales activities of these enterprises influence the same local ecosystem, characterized by specific and defined environmental aspects. Moreover, companies operating in one district often face similar environmental problems, because they dump the emissions from their production processes into the same receptacle: waste water that drains into the same river (eg the Bisenzio river that runs through the entire Prato textile area, or the Sarno in the Salerno tomato district) or solid waste that goes into the same landfill.

On the other hand, the high specialization of production and the usually very small size of enterprises (with all the implications in terms of limited availability of human, technical and financial resources) allows us to think of the district as an industrial area sufficiently homogeneous also in terms of production methods, degree of technology and organizational and managerial choices. The same technological and organizational matrix of the businesses in the district may show in common environmental problems that are related, for example, to the inefficiency and ineffectiveness of facilities to reduce pollution, to technology obsolescence, to inadequate structures for environmental management, cultural lag and so on.

Even relations with suppliers of equipment and components, according to the logic of "vertically integrated industry" that characterizes many districts, are often played at local level, thus also affecting the availability and appropriateness of the most innovative and advanced technological solutions for pollution prevention (think of the crucial role companies of the so-called mechano-ceramic play in the district of Sassuolo, as the almost exclusive repositories of technological know-how and, therefore, appointed to develop and propose new 'clean technology' to the ceramic businesses in the district).

A final aspect to highlight is the relationship with local stakeholders: for businesses in the district, interacting with the same community, the same institutions, the same local supervisory bodies means to deal with the same needs and requests concerning the quality of the environment. This is of fundamental importance if we consider that the significance of an environmental problem depends on the way in which it is perceived socially. The local dimension is a context where the relationship with company stakeholders is intensified, it becomes more straightforward (given the coexistence in the same area), more immediate (e.g. relationships with local institutions are more frequent than with national institutions), closer (just consider the number of residents employed by enterprises in the district). Besides, given the homogeneity of industrial activities, the physical proximity and frequent inability to attribute the environmental effects to any one production unit, enterprises in the district are considered by local partners almost as single entity.

The relational dynamics among companies and external stakeholders therefore become a crucial pressure factor to foster awareness on environmental issues within the district. By acting the same way and with the same incisiveness on a large number of similar businesses, it reinforces itself and strengthens its effects. For example, if the local population shows particular sensitivity to environmental issues, all enterprises in the district will undergo a high degree of examination from the public (which leads them to ensure continued compliance to regulations) and will be encouraged to use tools to enhance their environmental commitment to the local community.

Other important partners for companies in the district are local institutions. Sometimes companies interact with local authorities and supervisory bodies who are open to dialogue and willing to leave some room for negotiation, or with institutions that are particularly strict as regards law enforcement and extremely demanding on the compliance with obligations and deadlines. The different attitude of institutions can mitigate or amplify the context pressure, acting in the same direction for all firms in the district. Firms can be challenged with requests from local authorities that may focus on some environmental aspects (making them more problematic) or that may promote the application of certain environmental policy tools (e.g.: voluntary agreements at local level).

Local institutions may also prove to be particularly active in promoting common solutions (subsidiaries or consortium) to the most demanding and urgent environmental problems in the district, acting as a catalyst to encourage collaboration among businesses and promoting synergy in the commitment of human, technical and financial resources.

The local dimension represents an essential key in understanding environmental issues also because the same solution to environmental problems can be managed at district level. For example, the infrastructural equipment of a purification plant helps the industrial system in reducing the environmental impact. However, enterprises may find themselves having to directly invest in the installation of small treatment plants, which is known to result in a "scattered" distribution of facilities rather than in a systematic and consistent process.

Increasing awareness to environmental issues by the actors with whom the company interacts implies the need to meet certain "environmental questions".

This is especially significant for SMEs operating within an industrial district. In fact, efforts in the direction of environmental improvement by an individual company are here associated with new knowledge and with the onset of difficulties (the environment, as we have seen, is a challenge or new "turbulence") that once overcome constitute know-how that can be shared with other firms in the district. In this process of growth also appears/arises the need for support from (and relationships with) external actors, a need common to most SMEs, which fosters the development of new "answers" to the emerging needs/demands. In a territorial dimension the resulting "networking" takes peculiar forms, leading to the development of somewhat common solutions (i.e. based on sharing tangible or intangible resources) that are tied to the specific local environment in which businesses in the district interact.

Recent decades have shown the dynamic of those "common solutions" in industrial districts, connected to the different inputs and external forces that have enabled the development of strategies and tools to start up environmental management processes that could involve the whole district

4. Environmentally Equipped Industrial Areas (EEIA) Cluster

4.1. Cluster "Environmentally equipped production areas and eco-industrial parks: definition and boundaries"

"Industrial area" means an area with specific land uses, geographically limited, that is near or on the periphery of urban centers. The industrial area can affect one or more municipalities. "Environmentally equipped areas" were introduced in Italy by Legislative Decree no. 112/98 (the so-called "Bassanini Decree") that in Article 26 states: "The regions and the autonomous provinces of Trento and Bolzano govern, with their own laws, industrial areas and environmentally equipped areas, with infrastructures and systems necessary to guarantee the protection of health, safety and environment. The same laws govern the forms of single management of infrastructures and services in ecologically equipped areas by public or private actors... omission ... and procedures for land acquisition included in industrial areas,...omission .... The production plants located in ecologically equipped areas are exempt from the acquisition of permits concerning the use of services therein. The regions and autonomous provinces identify areas ... omission ...mainly choosing among already existing areas, zones or centres, even if partially or totally abandoned. The local authorities concerned participate in this identification procedure".

Hence, national legislation gives the individual regions the task to regulate the matter, giving some basic points of reference:

1. environmentally equipped areas have infrastructure and systems necessary to protect health, safety and environment;
2. environmentally equipped areas are characterized by forms of centralized management of infrastructure and services;
3. manufacturing plants located in environmentally equipped areas are exempt from the acquisition of permits concerning the use of services therein. [eliminato, era un copia e incolla che ripeteva la frase del punto 2).

A different approach must be used as regards the so-called "eco-industrial parks" (EIP), which are spread across Europe and the world and are in a way similar to Environmentally Equipped Areas, but are usually voluntary initiatives. Eco Industrial Parks as theorized by Lowe, Moran, and Holmes are communities of manufacturing and services firms linked by a common management, and they seek to improve their environmental, economic and social performances by collaborating when addressing environmental issues and using the resources (including energy, water and materials). This integrated approach aims to achieve collective benefits that exceed the sum of individual benefits each company would separately have from the optimization of its performances. The path to achieving this goal includes a new design or redevelopment of infrastructures, and planning of the production area, cleaner production, protection from pollution, energy efficiency and cooperation among enterprises.

The early theorists of the concept of industrial ecology as we consider it today were scientists Robert Frosch and Nicholas Gallopolus, who in an article published in Scientific American in September 1989 defined a new strategy for the manufacturing industry: "the traditional model of industrial activity, in which production processes generate products for sale and waste for disposal, must be transformed into a more integrated model: an industrial ecosystem. This system optimizes energy and raw materials consumption up to using residues from one process...to feed other processes".

Other important researchers (Tibbs, Allenby, Graedel, Lowe, Holmes, Moran) contributed to the development of the concept of industrial ecology. They developed industrial ecology into a discipline based on multidisciplinary contributions aiming at the improvement of the industry-environment relationship. In 1992 Tibbs, another important pioneer, held that "*in natural systems there is no waste, meaning something that can not be absorbed constructively elsewhere in the system" coming to a key concept which is closing cycles: "...making maximum use of recycled materials in new products, optimizing the use of integrated materials and energy, minimizing waste and recovering waste as raw materials for other processes*".

Thus industrial ecology considers the flow of matter and energy with the aim of significantly reducing the use of resources and pollution. It suggests the application to industrial systems and their processing/production cycles of the rules and principles that determine the functioning of non-human biological systems, of ecosystems that are characterized by symbiotic relationships and by the absence of the concept of waste. Every scrap is reintroduced into the cycle to generate energy or as raw material to start another process that is essential to maintaining the overall balance.

Both types of production area (APEA and EIP), therefore, aim at the so-called "closing cycles" of material, water and energy; they aim at sharing key environmental services (water, energy, waste) and at optimizing the organization of activities that have an impact on the environment.

The cooperative approach can be seen mainly in two basic aspects:

- The adoption of collective systems and infrastructures within the industrial area (e.g. purification plant, centralized area for storing waste, industrial water supply systems, power generators for the area);
- The identification of a single production manager that deals with common services within the production area (e.g. collective management of waste, energy, security).

4.2. Cluster "Environmentally equipped production areas and eco-industrial parks" in Italy

As mentioned above, Italian legislation lets the regions regulate the issue of environmentally equipped areas on their territory.

The spread of APEA in Italy depends on the choices promoted by each region. To date, the regions that have issued laws and regulations are:

- Abruzzo (R.G.D. October 10, 2003, No. 1122, "Leg. Decree March 31, 1998, number 112 - P.R.D October 20, 1998, No. 447 as amended by Presidential Decree March 31, 2000, No 440 Definition of the discipline of "Environmentally equipped areas").
- Calabria (Regional Law December 24, 2001. Number 38 "New legal system for Consortia for Areas, Centres and Industrial Development Zones);
- Emilia Romagna (Regional Law No 20/00 "General framework on the protection and use of the territory", Legislative Assembly Resolution No. 118/07 "Adoption of the Guidelines Act and technical coordination on the implementation of ecologically equipped areas in Emilia-Romagna).
- Liguria (Regional Council Resolution of December 28, 2000 No 1486 "Criteria, parameters and methods on the industrial areas and environmentally equipped areas)
- Marche (Regional Law n. 16/05 "Regulation of urban redevelopment and guidelines for the environmentally equipped production areas", and DGR n. 157 of 07/02/2005 "Guidelines for the ecologically equipped production areas (APEA) of Marche Region").
- Puglia (Regional Law January 31, 2003, No. 2 "Guidelines on actions for economic development, production activities, industrial areas and environmentally equipped areas).
- Tuscany (Tuscany Regional Law No. 87 of 22/12/2003 "Ecologically equipped production areas. Changes to the regional law December 1, 1998, No 87").

In other Italian regions in which there are no laws specifically dedicated to APEA, there are other standards that facilitate the environmental management of industrial areas or the creation of EIP. It is the case with regional laws governing consortia of industrial development which are suitable structures for the collective management of a number of issues in the area, including environmental issues (eg, Friuli Venezia Giulia, Sicily).

In other cases, although APEA characteristics are not regulated by regional standards, they are often referred to in regional planning documents (eg. DOCUP 2000-2006 Regione Piemonte).

An analysis of the related regional legislation shows that regulations concerning the management of industrial areas come from different disciplines, especially laws related to planning and environmental and production activities.

In some regions the choice to go towards an APEA is compulsory, since it was decided in urban planning (e.g. Emilia Romagna), while in others the choice is voluntary (e.g.Tuscany). Although the national law was enacted over 10 years ago, its application in different regions is not settled yet, but still under development.

APEAs could potentially become a very popular model of production area in Italy, but to date there are only few cases of full implementation.

In the various Regions there is the enactment of laws and technical regulations, the implementation of experimental projects and (albeit only recently) financing, but the reality of APEAs in Italy has not yet established itself in terms of actual implementation.

It is therefore difficult to identify APEAs in Italy. To date, the most advanced experiences are in Tuscany, where it is possible to identify 12 production areas involved in a qualification path towards APEA, and in Emilia Romagna, with a process of public funding for APEA which saw the application of 42 industrial areas.

This gradual development is also linked to the fact that the different regional regulations provide for environmentally equipped areas a very broad field of application, stating that issues such as waste management, water resources, transport and logistics, security, etc. should be dealt with, thus covering all environmental issues in a systematic way.

However, some environmentally equipped areas are precursors, i.e. industrial areas that implemented solutions fully in line with the contents of Legislative Decree No 112, 1998.

For these areas the definition 'eco-industrial parks' is more correct, since they are production areas that do not fully comply with regulatory requirements of APEAs, but where environmental management initiatives such as collective management of certain environmental problems (not all) were activated, or collective facilities were built.

These experiences represent case studies with established characteristics in the review of Italian good practices, while actual cases of APEA will be mostly developing initiatives.

4.3. Environmentally equipped production areas and eco-industrial parks Cluster in Europe

Eco industrial parks are not identical to the Italian environmentally equipped areas, but they certainly show strong similarities.

First, the reference cluster, identified in the industrial area intended as a geographically defined and limited area, with production activities; then the adoption of engineering and management solutions designed to reduce environmental impacts, in fact making the environment variable a lever for competitiveness.

Indeed, the main difference between APEAs and EIPs is that while the former are governed by technical and planning rules, often involving public entities, the latter are usually created on the basis of future economic gain.

The involvement of the public administration is also less frequent, although it still constitutes a major subject in many cases.

Eco-industrial parks are spread globally, and it is possible to find examples of success in Europe, North America and Asia. Although it is difficult to define the categories, it is interesting to highlight some elements that differentiate the European EIP from the American or Asian EIP.

Although there are exceptions, in general in European countries the initiator of the development process of the production area is a public actor which promotes the EIP as a solution to territorial problems. The involvement may be associated with a form of local governance. On the contrary, in Asia EIPs are usually linked to the economic value related to the application of the principles of industrial ecology, especially as concerns the issue of waste.

Another difference is related to the size of industrial areas (much bigger in Asia), which makes it easier to trade secondary raw materials. In Europe, collective solutions are usually linked to the sharing of financial and human resources (services, collective facilities).

The EIPs in North America are more similar to the European ones, although the public presence in developing the production area is lower.

A further difference between the eco-industrial parks in Europe and those outside Europe is that in Europe there are many technology parks that are classified as eco-industrial parks, i.e. areas that are not strictly involved in production, but which play a role in research, communication and promotion for the development of environmental technologies and sustainable management solutions in the production industry. It is difficult to gain a thorough knowledge of the distribution of eco-industrial parks in Europe, America and Asia since there are no institutional structures of reference.

The summary of the presence of eco industrial parks in the world was based on data collected from studies, networks and devoted sites.

A study by the University of Patras (Greece) in 2003 on the state of the art of eco-industrial parks in the world identified over 100 cases, of which 42% in the U.S., 36% in Europe, 11% in Asia and 6% in Canada. In Europe, eco-industrial parks are mainly concentrated in the North-West, especially in England, France and Scandinavia.

The geographic department of Hull University mapped a distribution of about 30 eco-industrial parks in Europe. It points out that in some cases they are not actual industrial parks but science and technology parks or initiatives and programmes aimed at sustainable management of production areas.

In a Finnish study on eco industrial parks conducted in 2006, data on the number of eco-industrial parks in Europe is lower and equal to 20, while North American EIPs are 36.

An investigation conducted by the Chinese 'National Commission for Development and Reform' (presented on the website of *Cleaner Production in China*) analyzed the Eco Industrial Parks initiatives undertaken in Asia. There are about 40 initiatives for environmental industrial networking or Environmental Management. These initiatives were launched in eastern Asia: China, India, Japan, Philippines, Malaysia, Taiwan, Vietnam, Thailand, Sri Lanka.

4.4. Features of the cluster approach in the management of environmental issues

The environmental performances of an APEA, both in consumption of non-renewable resources and in emission of pollutants in the air, water and soil, are based on three important aspects: urban planning, plant and infrastructure equipment, and management.

The search for performance excellence starts with the way in which the spaces within the industrial area are designed. The second aspect is based on using the best available techniques (eg. dual networks for the water cycle, own production of energy from cogeneration or from renewable sources) in line with national and EC legislation on Environmental Integrated Authorization (IPPC Integrated Pollution Prevention and Control) and on common spaces and facilities instead of individual ones (e.g. industrial sewage treatment plant for the entire area, common waste storage areas, centralized basins for the collection and treatment of stormwater).

The third aspect provides for the optimization of synergies already existing among the various businesses and the unified management of spaces and of centralized systems (e.g. provision of a grant for the recovery of waste among the firms located in the area, area mobility management, area energy management).

A collective management is the heart and engine of improving the environmental performance of the production area. Through the use of collective facilities and infrastructure it makes it possible to provide businesses with services that allow greater protection and environmental control and a reduction in costs.

The manager is generally an expression of the realities present on the territory, and it may be a public, private or mixed actor.

The single production manager is also responsible for the definition and implementation of an area environmental programme and for monitoring the environmental performance of the area. The Area Environmental Programme starts by analysing the existing issues and proposes solutions involving relevant local stakeholders (e.g. government, companies, trade associations, managers of public services).

Existing initiatives concern the activation of virtuous mechanisms among firms (sharing of human and technological resources, materials and energy flows in an industrial environmental perspective), the delivery of services to companies by the subject tha manages the area, the infrastructural and plant elements available to the production areas.

The following table shows the main solutions of good environmental management identified in the 32 analyzed cases (17 international cases[10] and 15 Italian cases[11])

---

[10] Eco-industrial park of Devens (USA); Burnside Industrial Park (Canada); Kokubo (Japan), Naroda Industrial Estate (India), LIK (Indonesia); Kalundborg (Denmark), Parc Industriel Plaine de l'Ain, Sphere EcoIndustrie D'Alsace, Syndival Lancadcres (France); Crewe Business Park, Sustainable Growth Park (U.K); Ecopark Hartberg (Austria); Vreten (Sweden); Value Park (Germany); S.Perpetua di Magoda, Parque Tecnologico de Reciclado Lopez Soriano.

[11] 1° Macrolotto Industriale di Prato, Z.I. Ponterosso - San Vito al Tagliamento (PN), Z.I.P. Padova, Z.I. Udine Sud, Z.I. Castello Lucento – Torino, Z.I. Valle del Biferno – Termoli, APO Ferrara, APO Ravenna, APEA Monte S. Vito (AN), APEA Pianvallico (FI), APEA Navicelli (PI), APEA Scandicci, APEA Ozzano (BO), APEA Ostellato (FE), APEA Colbordolo (PU).

RAW MATERIAL CONSUMPTION AND WASTE MATERIALS	
Good management of waste from construction activities	Recovery and safe disposal of waste generated during construction activities. The Italian experience demonstrates the success of the initiative when the monitoring activity carried out by the operator is accompanied by specific requirements in the Implementation Standards of Urban Planning.
Collective organization of waste disposal	It refers in particular to the special waste collection from businesses, which can be performed by the area operator or, more frequently, by third parties.
Promoting the principles of industrial ecology	Exchange activities among enterprises from the perspective of recovery and reuse of waste used as secondary raw materials. Usually there are specific agreements among a limited number of companies with synergic production processes.
SPREADING OF CLEAN TECHNOLOGIES	
Fostering of clean production technologies	Provision of research facilities; promotion of innovative solutions also through exhibition spaces; activation of pilot projects. Initiatives are generally aimed at creating of new jobs.
ADMINISTRATIVE SUPPORT FOR COMPANIES	
Interface between companies in the area and public administration	The area manager acts as mediator between businesses and the public administration with regard to administrative proceedings. The service can be simply informative (the manager contacts the local authority to obtain information and passes on said information to the interested companies) or it may concern the drafting of forms or the actual issuance of permits as delegated by the Municipality.
Environment info desk for businesses	Setting up an information service on environmental issues regarding regulatory obligations and as support for the processes of environmental certification.
Support to businesses for investment and funding	Activities carried out by the area manager for businesses as regards investments and research of grants and funding.
TRANSPORTATION	
Agreements with the provider of local public transportation	Study of traffic flows and routes from home to work, on the basis of which the single manager signs agreements with the provider of public transport to improve the service in terms of frequency and stops.
Car sharing and car pooling	Setting up a system of vehicles to stimulate the activation of a system of collective transport, as an alternative to the use of individual private vehicles. The initiative aims at reducing traffic in both commuting and missions.
TRAINING AND TECHNICAL SUPPORT	
Training on environmental issues	Courses for technical business personnel organized by the area manager, aimed at a proper application of environmental and

	safety regulations. Training and information activities for companies in the area offered directly or through third parties.
Eco – efficiency centers	Technical assistance and training to companies to improve their energy and water efficiency and in the production and disposal of waste.
SETTING UP OF BUSINESSES	
Environmental remarks in the selection of companies that wish to settle in the area	Generally carried out by drafting a questionnaire that the interested companies are required to fill in. The area manager is responsible for the selection as actor in charge of selling parcels of land or on behalf of the local authority.
Checking the companies status of compliance in terms of environmental requirements	Check up the relevant documentation prior to the settlement and visits to the site at the time of installation. This service is successfully carried out where companies have adopted an area Environmental Management System.
Envisaging environmental clauses in the contracts of sale for parcels	Provision of specific environmental clauses in contracts of sale for land parcels (e.g. when planning the areas for material handling or storage of waste) or request to adhere to the environmental policy.
Urban-environmental advantages	Removal of infrastructure costs, reduction of safety, health, and environmental taxes.
Landscape protection	Preserving the environment and integrating the business with the surrounding landscape through a centralized management of the development of the production area.
INNOVATIVE MANAGEMENT SOLUTION	
Implementation of environmental management system	Implementing an Environmental Management System complies with regulatory standards (ISO 14001 or EMAS) by the subject that manages the area. The EMS can provide a business involvement in varying degrees, for example in the adoption of common procedures or construction of a participatory environmental program. These elements also depend on the area of infrastructural facilities available
Creation of a territorial information system of the industrial	Development of a GIS on-line with all spatial information (geographical, cadastral and possibly environmental). The GIS is accessible by the companies settled and may represent a geo-referenced base that allows performance monitoring of the area. The management of the SIT is borne by the operator or company in charge of the area.
Agenda 21 meeting to define the action for cluster	Organise meetings within the local Agenda 21 to encourage the productive area to address environmental sustainability issues (energy supply, waste management, cleaning, transport and construction).

	Through working groups involving representatives of companies can address problems related to environmental issues and identify solutions that best suited for that territory.
Definition of an environmental action plan based on performance standards	Participation of local communities and businesses in programs of environmental action through the application of standards of environmental, social and economic performance and certification. The standards are defined by a committee made up of representatives of local production activities (trade associations) and citizens.

Table 4. Solutions for good environmental management

5. Cluster supply chain

5.1. Cluster "supply chain": definition and boundaries

The concept of this kind of cluster is not yet an 'autonomous' element in the economic literature, as is rather the case for other types of clusters (for example the large and consolidated strand of economic literature on *industrial districts*, or the most recent and innovative one on *environmentally equipped production areas*).

Although the meaning has not been conceptually developed yet, references to the 'supply chain' cluster are widespread and detectable in several contexts, both legal and academical.

Starting from Porter's remarks (1990), the *cluster theory* originally focused on identifying the key features of the 'cluster entity', with the aim of analyzing the ways in which its operation functioning mechanisms and its internal dynamics are able to determine a competitive advantage for businesses/industries belonging to the cluster itself. According to Porter, clusters are geographic concentrations of interconnected companies, specialized providers, service providers, and associated institutions in a particular field[12]. Firms located in the same area have the opportunity to operate with ease, in coordination along the value chain. The cluster is based not only on goods and *material* resources, but also on *intangible* resources such as development and exchange of knowledge, expertise, and relationships. These elements make a territorial area unique, an area in which human factor and knowledge make the difference when compared to any other area.

In other words, the supply chain is conceived as a geographically defined element within the cluster. This approach is confirmed by that part of literature aimed at investigating the connections between cluster theory, supply chain and supply chain management theory. This is a trend not developed theoretically and empirically yet, but which contributes to clarify some important aspects of the relationship between cluster and supply chain. Some authors observe that one of the key elements of Porter's cluster theory - the benefit deriving from knowledge exchange and cooperation between firms - is shared by the theorists of the

[12] *"A cluster is a geographically proximate group of interconnected companies and associated institutions in a particular field, linked by commonalities and complementarities"* (Porter, 1998).

supply management theory13 (De Witt et al., 2006; Mentzer et al., 2001). In fact, supply chain management aims at producing benefits for companies that are part of the chain through process integration and the set up of long term relationships (cooperation, trust) among companies. When companies belonging to the same supply chain operate in the same geographical context, they benefit from coordination and from the boost to competitiveness and innovation that comes from their geographical proximity (Mentzer et al., 2001).

To sum up, according to Porter the factors that characterize the cluster can be identified in geographical proximity, production specialization and in the interaction among the different actors in the cluster.

In this sense and for the purposes of our analysis it is relevant to note that proximity both upstream and downstream the supply chain is what facilitates interaction and promotes a continuous exchange of ideas and innovations.

Some studies aimed at providing empirical evidence of the existence of positive synergies, in terms of the impact on competitive performances that arises from belonging to a cluster. According to De Witt et al. (2001), the geographical proximity of firms belonging to a supply chain allows long-term competitive advantages that are more stable than those achieved by outsourcing to distant companies. Other studies highlight that the relational factors that characterize the relationship between buyers and suppliers belonging to the same cluster are able to enhance their performance in the long run (Noordewier *et al.* 1990; Corsten and Kumar, 2005). More specifically, the key elements that determine long-term competitive advantages are the *interdependence*, the mutual *trust* and *shared goals and commitments* that pervade the relationship among actors in the chain within a cluster, and the reduction of conflicts among the actors themselves (Ganeson, 1994, Doney and Cannon, 1997, Kumar et al., 1995).

A second important connecting element between cluster and supply chain refers to a particular connotation of the cluster, defined by the presence of a large multinational company surrounded by a 'halo' of suppliers ("A large demanding purchaser, such as a major multinational firm [...] surrounded by a 'halo' of suppliers", Johnston, 2003). In this respect, territorial location still plays a central role in the competitive dynamics of the cluster: the proximity of suppliers of a large enterprise allows for the development of agglomeration economies thanks to the direct or indirect links that are established between the economic activities upstream and downstream.

[13] In spite of the popularity of the *supply chain management* both in theory and in practice, there is no unanimous agreement on its meaning, also because its development in the managerial and academic fields is relatively recent. In his important work of 2001 (*'Defining Supply Chain Management'*), Mentzer defines the supply chain management, or management of the supply chain, *'a systemic, strategic coordination of the traditional business functions within a particular company and across businesses within the supply chain, for the purposes of improving the long-term performance of the individual companies and the supplu chain as a whole'* (Mentzer, 2001). With 'supply chain' he means *'a set of three or more companies directly linked by one or more of the upstream and downstream flows of products, services, finances, and information from the source to the ultimate customer'* (Mentzer, 2001).

More recently, the literature recognizes the need to introduce a new meaning of the word cluster alongside its more 'traditional' concept. Nowadays, the technological improvement of communication systems and of distribution networks at global level makes it possible to talk of 'virtual' clusters, where the element of geographical proximity is missing while emphasis is placed on features such as exchange and sharing of information and knowledge among the actors in the cluster (Johnston, 2003).

According to Rullani[14], this new type of cluster is characterized by an "evolved" form of proximity, not only physical but also virtual, among people and businesses that use technological mediation to develop 'close relationships' (easy, frequent, reliable, complex) even when they are physically distant from each other, thus replicating in a virtual space the benefits of proximity that in the past were a typical (and almost exclusive) characteristic of the territory.

The academic theory of virtual clusters is joined by their institutional recognition. In its working paper annexed to the EC Communication "Towards world-class clusters in the European Union: Implementing the broad-based innovation strategy" [COM(2008)652][15], the European Commission, while recognizing that most definitions of cluster focus on two factors - the concentration of one or more sectors in a given geographical area and the importance of networking and cooperation among enterprises and institutions - indicates that the spatial dimension of a cluster is variable and not necessarily limited to certain geographic boundaries, depending, among other things, on the ability and willingness of its actors to perform changes that are functional for the development and preservation of relationships that feed the cluster itself.

The definition of clusters developed by the OECD makes the "release" of the concept from the element of geographical proximity even more explicit, emphasizing once again the production and exchange of knowledge within the value chain as a key element of identity of the cluster "Clusters are characterised as networks of production of strongly interdependent firms, knowledge-producing agents and customers linked to each other in a value-adding production chain" (OECD, 1999).

A final element of analysis comes from recent literature aimed at defining the links between the concepts of cluster and supply chain in the context of today's 'knowledge-based economy". If it is true that the cluster theory in general takes a macroeconomic perspective in which the theories on supply chain refer to a purely microeconomic level, it is possible to recognize some key elements the two theories share and that can lead to the identification of significant similarities between the two concepts (Sureephong et al., 2008). In particular, the success of a cluster is often due to the presence of a Cluster Development Agent (CDA), an organizational entity that represents the different socio-economic actors in a cluster, acts as coordinator and facilitator of cooperative dynamics among the actors, promoting knowledge sharing, innovation, the ability to communicate and the mutual recognition and trust[16].

---

[14] Enzo Rullani, *Cluster: tendenze e scenari nell'economia globalizzata*, "Pattern of clusters evolutions" conference proceedings, Venezia.

[15] European Commission (2008b).

[16] Notice how this role and functions correspond to what in the context of E.U. policies are the tasks of the governing body of a cluster that is established when a cluster is 'institutionalized' (Cfr.: CE, 2002).

The literature recognizes a strong similarity between this role and the one that in the context of the supply chain management theory is played by the Supply Chain Facilitator (SCF), the person that guides the relationships among actors in a supply chain, and encourages the development of relationships and cooperation among them. If this role was originally played by the large multinational company, in the context of the current '"knowledge-based economy"' its function can be successfully carried out by other bodies - such as universities, associations, local institutions, *ad hoc* coordinating organizations - that belong to the same cluster as the socio-economic actors in the supply chain (Sureephong *et al.* 2008).

5.2. Supply chain cluster in Italy

As abovementioned, for the purposes of this analysis we should highlight how within this theoretical model it is possible to trace a first description of cluster as a type of "supply chain", relating to a local chain dimension, coinciding or included in the 'classic' industrial district, physically located in a given territorial area. The cluster supply chain is especially notable for the presence of a primary industry supported by several companies specializing in various phases of the industry. Thanks to territorial proximity and the presence of a socially cohesive community, these companies have the ability to easily operate by coordinating along the value chain (Sacco, Ferilli, 2006).

With respect to this first description, the literature emphasizes how clusters linked to local industries represent an area of strong identification and specialization in the economy of many Italian regions, although generally the cluster does not reach the regional dimension (Bardi, Bertini, 2005). In many contexts, some leading companies' ability to grow and export can foster local induced activities and is an example increasingly imitated by dynamic new small businesses, a catalyst of growth processes and territorial specialization. Examples of chains related to this meaning may be found in many traditional Italian industrial districts. Consider, for example, the area between Carpi and Reggio Emilia for knitwear and clothing, the area of Forlì and Cesena for the food industry, the Bassa Bresciana for shoe manufacturers, the furniture chain located between Matera and Puglia, etc.

More recently, the literature shows how in many cases this type of articulation of the cluster supply chain reached maturity during the 90s, and how the highest degree of complexity of the industry and the increased competition on international markets led to an expansion of the classical concept of district, which meant an area in which a single homogeneous production chain was focused. To respond to these changes, the new articulation of a cluster is as metadistrict, that is a territorial chain not concentrated locally, but that creates a widespread, sectorially specialized network - on a regional, multi-regional or national scale - with strong interaction/interdependence/competition among the business realities belonging to it. In many cases this connotation can be traced back to the development of products that identify the so-called 'Made in Italy' in some specific areas (typically the North East of the country and Tuscany), mainly focusing on sectors such as fashion, home, typical food products (e.g, buffalo mozzarella in Campania, bresaola of the Valtellina), light engineering (e.g. the biomedical field in Emilia-Romagna) (Rullani, 2002).

In this respect, the cluster supply chain is characterized by the different role of territorial contiguity of businesses with regard to cooperative and competitive dynamics that characterize it. Studies and remarks note how the metadistrict was established to overcome the old links among companies - where proximity is a prerequisite for the existence of relationships that can trigger processes of technological exchange and learning – by creating new links related to the development of new technologies and services necessary to keep those companies competitive. Globalization and new information technologies do not "disengage" companies from the territory, but help to provide the cluster with another operational dimension, in addition to and not as a replacement of the local dimension (Zucchetti, 2003).

The change of competitive scenarios also influence the *nature of relationships among companies within the supply chains and production chains*. While within the traditional districts of the '70s and '80s the relationships among companies along the supply chain are basically egalitarian, mostly informal and usually direct, beginning from the '90s these characteristics deeply change. The stability of the suppliers' relationship with a limited number of clients, typically located within the district, is no longer a 'dogma', i.e. the degree of 'mobility' of the system is increasing.

Furthermore, to support the internationalization and export activity aimed at enhancing the 'Made in Italy', territorial production chains sometimes adopt a form of 'twinning' among the original industrial districts. Aggregation, in these cases, also aims at improving the awareness of belonging to a quality industry, promoting high visibility of the territories, increasing transparency and credibility with all stakeholders - investors, tour operators, consumers, etc. - and especially at increasing cooperation among industrial districts within a sector at national level[17] (Fontana, 2007). Consider, for example, the chain of leather products for shoes and clothing in Tuscany, or the textile chain in Lombardy.

Finally, the literature recognizes that even in our country the previously defined *virtual clusters* are taking shape. According to Rullani[18], a virtual cluster may arise from very different evolutionary paths:

1. From *previously existent clusters* that learn to master new technologies, combining the advantages of the existing local network with the advantages deriving from long distance relationships and multi-territorial sharing;
2. From *medium enterprises*, once "plunged" in a regional system, that use new information technologies to extend their supply and distribution networks in the global circuit (without losing their roots in the area of origin);

- From *multinational companies*, that discover the importance of differences and specificities of each territory they have access to, eventually anchoring themselves permanently to certain territorial specializations. These specializations are enhanced

[17] An example is the 'agro-ichthyc-food industry' twinning between the districts of *San Daniele del Friuli* (famous for the prosciutto *Dop*), *Nocera Inferiore-Gragnano* (famous for its tomatoes and pasta), *Mazara del Vallo Co.S.Va.P.* (fisheries) and *Vulture* (specialized in wine, fruit and vegetables, olive oil, cheeses and dairy products) (Fontana, 2007).

[18] *Ibidem.*

and expanded through access to that global network of virtual proximity and coexistence made possible by the multinational company itself.

If, in general, our country sees setbacks in the rise of virtual clusters (mainly due to the difficulty for the traditional districts/clusters of SMEs to evolve rapidly[19]), the third path here outlined leads to the identification of a third type of supply chain cluster in Italy, defined in terms of **'trade mark' (or 'brand') industry.** The connotation, though not prevalent in the Italian production reality (characterized by a structure of small and medium enterprises), can be recognized in all those cases in which large companies and/or multinational companies or large chains (whether the property be domestic or foreign) operate on a national scale through a widespread and 'loyal' network of suppliers and subcontractors. The following section presents in detail this specific type of supply chain cluster, widely present in the European and international context.

5.3. Supply chain cluster in Europe

While in Italy the patterns adopted by the supply chain cluster are significantly influenced by the specificity of the economic and social fabric, in the rest of Europe chain clusters are generally configured as business combinations, often multi-sectoral, linked by supply relationships (of goods and services) at different levels with large leading companies and/or multinational corporations.

Due to globalization, in recent decades the European context has wirnessed the emergence of the concept of cluster supply chain in terms of virtual aggregation. As anticipated, this type developes around large companies that become a drive of the territorial development of some industrial areas at regional level, through processes of "virtualization" of the flow of information and of internationalization of the supply and distribution networks. These companies are able to stimulate and support the growth of competitiveness poles of global significance, within which operate production chains belonging to one or more related manufacturing sectors.

Many corporations belonging to different sectors, such as IKEA, Ericsson, ABB, H&M, Volvo, belong to this type of cluster (Sölvell, 2006).

Therefore, in this type of cluster the globalization of the supply chain and of innovation goes hand in hand with the strategic importance of the so-called *'local environments'*. While local markets seem to have exhausted their function of driving force for most goods and services, the local concentration of activities related to various production sectors (old and new) continues to act as a driver of (multinational) business innovation. In other words, the multinational companies' ability to compete is linked to their ability to act as *"insiders"* (through their branches or subsidiaries) within the most dynamic regional clusters, and to

---

[19] Among the main factors that hinder the 'virtual' development of Italian districts and clusters, the literature identifies chronic saturation of work environment (full employment), of infrastructures (congestion), of free space (crowding), and of environmental tolerances (with balanced under constant stress) (Rullani, ib.).

globally coordinate activities in order to connect in large-scale networks global markets and the innovative thrust deriving from local clusters[20] (Sölvell, 2002).

In summary, the literature recognizes in this type of cluster the persistence of some distinctive features of the supply chain cluster, except for the element of proximity, now redefined on a virtual basis. The specialization of production and the interaction among actors in the production chain continue to classify the cluster, developing through a virtualization of the flows of knowledge that stimulates the constant change and "upgrading" of products and services, thus creating the bases for advanced supply chains that are diversified geographically as well.

5.4. Features of the cluster approach in the management of environmental issues

This paragraph analyzes the characteristics of the "supply chain" cluster with reference to the elements that characterize its management from an environmental point of view. To this end we gathered a review of existing theoretical and empirical studies, oriented along the three trends of literature identified as priority areas of investigation about the connections between *cluster approach*, *environmental management* and *supply chain management*:

- The trend of the so-called **Green Supply Chain Management,** or the development of supply chain management practices in environmental terms and their connections with the application of Environmental Management Systems (EMS);
- Studies and experience related to the adoption and implementation of **strongly product-oriented EMS**, so-called **POEMS** (*'Product-Oriented Environmental Management Systems'*), mainly based on the need to identify, assess and manage the so-called "indirect environmental aspects" (introduced by EMAS II Regulation) related to the product;
- The literature on the application of the concept and methodologies of **Life Cycle Assessment (LCA) in line with the chain/supply chain.**

These three trends have in common (albeit with different meanings at times) the acknowledgement of the need to adopt an inter-organizational approach to environmental management based on coordination and co-operation among different actors in the sector and not necessarily linked the territory. This need is the culmination of an evolution in the application of environmental management tools, characterized by the progressive recognition and assimilation by the businesses of a management approach inspired by the so-called life cycle thinking[21], marked by a greater sense of awareness and responsibility towards the environmental impacts that their activities have outside the confines of the production site (Carnimeo et al., 2002).

---

[20] In literature such clusters are sometimes called *"Hollywoods"* with reference to the californian cluster, world leader in the *entertainment* industry.

[21] In summary, *Life Cycle Thinking* can be defined as a 'cultural' approach that aims at focusing all management aspects of a product through a single 'magnifying glass': its life cycle. Under this approach, the environmental impacts (actual or potential) generated during the life cycle should be considered in an integrated manner when designing, developing and managing a product. For further reading see: Carnimeo, Frey, Iraldo (2002).

As previously stated, the first trend refers to the so-called Green Supply Chain Management, or that part of the literature born by integrating and "contaminating" environmental management and supply chain management studies, and targeted at investigating supply chain approaches and management logic from an environmental point of view (Srivastava, 2007)[22]. This is a trend literature regards as underdeveloped although potentially fertile for investigation, and so far studied in depth only in relation to certain phases or activities of the supply chain *viewed individually* (eg. green design, green procurement, reverse logistics, etc.) (Srivastava, 2007; Sharfman *et al.*, 2009). Conversely, the same literature agrees in giving particular importance to studies and contributions that on the one hand account for the reasons for adopting environmental supply chain management practices, and on the other hand account for the benefits and difficulties related to the implementation of the same practices.

At the empirical level, studies have shown the existence of a wide range of internal and external factors that may push companies to expand the environmental management to upstream and downstream supply chain activities: from the need to respond to increasing *pressures from external stakeholders* (e.g. consumers or institutions), to the need to *ensure compliance to more stringent environmental regulatory requirements*, to reasons of a *strategic nature*, or related to the opportunity to gain a competitive advantage (Sharfman *et al.* 2009; Darnall *et al.* 2008; Nawrocka, 2008). Corbett and Decroix (2001) affirm that nowadays the need to extend environmental management practices to the supply chain to improve performances is so recurrent in studies it has become a 'mantra' ("We have heard [...] various versions of the 'mantra': 'the next step forward in environmental improvement lies in supply-chain coordination', Corbett and Decroix, 2001).

Important for the purposes of this analysis are those studies that investigate the reasons that push companies to extend the environmental management to the supply chain under a cooperative approach. In this context, the empirical contributions emphasize how - beyond the reasons more closely tied to a proactive and "'value-driven'" environmental management - the key element that pushes companies towards a logic of cooperation with suppliers is mainly due to the uncertainty of information that governs the nature and extent of environmental impacts associated with the production process upstream and downstream activities, and the complexity and difficulty of decision-making processes frequently created by this uncertainty (Vermeulen and Ras, 2006 ; Sharfman et al., 2009).

In the context of a relatively scarce literature it is significant to note that some recent studies focus on the relationship linking the adoption and implementation of an Environmental Management System and the development of supply chain management practices, with the main objective to investigate (i) if and what kind of correlation exists in the adoption of these management practices, that is (ii) if and to what extent the adoption of one practice influences the adoption of the other (Darnall et al. 2008; Nawrocka 2008).

The process of gradual 'opening' of EMS and their redefinition in inter-organizational terms is recognized by the literature as a significant innovation in environmental management

[22] Srivastava's work (2007) offers a very detailed review of GSCM literature.

practices, characterized by the gradual assimilation by enterprises of the management logic inspired by the so.called life cycle thinking. This trend is identified by some authors with the gradual extension of the objectives and scope of management systems to environmental issues related to the lifecycle of the product (or service). The result of this "integration" is called POEMS - Product Oriented Environmental Management Systems (Klinkers et al., 1999).

The links between environmental impacts related to production and product make the environmental management tools within the company ineffective by shifting the emphasis on the relationships the company itself has with the actors that influence those impacts. By this logic, the goal of the management system is not only to ensure the implementation of the company's environmental policy by governing the processes and internal resources, but also to manage relations with the outside world, to foster and promote dissemination of that policy to other actors that share it. In other words, the scope of implementation of the management system becomes that of the actions and interactions through which many actors manage the impacts related to the different stages of the product's lifecycle (Sharfman et al., 1997).

The literature on POEMS highlights the benefits and limitations of these tools. In fact, although taking into account the entire lifecycle can provide more opportunities to reduce the environmental impact associated with the products - either through actions on specific issues and joint efforts involving different stakeholders along the chain - (Sharfman et al., 2009), thus contributing to the achievement of tangible improvements in performance (van Berkel et al. 1999; Charter and Belmane, 1999; Brezet and Rocha, 2001), the process of 'opening up' the environmental management systems is not immediate or 'painless' for businesses.

At European level, some experiences of POEMS developed since the '90s. In the Netherlands the signing of voluntary agreements between various industries and local authorities has allowed the development and funding of some pilot projects for product-oriented EMS. A study of sixty Dutch companies shows that the most sensitive element in the development of such tools is the difficulty for organizations to obtain the necessary information from their suppliers, because of the other companies' unfamiliarity with POEMS and the little influence of single organizations, especially small ones, on the whole product chain. Other experiences have demonstrated the 'compatibility' of the approaches and the requirements of international and European standards on environmental management systems (ISO 14001 and EMAS) and POEMS.

In France, a project to develop product environmental management in the automotive sector involved 250 companies, mostly SMEs, with the aim of honing tools and methodologies for the development of POEMS in organizations belonging to the sector. In this case the main project stimulus was of a regulatory nature, in connection with the requirements of the European Directive End of Life Vehicle, which demands specific knowledge of the supply chain[23] (Andriola et al., 2003a).

---

[23] Directive 2000/53/CE.

In Italy as well several POEMS experiences were carried out in medium-large businesses, or under programme agreements with public institutions (Ardente et al., 2006; Andriola et al., 2003a). Overall, the Italian experiences and the experimental applications in Europe (mainly in the Netherlands, France and Denmark) have not led to a 'coding' of POEMS by any international or Community standard. However, it must be noted how these experiences have contributed significantly to the development of methodological and management tools aimed not only at accounting for the impacts associated with the lifecycle of the product, but also at involving the suppliers in the assessment and quantification dynamics of these impacts. Among these tools, the methodologies of Life Cycle Assessment (LCA) are predominant[24].

A vast literature deals with the technical and methodological aspects as well as with numerous case studies and methods of dissemination of the LCA in Italy and in other advanced countries. For the purposes of this analysis, it is particularly important to observe the studies that investigate the role and mode of application of the LCA in the context of the management of supply chain relationships and management. In this respect, the literature recognizes that the potential of a LCA is revealed mainly in the form of a methodology capable of triggering an inter-organizational and networking approach within the chain. In fact, the steps that make up the tool[25] imply the need to activate along the entire supply chain the information and communicational channels necessary to collect and process data and information that allow to quantify the interactions between the 'product system' analized and the environment (Pesonen, 2001, Lefebvre 2000, Krikke et al. 2004; Sarkis, 2001; Sroufe et al., 2000).

Finally, it is important to note that some studies go further in analyzing the relationship between LCA and supply chain. They apply the methodology of Life Cycle Assessment as an opportunity to "restructure" the supply chain, with the aim of improving the environmental performance associated with a specific "product system". It is essential to take into account a number of factors for the LCA to be able to effectively contribute to

---

[24] The most common definition of this method comes from the Society of Environmental Toxicology and Chemistry (SETAC): *"a LCA is a process to evaluate the environmental burdens associated with a product [...] by identifying and quantifying energy and materials used and wastes released to the environment to assess the impact of those energy and materials used and releases to the environment; and to identify and evaluate opportunities to affect environmental improvements. The assessment includes the entire life cycle of the product [...] encompassing, extracting and processing raw materials; manufacturing, transportation and distribution; use, re-use, recycling, and final disposal"* (SETAC, 1993). If the inter-organizational approach of the POEMS consists in considering the product-system as *"a network of operations linked together by flows of materials and energy [...] that ties activities and processes in different organizational contexts"* (Heiskanen, 2000), the use of a LCA proves itself extremely useful, since it can be aimed at finding the optimal solutions for the entire product-system, regardless of what would be preferable from an environmental point of view for each single process (or single organization) (Carnimeo *et al.*, 2000).

[25] The structure of the LCA derives from the scheme set up by SETAC in the early '90s with the aim of proposing a common approach for all the analyses carried out until then. This format is still the basic structure from which the subsequent changes and amendments derived. Internationally, the scheme proposed by SETAC was complemented by the ISO Regulations 14040 series, which rule the drafting of LCA studies. According to the ISO general regulation, the assessment of the life cycle must include the following steps: 1) definition of the objective and scope of the study, 2) inventory analysis, 3) impact assessment; 4) interpretation of results.

improve the environmental performances associated with the implementation of a product or service which combine multiple actors along the value chain (Hagelaar G. and van der Vorst J., 2001):

- The results of the LCA methodology are closely tied to the definition of the objectives and scope of this tool that are set at the startup of the evaluation process;
- The application of LCA in the perspective of the supply chain involves strong cooperation among all actors involved, not only in terms of trust and openness, but also of *transparency in the shared data and information* and of *consistency in their policies*.

In conclusion, it is possible to account for some empirical evidence related to the implementation of tools linked to *Life Cycle Thinking* in the logic of supply chain management. When LCA is integrated into the environmental management system in a logic of supply chain dynamics management, it is to be assumed that companies that use and promote this approach are able to affect the environmental impacts or influence the behaviour of actors that are external to the "boundaries" of the companies' organization. This is a pre-condition for the integration of product logic to be effective. If this pre-condition occurs, the use of LCA within the supply chain relationships provides important contact points, synergies and complementarity with an "extended" environmental management system. The above mentioned approach of "indirect environmental aspects" introduced by the EMAS II Regulation is well suited to be the "leverage" through which to introduce these tools.

The experiences in the analyzed scenario refer to the implementation of a LCA aimed at including *product-oriented* logic in an environmental management system. A first empirical evidence refers to the transition from a single company to a supply chain perspective, in the implementation of the Initial Environmental Analysis (as provided by ISO 14001 and EMAS) that crosses the company's boundaries and allows to identify and properly assess the indirect environmental aspects thanks to the adoption of the LCA. Although in practice the EMAS Regulation does not require the company to conduct a thorough LCA, it is clear that knowledge of these impacts is necessary, especially when it is functional to the identification of the most significant indirect environmental aspects. Understanding the impacts of product disposal, or of the product's packaging, may be decisive for a company that uses large distribution channels and targets/is addressed to the final market.

The same applies when a company uses a transport network for the delivery of its products to intermediate customers "spread" on the territory. This activity can be carried out as part of a chain. In the case of the local supply chain that drove the PIONEER project26, for example, a simplified LCA of paper products was implemented (streamlined or screening LCA, to use English terminology), useful to companies in the supply chain to identify

[26] The Life PIONEER Project (2003-2006) had the objective to define and experimentally implement a methodology based on the EMAS Regulation to the paper industry district of Lucca. The methodology has promoted a cooperative and integrated approach to the environmental management at local level, aimed at involving all stakeholders in actions to improve the environmental performance of the territory. For further reading, see: Frey and Iraldo, 2008 and the website www.life-pioneer.info

particularly relevant indirect aspects and to determine the connections among the various business activities of companies operating at different stages of the supply chain, in order to focus on critical points on which it was then possible to work with joint programmes and improvements.

LCA can be applied effectively in the supply chain environmental management also in the pursuit of ways to improve eco-efficiency. The so-called Life Cycle Costing (LCC), for example, provides guidance on how to integrate "conventional" accounting with an approach that allows to identify longer-term strategic opportunities and efficiency margins. An interesting example of application of LCC is that of the logistics of the large company Xerox (Bennet, James, 1999). Careful analysis of logistics costs by the company (related to product distribution and the recovery of remains to be reused) showed efficiency margins in its supply chain. However, in order to seize these opportunities some interventions were needed to design, manage and reorganize logistics flows which would mean a different allocation of costs and benefits among the different stakeholders (Xerox suppliers and customers).

Examples of these solutions were the internalization of the costs of packaging (including disposal) in accordance with the suppliers, and the standardization of packages so that they were adaptable to every product and, above all, reusable by customers to pack the product being replaced (and returned) at end of life. With an initial investment of 4-5 million dollars, Xerox estimated annual savings of $1.2, to which some "intangible" benefits (e.g. in the management and organizatin of logistics: handling of homogeneous packaging, reduction of operations' time, etc.) added up. Although the overall comparison gave a positive outcome, some phases of the chain showed an increase in costs. Had the company not analyzed costs and potential benefits in the different stages of the lifecycle using the LCC method and reasoning in terms of actions promoted and coordinated by Xerox in an integrated logic with the other industry players, they would have never independently decided to engage in the improvement programme.

There are other examples that show how attention to the product under a "supply chain" environmental management system driven by the logic of the life cycle can constitute a solid basis on which to build a strategy oriented to an "environmental" customer satisfaction27. An interesting case is that of Baxter International. After stating in its policy that "we will work with our clients to help them tackle their environmental problems", the company managed to translate this principle at the operational level by applying an approach heavily oriented towards the green supply chain management. Resolutely going beyond the boundaries of its organization, management at Baxter decided to take charge of the issues related to the disposal of waste that results from the use of intermediate products in the "downstream" activities. By this logic, a waste auditing service was activated with the aim to

[27] Although the concept of environmental *customer satisfaction* is not totally alien to the logic of a management system, it is usually related to social actors, since the main "customer" in environmental terms is the one that suffers the externalities of production. When this customer coincides with the traditional customer, however, the dynamics with which relationships can be managed differ from those that govern relationships with social *stakeholders*.

verify the needs in the waste management of their intermediate customers. The activity developed by this service has produced a set of suggestions/recommendations, then implemented through a thorough and extensive redesign of products and services in order to minimize their impact downstream of the production process (Fuller, 1999).

Another interesting case is that of the initiatives undertaken by a group of U.S. chemical companies in a "horizontal supply chain" (Elwood and Case, 2000) for the management of orders from customers in a logic of eco-efficiency. In this case, the needs of production efficiency go together with the attempt by the manufacturers to reduce the environmental impacts of their products in all stages of the lifecycle.

The examples here reported show some approaches that were made possible only through the cooperative relationship among the companies in the supply chain and, although not directly related to the application of LCA, they clarify that a product-oriented logic can effectively engage on the environmental management system of a company that operates as a producer or as a customer in any type of supply chain. This logic provides crucial support for the management of customer relationships, and bridges the gaps of the system when it comes to identifying the customers' needs, defining interaction modes, handling complaints and returns (for environmental reasons), reviewing the contract (which can include requirements concerning the product's impact) and measuring customers' satisfaction, which is essential in order to assess the environmental competitiveness and "green" marketing strategies implemented by the company itself.

In terms of marketing and environmental communication, there is another interesting example of application of product logic within a supply chain. It is the opportunity, now available and potentially very effective, to certify the environmental impact of a product of a "local" chain or a group of producers on the basis of an international scheme based on the ISO standard. The EPD international system (Environmental Product Declaration), currently managed by a body comprising representatives of some EU countries (including Italy) was born with the objective to certify the environmental performances of a product or service of a single firm.

Because of the evolution of the system and its gradual spreading (currently it counts more than 100 companies) there is the the need to promote products and services that come from an entire production system (i.e. a "cluster" or chain of companies) precisely because of environmental marketing objectives of a typical product or a product tied to a "brand". Thanks to the recent amendment of the EPD certification system groups of producers (from a district, a chain, a geographical area) were offered the opportunity to develop an *Environmental Product Declaration* that can enhance the excellent environmental performances of their "average" product.

6. Conclusion

SMEs are to be considered a crucial target if policy makers really want to pursue sustainable development. The environmental problem does not fully emerge if one considers individual firms (although in some cases there can be serious impacts on local environments and communities exerted by a single SME), but it pertains to their combined and cumulative impact across sectors. Therefore. these companies are responsible for a large share of business environmental impacts.

These observations have stimulated the development of the so-called "cluster approach" to manage the environmental issues of a large number of SMEs located in limited territorial areas. As above mentioned, this concept is not limited to the classical configuration of the industrial district (geographically confined industrial areas) but it also encompasses environmentally equipped production areas ("APEA"), or industrial parks up to the interactions along the supply chain.

For example, the Industrial District is a local system with the presence of a prevalent production activity carried out by a group of small independent firms highly specialized in different stages of the same production process. On the contrary, Production Areas represent an organizational model characterized by the territorial element but, unlike the District, they are concentrated in areas that are easier to define and circumscribe geographically, and do not necessarily show the presence of one or more specialised production sectors.

The networking approach allows enterprises to co-operate by identifying and assessing similar environmental aspects and by finding technological and operational solutions that can be applied to similar production processes and products, as well as by defining organisational structures suitable for the same kind of production cycles

In the APEA case, co-operation is facilitated by the 'physical contiguousness' and there are synergies both in improving the environmental impact on the same local eco-system, and in interacting and communicating with the same stakeholders (local population, authorities, etc.).

In some cases, a network was created among SMEs within a 'cluster' in order to foster information exchange and experience dissemination and to define and apply common solutions to similar environmental, technical and/or organisational problems, or to share environmental management resources (Iraldo & Frey, 2007). A specific kind of co-operation within a cluster of organisations takes place in the supply-chain: when a large customer is willing to support small suppliers in the EMS implementation process, then all the smaller organisations involved in the supply chain can greatly benefit from networking. This approach proved to be effective in some Member States such as Germany ("Konvoi" approach), Spain (co-operation in the tourism supply chain), Nordic Countries (Denmark and Sweden) but in particular in Italy, where by means of the so-called APO "Ambiti Produttivi Omogenei", it has shown its effectiveness in promoting the environmental compliance of SMEs.

Therefore, we collected some empirical cases to demonstrate how the innovative approach to environmental management called "cluster approach" can be an effective tool available to SMEs in order to improve their environmental performance and find innovative management solutions.

The goal for future research and experimental initiatives should be the development of the cluster approach and its structural inclusion in policy-making.

Author details

Francesco Testa[1,*], Tiberio Daddi[1], Fabio Iraldo[1,2] and Marco Frey[1,2]
[1]*Sant'Anna School of Advanced Studies, Pisa, Italy*
[2]*IEFE – Institute for Environmental and Energy Policy and Economics, Milano*

7. References

ADEME (2007). L'environnement et la maîtrise de l'énergie dans les PME, Paris.

Andriola L., Luciani R., Sibilio S. (2003b), 'I sistemi di gestione ambientale orientati al prodotto: POEMS, un nuovo strumento', Ambiente, n. 8

Bardo A., Bertini S. (a cura di) (2005), Dinamiche territoriali e nuova industria, Dai distretti alle filiere, V Rapporto della Fondazione istituto per il Lavoro, Maggioli Editore, Milano.

Becattini G. (1979). "Dal 'settore' industriale al 'distretto' industriale. Alcune considerazioni sull'unità di indagine dell'economia industriale", Rivista di economia e politica industriale, n. 1,

Calcagno D., Del Borghi A., Gaggero P.L., Sacerdote I. (2006), The contribution of LCA methodology to the environmental sustainability of cement production, 15th International Symposium on Mine Planning & Equipment Selection MPES, Torino - Italy, September 20 - 22, 2006

Cancila E., Bosso A. e Ottolenghi M. (a cura di), (2006), La gestione sostenibile delle aree produttive. Una scelta possibile per il governo del territorio e per il rilancio delle politiche industriali, Ervet, Bologna.

Carnimeo G., Frey M., Iraldo F. (2002), Gestione del Prodotto e Sostenibilità, Le imprese di fronte alle nuove prospettive delle politiche ambientali comunitarie e della IPP (Integrated Product Policy), Franco Angeli, Milano

Charter M., Belmane I., (1999), Integrated Product Policy (IPP) and Eco-Product Development (EPD), Journal of Sustainable Product Design.

Communication from the European Commission to the Council, the European Parliament, the european economic and social committee and the committee of the region "Towards world-class clusters in the European Union:Implementing the broad-based innovation strategy" (COM (2008) 652).

Corbett C. J., DeCroix G. A., (2001), Shared-Savings Contracts for Indirect Materials in Supply Chains: Channel profits and Environmental Impacts, Management Science, Vol. 47, n. 7

* Corresponding Author

Corsten D., Kumar N. (2005), Do suppliers benefit from collaborative relationships with large retailers? An empirical investigation of efficient consumer response adoption, Journal of Marketing, Vol. 69, n.3

Darnall N., Jolley G. J., Handfield R. (2008), Environmental Management Systems and Green Supply Chain Management: Complements for Sustainability?, Business Strategy and the Environment, n.18

De Witt T., Giunipero L. C., Melton H. L. (2006), Clusters and supply chain management: the Amish experience, International Journal of Physical Distribution & Logistics Management, Vol. 36, n 4.

Doney P.M., Cannon, J.P. (1997), An examination of the nature of trust in buyer-seller relationships, Journal of Marketing, Vol. 61, 1997.

ENEA - Progetto Life-Siam (2007), Linee Guida per l'insediamento e la gestione di aree produttive sostenibili [on line], disponibile su <http://www.life-siam.bologna.enea.it/> [data di accesso: 23 marzo 2009]

Fontana D. (2007), Distretti Italiani in quindici anni di vita di strada ne ha fatta, Amministrazione & Finanza, n. 22/2007, Inserto

Frosch R. e Gallopolus N., (1989), Strategies for manufacturing, Scientific American, 261: 94

Fuller D.A. (1999), New decision boundaries: the product system life cycle, in Sustainable Marketing, SAGE Publications, Thousand Oaks, London, New Delhi.

Ganeson S. (1994), Determinants of long-term orientation in buyer-seller relationships, Journal of Marketing, Vol. 58

Graedel T.E. and Allenby B.R., (2002), "Industrial Ecology", Prentice Hall, 363 http://www.chinacp.org.cn, Cleaner Production in China, [data dell'ultima consultazione: 25 marzo 2009]

Hagelaar G., van der Vorst J. (2001), Environmental Supply Chain Management: using Life Cycle Assessment to structure supply chains, Paper IAMA 2001, Sydney, Australia.

Heiksanen E, (2000), Managers'interpretation of LCA: enlightment and responsibility or conflusion and denial, Business Strategy and the Environment, Vol. 9

Johnston R. (2003), Clusters: A Review, The Australian Centre for Innovation Limited.

Lowe, E.A., Moran S.R. and Holmes D.B. (1996). Fieldbook for the development of Eco – Industrial Park, final report, Washington

Marceau J., Dodgson M., (1999) Systems of Innovation, Paper N. 1, Innovation Summit, Department of Industry, Science and Resources, Canberra

Marshall A., (1980). Principles of Economics: an Introductory Volume. London.

Marshall C. (1998). Report for HM Treasury.Economic Instruments and the Business Use of Energy. The Stationary Office, London.

Mentzer J.T., DeWitt W., Keebler J.S., Nix N.W., Smith C.D., Zacharia, Z.G. (2001), "Defining supply chain management", Journal of Business Logistics, Vol. 22 No. 2.

Morikawa M., (2000), Eco–industrial development in Japan, Indigo Development Center

Nawrocka D. (2008), Inter-Organizational Use of EMSs in Supply Chain Management: Some Experiences from Poland and Sweden, Corporate Social Responsibility and Environmental Management, n. 15

Noordewier T.G., John G., Nevin J.R. (1990), Performance outcomes of purchasing arrangements in industrial buyer-vendor relationships, Journal of Marketing, Vol. 54

Pesonen H-L. (2006), Multinational Corporations and Clusters, Vinnova Conference Proceedings "Challenges to National Innovation Systems in a Globalizing World", Stockholm, 19th January 2006.

Porter M. E., (1998). On Competition, Harvard Business School Press, Boston.

Porter, M.E. (1990). The Competitive Advantage of Nations, Macmillan, London.

Regional Council of Etela – Savo, (2006), Eco Industrial Parks, background report for the eco – industrial park project at Ratasalmi – Interreg IIIB.

Rullani E. (2002), Dallo sviluppo per accumulazione allo sviluppo per propagazione: piccole imprese, clusters e capitale sociale nella nuova Europa in formazione, East West Cluster Conference Proceedings, 28-31 Ottobre 2002, OECD, LEED, Udine.

Sacco P., Ferilli G. (2006), Il distretto culturale evoluto nell'economia post industriale, Università Iuav di Venezia, DADI Dipartimento delle Arti e del Disegno Industriale, WP WORKING PAPERS DADI/ WP_4/06.

Schumpeter J.A. (1971), Teoria dello Sviluppo economico, Sansoni, Firenze.

Sharfman M.P., Ellington R.T., Meo M. (1997), The next step in becoming "green": life-cycle oriented environmental management, Business Horizons, vol. 40

Sharfman M.P., Shaft T. M., Anex R. P. Jr (2009), The Road to Cooperative Supply-Chain Environmental Management: Trust and Uncertainty Among Pro-Active Firms, Business Strategy and the Environment, n. 18.

Sinding K. (2000), Environmental management beyond the boundaries of the firm: definitions and costraints, Business Strategy and the Environment, Vol. 9, n. 2

Sölvell Ö. (2002), The Multi-Home Based Multinational – Combining Global Compentencies and Local Innovativeness, Institute of International Business – Stockolm School of Economics

Srivastava S. K. (2007), Green supply-chain management: A state-of-the-art literature review, International Journal of Management Reviews, Vol. n. 9, Issue 1

Sureephong P., Chakpitak N., Buzon L., Bouras A. (2008), Cluster Development and Knowledge Exchange in Supply Chain, The Proceeding of International Conference on Software Knowledge Information Management and Applications (SKIMA 2008), Katmandu, Nepal.

Tibbs B.C, (1992), Industrial Ecology: an Environmental Agenda for industry, Whole Earth Review, 4 - 19

Van Berkel R., van Kampen M., Kortman J. (1999), Opportunities and constraints for Product-oriented Environmental Management Systems (P-EMS), Journal of Cleaner Production, n.7(6)

Vermeulen W. J. V., Ras P. J.,(2006),The Challenge of Greening Global Product Chains: Meeting Both End, Sustainable Development, vol. 14.

Zucchetti S. (2003), Una nuova generazione di distretti industriali, Impresa & Stato.

Chapter 15

Beach Erosion Management with the Application of Soft Countermeasure in Taiwan

Ray-Yeng Yang, Ying-Chih Wu and Hwung-Hweng Hwung

Additional information is available at the end of the chapter

http://dx.doi.org/10.5772/45883

1. Introduction

The total length of Taiwan's shoreline is approximately 1,100 kilometers including sand, rock, cliff, gravel and reef coasts (see Figure 1). Almost half of the shoreline has been protected by seawalls. From the viewpoint of shore protection in coastal area, these seawalls actually play an important role of costal protection that prevents people and infrastructure from coastal disasters. Furthermore, detached breakwater and groyne are built to protect the coastal area with serious erosion problem. These efforts made our land safe over the last fifty years to some extent. Due to the martial law, it was not so easy for people to walk or visit near the coastal area in Taiwan during 1949 to 1987. However, after 1987 people gradually valued the coastline for environmental protection and, recreational use as well as the economic activity. The purpose of the coastal protection is diversified by these new demands. In this study, we will introduce environmentally and user-oriented coastal protection works as well as technically sound creditable coastal protection works to meet these new trends. Therefore, the purpose of this study is to evaluate on how to join soft solution strategies into current shore protection system throughout Taiwan's coast. Moreover, a feasible application for hard solutions complemented by soft issues for beach erosion management has also been evaluated. In Taiwan, beach erosion has become more serious in the recent past. The time for this beach erosion to become apparent chiefly depends on how fast the rate of longshore sediment transport decreased from the up-coast area and on the river sediment supply. Many industrial, commercial and fishery harbor construction projects were also observed to have disturbed the continuity of littoral sediment transport, and lead to the retreat of the shoreline in the downcoast area. However, sufficient knowledge on nearshore hydrodynamic forcing (incoming wave energy, wave-induced currents in the surf zone and tidal range), sediment transport processes and morphological features along coasts, will be helpful to the improvement of shore protection work. Countermeasures for beach erosion control should depend on the local conditions of hydrodynamic forcing characteristics, littoral sediment transport and various morphologies. Therefore better applications of the various soft methodologies available for beach erosion management will be proposed in this study.

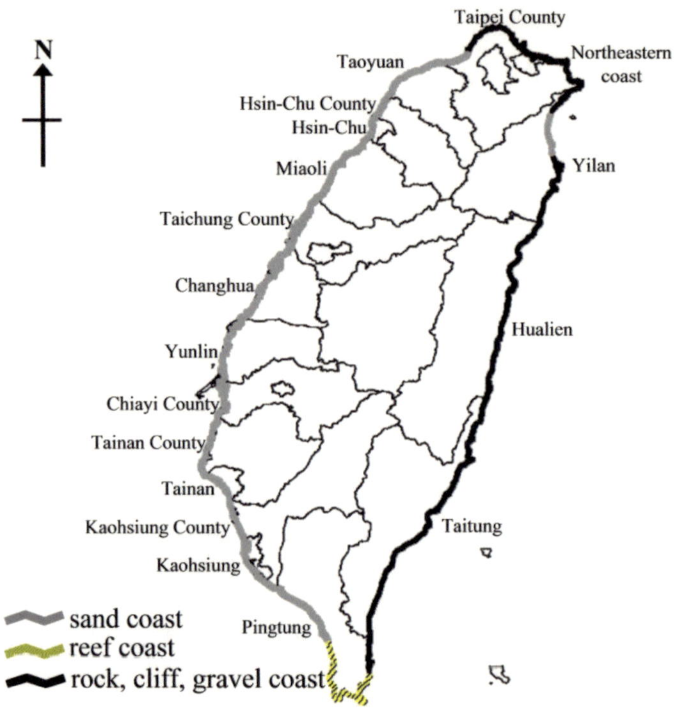

Figure 1. Different types of coasts in Taiwan

2. Evaluation of soft engineering structures

The protection of beaches against erosion has always been an important aspect of coastal engineering works in Taiwan. History is replete with the loss of valuable coastal lands such as beaches, reclamation areas, harbors, and other valuable coastal property to erosion induced by sea encroaching. On the other hand, there are also some cases of harbors being abandoned because of infilling by sediments, which is quite a different coastal engineering problem but a significant one. Erosion control measures should incorporate a reduction in the cause of the beach erosion where possible, when, for example, the erosion is caused by human activities along the coastal areas such as hard engineering structures or harbors. For each coastal erosion mitigation measure, it is important to know how they work. In fact, some of the mitigation schemes are able to reduce the wave energy at the shoreline, or simply provide a sacrificial beach, whereas others try to impede the long shore transport of sand. In a particular situation, one mitigation scheme will work better than some others owing to the difference in operation. In fact, some methods will fail in one situation and do very well in others. Therefore, in order to mitigate the erosion problems due to the current hard shore protection system around Taiwan coast, detailed evaluation of various feasible soft engineering structures should be conducted in advance. The evaluation of soft countermeasure includes beach nourishment, near shore disposal berms, geosystems, artificial oyster reef, fluid-elastic sheet and aquatic vegetation.

2.1. Beach nourishment

Beach nourishment is the mechanical or hydraulic placement of sand on the beach and/or shoreface to advance the shoreline or to maintain the volume of sand in the littoral. It is a soft protective and remedial measure that leaves a beach in a more natural state than hard structures and preserves its recreational value. Of the many remedial measures for beach erosion, beach nourishment is the only approach that introduces additional sand sources into the coastal system. Without the construction of coastal structures, beach nourishment seldom causes damage to the landscape, and can flexibly responds to changes of the littoral environment. Beach nourishment, with its expected widening of beach, is used to accomplish several goals as follows: formation of additional recreational area; land reclamation; maintenance of shoreline; reinforcement of dunes against breaching; protection of coastal structures; reduction of the wave energy near shore and creation of a sacrificial beach to be eroded during a storm; provide, in some cases, environmental habitat for endangered species.

Sand nourishment can be carried out at various locations in the profile or along the shoreline. The options of nourishments in cross-shore profile are shoreface (underwater nourishment or profile nourishment), dune zone (landward and seaward above dune toe), beach and swash zone. Leonard et al., 1990 evaluated 155 beach nourishment projects in the U.S.A. In all, about 300 million m^3 of sand was placed along 700km of shoreline (470km along Atlantic coasts, 180km along Gulf coasts and 50km along Pacific coasts). In 1996, Rijkswaterstaat also evaluated nine nourishment projects (volumes between 50 and 100 $m^3 / m / yr$; sand size between 0.15 and 0.3mm) carried out along the coasts (tidal range of about 2m) of the Netherlands in the period 1975-1994. Several characteristics, including ratio of design nourishment volume and required volume to compensate annual erosion volume in active zone before nourishment and after nourishment, were analyzed in his evaluation. Leo C. van Rijn, 1998 summarized the sand nourishment characteristics under micro and meso-tidal conditions in great detail from five projects (Delft Hydr., 1987; Dette & Raudkivi, 1994; Mϕller, 1990; Rijksw, 1996; Work & Dean, 1991). From his result, it shows that beach nourishment can be mostly utilized on coastal areas of low or moderate wave energy with micro-tidal condition. Meanwhile, the three basic elements including the eroded area, the borrow area and the transportation/ dumping methods should be investigated in detail when sand nourishment is applied to beach erosion control. Dean, 1986 recommended a mitigative approach for armoring on an eroding coastline that calls for the placement of sand annually in the amount that has been prevented from entering the system by the armoring structure. This approach maintains a more natural littoral system. Often, the nourishment scheme is remedial rather than preventive (Hamm et al., 1998). In summary, beach nourishment is the approach that directly addresses the deficit of sand in the coastal system without at least the potential of causing adverse effects on adjacent property. Bridges & Dean, 1996 concluded that beach nourishment is the most benign and acceptable approach to beach erosion mitigation. However from the new demands of shore protection now in Taiwan, beach nourishment can not be the only option of beach erosion control.

2.2. Near shore disposal berms

Open-water disposal of dredged material has been practiced worldwide for over sixty years. The initial attempts have arisen from the search for a beneficial use of the large amounts of dredged material obtained from navigation channel maintenance operations. The removed material which varies in size and quantity has been placed in nearshore disposal sites seaward of the surf zone. A major cost savings often accrues if beach fill material can be placed offshore rather than on the beach in the expectation that natural processes will move the material to the beach. The performance of underwater berms has been investigated both in the laboratory and through field monitoring programs. Hands, 1991 provided a thorough review of the behavior of 11 berms and their performance. Furthermore, Otay, 1994 presented a summary of submerged berms and their characteristics, including whether they were judged to be stable or migrated. Of the berms placed to benefit the landward beaches, possible designs could be a feeder berm, in which sand would be transported to the beach from an active berm or as a stable berm that causes damping of the waves and thus sheltering of the landward beach. In his research, Otay also described the monitoring results of an underwater berm placed off Perdido Key, Florida. Monitoring included repetitive beach profiles and wave measurements. His result showed that the berm had exerted a stabilizing effect on the beach leeward of the berm.

2.3. Geosystems

Geosystems (tubes, containers) have already found various applications in coastal engineering. The tubes and containers are mainly applicable for construction of groynes, perched beaches, and offshore breakwaters, and as bunds for reclamation works. Application of these systems has executed by a number of projects in the Netherland, Germany, Japan and U.S. Some information on U.S. experience with geotubes can be found in Fowler et al., 1995 including the application of geotubes for dewatering of contaminated maintenance dredged material. Geosystems have much applicability in erosion control, water control (small weirs and reservoirs), flood control, etc. For example, breakwaters made of sandbags, geotubes, etc, have been used successfully in the United States of America under conditions for low tidal range and low wave activity (Krystian, 2000). Under gentle wave climates such structures may not only attenuate waves, but can also encourage the accretion of sediment between them and the shore. Geotube can also be used to assist in dike, groin and breakwater construction. Krystian summarized the examples of application of geotube, as dune reinforcement, core of breakwater, and bunds for dike construction, from a number of projects executed in the Netherlands and Germany. The main advantages of geosystems in comparison with more traditional methods (rock, concrete armor units, block mats, asphalt, etc.) are: a reduction in work volume, execution time and cost, the use of local materials, low-skilled labour and locally available equipment. However, until now, geosystems were mostly applied as temporary structures. The reason for that was their relatively low resistance to the loading of waves and currents, the lack of proper design criteria, and a low durability in respect to UV-radiation and vandalism.

2.4. Artificial oyster reef

There is increasing interest in oyster reefs used to restore eroding coastlines. Occasionally, subtidal oyster reefs can be found offshore. These immense natural submerged breakwaters protect the beaches from storms and wave erosion by dissipating wave energy. The study of how artificial and natural reefs have protected shorelines has been conducted by Hamaguchi et al., 1991. They investigated the effects of an artificial reef on the Niigata coast in Japan. It was found that a significant amount of sand was deposited landward of this artificial reef. This reef was developed to mimic the effect of the natural coral reefs in the area. There has been an effort to find different methods of restoring oyster reefs in various estuaries around the world. O'Beirn et al., 2000 conducted the experiments by using oyster shell, concrete, and rubber tire chips as oystercultch material. A structure termed an "oysterbreak" was designed to stimulate the growth of biological structures in an optimal shape to serve as submerged breakwaters (Foret, 2002). Oysterbreak can form immense structures that can protect shorelines and coastal communities by reducing wave energy.

Currently, mineral accretion amelioration on gabion that was filled with oyster cultch & rock to form a new biological unit has been investigated in field experiment by Hwung et al., 2008. It is hoped that this combination of oyster cultch, mineral accretion and cage meshed into berm breakwater can improve the toe revetment and berm advance, and simultaneously enrich the local environment to a higher level.

2.5. Fluid-elastic sheet

Fluid-plate hydro-elastic interaction problems have been of common interest for a long time because of their engineering applications. During the past decades, for instance, there has been a gradual increase in interest in the use of flexible plates or membranes as alternative effective inexpensive wave barriers in a beach zone. Currently, developing of the new design of floating wave breakers in a beach zone using coating of the sea surface by an elastic plate, which absorbing the energy of sea waves, is investigated by Hwung et al., 2008. A properly designed horizontal flexible membrane can be a very effective wave barrier and its optimal design can be found through a comprehensive parametric study using the experiments, theory and computer programs developed. In particular, the membrane is light and rapidly deployable; thus, it may be an ideal candidate as a portable temporary breakwater. Since a horizontal membrane does not directly block incoming waves, the transmitted and motion-induced waves need to be properly cancelled for it to be an effective wave barrier.

2.6. Aquatic vegetation

Aquatic vegetation provides important ecosystem services to coastal marine systems. They influence their environments through wave attenuation, the stabilization of sediments, increased setting of the suspended particulate particle and nutrient cycling. For environmental and esthetic purposes, projects on natural development of wetlands and restoration of river basins toward natural development have been promoted recently. The growth of vegetation in these areas is favored. Such vegetation increases the resistance of the watercourse, leading to an increase in water depth and reduction of flow velocity. In estuarine and coastal areas with vegetation, in addition to freshwater flow upstream, waves and tidal currents exist and will play a significant role in the hydrodynamics and mixing processes. Waves over vegetation will be attenuated due to the resistance offered by the vegetation. The bidirectional nature of wave motion will increase the mixing between the water column and that within the vegetation (Li & Yan, 2007). Wave motion tends to be highest in the shallow waters where, in combination with tidal currents, water movement imposes a shear stress on bottom sediments. If bottom shear stress exceeds a critical value, sediment will be resuspended, increasing turbidity and light attenuation (Wright, 1995).

For waves propagating over vegetation, Kobayashi et al., 1993 developed an analytical model to predict wave attenuation over vegetation by assuming an exponential decay of incoming regular waves. Vegetation meadows can reduce suspended sediment concentrations; friction from vegetation leaves reduces current velocity and attenuates waves, thus reducing the stress on bottom sediments, decreasing resuspension, and promoting sediment settling within the vegetation bed (Fonseca & Cahalan, 1992; Rybicki et al., 1997). Vegetation beds may also increase particle settling shoreward of the bed (Chen et al., 2007).

3. Hydrodynamic energy and morphology classification of Taiwan coast

The hydrodynamic and morphological processes in the coastal zone are governed by two primary phenomena, namely, winds and tides. The winds are directly responsible for the transport of sand on the dry beach and for the generation of waves, currents and water-level fluctuations, while the tides express themselves in a periodic rising and falling of the water and in tidal currents. Therefore, coastal classification based on hydrodynamic energy was presented by Davis & Hayes, 1984. The classification is shown in Table 1. The wave climate is generally characterized, as: low wave energy, if annual mean significant wave height at edge of surf zone (say, depth of 6m) is $H_{s,am}<0.6$m; moderate wave energy, if $H_{s,am}$ between 0.6m and 1.5m; high wave energy, if $H_{s,am}>1.5$m.

However, tides are classified as micro-tidal, if the tidal range (TR) <2m, meso-tidal for TR between 2m and 4m and macro-tidal for TR>4m. Furthermore, the relative strength of tide-induced (tidal range TR) and wave-induced forces (mean annual nearshore wave height H) acting in coastal system, the following classification may also be given as: wave energy-dominated coasts (TR/H=0.5 to 1.0), tide energy-dominated coasts (TR/H>3); mixed energy coasts (TR/H=1 to 3).

Regarding the long-term marine observation data, we refer to the research reports analyzed by Tainan Hydraulics Laboratory [THL], 2002 and then summarize the hydrodynamic energy classification of Taiwan coast shown as Figure 2 and Figure 3. Therefore based on Davis and Hayes's classification, the results show that northeastern coast, east coast and south coast of Taiwan belong to micro-tidal coast. However, Yun-Lin, Changhua (mid-western coast) and north coast are meso-tidal coast. As mentioned about classification of wave energy coast, the coast from Taipei county to Hsinchu county (northwestern coast), and coast between Yun-Lin county and Tainan county (west coast) are moderate wave energy coast.

Wave Energy	low	moderate	high
	Hs,am<0.6m	0.6m <Hs,am<1.5m	Hs,am>1.5m
Tidal Energy	micro-tidal	meso- tidal	Macro- tidal
	TR<2m	2m<TR<4m	TR>4m
Coastal classification based on hydrodynamic energy			
Wave Energy-dominated			TR/ Hs,am=0.5~1
Tide Energy-dominated			TR/ Hs,am>3
Mixed Energy			1<TR/ Hs,am<3

Table 1. Hydrodynamic forcing in the coastal zone

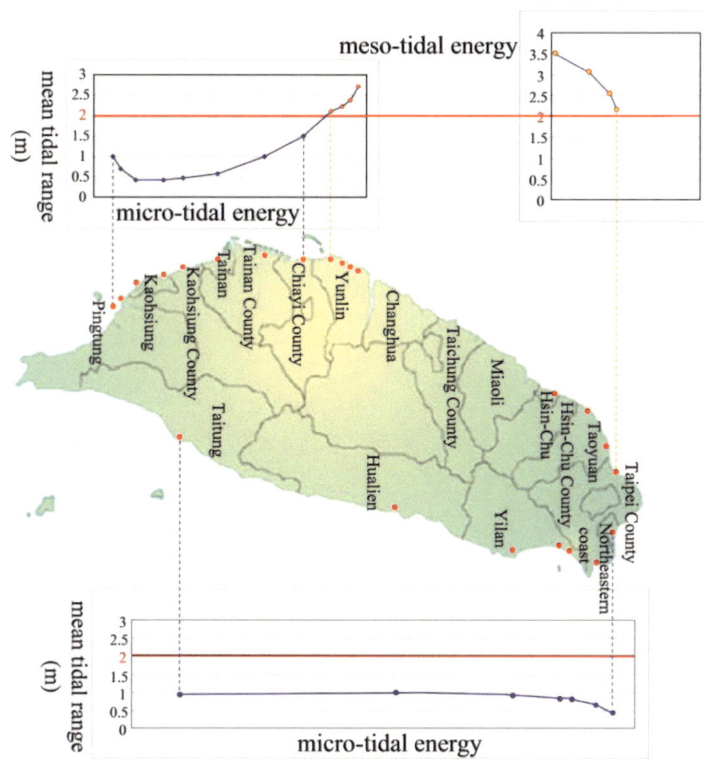

Figure 2. Tidal energy classification of Taiwan coast

However, southwestern coast from Tainan city, Kaohsiung to Ping-Tung county belong to low or moderate wave energy coast. Regarding to the east coast of Taiwan, besides partial coast of Taitung county is moderate wave energy coast, most parts of the east coast, (Hualien, Yi-Lan) and northeastern coast, are high wave energy coasts. For morphology classification, the slopes of beach profiles around Taiwan coast are shown in Figure 4. Due to these hydrodynamic energy and morphology classification of Taiwan coast, a strategic management proposal can be made to integrate soft countermeasure into the current hard shore protection system around Taiwan coast.

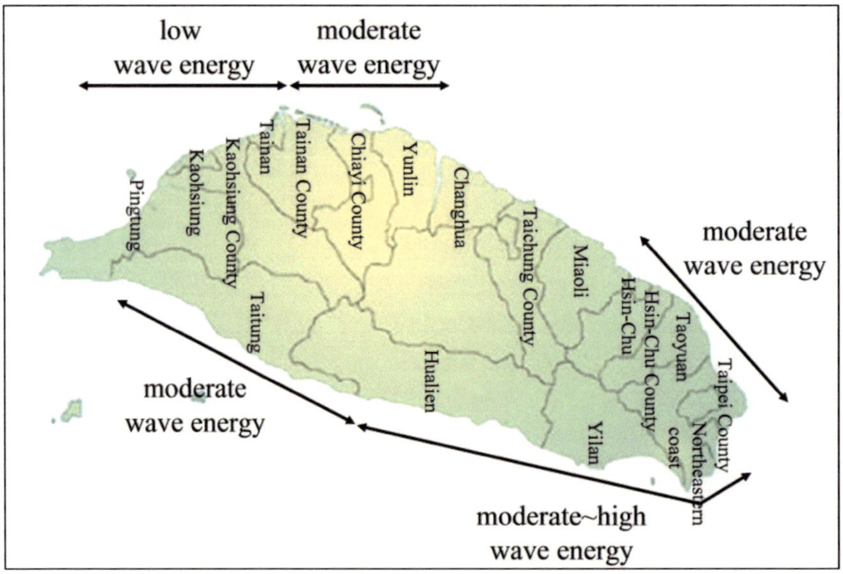

Figure 3. Wave energy classification of Taiwan coast

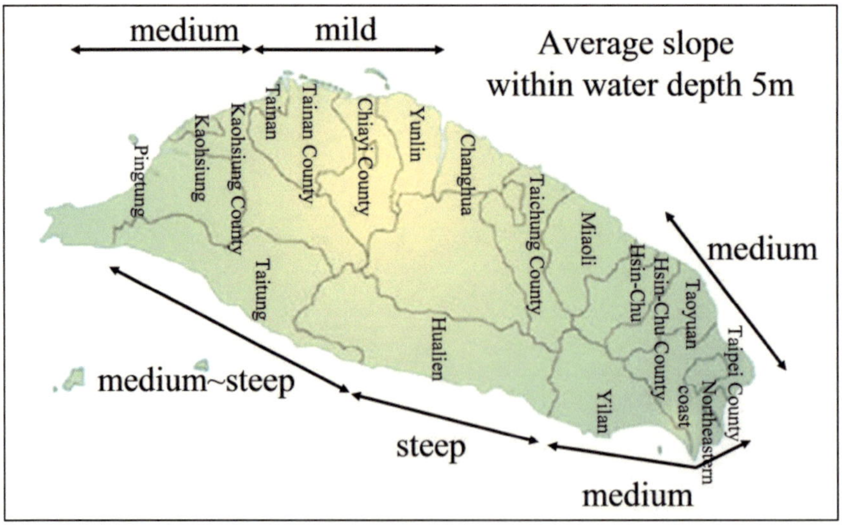

Figure 4. Morphology classification, the beach slopes around Taiwan

4. Discussion on application of combining hard and soft solutions for beach protection in Taiwan

Water Resource Agency, Taiwan, has claimed nearly 95 percent of fulfillment on coastal protection after more than 30 years' efforts. However, under the disaster prevention demand in earlier days, coastal protection has been accomplished mostly be lining up the hard engineering structures such as seawalls, groyne and armour units along the coastlines. Until now, the constriction of the seawalls for shore protection does not always work well on each coast around Taiwan. Some shores still get eroded seriously with seawalls being damaged partially. Meanwhile, the coastal engineering development in Taiwan currently has changed the previous strategy, only focused on shore protection, to a new one taking into consideration several aspects like safety, economy, construction, recreation, landscape and ecology. According to this new trend, the current shore protection system around Taiwan can be properly mended by soft countermeasures, however, the specific characteristics and requisite at each local site should be taken into consideration. Therefore, we have divided Taiwan coast into several categories based on the collected long-term observation information such as geological characteristics, hydrodynamic forcing, the intensity of beach erosion and shore protection. Then, the national beach management and protection problems were evaluated and the solutions for shore protections and further improvements were proposed.

After evaluation of a number of soft solution results, it is indicated that beach nourishment is a natural and popular soft shore protection technique that has been applied worldwide recently. This method can be utilized on the coastal areas under low or moderate wave energies with mild or moderate bottom slope for engineering purpose. However from the new demands of shore protection now in Taiwan, beach nourishment can not be the only option of beach erosion control. Another soft solution should be also taken into consideration for integration. Based on these criteria and the categories of Taiwan coast, the countermeasure of integrating soft solution into current hard shore protection system around Taiwan coast will be proposed as follows:

Because the north coasts in Taiwan are meso-tidal and moderate wave energy coasts, and their beach profiles are of mild slope, the eligible improvement criteria for current shore protection scheme are based on headland control strategy plus sand nourishment. However, volume of sand, sand size, beach and swash zone placement should be taken into account in the works of beach nourishment.

On the other hand, many barrier islands in the offshore of Yun-Lin, Chia-Yi and Tainan coast, can be treated as natural offshore breakwaters. In fact, those barrier islands can form a defense line of low-lying coastal plains and back-barrier basin against storms attacking these areas. Therefore, the shore protection strategy for these areas should be focused on how to protect these barrier islands. However, most important for the formation and maintenance of barrier islands and inlets is the relative strength of the wave processes and of the tidal processes. In order to protect these barrier islands, detailed field investigations on sediment supply (sources and sinks), hydrodynamic forces (waves, tides and rate of sea level change) and geomorphic setting (shoreface profile shape, sub-strata composition) should be carried out in advance. Meanwhile, oyster cultivation is an important fishery industry in the coast of these areas. Thus, there is large volume of oyster cultch in these coasts. Therefore, one new shore protection technology using mineral accretion technique is proposed by Hwung et al., 2008. Regarding to this new technology, mineral accretion (an advance on cathodic protection) amelioration combined with oyster cultch and rock is used to form the new biological unit in order to enhance the efficiency of anti-rusting and function of shore protection. Several field experiments of this new shore protection technology have been done on Chigu Lagoon in Tainan County. It is hoped that this new shore protection technology can be successfully applied to coastal engineering in the near future.

Offshore sills or breakwaters have proved to be much effective when used in combination with beach nourishment schemes. The retention capacity of a perched beach not only helps to reduce wave attack but is clearly beneficial from a recreational point of view. Since many offshore breakwaters already exist in the coasts of Kao-Shung and Pin-Tung counties, sand sources can be filled in the region between offshore breakwaters and sea dikes. The expanding beach faces will be helpful for wave damping and sightseeing. However, as many successful fisheries exist in the coastal regions of Kao-Shung and Pin-Tung Counties, when the beach nourishment is taken as the shore protection method in these areas, the influence of beach nourishment to coastal fisheries should be taken into considerations.

As for steep beaches and high wave energy, such as those of the east coast region, more specific parameters should be taken into account. For example, the erosion problem at the Tou-Chen beach in Yi-Lan County can be remedied by headland control strategy plus sand nourishment. The Tou-Chen beach now is defended only by a seawall and some short groynes. The headland control strategy can be based on reconstructing two long arc-shaped groynes with a submerged breakwater to support the recreational beach fill in front of seawall. Meanwhile, for the purpose of recreational activity, the fluid-elastic shirt can be applied as a portable temporary breakwater in this beach zone for wave damping. However in order to mitigate the erosion of gravel beaches in Hualien and Tai-Tung counties, the beach nourishment of mixed grain sizes is to become an alternative solution. Since there is little experience on the movement of gravel on steep coast of Hualien and Tai-Tung counties, a comprehensive field investigation should be done on the mechanism of sediment movement especially for action of typhoon wave. Moreover, a detailed physical model study about the effect of dynamic nourishment as a countermeasure against erosion should also be conducted before gravel nourishment can be carried out on the Taiwan eastern coast.

5. Experimental application study on integration of soft solution into current hard shore protection system

Two local sites in the southwestern Taiwan coast and one biggest offshore barrier island in Taiwan are selected for the experimental application study on integration of soft solution into current hard shore protection system. These in-situ experimental studies are therefore designed to improve the security as well as to involve ecological and scenic remediation for the beach erosion problem.

5.1. Ching-Tsao-Lun coastal area

With the length of 5 kilometer, Ching-Tsao-Lun coastal area is located between Zeng-Wun River mouth and Lu-Erh-Men River mouth at the middle section of Tainan coast. Without manmade intrusion, Zeng-Wun River mouth was once a natural estuary for more over three decades ago. Unfortunately, nowadays Ching-Tsao-Lun coastal area was invaded by constructing concrete and pebble dikes or other artificial protection. Therefore, the beach in front of Ching-Tsao-Lun concrete dike had been eroded ten years ago (Figure 5 and Figure 6). In order to have an overall study on rebirthing Ching-Tsao-Lun coastline, the in-situ study (Figure 7) focuses not only on safety evaluation of remedial Ching-Tsao-Lun sea dike but also on low to reform the beach back with ecology evaluation (Liou et al., 2007). Furthermore, by joining government resource and local manpower together, advance coastal management will enlighten the environment and landscape of Ching-Tsao-Lun coast again. The overall coastal protections and environment rebirths for Ching-Tsao-Lun coastal are listed as follow:

The coastal area around Ching-Tsao-Lun concrete dike:

1. Short-term goal: If rebuild is required, dike section adjustment could be put into consideration.
2. Long-term goal: Three schemes with beach nourishment and offshore breakwater are proposed for beach rebirth. Physical remediation will be half the way whiles the beach reversible.

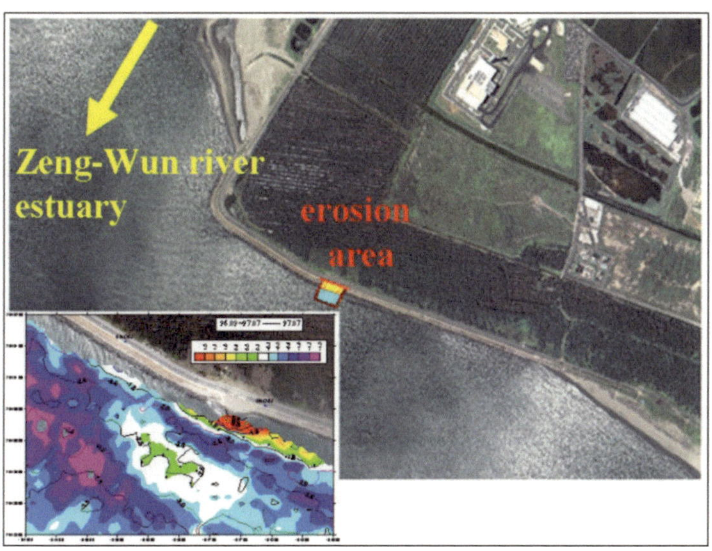

Figure 5. The erosion area in the Ching-Tsao-Lun coastline

The coastal area around Ching-Tsao-Lun pebble dike:

1. Apply "Vegetation Evolution Method" to enhance vegetation diversity and landscape.
2. Advance coastal management and use floating logs to enlighten the environment and coastal protection.

The natural beach zone:

1. Short-term goal: Build wind fence with planting to enhance the environment and landscape.
2. Long-term goal: Build artificial sand dune and increase dune elevation by using local floating logs, oyster cultch or dredged sedimentation form Lu-Erh-Men River.

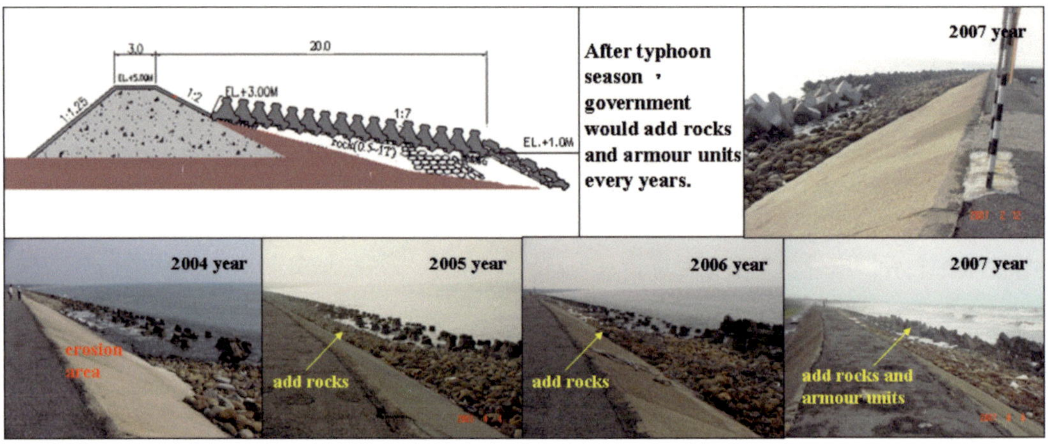

Figure 6. The erosion beach in front of Ching-Tsao-Lun concrete dike

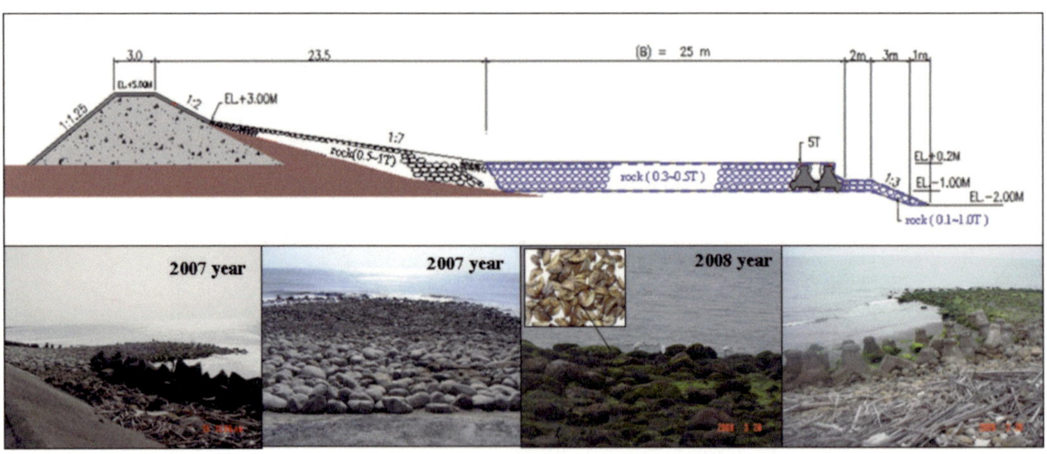

Figure 7. In-situ study on beach erosion control in the Ching-Tsao-Lun area

5.2. Shuang-Chun coastal area

With a beautiful sight and a mangrove ecosystem preservation area, Shuang-Chun coastal area (Figure 8) locates between Ba-Jhang River mouth and Ji-Shuei River mouth at the northern section of Tainan coast. Historical evolution of the coastal morphology shows that one third of the Shuang-Chun coastline at northern section has revealed shoreline retreat problem in the recent years. It can be shown by the evolution of coastline, satellite photos and aerial photographs from 1993~2002 (Figure 9, Figure 10 and research reports by THL, 2004).

Figure 8. The location of Shuang-Chun coast between two river mouths

After analyzing the long term field investigation data, the hydrodynamic characteristics of Shuang-Chun coast are moderate wave energy coast and micro-tide condition. The dominant hydrodynamic characteristics (incoming wave climate, tidal range) and local morphological information are considered for the proposed countermeasure to control beach erosion. Based on parabolic bay orientation on equilibrium shape (Hsu & Evan, 1989), the final select countermeasure is show in Figure 11 and Figure 12. The design is to establish three offshore breakwaters with a southwest stretched breakwater as a down-coast artificial headland on the existing seawall of Ba-Jhang River mouth (Yang et al., 2004). All the new breakwaters are set up by geobags filled with sand from the northern deposited area of Ba-Jhang River mouth. Behind the shelter areas of three offshore breakwaters, oyster booth is used as wave energy dissipation and sand trapping. Meanwhile, the existing dune is reinforced by dune zone nourishment. Monitoring of this experimental application study has still been conducted from 2008~2012. The final results of this proposed countermeasure will be verified by measurements of hydrodynamics and topography, sand sampling and monitoring of geosystem and oyster booth for their function of anti-damage and suitably applied environment.

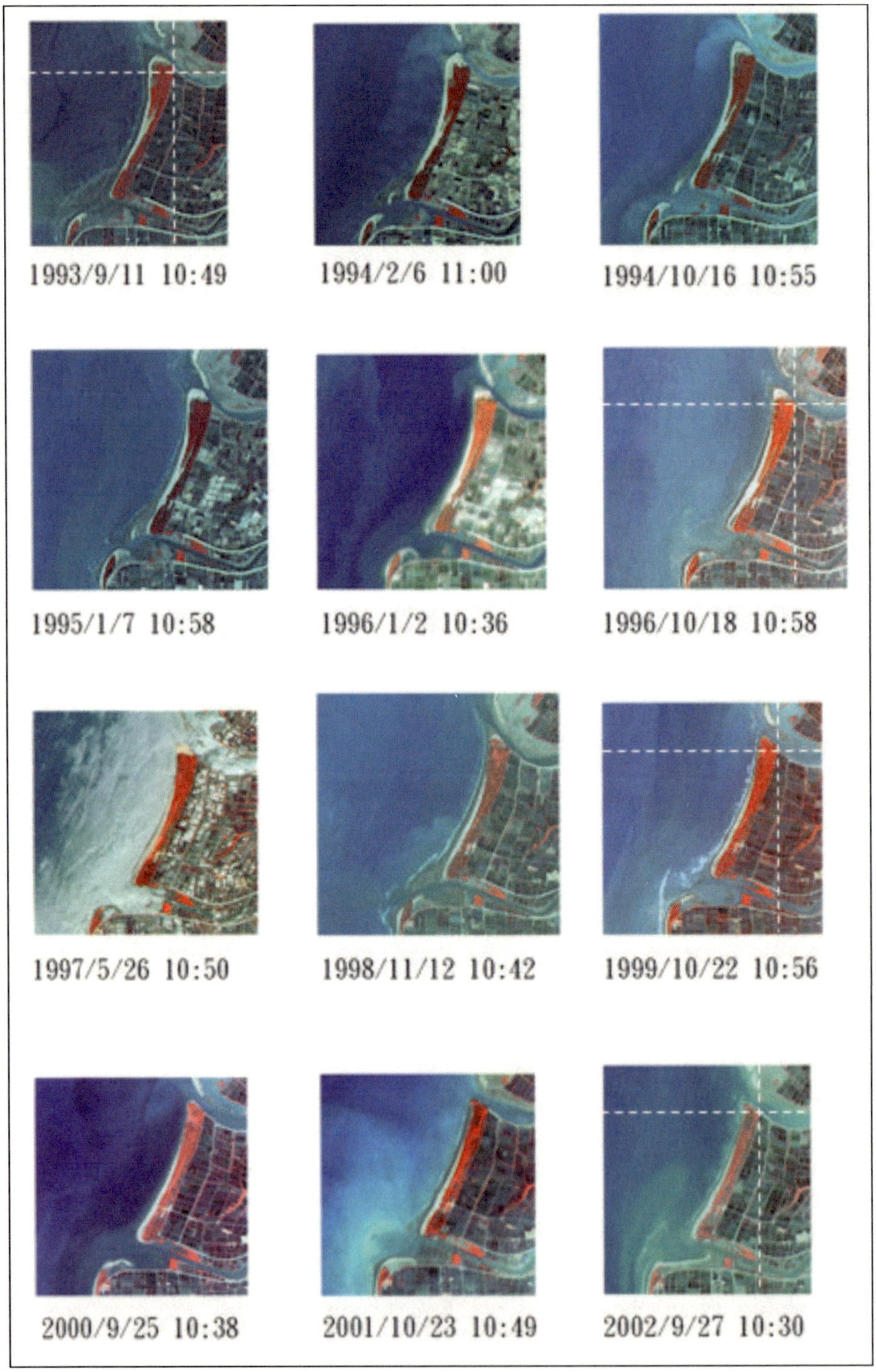

Figure 9. The satellite photos of Shuang-Chun coast from 1993~2002

Figure 10. The comparision of Shuang-Chun coastal morphology between 1993 and 2002

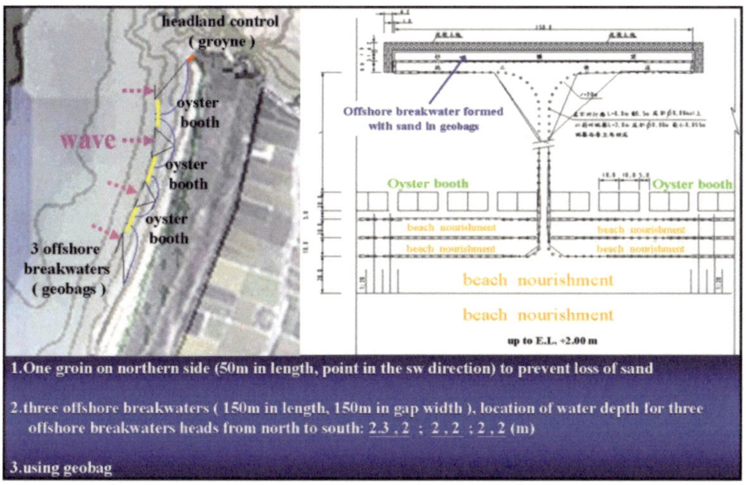

Figure 11. The final sketch of shore protection countermeasure for Shuang-Chun coastal area

Figure 12. Beach nourishment and geobag application in experimental study of the Shuang-Chun

5.3. Wai-San-Ding barrier island

Barrier islands as their name implies, they form a protective barrier between coastal shorelines and wave action that originates offshore. Barrier islands are also ecosystems that border coastal shorelines and physically separate the offshore oceanic province from inshore wetlands, bays and estuaries. Coastlines fronted by barrier islands also include some of the greatest concentrations of human populations and accompanying anthropogenic development in the word. The native vegetation and geological stability of these ecosystems are coupled and vulnerable to erosion events, particularly when also disturbed by development.

The Wai-San-Ding barrier island (Figure 13), protruding at about a forty-five-degree angle from the natural trend of the mainland shoreline at the mouth of the Peigang Shi River, is the largest remaining barrier island off the Taiwan coast. The overall length of this barrier is twenty kilometers, and her area is around two-thousand hectares during the Mean Water Level (M.W.L.). Much of the island shoreline is investigated to have been eroding at a rate of 50m~60m per year in recent years. Furthermore, this island holds some sort of "land speed" with continuing 0.2 degree/year counter-clockwise rotation to migrate southeastware to the mainland shoreline and gradual submerging into the sea. The Wai-San-Ding barrier island located on the southwestern Taiwan, is normally treated as natural offshore breakwater. In fact this biggest barrier island can form a defense line of low-lying coastal plains and back-barrier basin against storms attacking the southwestern coastal area in Taiwan. However, the erosion problem of the Wai-San-Ding barrier island has become more serious in the recent past. Therefore, how to protect this barrier island is always an important issue both from the consideration of coastal hazard and sustainable environment in Taiwan.

The objective of this experimental application study is to find the suitable measure for mitigating the existing erosion problem of the Wai-San-Ding barrier island. After collecting enough hydrodynamic and morphodynamic data from the long-term field investigation, the erosion mechanisms of the barrier island were analyzed in detail. Figure 14 shows that the time for this beach erosion to become apparently chiefly depends on how fast the rate of longshore sediment transport decreased from the up-coast area and on the river sediment supply. Meanwhile, run up mechanism under various waves, storm surge and overwash threshold on sand barrier during typhoon are also the important factors to be investigated. However, sufficient knowledge on nearshore hydrodynamic forcing, sediment transport processes and morphological features along this offshore barrier island, is helpful to the countermeasure control work. Based on the analysis of the erosion mechanisms, consideration of some measure options were proposed and firstly simulated by numerical model to find the two better solutions. Then two better applications (Figure 15) of the various soft methodologies available for the beach erosion control were proposed after numerical model analysis and further investigated by physical model test in the Near-shore Wave Basin (NSWB, 150x60x1.5m) at the Tainan

Hydraulics Laboratory (THL), National Cheng Kung University (NCKU), Tainan, Taiwan to validate their effect. The results showed that soft groins in the downstream and submerged artificial berms in the midstream are the effectively integrated measure to mitigate the continuing erosion problem of the Wai-San-Ding barrier island. Meanwhile, the plant evolution method and oyster cultch with aquatic vegetation were also proposed to apply in mitigation of wind sand transport and stabilization of sand dune. In order to protect this offshore barrier island, the more detailed field investigations on sediment supply (source and sinks), hydrodynamic forces (waves, tides and rate of sea level change) and geomorphic setting (shoreface profile shape, sub-strata composition) should be continuously conducted. Furthermore, the in-situ experimental study based on two proposed countermeasures is suggested to apply in improving the security as well as to involve ecological and scenic remediation for the erosion problem of Wai-San-Ding barrier island.

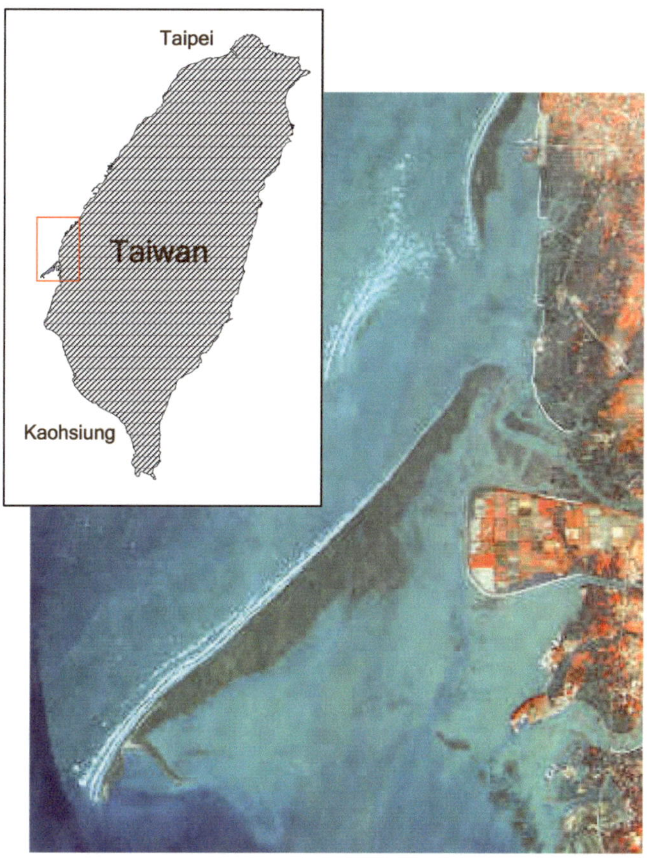

Figure 13. Studied area Wai-San-Ding Barrier Island image from satellite SPOT(2001)

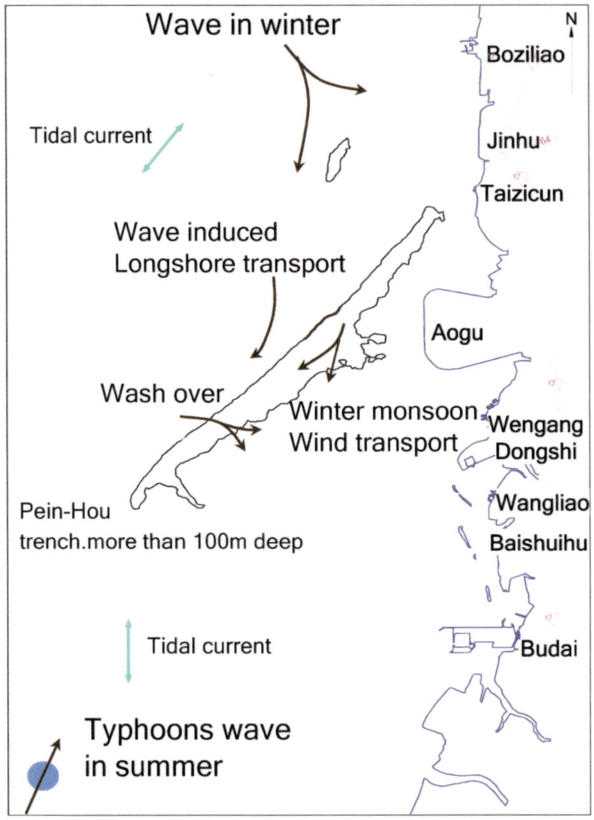

Figure 14. Hydrodynamic forcing characteristics, littoral sediment transport and morphology dynamics of the offshore barrier island

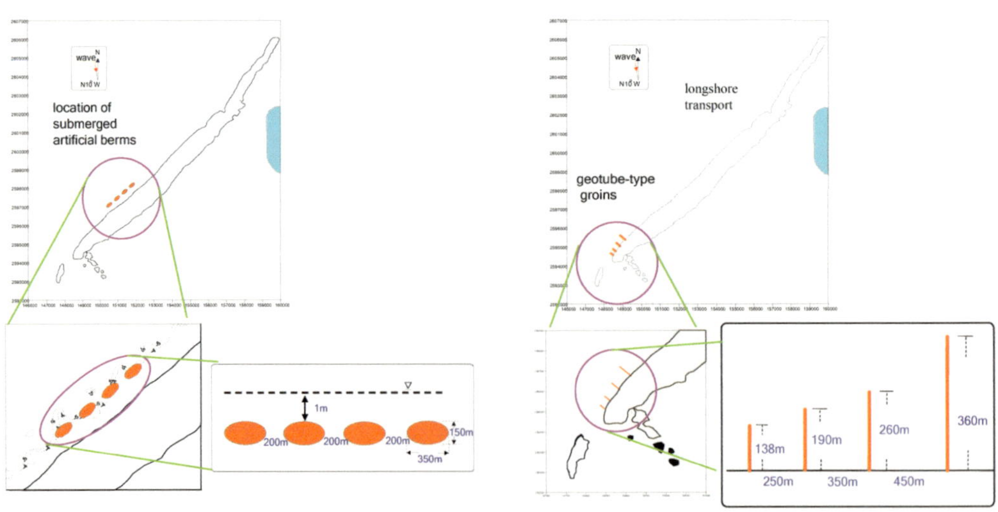

(a) submerged artificial berms (b) soft groins used by geotube

Figure 15. Two better countermeasures for mitigation the erosion problem of Wai-San-Ding Barrier Island

6. Conclusion

The efficiency and productivity of the infrastructure facilities is strictly required nowadays owing to escalating fiscal deficit of the government. We have questioned ourselves about what the people in Taiwan expect on coastal protection work for the next decade; moreover, is it worth applying soft solution instead of keeping the previous reinforced concrete revetment under this tight budget condition? The objective of this study is therefore to present various soft solution strategies available for beach erosion control in the hope of providing better efficiency and cost-effectiveness as well. The results also reveal that the current shore protection system around Taiwan can be properly controlled by beach nourishment. However, the specific characteristics at each local site should be taken into consideration. Accordingly, we divide Taiwan coast into categories based on the collected information such as geological characteristics, hydrodynamics, and the intensity of beach erosion. The national beach management and protection problems will therefore be evaluated followed by the offering of resolutions for shore protection and further improvements. For the purpose of beach erosion management, we also have completed collecting and analyzing coastal data around Taiwan and constructed a database as well a geographic information system (see Figure 16) as reference. Related units of coastal management agency, in Taiwan, are permitted to log on to and use the system via the Worldwide Web with an authorized username and password. The actual locations and related information of the current shore protection constructions with suitable principle and countermeasure of the future beach erosion control around Taiwan can be obtained via this geographic information. It is helpful for future reference of beach erosion management for the governmental agency in charge of shoreline policies.

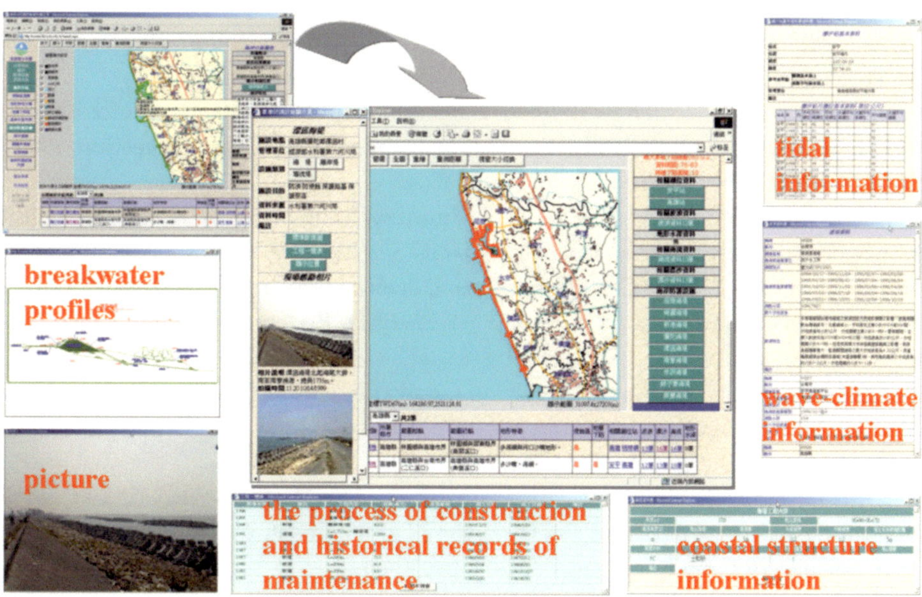

Figure 16. GIS information system of Taiwan coast

Author details

Ray-Yeng Yang and Ying-Chih Wu
Tainan Hydraulics Laboratory, National Cheng Kung University, Taiwan
Department of Hydraulic and Ocean Engineering, National Cheng Kung University, Taiwan

Hwung-Hweng Hwung
Department of Hydraulic and Ocean Engineering, National Cheng Kung University, Taiwan

Acknowledgement

The authors wish to acknowledge the generous support from Water Resources Agency, Ministry of Economic Affairs and National Science Council through the project "KUN–SHEN" of ANR (France) and NSC (Taiwan) joint funding initiative -- grant no. NSC-100-2923-E-006-001-MY3.

7. References

Bridges, M. & Dean, R.G. (1996). Erosional Hot Spots: Characteristics and Causes, *Proceedings 10th National Conference on Beach Preservation Technology*, Florida Shore and Beach Preservation Assoc.

Chen, S.N.; Sanford, L.P.; Koch, E.W.; Shi, F. & North, E.W. (2007). A Nearshore Model to Investigate the Effects of Seagrass Bed Geometry on Wave Attenuation and Suspended Sediment Transport, *Estuaries and Coasts*, Vol.20(2), pp.296-310.

Davis, R.A. & Hayes, M.O. (1984). What Is a Wave-Dominated Coast ? , *Marine Geology*, Vol. 60, pp. 313-329.

Dean, R.G. (1986). Coastal Armoring: Effects, Principles and Mitigation, *Proceedings 20th Intl. Conference Coastal Engineering*, pp. 1943-1857, ASCE, Taipei, Taiwan.

Delft Hydraulics (1987). *Manual on Artificial Beach Nourishment*, Delft, The Netherlands.

Dette, H.H. & Raudkivi, A.J. (1994). Beach Nourishment and Dune Protection, *24th ICCE*, pp. 1007-1022, Kobe, Japan.

Fonseca, M.S. & Cahalan, J.A. (1992). A Preliminary Evaluation of Wave Attenuation by Four Species of Seagrass, *Estuarine, Coastal and Shelf Science*, Vol.35, pp.565-576.

Foret, J. (2002). Role of Artificial Oyster Reef Development in the Restoration of Coastal Louisiana, *6th International Conference on Shellfish Restoration*, Charleston, SC, USA, NOAA/Sea Grant.

Fowler, J.D.; Toups, Ch. & Gilbert, P. (1995). Geotextile Contained Contaminated Dredged Material, Marina del Ray, Los Angeles and Port of Oakland, California, *Proceedings 14th World Dredging Congress (WODA)*, Amsterdam.

Hamgauchi, T.; Uda, T.; Inoue, C. & Igarashi, A. (1991). Field Experiment on Wave-Dissipating Effect of Artificial Reefs on the Niigata Coast, *Coastal Engineering in Japan*, Japan Society for Civil Engineers, Vol. 34, pp.50-65.

Hamm, L. et al. (1998). *Beach Fills in Europe; Projects, Practices and Objectives,* Book of Abstracts, 26th ICCE, Copenhagen, Denmark.

Hands, E.B. (1991). Unprecedented Migration of a Submerged Mound off the Alabama Coast, *Proceedings 12th Ann. Conference Western Dredging Assoc.*, pp.1-25.

Hwung, H.H.; Huang, H.Y.; Wu, Y.C.; Liou, J.Y. & Liu L.L. (2008). Mineral Accretion Technique during Biological Attachment In-Situ, *30th Ocean Engineering Conference in Taiwan,* pp.553-558, National Chiao Tung University, *Taiwan*. (in Chinese).

Hwung, H.H.; Yang, R.Y. & Igor V.S. (2008). Sea Wave Adaptation by an Elastic Plate, *Proceedings 18th International Offshore (Ocean) and Polor Engineering Conference*, pp.296-302, Vancouver, Canada.

Hsu, J.R.C. & Evans, C. (1989). Parabolic Bay Shapes and Applications, *Proceedings Instn. Civil Engers.*, Part 2. London: Thomas Telford, Vol.87, pp.557-570.

Kobayashi, N.; Raichle, A.W. & Asano, T. (1993). Wave Attenuation by Vegetation, *J. Waterway, Port, Coastal, Ocean Engineering,* Vol.119(1), pp.30-48.

Krystian, W.P. (2000). Geosynthetics and Geosystems in Hydraulic and Coastal Engineering, A. A. Balkema, Rotterdam, Netherlands.

Leo C. van Rijn (1998). *Principles of Coastal Morphology*, Aqua Publications.

Leonard, L.A. et al. (1990). A Comparison of Beach Replenishment on the U.S. Atlantic, Pacific and Gulf Coasts, *Journal of Coastal Research*, SI6, pp.127-140.

Li, C.W. & Yan, K. (2007). Numerical Investigation of Wave-Current-Vegetation Interaction, *Journal of Hydraulic Engineering*, Vol.133, No.7, pp.794-803.

Wright, L.D. (1995). Morphodynamics of Inner Continental Shelves, Boca Raton, Florida :CRC.

Liou, J.Y.; Huang, H.Y.; Kuo, C.H.; Shieh, C.T. & Chiang, W.P. (2007). An Amelioration Study upon Ching-Tsao-Lun Dike in Tainan Coast, *Proceeding of the 29th Ocean Engineering Conference in Taiwan* (in Chinese).

Mφller, J. T. (1990). Artificial Beach Nourishment on the Danish North Sea Coast, *Journal of Coastal Research*, SI6, pp. 1-10.

O'Beirn, F.; Luckenbach, M.; Nestlerode, J. & Coates, G. (2000). Toward Design Criteria in Constructed Oyster Reefs: Oyster Recruitment as a Function of Substrate Type and Tidal Height, *Journal of Shellfish Research*, Vol.19, No.1, pp. 387-395.

Otay, E.N. (1994). *Long-Term Evolution of Nearshore Disposal Berms*, Ph.D. dissertation, Dept. of Coastal and Oceanographic Engineering, University of Florida.

Rijkswaterstaat (1996). *Evaluation of Sand Nourishment Projects along the Dutch Coast 1975-1994(in Dutch)*, Report RIKZ 96.028, The Hague, The Netherlands.

Rybicki, N.B.; Jenter, H.L.; Carter, V.; Baltzer, R.A. & Turtora, M. (1997). Observations of Tidal Flux between a Submersed Aquatic Plant Stand and the Adjacent Channel in the Potomac River near Washington, D. C., *Limnology and Oceanography*, Vol.42(2), pp.307-317.

Tainan Hydraulics Laboratory Technical Report. (2002). *The Research of New Shore Protection Technology (3/4)*, National Cheng Kung University, Taiwan, Bulletin No.285. (in Chinese).

Tainan Hydraulics Laboratory Technical Report. (2004). *The Research of New Shore Protection Technology (4/4)*, National Cheng Kung University, Taiwan, Bulletin No.312. (in Chinese).

Work, P.A. & Dean, R.G. (1991). Effect of Varying Sediment Size on Equilibrium Beach Profile, *Coastal sediments*, Seattle, USA.

Yang, R.Y.; Wu, Y.C.; Liou, J.Y. & Hong, C.S. (2004). A Research of Field Case Protection on Shuang-Chun Coastline, *Proceeding of the 26th Ocean Engineering Conference in Taiwan*, pp. 691~698 (in Chinese).

Chapter 16

The Economic and Financial Feasibility of a Biodigester: A Sound Alternative for Reducing the Environmental Impact of Swine Production

Antonio Zanin and Fabiano Marcos Bagatini

Additional information is available at the end of the chapter

http://dx.doi.org/10.5772/47428

1. Introduction

The modern landscape is one where environmental impact encroaches upon our quality of life. The search for viable technologies which both alleviate and lessen environmental pollution has become a priority, especially in the arena of production. Business, as with society, is now focused on minimizing environmental degradation, reviewing its strategies, structures and responsibilities. Tinoco (2001) explains that this modern responsibility is defined by the environmental and social demands of our day, where it is not merely a question of profit but also social conscientiousness. Business is now geared towards the interests of society, where environmental policy features high on its agenda.

It seems that human being has the tendency to risk his existence and wellbeing when the environmental impact of business proves a constant disruption to our natural world. Rural swine production and the environmental dangers it poses, is one example: waste, in remaining exposed releases methane gas into the atmosphere. As Brilhante and Caldas (1999) on this point, over the last few decades the dissipation of gasses has been affected by such practices, resulting in an increased concentration of carbonic gas (CO^2); methane (CH4); chlorofluorocarbons (CFCs), nitrous oxide and atmospheric ozone. As we now know, these gasses disrupt the energetic equilibrium of the Earth's atmosphere, and by consequence, our climate system.

Swine production has contributed substantially to Brazilian trade; it has received large investment incentives from genetics and other technologies, in order to provide and ensure a quality product. Panty (2008) highlights, that Brazil and its State of Santa Catarina have a model integrating both industry and producer, where specialists and rural farmers are well attuned to the competitive advantages of swine production.

Rural producers of swine do however face a lack of financial assistance in their search for an environmentally friendly solution. But an alternative is available, one able to reduce the endangerment of even more natural resources. Various "techniques" bringing to light and lessening the environmental effects of swine waste have been developed and put into practice. For instance, the process of biodigestion transforms methane gas into carbon dioxide, lessening environmental impact. The implementation of a biodigester means that waste can be reused and transformed into a renewable resource, proving an important mechanism for both business (finance) and society (the environment) alike. In other words, the biodigester is today's alternative solution for the rural farming industry, capable of minimizing the environmental effects of swine production. Waste on each farm, can be ably "re-directed" benefiting financial return; and where the farmer's quality of life is enhanced economically and financially, so too is that of the population at large who rely on the natural world for sustainability and survival. Indeed, we not only have the importance of increased productivity and the ensured success of new markets, but of being alert to a challenging future of sustainability and social responsibility.

Our research assesses the financially and environmental viability of installing a biodegster on a farming property. We divide our present study into 7 subsections, not counting the introduction. Beginning with *Environmental Management* we seek to contextualize the discourse and importance of our research by drawing upon recent literature in order to underline the contemporary shift in practice – namely the move towards more environmentally sensitive and socially sympathetic business strategies. Environmental strategy is vast becoming the defining characteristic of market competitiveness, and the success of a business to be environmental conscious and aware in its strategy and decision making, will prove the measure of its market edge and value. Our third section, *Swine Production and the Environment*, introduces the readership to the unique global positioning of Brazilian swine production and the very real possibilities of introducing Biodigestion as a viable and important measure ensuring both environmental integrity as well as cost effectiveness for the farming business. Section four, *Biodigesters*, introduces anaerobic biodigestion as the natural mechanism for both farming and environmental integrity. We seek to detail this anaerobic process and set out the processes and procedures of installing a functional biodigester on farming properties. Our sections on *Methodology* and the *Interpretation of Data* introduce the reader to the sampling method and data specifics of the herd studied. We then follow on with a section dealing with initial investment into the biodigester project, detailing the projected return and revenue, this, qualifying such investment as both environmentally timely and financially cost-effective for the business.

We conclude our study by underlining the necessity and urgency of biodigestion for swine producers, this, supported by the modern context of social conscientiousness and the benefit of sound financial return in line with our research projections.

2. Environmental management

The more our environment is damaged, the more our planet earth is compromised. We are witnessing, and experiencing, the progressive extinction of fauna and floral species, the pollution of groundwater and global warming.

In the face of these threats, is environmental management which aims to minimize environmental impact and maintain the wellbeing of people by redefining practices, processes and procedures for private, public and rural life. Good, proper working conditions and environmentally sound products complying with environmental laws and regulations fall to the responsibility of environmental management, as does the proper handling of waste produced in rural areas and the legal remit within which organizations can operate. This said, environmental management seeks to strike the legal and ethical balance between quality, productivity and competitiveness with the minimization of environmental degradation. Thus, for Moura (2000), the workings of environmental management involve putting specialized concepts and management techniques into environmental practice.

An environmental management system also signals greater competitiveness for the business which can equally retain and attract modern, learned consumers, whilst meeting the growing demands of external markets. Barbieri (2007) tells us that environmental management is defined by its administrative guidelines and key activities such as the planning, direction, control, and allocation of resources. Its main objective, furthermore, is to achieve positive, environmental results by reducing and eliminating the damage caused by human practices, or indeed to prevent such damage from even arising now and in the future.

This is why companies are now looking to develop and implement environmental management at the core of their operations in order gain an environmental advantage over market competitors, in strict accordance with the principles of sustainable development. Thus in retaking the concept of Barbieri, Tinoco and Kraemer (2004), we can assert that the role of environmental management is to effectively minimize and eliminate the environmental risks of private, public and rural businesses.

Companies can therefore meet the demands of the environment (and likewise of society), by tallying the expenditure of resources with legislation, restoring the natural resources extracted from the environment.

In this context, environmental management underlines the importance of environmental certification and accreditation which aims to help companies engage with, and commit to, the environment. Ribeiro (2006), for example, states that it is necessary to determine the particular strategy of implementing guidelines so as to more broadly define a company's environmental status and profile. Here, tools such as economic planning can be implemented into environmental programs that seek to change the current management system. Such programs, moreover, must be constantly checked through environmental audits. Yet aside from the virtues of the environmental management model, there is a constant need for businesses to be aware of the rapid changes which are occurring in information technology and strategic cost initiatives.

Indeed, models are constructed using the concepts defined by companies to guide and achieve goals. Barbieri's findings (2007) underline that the adoption of a model is critical because activities can be developed by different people at different times, in different places and through different ways of perceiving and positioning crucial issues. Companies can create their own environmental management models or take advantage of the various generic models that have been with us since the mid-1980s.

We see then that companies are finally coming into the age of environmental awareness, where standards and environmental legislation feature high on the agenda of those wanting to be maintain a market edge, nationally and internationally. We have a new series of standards defining our environmentally sensitive and fragile modernity, where the environmental status of a particular company has become the internationally accepted standard, seal and guarantee.

3. Swine production and the environment

For Marion (2002 p. 24) rural business, "explores the productive capacity of soil through the cultivation of land, as well as breeding and processing of certain agricultural products". The author classifies such rural activity as (1) agricultural; (2) zootechnical; and (3) agro-industrial. Similarly, Araújo (2003, p. 31) points out that, "agro-industries are businesses defined by the processing, handling and transformation of natural agricultural products into commercially packaged goods." Such businesses – deemed "agro-industries" by Marion (2002) and Araújo (2003) – are those which transform the agricultural/zootechnical product: the process of breeding, raising and slaughtering pigs for example, is for the purpose of transforming and commercializing derivatives. In a strictly economic context, the swine industry plays an important role in the movement of the food and supply chain. Sobestiansky (1998) shows then, that the modern swine industry is primarily focused on the production of pigs for slaughter and / or the breeding of livestock.

One of the most evident changes occurring both worldwide and in the Brazilian pig industry is the linear trend where a decrease in the number of production systems runs parallel to an increase in the number of system matrices. In terms of international agriculture, Brazil has the specialized workforce capable of producing technology that ensures a competitive advantage – it is a country fit to compete on equal terms with any other in the agricultural business, heavily investing in research and production strategies. The work of Gonçalves and Palmeiras (2006) shows us that the Brazilian swine industry has received greater international attention for its advantageously competitive edge: swine production in Brazil has lower costs than its major worldwide competitors; its system of production is vertically integrated (meeting agro-industrial demand); foodstuffs and basic grains such as soybeans and corn are plentiful, and there is technological investment.

In retrospect, market realities and the accelerated growth of a global economy meant that agro-industrial businesses had to seek improvements and overhaul their organizational structures in order to guarantee market presence and competitiveness. The swine industry was by no means an exception to this trend, requiring both injections of investment into its processes and new facilities. As Leite (2008) makes clear, swine production underwent change because technological innovation became the rule-of-thumb for a new generation of engineers focusing on the economic and environmental viability of such practices. Zuin and Alliprandini (2006, p. 255) similarly assert that, "over the years innovation and invention proved powerful tools in achieving better efficiency in farming systems." An equally important factor to this is the partnerships established between pig farmers and agro-industrial businesses, which have ensured the commercialization of production and added market value to the final product. Yet Mior (2005) equally points out that at the close of the 1990s in the western region of Santa Catarina, partnerships between agro-industries and swine producers heralded a new epoch in pig production, by reducing the number of contracts at the same time as increasing production.

According to findings from the Associação Catarinense de Criadores de Suínos (2011), Brazil has 2,460 matrices for the housing of swine, where production sprang from 2,708 million tons in 2003 to 3,240 million in 2010 – this, an increase of 19.65% in merely five years, signaling market growth and the importance of swine in the supply chain. Yet the concentration of pig herds on small rural farms nevertheless carries environmental impacts owing to the vertical model of production adopted by Brazil, which is characterized by partnerships with industry.

Generally, animal waste is treated in liquid form: water runs into deposits which are stored and the soil then used as organic fertilizer (EMBRAPA, 2008). This model coupled with the growth of swine production in Brazil heightens the risks of environmental degradation whereby the measures for treating swine waste are not only costly but require constant precision – mishandled waste can lead to water, soil and air pollution with both an unwelcome stench and mosquitoes.

In addition to industry and urbanization (domestic sewage), swine production is monitored by regulatory bodies and environmental agencies, so much so that the law clearly identifies the environmental dangers of such practices (GUIVANT e MIRANDA 2004). This is due to the large number of contaminants found in effluents which represent a potential source of air, water and soil contamination. Indeed, due to the high concentration of livestock at these rural sites, swine waste can easily exceed the capacity of local ecosystems, potentially disrupting the natural environment and human health through organic matter; nutrients; pathogens; odors and microorganisms generated in the atmosphere (PEREIRA, DEMARCHI e BUDINO, 2009).

A series of requirements can therefore be provisionally laid out, aimed at preventing and correcting increased environmental degradation. Among these we can list:

1. The need to maintain a permanent boundary of preservation set at a distance of 30 meters, with distances between dwellings and settlements of at least 300 meters, and distances to roads at least 50 meters (Bezerra, 2005).
2. The need to prohibit and monitor the dumping of waste and/or effluents from any polluting source, including waste from livestock, into Class I rivers intended for domestic supply. Such material may be released directly or indirectly into Class II and III rivers only after appropriate treatment and having satisfied the conditions, standards and requirements set forth by government Decree (RESOLUÇÃO CONAMA nº 357, from 17th March, 2005).
3. The need to research ways of combining the use of waste for crops (fertilizers), or for the production of energy. This would reduce the degree of environmental pollution in line with the realities faced by rural farmers. Given this context, some environmental problems could be solved if environmental measures for swine production were effectively researched and put into practice. However, not all swine producers have the awareness or financial resource to treat waste correctly.

4. Biodigesters

Seganfredo (1999) notes that the continuous use of large quantities of swine waste as fertilizer has proven environmentally detrimental, not only in terms of air pollution but in terms of the progressive accumulation of nutrients in the soil and the presence of excess nitrates in water. Likewise Sampaio et al. (2010) show that the mismanagement of remaining waste water can lead to an excess of pig manure in the soil (depending on the capacity for absorption some of these nutrients may lead to water contamination).

Nogueira (PALHARES, 2008) therefore does well to remind us that in 1086 the Englishman Humphrey Davy identified a gas rich in carbon and carbon dioxide resulting from the decomposition of manure in humid places. Released into the atmosphere, the gas attacks the ozone layer and causes global warming. Jordan (2005) states, moreover, that methane - one of the gases produced by the degradation of waste – is twenty-one times more volatile than carbon dioxide (CO2).

It is precisely here that the process of anaerobic biodigestion can prove the necessary environmental solution in that it destroys pathogenic organisms and parasites. In this way the treatment of waste by such means carries great advantages, transforming harmful gasses into a source of energy (bio-gas) *for the better*. In addition to this, solid matter decanting in the bottom biodigesting tank acts as biofertilizer, with liquid matter the (treated) mineralized effluent. Nogueira (1986) further points out that such a process offers multiple advantages. The production of fuel gas; the control of water pollution and odor; the elimination of pathogens from organic matter and the preservation of fertilizer are the immediate benefits of such waste removal. We can likewise emphasize that anaerobic digestion helps to minimize negative environmental impact, at once reducing relative risks and improving quality of life issues.

A digester is essentially built from a tank sealed with canvas suitable for storing waste. The process of biodigestion occurs in rotating fashion where organic matter enters through the side and anaerobic fermentation taking place without the presence of air. The result of this process is the transformation of methane into carbon dioxide.

The biogas storage balloon is used for the intake and utilization of highly corrosive gases which require beneficiation treatment. Such processing is essential and increases their efficiency, for without it the use of biogas is not recommended. The process has specific stages of washing; cooling and compression. Finally, there is the intake of biogas by the storage (balloon), as evidenced in Figure 2.

The liquid effluent resulting from this process then exits the digester for maturation ponds wherein the release of gases is completed: even after waste digestion in the machine, some substances still remain and must be released into the atmosphere so that the liquid can be used for fertilization.

As Nogueira explains (1986) waste is transformed by agitation which distributes the substrate and bacteria, efficiently using the volume of the biodigester and reducing/eliminating supernatant scum matter. In order to have a guaranteed and precise process, agitators must be inserted into the biodigester in order to correctly agitate the substances necessary for transforming waste. It is also important to maintain the biomass at a heated temperature within the biodigester, for as Nogueira (1986) further points out, the biodigester has to be constructed and set up below ground-level, since this depth serves as a thermal insulator. Temperature plays an important part and it is advisable to ensure internal or external heat (of course depending on the agricultural needs of the producer) because bacteria can reproduce in this way, thereby transforming waste into biogas.

Figure 1. The biodigester set up on the property where research findings were collated.

Figure 2. Biogas storage balloon

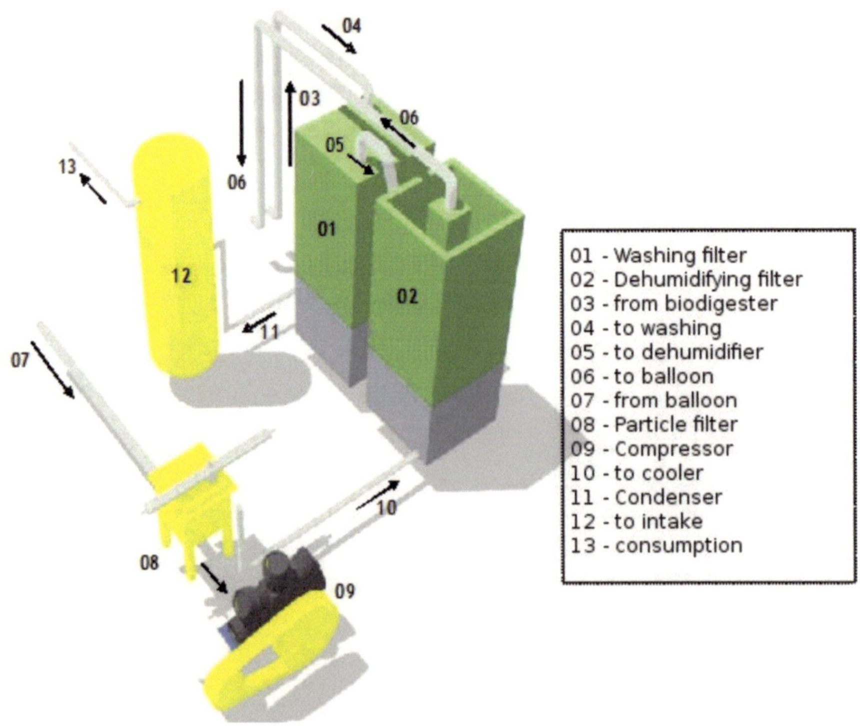

Source: Gter Energias Renováveis.

Figure 3. The biogas treatment process prior to intake.

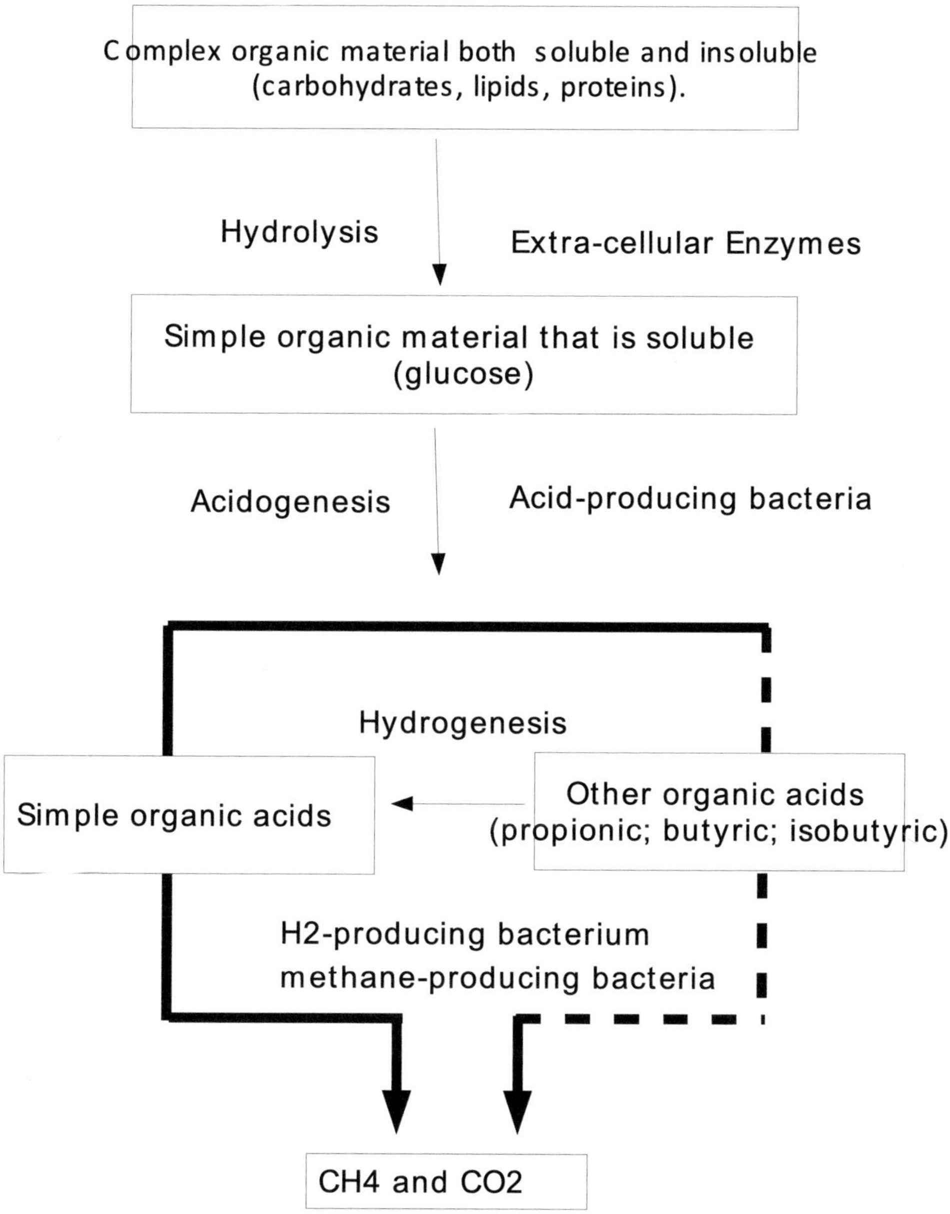

Source:http://www.biodieselbr.com/i/energia/renovavel/biogas/processo-biodigestor.gif.
Accessed: 18/5/08.

Figure 4. The metabolic stages of anaerobic digestion occurring in the biodigester: the transformation of organic materials into gases used by farms, means that the environment is benefitted through the reduction of harmful methane gas emission.

The introduction of biodigesters to rural swine production can thus equally benefit producers and the environment, reducing and possibly eliminating risks of environmental degradation whilst ensuring the quality of life required by human being.

5. Methodology

Our present study focuses on a rural farm chiefly producing swine. Research was carried out on a farming property with a herd of 5, 362 pigs. All processes and procedures undertaken at the farm are officially approved and licensed by the governing environmental agency. We also investigated whether the implementation of a biodigester is an economically and financially viable alternative: where there is the appropriate treatment of swine waste from production and the financial return of investment for the farmer.Data was solicited by semi-structured interview, with the additional analysis of documentation. The methodology of our study consisted of exploratory (qualitative) research, case studies of site procedures as well as quantitative research. Collated data was then tabled, and calculations / financial projections were made for the economic viability of investment.

For the sake of anonymity, "Business A" is used to designate the slaughterhouse and "Business B" the company which invested funds into the biodigester.

Calculations for the economic and financial viability of investment drew upon data collected in the first half of 2009. These data were tabled and are the basis for projections. In January 2012, we revisited the businesses which had participated in the research study in order to measure the strength and validity of the 2009 projections. We found that the generation of biogas and the financial return were in line with our 2009 forecasting/calculations. Adjustments were made in order to ensure a greater reliability for the return of investment.

6. Interpretation of data

The object of our study is Farm III and its 2 nuclei. The farm provides raw material to Business A. Our study seeks to measure the economic and financial feasibility of installing a digester and the proper treatment of swine waste this promises. All necessary data was therefore collected from Farm III/Nuclei I and II.

Farm III currently has a herd number of 5,362 pigs between nuclei I and II. It has a legally authorized waste treatment system. Each nucleus on the site has six pools and a Geomembrane Biodigestor of HDP (High Density Polyethylene) and LLDP (Linear Low Density Polyethylene) of 1.00 mm with a volume capacity of 800m ^{3} in nucleus I and 1400m ^{3} in nucleus II. Each nucleus has a homogenization tank which heats waste prior to it entering the biodigester. This tank has a volume capacity of 75m ^{3} in nucleus I and 150m ^{3} in nucleus II. There is also a homogenization pump in each nucleus with a power of 5 hp.

7. Initial investment in the project

Installing a biodigester requires building a workable structure. Costing included reservoirs of clean biogas; biogas cleaning kits and labor. It was also necessary to purchase equipment such as biogas dryers; piping; compressors. Table 1 presents the total costing of initial investment necessary for installing a biodigester.

Description	Qty	Value (R$)
Biodigester 1.400 m^3	1 und	131.700,00
Biodigester 900 m^3	1 und	102.426,00
Earthwork/excavation	N/D	37.000,00
Machinery room	2 und	26.000,00
Biogas cleaning kit	2 und	49.640,00
Biogas dryer set	N/D	38.960,00
Reservoirs of clean biogas	3 und	54.000,00
Pipe connections	N/D	12.000,00
Compressor	2 und	24.000,00
High pressure network	N/D	9.600,00
Pipeline	4 km	41.735,00
Pipeline excavation	4 km	22.000,00
Control fittings	N/D	28.716,00
Command table	2 und	24.800,00
Burners / heating equipment	N/D	32.321,00
Unforeseen costs	N/D	15.873,00
Total		**650.771,00**

Source: data collected from Business B.

Table 1. Initial investment.

7.1. Revenue generated by investment

Business B is responsible for funding the biodigester. It receives a return through the use of biogas consumed in the slaughterhouse and the singeing of swine. Values are invoiced using the following calculation:

$$CBG(R\$) = \frac{m^3\,\text{biogas} \times \text{cost per kg of GLP} \times 0.80}{2.3}$$

KEY: CBG = value of biogas consumed
m^3 biogas = quantity of biogas consumed;
cost per kg of GLP (liquefied petroleum gas);
0.80 = eighty percent of the value of LPG (as per the binding contractual clause between investor and slaughterhouse);
2.3 = conversion factor for energy equivalence GLP/biogas.

Table 2 shows the net income accrued over 2006; 2007; 2008; 2009; 2010 and 2011, and the projected income for 2012; 2013; 2014 and 2015. The projections were made by calculating the average between 2009, 2010, 2011 and applying a 5% growth estimate (subsequent years are projected to grow by 5% when compared to previous years).

Year	Revenue
2006	94.922,09
2007	115.993,24
2008	107.239,10
2009	111.354,05
2010	116.921,75
2011	122.767,84
2012	122.865,27
2013	129.008,54
2014	135.458,96
2015	142.231,91

Source: Data collected from Business B.

Table 2. Generated and projected revenue.

In 2006 we see that the investing company had revenues of R $ 94,922.09 (ninety four thousand, nine hundred and twenty-two Reals, nine cents). This, representing a nominal return of 14.95% from the value invested. The rate of inflation for 2006, using the IPCA index (adopted by the Brazilian government as the official measure of inflation) was 3.14% (according to the Central Bank of Brazil) and stands at 11.45% as the effective rate of return for that year.

In 2007, the investing company achieved a total revenue of R $ 115,993.24 (one hundred and fifteen thousand nine hundred and ninety-three Reals, twenty-four cents), representing an 18.27% nominal return on the value invested. The IPCA inflation rate for 2007 stood at 4.46% (according to the Central Bank of Brazil), with 13.22% as the effective rate of return on investment for the period.

In 2008, total revenue accrued by the company amounted to R $ 107,239.10 (one hundred and seven thousand, two hundred and thirty-nine Reals, ten cents), reaching a nominal return of 16.89% on invested capital. IPCA (according to Central Bank of Brazil) registered inflation at 5.90%, thus qualifying 10.38% as the effective return of investment.

In 2009 the total net revenue obtained by the company amounted to R $ 111,354.05 (one hundred and eleven thousand, three-hundred and fifty-four Reals, five cents), reaching a nominal return of 17.54% on invested capital. Using IPCA as an index (and according to the Central Bank of Brazil) inflation stood at 4.31%, thus qualifying a 12.68% effective return of investment.

2010 had total revenues of R $ 116,921.75 (one hundred and sixteen thousand, nine hundred and twenty-one Reals, seventy-five cents), reaching a nominal return of 18.42% on invested capital. In 2010 IPCA (according to the Central Bank of Brazil) pegged inflation at 5.91%, this putting the effective return of investment at 11.81%.

In 2011 net revenue was R $ 122,767.84 (one hundred and twenty-two thousand, seven hundred and sixty-seven Reals, eighty-four cents), reaching a nominal return of 19.34% on invested capital. Inflation stood at 6.50% according to IPCA (and the Central Bank of Brazil), with the effective return of investment at 12.06%.

It appears that the annual return of savings (investment considered low risk and yield equal in all financial institutions in the country) for the years 2006; 2007; 2008; 2009; 2010 and 2011 was 8.41%; 7.80%; 7.74%; 7.09%; 6.81% and 7.50% respectively. The rate of cumulative nominal return on investment in question was higher in all the periods analyzed when compared to the accumulated rate of return of savings. From a purely financial viewpoint, such investment is attractive in terms of yields.

Year	Initial Investment	Net revenue	Return
2006	-650.771,00		-650.771,00
2006		94.922,09	-555.848,91
2007		115.993,24	-439.855,67
2008		107.239,10	-332.616,57
2009		111.354,05	-221.262,52
2010		116.921,75	-104.340,77
2011		122.767,84	18.427,07
2012		122.865,27	141.292,34
2013		129.008,54	270.300,88
2014		135.458,96	405.759,84
2015		142.231,91	547.991,75

Table 3. *Payback and return of investment.* The payback method of analysis presents to shareholders the necessary time to recover the financial resources invested in the project. Source: authors.

One of the weaknesses of payback strategy is the depreciation of monetary value over time. This said, a reversal seems to have occurred between 2010 (negative) and 2011 (positive). In terms of our research and its specific scenario, a recovery of initial investment kicks in at about five years and 11 months. The projection of discount rates for the following years was calculated by using the simple average of yields recorded over the previous three years, this being approximately 7.13%.

Year	Initial investment	Net revenue	Tax deduction	Net income	Deducted return
2006	-650.771,00				-650.771,00
2006		94.922,09	8,41%	87.558,43	-563.212,57
2007		115.993,24	7,80%	99.814,85	-463.397,72
2008		107.239,10	7,74%	85.747,65	-377.650,07
2009		111.354,05	7,09%	84.666,25	-292.983,82
2010		116.921,75	6,81%	84.107,70	-208.876,12
2011		122.767,84	7,50%	79.548,84	-129.327,29
2012		122.865,27	7,13%	75.866,74	-53.460,55
2013		129.008,54	7,13%	74.358,33	20.897,78
2014		135.458,96	7,13%	72.879,90	93.777,68
2015		142.231,91	7,13%	71.430,88	165.208,56

Source: authors.

Table 4. *Deducted* return: this table shows the calculation of discounted payback. The discount rates applied to net annual income were from savings accounts checked for the years 2006, 2007, 2008, 2009, 2010 and 2011.

Considering this data, a sign inversion occurs between the projected years 2012 (negative) and 2013 (positive), pitching the recovery of initial investment at roughly 7 years and 9 months. Using the information in Table 4 as a reference point, that is, the initial investment and net income both generated and projected, it is also possible to calculate the internal rate of return for the project – this being 12.16% per year.

Financial analysis reveals the benefits of investing in a biodigester. The farm reported a total revenue of R $ 650,771.00, with a net income of R$ 94.922,09 for 2006; R$ 115.993,24 for 2007; R$ 107.239,10 for 2008; R$ 111.354,05 for 2009; R$ 116.921,75 for 2010 and R$ 122.767,84 for 2011. For 2012 net income is projected to be R $ 122,865.27 with 5% real annual growth applied.

The payback period of investment recovery stands at 5 years and 11 months, and the discounted payback at 7 years and 9 months. Comparing the annual return of investment to the cost of savings, the rate of return on investment yielded higher gain over all periods analyzed, this, confirming the economic and financial viability of such a project. Financial and economic optic thus evidences the viability of financial investment into the project. This, combined with the environmental benefits, marks the entrepreneurial quality of such an enterprise.

8. Conclusion

The goal of our study was to demonstrate the relationship between swine production and the environment – where the former can be developed through new, innovative technologies, the latter is not impervious and remains vulnerable. There is a need to consider the negative environmental effects of such practices, highlighting the urgency for due care and attention in waste management and business strategy.

The current market calls for fundamental changes in the economic situation and nature of organizations as well as flagging demands for a new benchmark in business management. An innovative approach to the way new realities are both understood and dealt with is a modern requirement.

There is nevertheless a lack of accurate information about the chemical concentration of swine waste. Alternatives remain limited and the rationale underdeveloped. Ongoing research is required into the suitability of different soil types and crops receiving fertilizer. Likewise, both the short and long-term environmental effects of swine waste need to be studied and known. What we do know is that the indiscriminate disposal of swine waste in natural environments, runs the elevated risk of contaminating soil; water supplies; rivers; effluents and the air itself – this, a condonable practice directly affecting the health of rural and urban communities.

Any waste distribution system then, must take into account the cultural and economic realities of farmers and local producers of swine. It is essential to raise the awareness of farmers and society alike, bringing to their attention issues of waste pollutants, as well as the benefits of implementing a technology that combines the agronomic use of manure as fertilizer, providing the economy with greater input and systems which minimize the effects of pollution.

As we previously asserted, our modern scenario is one requiring a social conscientiousness for the ecosystem is on the verge of total collapse. It is where modern man is faced with the urgency to change his world-view through sustainable business practices, igniting a change of values and a new direction in operating systems which engage with sustainable development and environmental preservation.

According to Freitas (2008), a great many experts fear that if emissions of greenhouse gases (mainly carbon dioxide, methane and nitrous oxide) continue to increase, then the planet's temperature will rise and the results will be drastic if not unimaginable. Formidable changes in our climate with extreme cold and elevations of wetland will lead us to experiencing periods of drought; fertile farmland will not too far into the future succumb to desertification; incidences of severely destructive storms, tornadoes, hurricanes or typhoons will be frequent; the dwindling and complete loss of floral species and fauna in different parts of the natural world will be commonplace, as will the melting of ice caps and the consequent increase in global sea levels. With the unbridled increase in global warming, mainly due to high levels of carbon dioxide and methane, serious consequences can already be felt on a global scale, threatening the survival of the earth's inhabitants.

The aim of our study is to demonstrate the benefits generated by installing a biodigester as a financially and economically viable alternative in the management of swine waste. Given this, biodigester technology and its environmental value, is more than a mere hypothesis but a tried and tested means of treating and reusing detrimental waste material – this, important for both swine producer and society. Besides being an alternative source of renewable gas, such a technology reuses and recycles, improving and maintaining soil without jeopardizing environmental standards and human wellbeing.

Author details

Antonio Zanin and Fabiano Marcos Bagatini
Universidade Comunitária da Região de Chapecó – UNOCHAPECÒ, Brazil

9. References

Associação Catarinense De Criadores De Suínos. Available at: http://www.accs.org.br/dados_ver.php?id=4 Accessed on 10/7/2011.

Araújo, Massilon J. Fundamentos do agronegócio. São Paulo: Atlas, 2003.

Barbieri, José Carlos. Gestão ambiental empresarial: conceitos, modelos e instrumentos. 2. ed. Atual. Ampl. São Paulo: Saraiva, 2007.

Bezerra, Severino Antunes. Gestão ambiental da propriedade suinícola: Um modelo baseado em um biossistema integrado. Revista Ciências Empresariais da UNIPAR, Toledo, v.6, n.2, jul./dec., 2005.

Brasil. Resolução CONAMA nº 357, 17th March, 2005. Ministério do Meio Ambiente. Available at: http://www.mma.gov.br/port/Conama. Accessed on 17/2/ 2012.

Brilhante, Ogenis Magno; CALDAS, Luiz Querino de A. Gestão e avaliação de risco em saúde ambiental. Rio de Janeiro: Fiocruz, 1999.

Embrapa. Licenciamento ambiental para criação de suínos. Available at: http://hotsites.sct.embrapa.br/diacampo/programacao/2008/licenciamento-ambiental-para-criacao-de-suinos. Accessed on: 30/1/2012

Gonçalves, Rafael Garcia; PALMEIRA, Eduardo Match. Suinocultura Brasileira. Observatório de Economia Latino americana, Número 71, 2006. Available at: http://www.eumed.net/cursecon/ecolat/br/06/rgg.htm. Accessed on 5/4/2011.

Guivant, Julia S.; MIRANDA, Cláudio R. de. Desafios para o desenvolvimento sustentável da suinocultura: uma abordagem multidisciplinar. Chapecó: Argos, 2004.

Jordan, Danielle. Dejetos de suínos e outros animais podem ser convertidos em créditos de carbono. Homologa.ambiente.sp.gov.br/proclima, September, 2005. Available at: //http://noticias.ambientebrasil.com.br/exclusivas/2005/12/09/22129-exclusivo-dejetos-de-suinos-e-outros-animais-podem-ser-convertidos-em-creditos-de-carbono.html. Accessed on: 2/3/2012.

Leite, João Guilherme Dal Belo. Inovações Tecnológicas na Suinocultura Sul – Brasileira sobre a Ótica da Sustentabilidade. ENGEMA- Encontro Nacional de Gestáo Empresarial e MeioAmbiente (National Congress for Business and Environmanetal Management), 9th- 12th November, 2008. Porto Alegre, Brazil.

Marion, José Carlos. Contabilidade Rural: Contabilidade Agrícola; Contabilidade Da pecuária; imposto de renda pessoa jurídica. 7. ed. São Paulo: Atlas, 2002.

Mior, Luiz Carlos. Agricultores Familiares, Agroindústrias e redes de desenvolvimento rural. Chapecó: Argos, 2005.

Moura, Luiz Antonio de. Qualidade e gestão ambiental: sugestões para implantação das Normas ISSO 14.000 nas empresas. 2ª ed. São Paulo: Editora Juarez de Oliveira, 2000.

Nogueira, Luiz A. Horta. Biodigestão: a alternativa energética. São Paulo: Nobel, 1986.

Palhares, Julio César Pascale. Biodigestão anaeróbia de dejetos de suínos: aprendendo com o passado para entender o presente e garantir o futuro. Available at: http://www.infobibos.com/Artigos/2008_1/Biodigestao/Index/.htm. Accessed on: 9/1/2012.

Panty, Eliana. I simpósio Brasil Sul de Suinocultura. Agromais, Chapecó, n. 9, p. 66, April/May, 2008.

Pereira, E.R.; Demarchi, J.J.A.A.; Budiño, F.E.L. A questão ambiental e os impactos causados pelos efluentes da suinoculturaa. 2009. Article in Hypertext format. Available at: http://www.infobibos.com/Artigos/2009_3/QAmbiental/index.htm. Accessed on: 30/1/2012

Queiroz, Timóteo Ramos (Org.). Agronegócios: gestão e inovação. São Paulo: Saraiva, 2006. p. 251-280.

Ribeiro, Maisa de Souza. Contabilidade ambiental. São Paulo: Saraiva, 2006.

Sampaio, Silvio C; Fiori, Marciane G.; Opazo, Miguel A. U.; Nóbrega, Lúcia H. P. Comportamento Das Formas De Nitrogênio Em Solo Cultivado Com Milho Irrigado Com Água Residuária Da Suinocultura. Eng. Agríc., Jaboticabal, v. 30, n1, jan-feb. 2010.

Sasaki, Yosuke; Saito, Hikari; Shimomura, Ai; Koketsu. Consecutive reproductive performance after parity 2 and lifetime performance in sows that had reduced pigs born alive from parity 1 to 2 in Japanese commercial herds. Livestock Science. School of Agriculture, Meiji University, Kanagawa 214–8571, Japan, 2011.

Seganfredo, Milton Antonio. Os dejetos de suínos são um fertilizante ou um poluente do solo? Cadernos de Ciência e Tecnologia, Brasília, v. 16, n. 3, p. 129-141, Sept./Dec. 1999. Available at: http://www.uov.com.biblioteca_arquivos/Curso55-3.pdf. Accessed on: 4/2/2012.

Silva, Benedito Albuquerque da; Robles Junior, Antonio. Modelagem de custos das atividades integradas de suinocultura: cria, recria e engorda. XVI Congresso Brasileiro de Custos – Fortaleza, Ceará, Brasil, 3rd -5th November, 2009. Annals.

Sobestiansky, Jurÿ; et al. Suinocultura intensiva: produção, manejo e saúde do rebanho. Brasília: Embrapa, 1998.

Tinoco, João Eduardo Prudência. Balanço Social: uma abordagem da transparência e da responsabilidade pública das organizações. São Paulo: Atlas, 2001.

Tinoco, João Eduardo Prudêncio; KRAEMER, Maria Elisabeth Pereira. Contabilidade e Gestão Ambiental. São Paulo: Atlas, 2004.

Zanin, Antonio; Bagatini, Fabiano Marcos; Pessato, Camila Batista. Viabilidade econômico-financeira de implantação de biodigestor: uma alternativa para reduzir os impactos ambientais causados pela suinocultura. Custos e agronegócio on line, ISSN 1808-2882, Recife, volume 6, número 1, p.1-161, Jan/April, 2010. Available at: http:// www.custoseagronegocioonline.com.br. Accessed on: 12/3/2012.

Zuin, Luíz Fernando Soares; AlliprandinI, Dário Henrique. Gestão de inovação na produção agropecuária. In: Zuin, Luiz Fernando Soares.

Chapter 17

Overview of Environmental Management by Drill Cutting Re-Injection Through Hydraulic Fracturing in Upstream Oil and Gas Industry

Mansoor Zoveidavianpoor, Ariffin Samsuri and Seyed Reza Shadizadeh

Additional information is available at the end of the chapter

http://dx.doi.org/10.5772/45828

1. Introduction

For the reason of worldwide increased activities of upstream oil and gas industry for future energy demands which will be associated with more waste generation, zero discharge is considered an environmentally friendly approach of complying with environmental legislations. Drilling is one of the major operations in upstream oil and gas industry that can potentially impact the environment through generation of different types of wastes. The drilling process generates millions of barrels of drilling waste each year; primarily used drilling fluids and drill cuttings especially oil-contaminated drill cuttings. In the early years of the oil industry, little attention was given to environmental management of drilling wastes. The rapid development of drilling operation in order to fulfill the global energy demands and so the drilling environmental regulatory requirements have become stricter, drilling and mud system technologies have advanced, and many companies have voluntarily adopted waste management options with more benign environmental impacts that those used in the past. Moreover, it is crucial to find out why drilling wastes are important nowadays, how they generated and by which means those waste could be disposed off with higher efficiency and acceptable HSE and economically concerns. Drill Cutting Re-Injection (DCRI) is one of the processes that developed as an environmentally friendly and zero discharge technology in upstream oil and gas industry.

A variety of oil field wastes are disposed of through injection, such as produced water that re-injected through tens of thousands of wells for enhanced recovery or disposal. Other oil field wastes that are injected at some sites include work over and completion fluids, sludge, sand, scale, contaminated soils, and storm water, among others. The focus of this chapter is injection of wastes related to the drilling process, which involve processing cuttings into small particles, mixing them with water and other additives to make slurry, and injecting it into a subsurface geological formation at pressure high enough to fracture the rock. DCRI has been given other terms by different authors such as fracture slurry injection, grind and inject, and drill cuttings injection.

The most critical aspect in waste injection through hydraulic fracturing (HF) in upstream oil and gas industry, which is DCRI, will be reviewed in this chapter. The subject of this chapter, DCRI, is a specialized area in upstream petroleum industry; even though many brilliant papers presented on various environmental areas, overview papers that present a context for those more specific studies are needed. This chapter will presents in an effort to review the environmental management of DCRI in upstream petroleum industry. The aims are firstly, to review the drilling process and different types of drilling fluid. Afterwards, because it's considered as a key in identifying containment formations to prevent waste migration to water resources and environment in DCRI operations, HF technology will be introduced in the second part of this chapter. Finally, after reviewing the essential parts of DCRI, drilling wastes and HF, the nature of DCRI and its role in environmental management will be presented in details.

2. Overview of drilling operation

Oil and gas wells are drilled to depths of several hundred to more than 5,000 meters. Figure 1 shows a schematic of typical drilling rig, which uses a rotating drill bit attached to the end of a drill pipe. Drilling fluids (muds) are pumped down through the hollow drill pipe, through the drill bit nozzles and up the annular space between the drill pipe and the hole. Drilling mud mixture is particularly related to site and hole condition; it used to lubricate and cool the drill bit, maintains pressure control of the well as it is being drilled, and helps to removes the cuttings from the hole to the surface, among other functions. In fact, the technology of mud mixing and treatment has been recognized as a source of pollutants.

Mud and drill cuttings are separated by circulating the mixture over vibrating screens called shale shakers. As the bit turns, it generates fragments of rock (cuttings), which will be separated from the mud by shale shakers that will moves the accumulated cuttings over the screen to a point for further treatment or management. Consequently, additional lengths of pipe are added to the drill string as necessary. As a common practice in drilling of oil and gas wells, when a target depth has been reached according to the drilling plan, the drill string is removed and the exposed section of the borehole is permanently stabilized and lined with casing that is slightly smaller than the diameter of the hole. The main function is to maintain well-bore stability and pressure integrity. (Three sizes of casing depicted in Figure 1). Cement is then is pumped into the space between the wall of the drilled hole and the outside of the casing to secure the casing and seal off the upper part of the borehole. Each new portion of casing is smaller in diameter than the previous portion through which it is installed. The final number of casing strings depends on the total depth of the well and the sensitivity of the formations through which the well passes. The process of drilling and adding sections of casing continues until final well depth is reached.

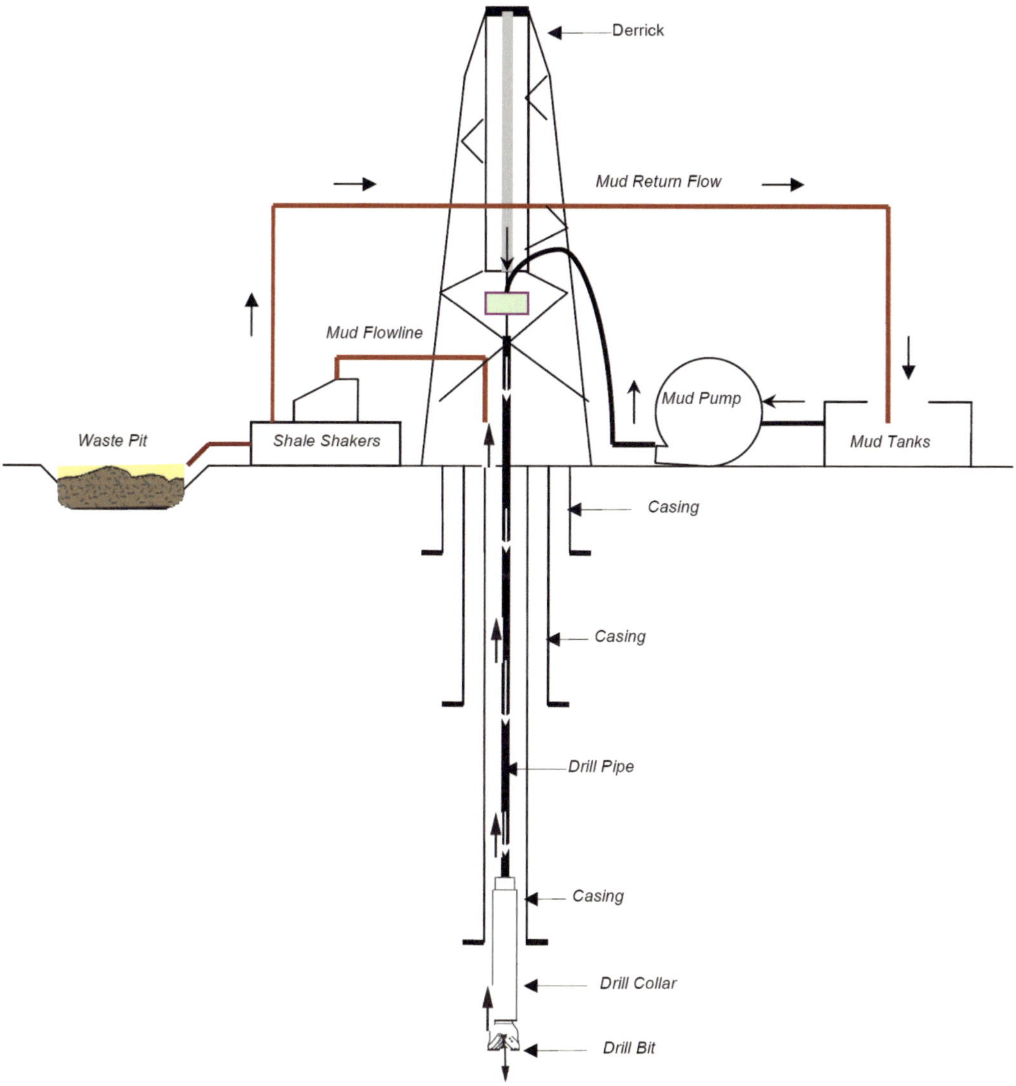

Figure 1. A schematic of a drilling rig (not to scale).

Two primary types of wastes are generated in drilling of oil and gas wells; drill cuttings and drilling fluids. Most drilling fluids contain bentonite clay, water, barite, specialized additives, and some types of muds also contain hydrocarbons. Large volumes of drilling muds are stored in aboveground tanks or pits. The liquid muds pass through the screen and are recycled into the mud system, which is continuously treated to maintain the desired properties for a successful drilling operation. Depending on the depth and diameter of the well bore, the volume of drilling wastes generated from each well varies; typically, several thousand barrels of drilling waste are generated per well. Figure 2 is a demonstration of the generated drilling waste from a 2400 meters well depth that comprises of four different borehole sizes.

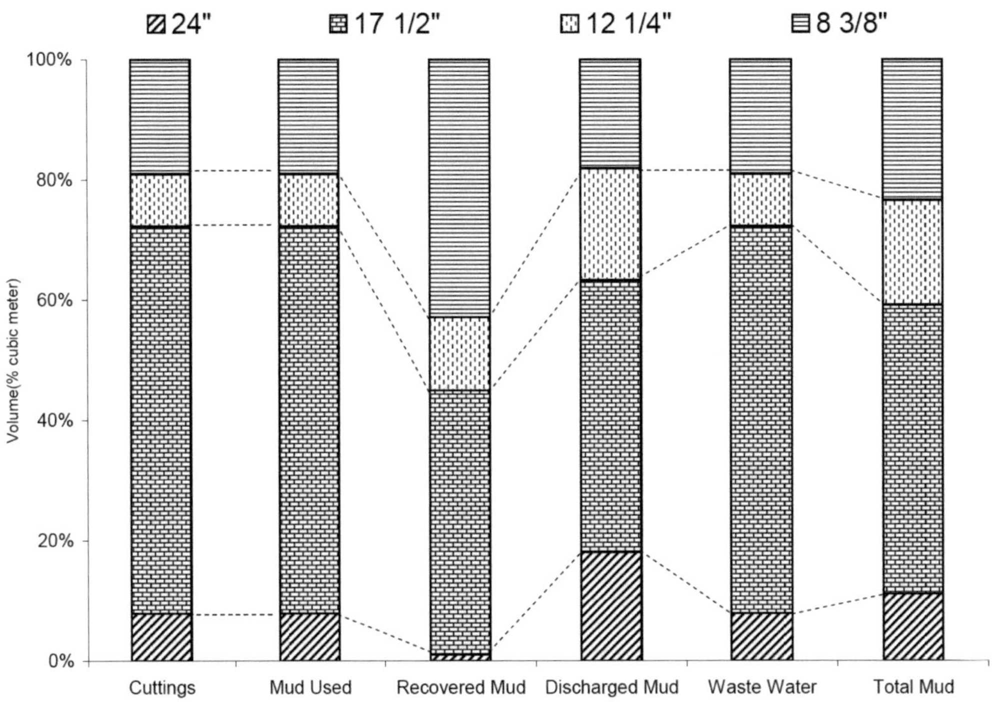

Figure 2. A typical drill cutting and mud volumes for a 2400 meters well depth.

The generation of wastes from drilling fluid and drill cuttings could be recognized at different stages of the drilling operation. When drilling at the first few hundred meters to run conductor casing or surface casing, higher quantities of cuttings are produced; that's because borehole diameter is the largest during this stage. Substantial waste fluid must be handled when drilling deep wells that encountered shale's and/or unstable formations. So, oil based muds (OBMs) is utilized to overcome those problems which will be mixed with other drilling fluids in waste pit and disposed to the environment. Furthermore, higher volume of wastes must be displaced in the completion phase of drilling operation which is replaced by completion fluids and equipment. Physical condition of a waste pit during and after drilling operation is illustrated in Figures 3 and 4, respectively. More details could be found by Shadizadeh and Zoveidavianpoor, (2008).

Figure 3. Mud pit condition during drilling operation.

Figure 4. Mud pit condition after drilling operation.

3. Environmental impacts of drilling muds

In upstream petroleum industry, drilling is the major operation that can potentially impact the environment. Drilling operation generates a significant volume of wastes. The composition of drilling fluid constituents is depicted in Table 1. Environmentally responsible actions require an understanding of the characteristics of these wastes and how they are generated in order to minimize their environmental impacts by known environmental protection methods. In this section, environmental impacts of a drilling mud will be presented along with a case study on mud pit samples for heavy metals (Cd, Cr, Ni, and Al) concentrations during and after the drilling operation. For more details please consult Shadizadeh and Zoveidavianpoor, 2008 and 2010.

Elements	Water	Cuttings	Barite	Clay	Chrome-lignosulfonate	Lignite	Caustic
Aluminum	0.3	40,400	40,400	88,600	6,700	6,700	0.013
Arsenic	0.0005	3.9	34	3.9	10.1	10.1	0.039
Barium	0.01	158	590,000	640	230	230	0.26
Cadmium	0.0001	0.08	6	0.5	0.2	0.2	0.0013
Chromium	0.001	183	183	8.02	40,030	65.3	0.00066
Cobalt	0.001	183	183	8.02	40,030	65.3	0.00066
Copper	0.0002	2.9	3.8	2.9	5	5	0.00053
Iron	0.003	22	49	8.18	22.9	22.9	0.039
Lead	0.5	21,900	12,950	37,500	7,220	7,220	0.04
Magnesium	0.003	37	685	27.1	5.4	5.4	0.004
Mercury	4	23,300	3,900	69,800	5,040	5,040	17,800
Nickel	0.0001	0.12	4.1	0.12	0.2	0.2	5
Potassium	0.0005	15	3	15	11.6	11.6	0.09
Silicon	2.2	13,500	660	2,400	3,000	460	51,400
Sodium	7	206,000	70,200	271,000	2,390	2,390	339
Strontium	6	3,040	3,040	11,000	71,000	2,400	500,000
Cobalt	0.07	312	540	60.5	1030	1030	105

Table 1. Elemental composition of drilling fluid constituents (ppm) (Bleier et al., 1993).

A potential source of heavy metals in drilling fluid is from crude itself. Crude oil naturally contains widely varying concentrations of various heavy metals. In the selected well a combination of water based muds (WBMs) and OBMs had used. As shown in Table 2, the major components of WBMs in the investigated site were barite, salt, starch, bentonite, and lime. The metals of greatest concern, because of their potential toxicity and/or abundance in drilling fluids, include chromium, cadmium, and nickel (Neff, 2002). Some of these metals are added intentionally to drilling muds as metal salts or organometallic compounds. Others are present as trace impurities in major mud ingredients, particularly barite and bentonite. One of the major drilling mud additives used in both WBMs and OBMs in the investigated well is barite. The amount of barite used in the investigated well as shown in Table 2 is 702 tonnes. Barite contains variable amounts of heavy metals and it is the main source of heavy metals in the investigated site. Metals concentrations in mud pit of selected well during and after drilling operation are presented in Figure 5. Chromium concentration was detected in the samples at 0–0.08 ppm. Other heavy metals were also at high levels and showed significantly higher values specially by using OBMs: cadmium 0–0.006 ppm, nickel 0–0.024 ppm, and aluminum 0–341 ppm. However, these heavy metal levels are generally above toxic levels. As shown in Figures 5, the concentrations of cadmium, chromium, and nickel increased progressively in the fourth sampling periods because of the contamination of the mud pit with OBMs that was initiated in the fourth sampling period. Concentration of aluminum increased from the first to the third sampling periods, whereas in the fourth period it shows decreased values from 0.05 ppm to 0.006 ppm. Aluminum was not observed in the fifth and sixth sampling periods but maintained an increased value from the seventh to the end of the sampling periods. In the entire study area, chromium levels ranged from 0 to 0.08 ppm but no concentration was observed after the seventh period of the sampling. This can be explained by the storm runoff water at the investigated well site that washes away all these wastes, especially in the mud pits to other locations or seepage from the discharge pits into the surrounding soils. The statistics of the investigated heavy metals are shown in Table 3.

	Properties	**24″ hole ≅ 60 m**	**17½″ hole ≅ 1510 m**	**12¼″ hole ≅ 2158 m**	**8½″ hole ≅ 2330 m**
Mud Properties	Mud system	WBM	WBM	WBM	OBM
	pH	10-10.5	10.5-9.8	8-10	9-9.5
	Average salt concentration (mg/l)	2000	185600	297600	380100
	Average calcium concentration (mg/l)	464	2404	3320	231
	YP	11	4-7	6-78	19-27
	PV	35	5-10	8-58	8-12
	Initial Gel	22	3-6	1-13	2
	10 Min. Gel	30	4-8	2-6	3
	Mud lost @ unit (bbl)	0	2588	1252	802
	Density (pcf)	70-62	68-79	79-146	69.5
	Barite (t)	0	27	674.4	0
Mud Material	Salt (t)	2	166	168	15
	Starch (sx)	0	30	727	0
	Bentonite (t)	160	750	0	0
	Lime (sx)	123	69	222	130
	CMS H.V (sx)	0	0	0	17
	IRSATROL(sx)	0	0	0	140
	Diesel (bbl)	0	0	0	615

Note: YP=yield point; PV=plastic viscosity; bbl=barrel; pcf=pound per cubic feet; t=ton; sx=sacks

Table 2. Drilling fluid used in the selected well (Shadizadeh and Zoveidavianpoor, 2010).

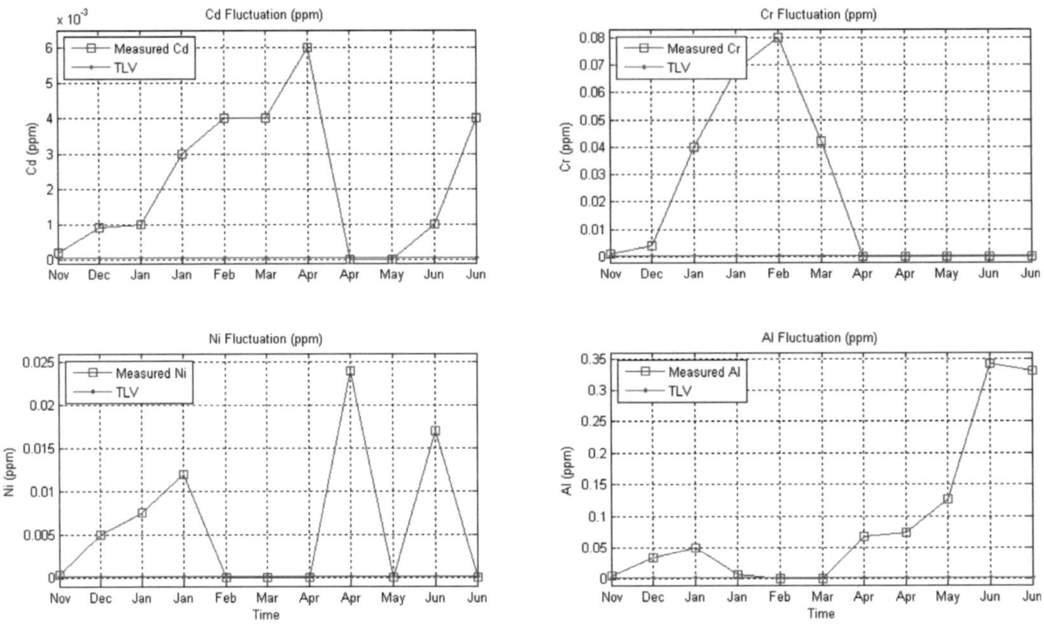

Figure 5. Heavy metals fluctuation during and after drilling operation.

Statistics	Heavy Metals (ppm)			
	Cd	Cr	Ni	Al
Max	0.0060	0.0800	0.024	0.341
Mean	0.0022	0.0214	0.005991	0.09396
Median	1.0000e-003	1.0000e-003	0.0003	0.05
Mode	0.0040	0	0	0
Std	0.0021	0.0306	0.008349	0.1255
Range	0.0060	0.0800	0.024	0.341

Table 3. Heavy metals statistics in the case study

4. Potential effects on natural resources, and minimization strategies

Drilling wastes can harm ecosystems, plants, and animals and cause health problems in humans. Many materials that are released into reserve mud pits also release drilling wastes into the environment, which calls for public awareness as well. When released heavy metals are discharged into unlined pits the toxic substances in the pits can leach directly into the soil and may contaminate groundwater. Additionally, there is no evidence of zero discharge in lined pits. In contrast to most organic pollutants, trace metals are not usually eliminated from aquatic ecosystems by natural processes due to their non-biodegradability. Both toxic and nontoxic heavy metals tend to accumulate in bottom sediments, from which they may be released by various processes of remobilization. Frequently, these metals can move up the biological chain, eventually reaching humans, where they can cause chronic and acute ailments (Ankley et al., 1993). As presented in the previous sections, routine drilling wastes such as drilling muds and cuttings contain a variety of toxic chemicals; they are known to be hazardous to wildlife, livestock, and human health. If pollutants from oil well drilling build up in the food chain, people who consume those natural resources from the contaminated drilled well area could be at risk of health problems such as genetic defects and cancer. For environmental protection, different strategies are considered; (1) restoring the well site to its natural state after drilling, (2) let the liquid to be evaporated, (3) Bioremediation, (4) multi-pit system, and (5) DCRI, which is the focus of this chapter. Because DCRI deal with the initiation and propagation of a fracture in a rock matrix by means of hydraulic pressure, HF will briefly be discussed in the next section.

5. Hydraulic fracturing

Initially, fracturing was a low technology operation consisting of the injection, at low temperature, of a few thousand gallons of napalm into low-pressure reservoirs. Substantially, HF has evolved into a highly engineering and complex procedure. As a technology has improved, so has the number of wells, formations, and fields that can be successfully fractured, increased. The development of high pressure pump units, high strength proppant, and sophisticated fracturing fluids, has meant that deep, low permeability, high temperature, reservoirs can now be fractured (Veatch et al., 1989). This technology is a well-known process, which was originally applied to overcome near wellbore skin damage (Smith, 2006). Since then, it has been expanded to such applications as (1) reservoir stimulation for increase hydrocarbon deliverability, (2) increase drainage area, and decrease pressure drop around the well to minimize problems with asphaltene and/or paraffin deposition, (3) geothermal reservoir recovery, (4) waste disposal, (5) control of sand production, (6) to measure the in-situ stress field and (7) heat extraction (geothermal energy) from deep formations. Obviously, there could be other uses of HF, but the majority of the treatments are performed for the mentioned reasons. HF has made significant contributions to the petroleum industry since its inception (Veatch et al., 1989). By 2009 HF activity has increased 5-fold compared to the investment of a decade earlier and has become the second largest outlay of petroleum companies after drilling (Economides, 2010).

HF is the pumping of fluids at high rates and pressures in order to break the rock. A typical chart of fracturing which shows the common treatment stages is shown in Figure 6. The operation begins with injection of a mixed acid and water named Pre-pad. A mixture of water and a polymer, named Pad, will follows. The fracture will initiated in this stage but contains no proppant. To make the fracture open for fluid flow, a mixture of proppant and the fracturing fluid, which called Slurry will have injected. For more details please consult Daneshy, 2010.

As it clear from section 2, the need has been arises to treat/manage the drill cuttings toward zero discharge by utilization of HF.

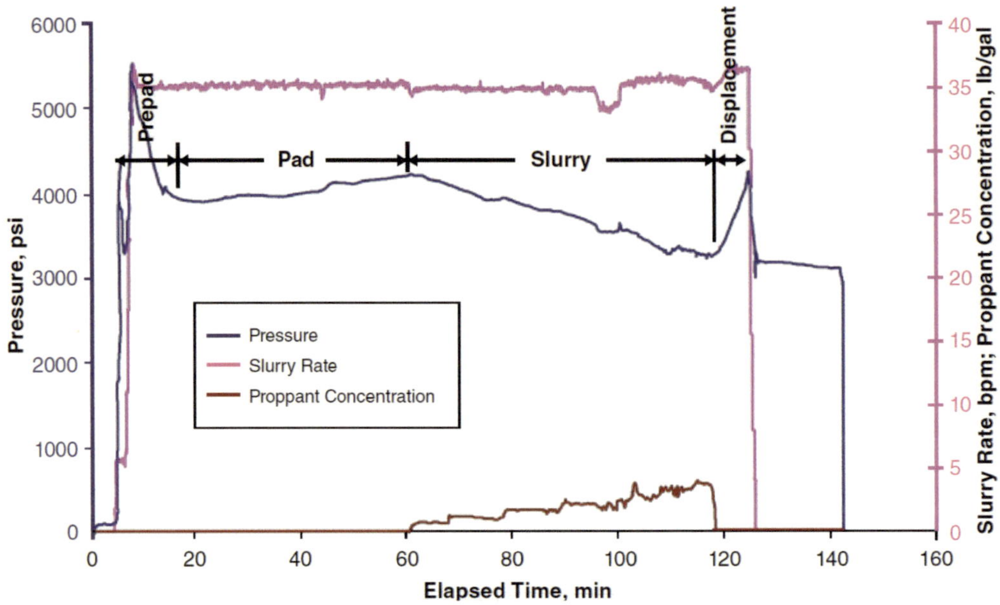

Figure 6. A typical fracturing chart illustrates the steps to HF a well (Daneshy, 2010).

There are both similarities and distinct differences between HF and DCRI which shown in Table 4. More details could be found from Arthur (2010).

Issue	Drill Cutting Re-injection	Hydraulic Fracturing
Target interval	Non Reservoir	Reservoir
Pumping period	Long-term	Short-term
Pumping pressure	Fracture	Fracture
Slurry mixture	Cuttings and fracturing fluids	Proppant and fracturing fluids
Fracture containment study	Essential	Essential

Table 4. Comparison between DCRI and HF.

6. Waste management by DCRI

6.1. An overview

Even though the generation of drill cuttings is a certain result of drilling, those wastes can be treated and/or managed in a number of ways. A summary chart on different drilling wastes management options are presented in Figure 7. As mentioned earlier, the focus of this chapter will be on DCRI.

Valuable literature available regarding the disposal options including: lessons learned concerning biotreating exploration and production wastes (McMillen et al. 2004), successful cases of fixation (Zimmerman and Robert, 1991), converting cuttings into a valuable sources by using vermicomposting (Paulse, 2004), and thermal treatment (Bansal and Sugiarto, 1999).

As summarized in Figure 7, environmental management of drilling wastes may be categorized in three options; waste minimization, recycle/reuse, and disposal. The first and second options are not addressed here. Table 5 shows a comparison among disposal methods which may classified into fixation, thermal treatment, DCRI, and bioremediation/composting. Among the four methods for disposal option that may be considered when deciding on waste management options, the focus of this chapter is on DCRI.

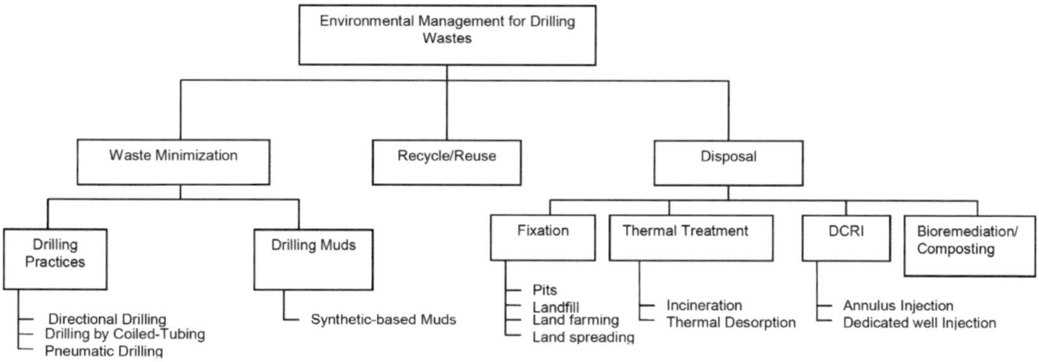

Figure 7. Different approaches in environmental management for drilling wastes

Comparative assessments on alternative disposal options are outlined in Table 5. As clearly shown, environmental impacts and safety risks, which are the most important factors among others, have low level degree and therefore its vulnerability as the best option increases to be adopted as the environmentally friendly drilling waste disposal process. in addition to zero discharge, other advantages of DCRI include; no transportation concerns, no future cleanup responsibilities by the operator, full control over the waste management process, world wide applicability, and its favorable economics. According to Reddoch, (2008): "DCRI is simply the lowest cost, easiest course of action for most drilling operations."

Comparison Factors	Fixation	Thermal Treatment	DCRI	Bioremediation/ Composting
Environmental Impact	Low	High	Low	Medium
Cost	\$9-10/bbl [a]	\$90/metric ton [a]	\$5/bbl [b]	\$500/cubic meter [c]
Safety Risks	High	High	Low	Medium
Technical	Low	Medium	High	Medium

$1m^3$=6.29 bbl (US bbl); 1 metric ton=7.1 bbl (for an oil with 0.88 specific gravity)
[a]Bansal and Sugiarto (1999); [b]Reddoch (2008); [c]McMillen and Gray (1994)

Table 5. Qualitative and quantitative comparison in disposal approaches

The question is raised that what is the relationship between environmental management and DCRI? It's clear that DCRI process will maintain waste containment in a target interval with zero discharge and consequently low HSE risks. Other goals such as cost management and asset management are not covered in this chapter. For more details please consult Bruno et al. (2000).

We can visualize DCRI to loss of circulation of drilling fluids in conventional under balanced drilling operation. Also, it's quite similar to HF operation, because we need to propagate the fractures in the selected horizon and this goal will be achieved by utilization of fracture propagation models which conventionally employed in HF treatment.

Cuttings may be re-injected into the annulus of a well being drilled or into a dedicated well. In annulus injection, cutting would be stored until the desired formation is reached. Whereas in dedicated disposal well, one or more dedicated disposal wells would be drilled and drill waste systems put in place in those wells. A schematic of both types of DCRI is shown in Figure 8.

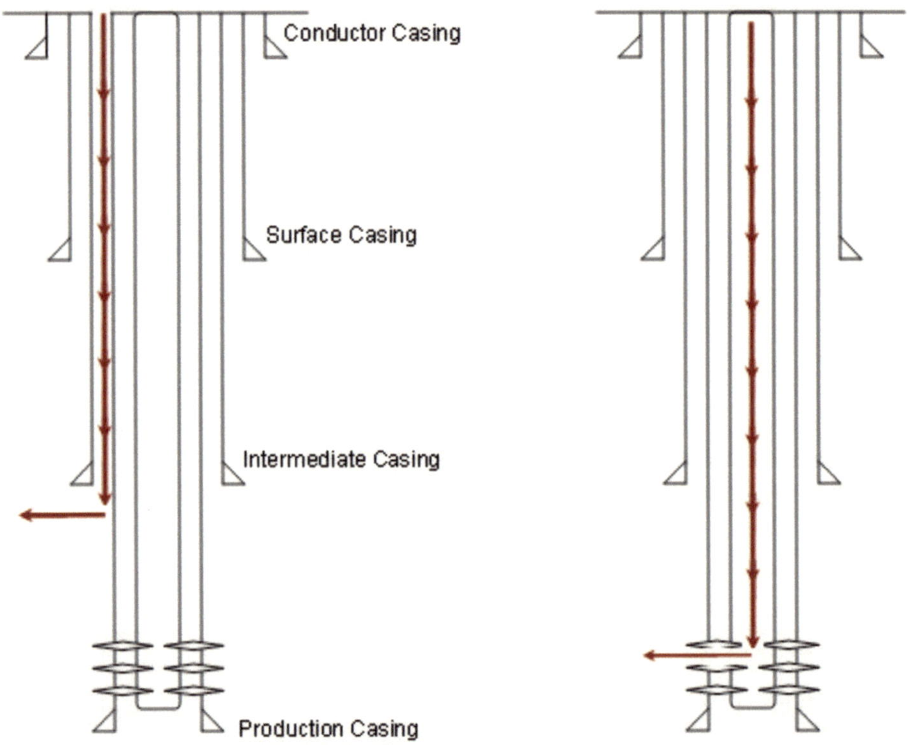

Figure 8. Two major types of DCRI; annulus injection (left) and dedicated well (right).

Drill cuttings may be injected into subsurface geological formations at the drilling site, offshore or onshore and would provide a complete disposal solution. Its worth to note that onshore operations have a wider range of options than offshore operations.

Readers may be asks why this process is called drill cutting re-injection? That's because drill cuttings will be returned back to their origin, deep beneath the Earth's surface.

A sketch of basic setup and flow of DCRI process is shown in Figure 9. Drill cuttings and other oilfield wastes are slurried by being milled and sheared in the presence of water. The resulting slurry is then disposed of by pumping it into a dedicated disposal well, or through the open annulus of a previous well into a fracture created at the casing shoe set in a suitable formation.

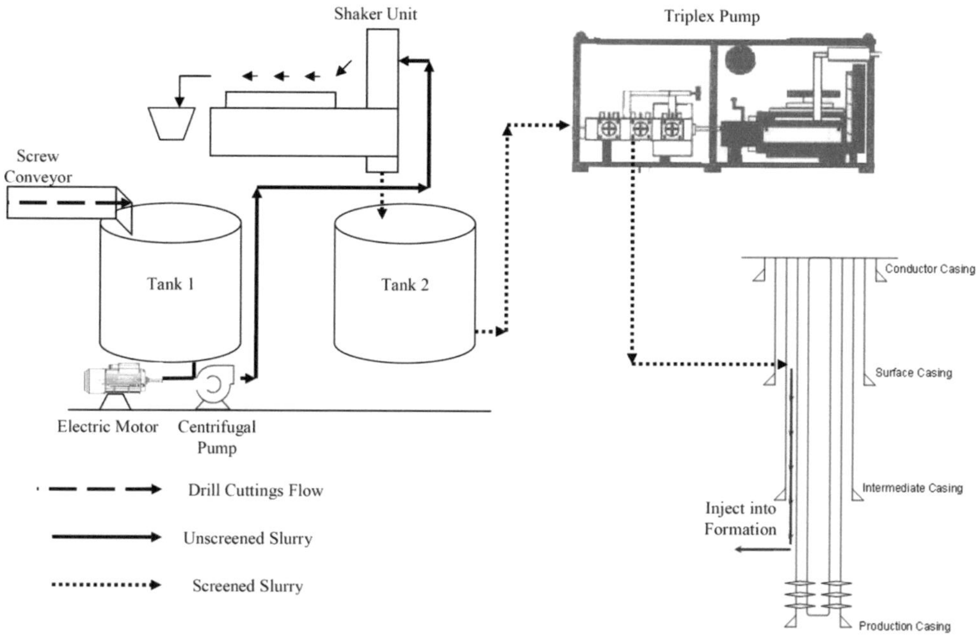

Figure 9. A sketch of basic setup and flow of DCRI

6.2. A case study

In addition to the drill cuttings and drilling fluids, various waste streams need to be handled and disposed of properly include: produced water, contaminated rainwater, scales, and produced sand. DCRI provides a secure operation by injecting cuttings and associated fluids up to several thousand meters below the surface into hydraulically created fractures. In order to guarantee containment within the selected underground formation and perform sufficient design of surface facilities, simulations are performed for the anticipated downhole waste domain.

In this regard, a feasibility study was performed to show the possibility of DCRI in Ahwaz oilfield located in southern Iranian oilfields. The possibility of annular injection and dedicated injection wells was investigated in this study. The objectives were to (1) estimate the volume of drilling waste produced from drilling of each wellbore of the field, (2) select the most appropriate disposal formation in the field, and (3) determine whether the drill wastes can be safely injected into a dedicated well or annular space. Numerous scenarios were considered in the feasibility studies to ensure safe containment of any injected drilling waste. More details could be found by Shadizadeh and Zoveidavianpoor, (2011).

The volumes of drill cuttings and muds, type of utilized mud, and geological information are shown in Table 6. The required data to conduct this study is depicted in Table 7.

Depth (m)	Formation Name	Column	Setting depth (inch≅m)	Lithology	Hole Size (inch)	Cutting volume (bbl)	Mud volume (bbl)	Mud type
1550	Aghajari	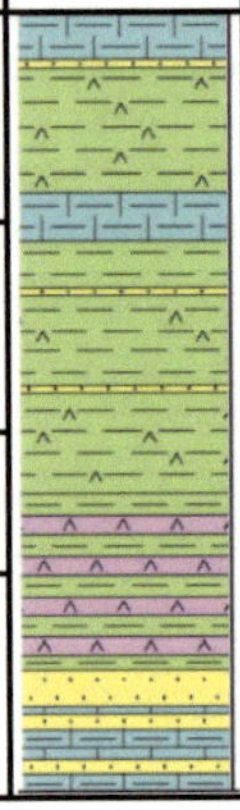	18 5/8 ≅ 60	Marl with Sandston bonds	26	132	4400	WBM
1660	Mishan		13 3/8 ≅ 1767	Marl with Limestone basement	17 1/2	2040	2800	WBM
2332	Gachsaran		9 5/8 ≅ 2337	Marl, Salt, Anhydrate.	9 5/8	219	3500	WBM
3590	Asmari		7 ≅ 3594	Limestone with Sandstone	8 1/2	73	800+400	WBM+ OBM

Table 6. Generalized geologic data along with drill cuttings and mud volumes.

Required data	Description
Injection batch volumes and injection rates	Injection of the slurry is often conducted intermittently in batches into the selected disposal formation, followed by a period of shut-in. depending upon the batch volume and the injection rate, each batch injection may last from less than an hour to several days or even longer.
Minimum in situ stress	Most important in fracture simulation that controls fracture-height growth, fracture azimuth and vertical and horizontal orientation, fracture width, treatment pressures, fracture conductivity, and wastes containment in disposal horizon.
Pore pressure	Very critical parameter to planning and carrying out successful DCRI, because the stress state of the poroelastic medium is directly influenced by pore pressure or reservoir pressure.
Young's modulus	Is the ration of longitudinal stress to longitudinal strain, which has significant effect on fracture geometry, especially on fracture width
Poisson's ratio	Is a measure of the compressibility of material perpendicular to applied stress that has significant effect on fracture geometry
Casing setting depths and injection point	The target which the slurry has to be injected via annulus or dedicated well.
Fluid leak-off data	Means the leaking of fluid from the surface of a fracture into the surrounding rock formation. It's an important parameter controlling the size and geometry of the hydraulically induced fracture.
Slurry rheology	The study of the deformation and flow of matter, that crucial for maintaining zonal isolation.
Fracture toughness	Is an important parameter in fracture modeling and is a measure of a material's resistance to fracture propagation

Table 7. Explanation of required data for DCRI simulation.

In particular, the expense of DCRI requires that the operator knows how the formation will respond to treatment, and whether the treatment design such as selection of pump rates, fluid rheology, accurate rock mechanic properties, pumping schedule and fracture propagation model, will create the intended fracture.

Most 2D models are based on three common models entitles Perkins-Kern-Nordgren (PKN), Khristianovic-Geertsma de Klerk (KGD), and Radial models. The first and second models which assume constant height, are appropriate when the stress contrasts are high between the pay layer and neighboring formations and these contrasts follow lithologic boundaries. For Radial model, its better works in a setting where the fracture grows in a formation of homogeneous stress and mechanical properties so that fracture height is small compared to formation layer thickness. A brief comparison among 2D models is listed in Table 8.

The main advantage of a more advanced method such as pseudo 3D (P3D) over 2D models is that it does not require estimating fracture height, but it does require input of the magnitude of minimum horizontal stress in the zone to be fractured and in the zones immediately above and below.

Model Name	**Plan View**	**Cross Section View**	**Pressure-Time Trend**	**Description**
PKN	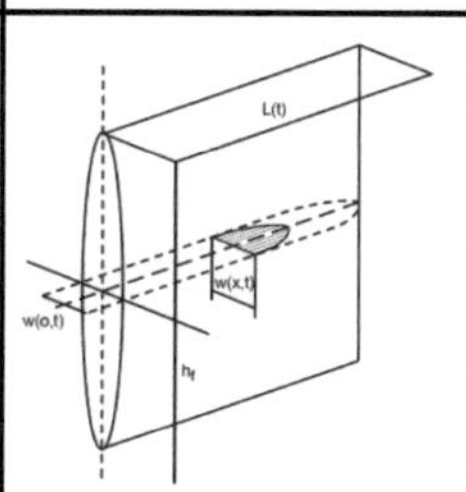	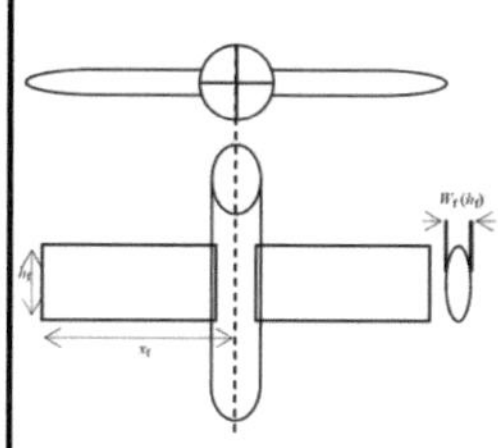	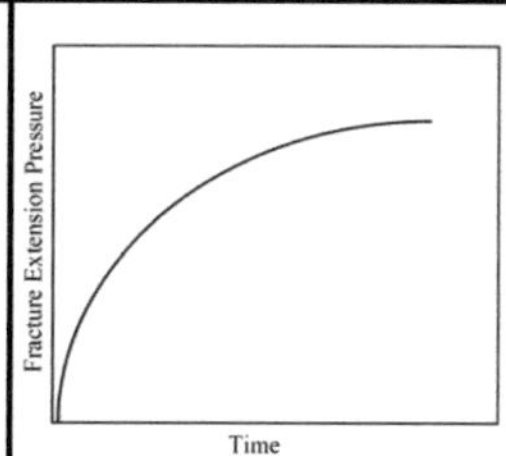	Cross section= Elliptical Width ∝ height Width < KGD Length > KGD Suitable when: length>height
KGD	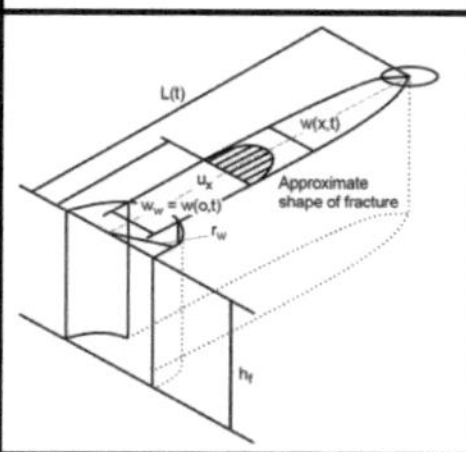	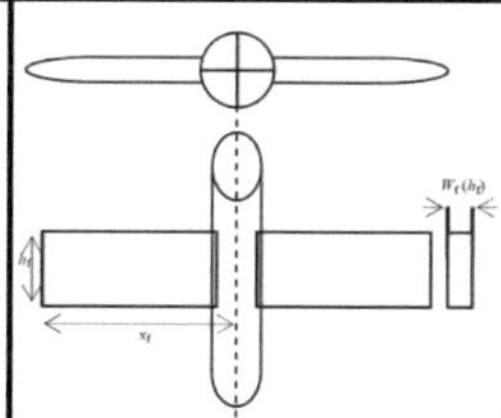	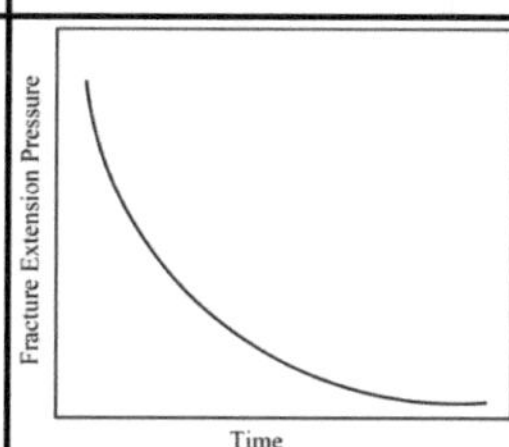	Cross section: Rectangular Width ∝ height Suitable when: length<height
Radial	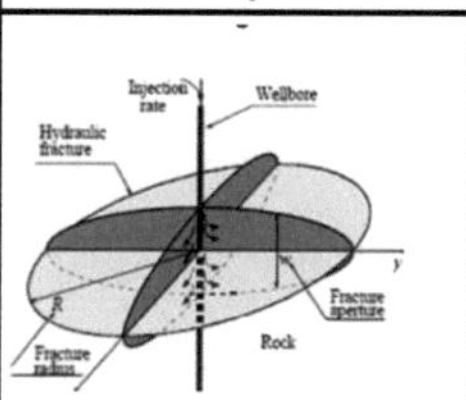	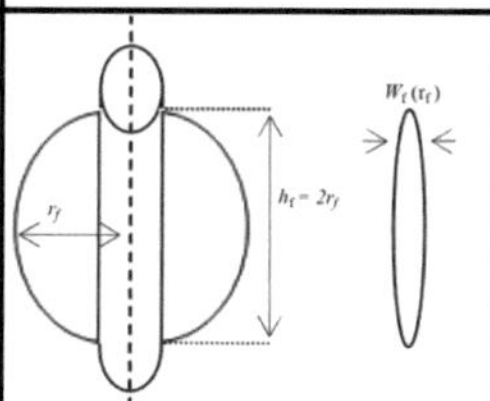	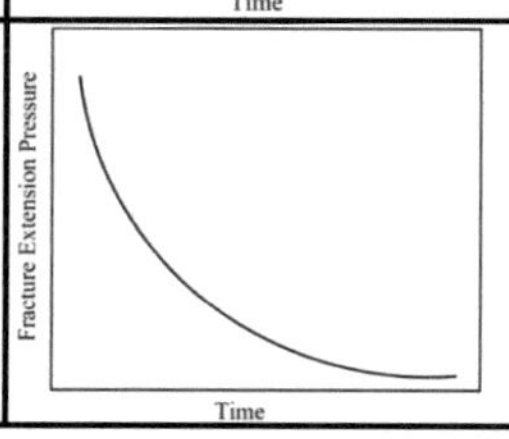	Cross section= Elliptical Suitable when: length=height

Table 8. Comparison of 2D fracture models

6.2.1. Simulation study

Based on the petrophysical logs, from lithological point of view, the relevant formations are fairly marl, sandstone and limestone with an average rock density 2.33gr/cm3. The vertical stress was calculated by integrating the available bulk density with respect to depth. Vertical stress gradient is calculated as Eq. (1):

$$\sigma_v = 0.433\rho_{OB} = 0.433 \times 2.33 = 1 psi / ft \quad (1)$$

The values of minimum horizontal stress of Aghajari, Mishan, and Gachsaran formations were 1693, 3847, and 4489, respectively which calculated from Eq. (2) is:

$$\sigma = \frac{\upsilon(\sigma_v D - 2p) + p}{1 - F\upsilon} \quad (2)$$

Elasticity of the formations is determined with the sonic log. Table 9 lists the values of the static elastic Young's modulus, Poisson's ration, leak-off coefficient for the different formation zones shown in Table 5. These values are based on the dynamic elastic Young's module obtained from sonic and density logs. Static elastic Young's module values are often two times smaller than dynamic values derived from sonic logs. The elastic Young's module values that are listed in Table 9 are arbitrarily one-half of their dynamic equivalents. The larger than usual values were used in the analysis for these shallower formations.

Zone Name	Zone Height (ft)	Poission's Ratio*	Pore Pressure* (psi)	Fracture Gradient* (psi/ft)	In-situ Stress (psi)	Young's Modulus (MM psi)	Leak-off Coefficient (ft.min $^{-0.5}$)	Toughness (psi.min $^{-0.5}$)
Aghajari	5250	0.29	1050	0.650	1693	2	0.00081	1000
Mishan	330	0.31	2567	0.714	3847	2	0.00087	1000
Gachsaran	330	0.36	2878	0.780	4489	2	0.00089	1000

Table 9. Formation properties used in fracture simulations.

Slurry rheology design did not performed in this paper and is beyond the scope of this article; however by considering the cuttings brought out of the wellbore and the drilling muds used in Ahwaz oil field, a reasonable result was earned of rheology characteristics of the injection slurry. It was assumed that the cuttings slurry with final rheological condition would behave in a manner similar to the drilling muds used in Ahwaz oil field. Slurry and solid properties are selected from past DCRI operation in literature (Abou-Sayed et al., 2002), which is also near the nature of selected drilling fluids and cuttings lithology of the Ahwaz oilfield and are presented in Table 10.

Density	1.26 SG
Particle Loading	80/100 mesh proppant at a consternation of 2 PPG
Apparent Viscosity	161 cp ≅ 170 1/S
Non-Newtonian power law indices	N=0.26; k=0.15

Table 10. Physical properties of injected cuttings slurry.

For the scenario of casing injection into a dedicated injection well, the intermediate casing can be set on top of Gachsaran formation. The casing is assumed to perforate at a depth about 50 m under the Aghajari formation and the center of the Mishan formation. The initial fracture is assumed to be at the center of the perforated interval.

6.2.2: Simulation results

After determining all required data, a fracture geometry model was selected for use in the simulation. As described previously, the dedicated injection mechanism is more suitable for the Mishan formation because it is deep enough and consists of limestone lithology in a base that is appropriate for reinjection. In each case, the geometry reported indicates the maximum fracture achieved when slurry is pumped continuously. The simulation study is represented for both dedicated wells that consist of two cases and annulus injection well mechanisms.

Dedicated Well Injection Mechanism

Two cases will be presented in this section, which differs in the magnitude of two parameters; Young's modulus and leak-off coefficient.

Case 1: For a case like Ahwaz oilfield in which the vertical distribution of the minimum in situ stress is uniform, a circular fracture is expected. The formations had Young's modulus and leak-off coefficients as shown in Table 9. For this simulation, the fracturing would initiate from the Mishan formation and broke through the Aghajari formation but was still 4,700 ft. below the surface when 50,000 bbl of slurry had been injected continuously. Table 11 summarizes the results of this simulation. Figures 10 and 11 show predicted the fracture shape plot after injection of 50,000 bbl continuously at 5 bbl/min.

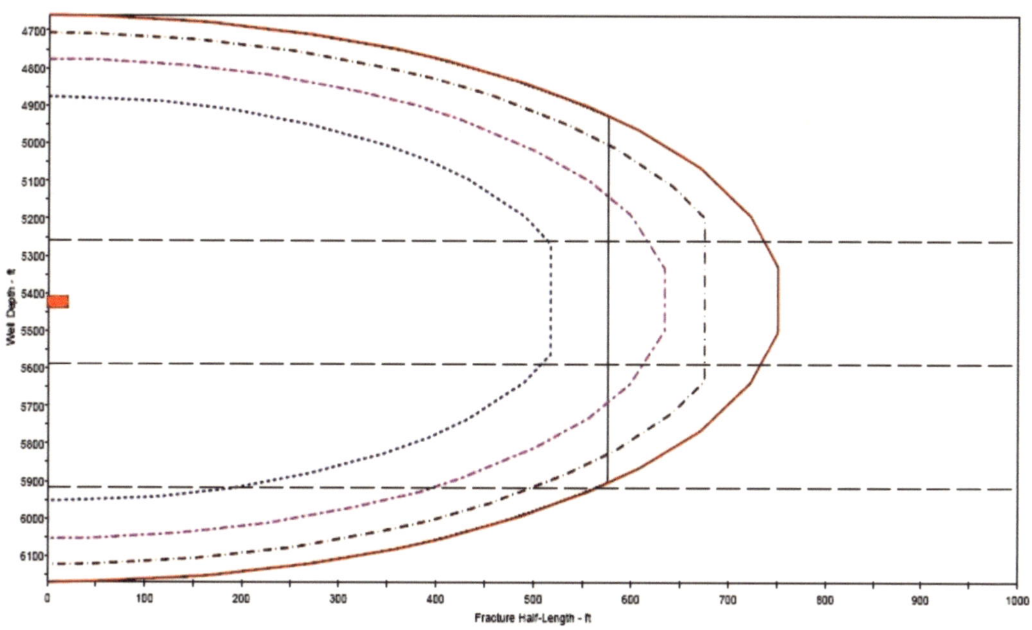

Figure 10. Fracture geometry history- Radial model (case 1).

Case 2: The formations were assumed to have Young's modulus that was twice those listed in Table 9. Also, the leak-off coefficient for formations used was specified as one half of the value listed in Table 9. This extremely large modulus and small leak-off resulted in a much larger fracture. Consequently, this is a very conservative analysis. Even for this very conservative case, the fracture that broke through the Aghajari formation was still 4,550 ft. below the surface when almost 50,000 bbl of slurry had been injected continuously at 5 bbl/min. Table 11 summarizes the results of the fractures created.

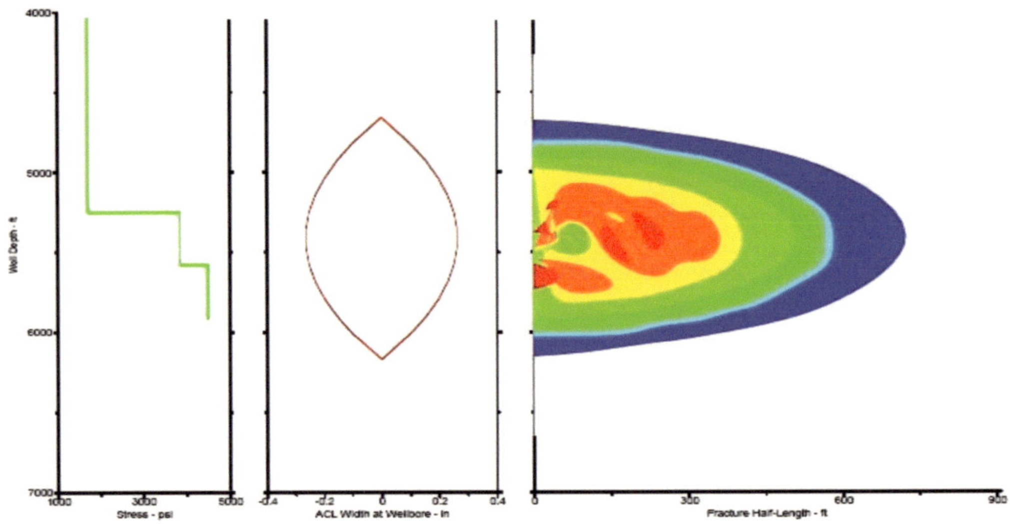

Figure 11. Fracture profile and cuttings concentration- Radial model (Case 2).

Parameters	Case 1	Case 2
Slurry volume (bbls)	50000	50000
Fracture half-length (ft)	576	795
Fracture width at well (in)	0.276	0.237
Net pressure (psi)	71	89
Max surface pressure (psi)	1755	1807
Shut-in time (hrs)	13	26

Table 11. Simulation's results of dedicated well injection.

Annulus Well Injection Mechanism

Annulus injection is only possible if the annulus of an intermediate casing string in an existing well is open to a suitable subsurface formation and this well satisfies a range of screening criteria. The allowable injection pressures for annulus injectors are often lower than the allowable pressures for dedicated wells because of casing burst and collapse limitations for annulus injectors. By considering the lithology and casing design of Ahwaz oilfield, it is concluded that the planned slurry injection would occur in an 18 5/8-in./13 3/8-in. annulus. Other annuli are not possible for injection because they are open to unsuitable subsurface formations. To prevent the upward migration of injected wastes to the surface, the 18 5/8-in. casing string should set at about 1,000 ft. and cement back to the surface, and the 13 3/8-in. string should cement back to 1,500 ft. below the previous casing shoe. This provides a window across the Upper Miocene marl and sandstone of Aghajari formation. For this simulation, the fracturing initiated from the Aghajari formation and grew toward the surface but was still 500 ft. below the surface when 15,000 bbl of slurry had been injected continuously. Table 12 presents the different parameters of the fracture created. Figure 12 shows the predicted fracture shape plot after injection of 15,000 bbl continuously at 5 bbl/min.

Parameters	Radial Model
Slurry volume (bbls)	15000
Fracture half-length (ft)	230
Fracture width at well (in)	643
Net pressure (psi)	2.39
Max surface pressure (psi)	968

Table 12. Simulation results of annular well injection.

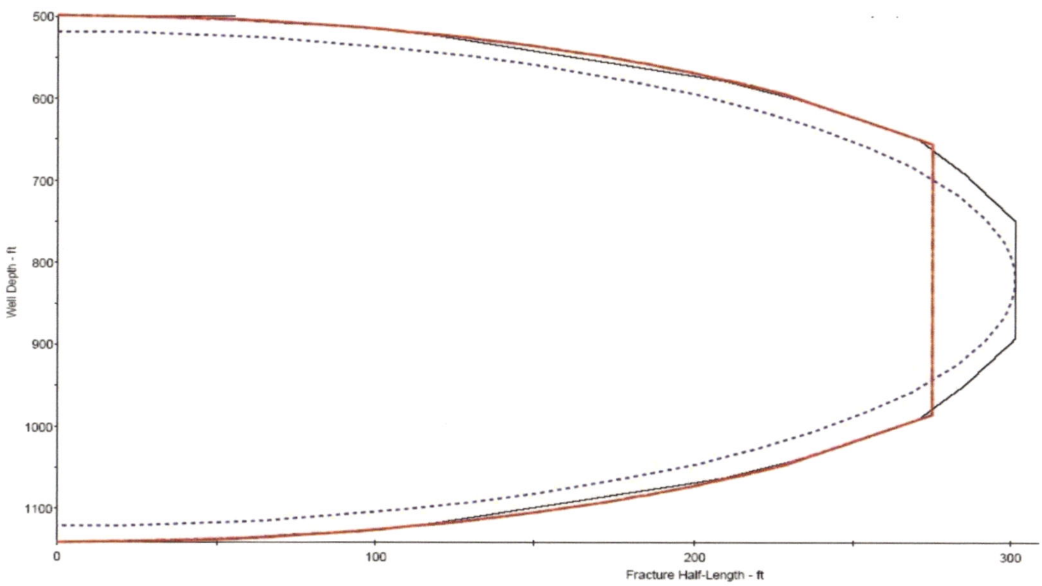

Figure 12. Fracture geometry history- annular injection well (Radial model)

7. Discussion and conclusions

Assessment of environmental impacts of drilling operations and searching for the methodologies to protect nature and resources against negative impacts has become an interesting topic during the last thirty years in upstream petroleum industry. The necessity of environmental management in drilling operation, lessons learned, and a brief list of mitigation options from wastes generated by drilling operations in a southern Iranian oilfield were documented previously (Shadizadeh and Zoveidavianpoor, 2008, 2010). Most of the drilling wastes sources in the oilfields are OMBs and oily cuttings associated with them. Unfortunately, lack of demanding regulations regarding drilling waste discharge leaves room for drilling companies to leave the waste in the nature without treating them (Shadizadeh and Zoveidavianpoor, 2008, 2010). This chapter tried to study the possibilities of waste prevention and zero discharge by utilization of serviceable methods in drilling well sites. So, the feseability study of DCRI at Ahwaz oilfield was initiated and conducted to fulfill the needs of growing upstream petroleum industry in Iran. This article focuses on the design aspect of the technology. Design guidelines are given to include data required for project planning, injection scheme (annulus versus dedicated well) selection, injection well and disposal formation identification, subsurface fracturing simulation, and waste containment. Operational procedures such as slurry rheology were the area of investigation in this study; however, it was determined as input data for simulation that has conformity with the nature of selected drilling fluids and cuttings lithology of the Ahwaz oilfield. Well design requirements and estimation of disposal capacity in each of the injection schemes was performed. This study shows that the DCRI study at Ahwaz oilfield is practical by considering some potential risks involved in any DCRI job. It was determined that by using HF technology, drilling wastes could be reinjected to the Mishan formation or even a shallow formation such as the Aghajari formation without propagation of the fractures to the surface or near wellbores. The thickness of the Aghajari formation provides an appropriate barrier to upward growth of DCRI at the Mishan formation through a dedicated injection well. A dedicated injection well is more typical of longer-term, permanent injection operations and is more common onshore (Keck, 2002). It is simulated that a large amount of drilling waste can be safely injected to Mishan formation. The maximum surface pressure required to inject the slurry is in a range of 1,500 to 2,000 psi, which is completely reasonable with the current surface facilities. The propagation of the fracture to the surface showed to be efficient and safe in the two cases performed in the dedicated well injection scheme. The simulation results confirm that the drilling wastes produced from each wellbore could be injected through annulus of the same wellbore while drilling. The selected annulus for annular reinjection in Ahwaz oilfield is not very favorable because the injection point is close to the surface. As described before, other annuli are not suitable due to abnormal pressure or hydrocarbon bearing. The annular reinjection at Ahwaz oilfield has many serious risks that need a careful job planning. However, the amount of drilling wastes from a typical wellbore is not high and the simulations confirm that 15,000 bbl wastes from a typical wellbore can be injected without serious danger. Advantages and disadvantages of annular and dedicated well injectors are presented in Abou-Sayed and Guo (2001).

It should be noted that the simulations represent upper-bound predictions of the fracture geometry because low leak-off and high Young's modulus is assumed in different formations. In reality, even a very limited change in the amount of fluid leak-off, coupled with intermittent batch injection of slurry, would result in a significantly reduced fracture area. The analyses confirm the integrity and suitability of the injection operations and ensure safe application of this technology at Ahwaz oilfield.

Author details

Mansoor Zoveidavianpoor and Ariffin Samsuri
Universiti Teknologi, Faculty of Petroleum & Renewable Energy Engineering, Malaysia

Seyed Reza Shadizadeh
Petroleum University of Technology, Abadan Faculty of Petroleum Engineering, Iran

Acknowledgement

The authors of this chapter would like to express thier gratitude to University Teknologi Malaysia due to the valuable supports during this study.

8. References

Abou-Sayed, A. S., Quanxin G., McLennan, J. D. and Hagan, J. T. (2000). Case studies of waste disposal through hydraulic fracturing. Presented at the Workshop on Three Dimensional and Advanced Hydraulic Fracture Modeling, held in conjunction with the Fourth North American Rock Mechanics Symposium, July 29, 2000, Seattle.

Abou-Sayed, A. S., and Guo, Q. (2001). Design considerations in drill cuttings re-injection through downhole fracturing. Paper IADC/SPE 72308 presented at the IADC/SPE Middle East Drilling Technology, Bahrain. October 22–24.

Ankley, G. T., Mattson, V. R., Leonard, E. N., West, C. W., and Bennett, J. L. (1993). Predicting the acute toxicity of copper in freshwater sediments: Evaluating the role of acid-volatile sulfide. Environ. Toxic Chem. 12:315–320.

Arthur, J.D., "A Comparative Analysis of Hydraulic Fracturing and Underground Injection", Presented at the GWPC Water/Energy Symposium, Pittsburgh, Pennsylvania. September 25-29, 2010.

Bansal, K. M. and Sugiarto, (1999). Exploration and Production Operations-Waste Management A Comparative Overview: U.S. and Indonesia Cases. Paper SPE 54345 presented at the SPE Asia Pacific Oil and Gas Conference, Jakarta, Indonesia, April 20-22.

Bleier, R., Leuterman, A. J. J., and Stark, C. (1993). Drilling fluids making peace with the environment. J. Petrol. Tech. 45:6–10.

Bruno, M., Reed, A. and Olmstead, S. (2000). Environmental Management, Cost Management, and Asset Management for High-Volume Oil Field Waste Injection Projects. Paper SPE 51119 presented at the IADC/SPE Drilling Conference held in New Orleans, Louisiana, 23–25 February 2000.

Daneshy, A. (2010). Hydraulic Fracturing to Improve Production. The Way Ahead: Tech 101. 6(3), 14-17.

Economides, M. J. (2010). Design Flaws in Hydraulic Fracturing. Paper SPE 127870 presented at the SPE International Symposium and Exhibition on Formation Damage Control. 10-12 February. Lafayette, Louisiana, USA.

Keck, R. G. (2002). Drill cuttings injection: A review of major operations and technical issues. Paper SPE 77553 presented at the SPE Annual Technical Conference and Exhibition, San Antonio, TX, September 29–October 2.

McMillen, S. J. and Gray, N. R. (1994). Biotreatment of Exploration and Production Wastes. Paper SPE 27135 presented at the Second International Conference on Health, Safety & Environment in Oil & Gas Exploration and Production, Jakarta, Indonesia, January 25-27.

McMillen, S. J., Smart, R. and Bernier, R. (2004). Biotreating E&P Wastes: Lessons Learned from 1992-2003. Paper SPE 86794 presented at the 7th SPE International Conference on Health, Safety and Environment in Oil and Gas Exploration and Production, Calgary, Alberta, Canada, 29-31 March 2004.

Neff, J. M. (2002). Bioaccumulation in Marine Organisms. Effects of Contaminants from Oil Well Produced Water. Amsterdam: Elsevier Science Publishers. ISBN: 978-0-08-043716-3. 452 pages

Paulse, J. E., Getliff, J. and Sørheim, R. (2004). Vermicomposting and Best Available Technique for Oily Drilling Waste Management in Environmentally Sensitive Areas. Paper SPE 86730 presented at the 7th SPE International Conference on Health, Safety and Environment in Oil and Gas Exploration and Production, Calgary, Alberta, Canada, 29-31 March 2004.

Reddoch, J. (2008). Why cuttings reinjection doesn't work everywhere—or does it? World Oil, January 2008, 69-70.

Shadizadeh, S. R. and Zoveidavianpoor, M. (2010) 'A Drilling Reserve Mud Pit Assessment in Iran: Environmental Impacts and Awareness', Petroleum Science and Technology, 28: 14, 1513-1526. ISSN: 1091-6466 print/1532-2459 online

Shadizadeh, S. R., and Zoveidavianpoor, M. (2008). Environmental Impact Assessment of Onshore Drilling Operation in Iran, Abadan, Iran: Petroleum University of Technology.

Shadizadeh, S. R., Majidaie, S. and Zoveidavianpoor, M. (2011) 'Investigation of Drill Cuttings Reinjection: Environmental Management in Iranian Ahwaz Oilfield', Petroleum Science and Technology, 29: 11, 1093-1103. ISSN: 1091-6466 print/1532-2459 online.

Veatch, Jr. R. W. Moschovidis, Z. A. (1989). An Overview of Recent Advances in Hydraulic Fracturing Technology. Paper SPE 14085 presented at the International Meeting on Petroleum Engineering. 17-20 March. Beijing, China.

Zimmerman, P. K. and Robert, J. D. (1991). Oil-Based Drill Cuttings Treated by Landfarming. Oil & Gas Journal, 89(32): 81-84.

Chapter 18

A Geologic and Geomorphologic Analysis of the Zacatecas and Guadalupe Quadrangles in Order to Define Hazardous Zones Associated with the Erosion Processes

Felipe de Jesús Escalona-Alcázar, Bianney Escobedo-Arellano, Brenda Castillo-Félix, Perla García-Sandoval, Luz Leticia Gurrola-Menchaca, Carlos Carrillo-Castillo, Ernesto-Patricio Núñez-Peña, Jorge Bluhm-Gutiérrez and Alicia Esparza-Martínez

Additional information is available at the end of the chapter

http://dx.doi.org/10.5772/45852

1. Introduction

In the State of Zacatecas, Mexico (**Figure 1**), the environment is not usually taken into account as a critical variable for the urban growth and development planning. The expansion of the cities of Zacatecas and Guadalupe in the study area is merely based on the land use change according to the Urban Development Code[1] and the Urban Development Plan 2004-2030 [2]. The Code states the urban growth policies which apply to the whole state; whereas the Plan is a compilation of documents related to the urban growth tendency, population distribution, and basic population services (i.e. water supply). Regardless of the scale, the criteria for the land use change policies are unclear. The Urban Development Plan [2] suggests avoiding urban growth towards areas geologically and topographically unstable and those with flood potential. In every case, the slope should be less than 30°. There are no available maps that indicate where these areas are located, so the criteria for land use will remain unclear.

Although the geology and geomorphology are mentioned [2], their value as critical variables is not taken into account in practice for any purpose. The geologic and geomorphologic variables defined [3] are only indicated, but not located on a map for planning development. Moreover, it is suggested [2] that there is a necessity for a detailed mapping of the geomorphic agents defined [3].

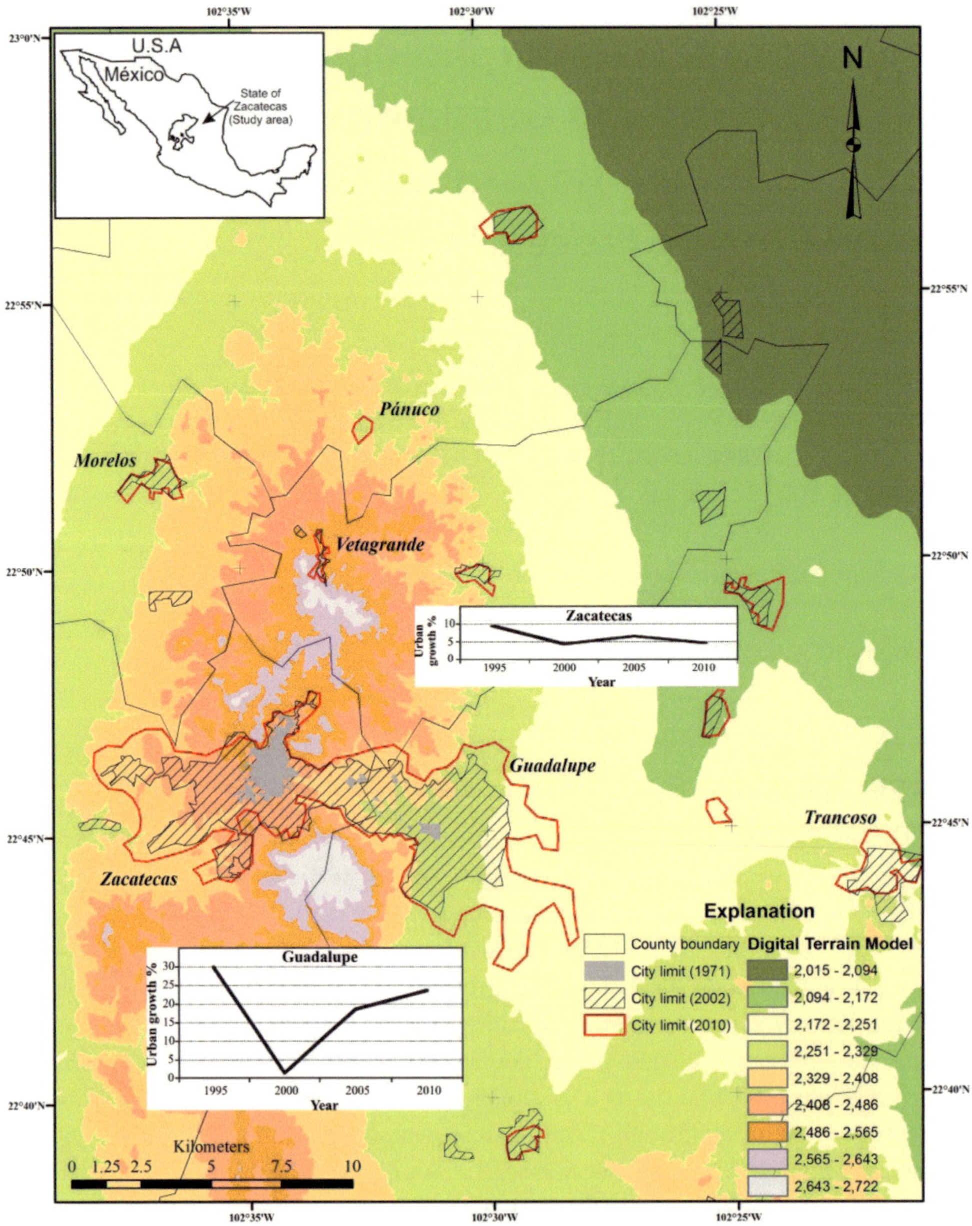

Figure 1. Digital Terrain Model of the study area. The area includes six municipalities whose names are in bold-italic. The largest cities are Zacatecas and Guadalupe. The graphs show the population growth during the last 20 years [8-10], Guadalupe is among the fastest growing municipalities in Mexico (~25% in 2010). The 1971 city limit was obtained from INEGI [11-12]; the growth of these cities until 2002 is a compilation made by the first author; while for 2010 it was a combination between a Google Earth image (August 5th, 2009) and field work. Inset is the location of the State of Zacatecas and the study area.

Due to the land use change and the subsequent landscape modification, the erosion processes are currently becoming active in places usually considered to be stable. The effects are: fractured streets, roads, and houses, voids under the streets and buildings, slope instability, and rock blocks falling next to the roads themselves (**Figure 2**). The primary erosion agent is rainwater. The geomorphic processes are slow but steady contributors, due to the semiarid climatic conditions [3]. The average annual precipitation is 500 mm/yr; while the average annual temperature is 22°C. Therefore, the erosion processes develop slowly, but successfully. Until now, the hazardous zones have only been reported in specific places; the effects of which are the mass removal under houses or streets with active fracturing and, in worst case situations, their slow collapse (**Figure 2**). The exact location of these areas has not been taken into account as a social, economic or environmental problem. The natural hazards recognized by the authorities are mostly related to the mining industry and their products (i.e. mine tailings, open pits); on the other hand, the erosion and its effects [2] are considered of minor importance. The landscape, geology, land use, soil cover, and their modifications are barely considered to be serious enough to promote the development of dangerous areas once the original conditions are changed.

Figure 2. Images showing the erosion hazards associated with different lithology: a) and b) are in the granitic facies of the Zacatecas Conglomerate; c) moderately consolidated facies of the Conglomerate; and d) deformed lava flows from Las Pilas Complex. Yellow arrows point toward the first attempt of damage repair. Purple arrows are the sites where a second repair phase has been attempted; while the green arrow shows a third repair phase. The red arrows point to the voids created due to the soil removal by the erosion processes.

For several years now, attempts have been made in order to identify and describe the geomorphic processes acting in the Zacatecas and Guadalupe cities and their relationship with geology and geomorphology [3-7]. In this paper we define the hazardous zones by means of a GIS analysis that integrates the geologic and geomorphologic mapping combined with the digital slope modeling, land use, and soil type.

1.1. Local geology overview

The stratigraphic sequence of the study area is composed by units ranging from the Early Cretaceous Period to Present Age (**Figure 3**). The oldest unit is the Early Cretaceous Zacatecas Formation[13]. The Zacatecas Formation is made up by Greenschists Facies metamorphic rocks whose protoliths are wacke, mudstone, limestone, as well as, interstratified conglomerates with lava flows and tuffs. The whole sequence is cut by dikes, sills, and dioritic laccoliths. Its upper contact is transitional with the Early Cretaceous Las Pilas Complex. This unit is composed of mafic lavas with pillowed and massive structure, commonly foliated and deformed (**Figure 4a and b**). Lava flows contain interbedded wacke, greywacke, mudstone, and minor tuff and limestone. It is considered that laccoliths and dikes of the Las Pilas Complex are part of the same volcanic sequence. Moderate to intense hydrothermal alteration affects the entire Mesozoic record. Both, the Zacatecas Formation and the Las Pilas Complex are known as the Zacatecas Group. This group shows the results of a Late Cretaceous compressive deformation stage and at least five extensional deformation phases that have occurred from the Oligocene to Recent [16]. The association of deformed structures with faulting and steep slopes promotes rock falling.

The Las Pilas Complex is unconformably covered by the Eocene Zacatecas Conglomerate that crops out in the Zacatecas and Guadalupe cities. This conglomerate is composed of five facies named according to their clast-rich abundance and their physical characteristics (**Figure 4c and d**). The sandstone-rich layers are more easily erodible than the conglomeratic ones. This characteristic promotes differential erosion processes that, in combination with moderate to steep slopes, favors the generation of hazardous areas. The conglomerate was deposited in a WNW-ESE fault-bounded basin whose deformation is less intense than the one showed by the Zacatecas Group.

At the top of the stratigraphic column the Zacatecas Conglomerate is in transitional contact with the Oligocene-Miocene Volcanic Sequence that is composed of interbedded ash-flow and air-fall tuffs, breccias, and rhyolitic flows and domes [3, 17]. They commonly develop cliffs that, when associated with steep slopes, encourage the falling of the rocks. The cliffs next to the main faults and steep slopes are also suitable areas for the rock falling process to happen.

1.2. Geomorphic features and unconsolidated deposits

The study area belongs to the Basin and Range extensional province [16] that is formed by NNW-SSE normal faults forming horsts and grabens. The main geomorphologic feature is the Sierra de Zacatecas that is oriented NNW-SSE (**Figure 5**) bounded to the west by the

Calera Valley and to the east by the El Palmar Valley. A minor range, Sierra de Tolosa, is located to the east; both ranges are separated by the El Palmar Valley.

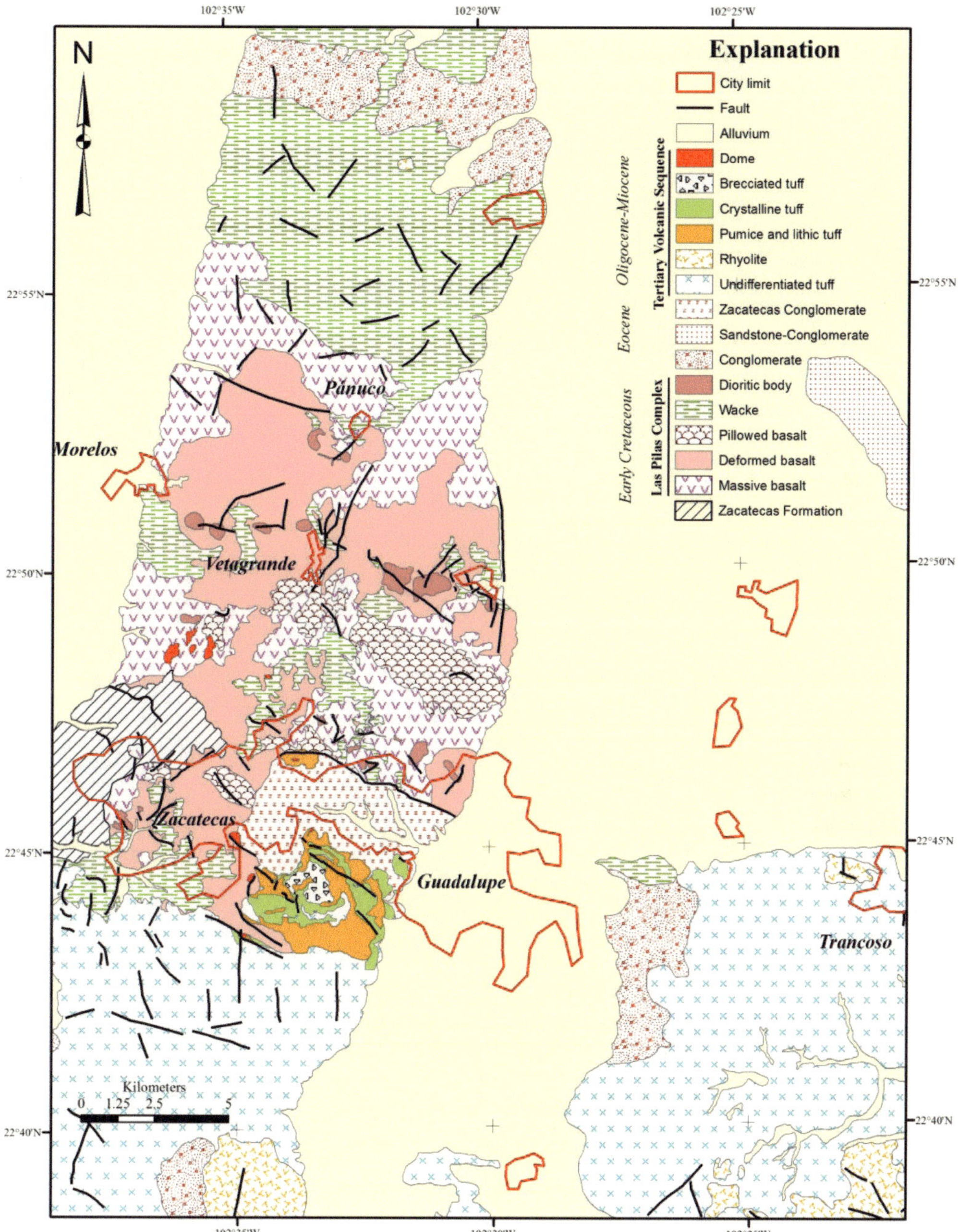

Figure 3. Geologic map of the Zacatecas and Guadalupe quadrangles. Detailed mapping from the Mesozoic sequence is taken from [13-14], while the Tertiary Volcanic Sequence is from [3, 15]. The names are from the municipalities of the study area.

Figure 4. Images showing the main physical characteristics of lithological units that crop out in the Zacatecas and Guadalupe quadrangles. The Las Pilas Complex can be deformed (a) or massive (b), while the Zacatecas Conglomerate can be moderately consolidated (c) with differential erosion (d).

The main erosional agent in the area is rainwater; it affects unconsolidated to moderately consolidated sediments either from slope or fluvial deposits, as well as alluvial fans. These geomorphic features occur in the neighborhood of the present city limit (**Figure 5**). The slope deposits are made of interbedded sandstone and conglomerate. The sandstone thickness is less than 50 cm, whereas the conglomerate thickness varies from a few centimeters up to 30 meters. Commonly, both show normal gradation and, sometimes there are interbedded pumice and/or lithic tuffs. Since the sandstone more easily eroded than the conglomerate, it generates a differential erosion process (**Figure 4d**).

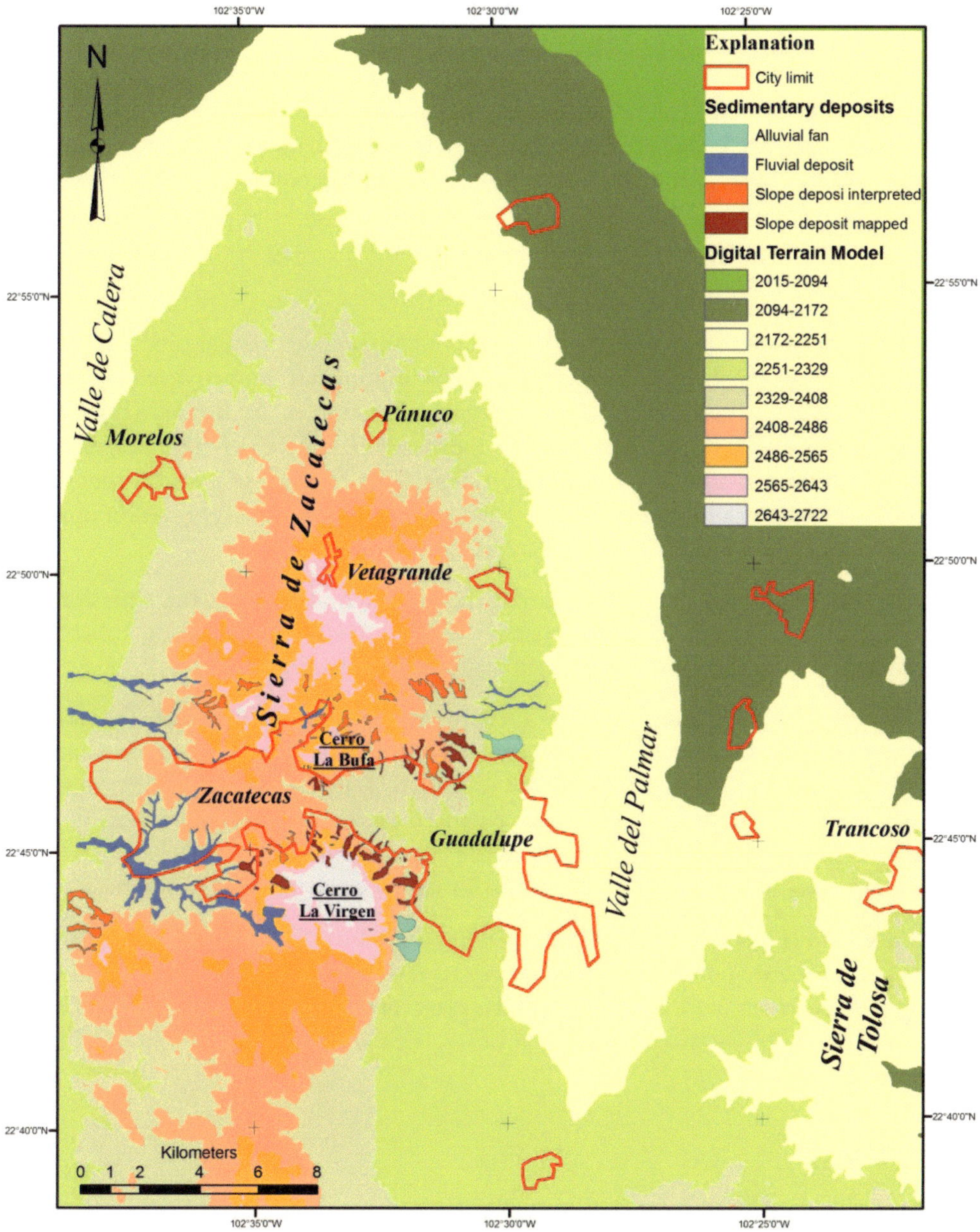

Figure 5. Digital Terrain Model showing sedimentary deposits that are in the neighborhood of the city limit of Zacatecas and Guadalupe. Horizontal bold-italics names are the municipalities; while in italics are the ranges and valleys. The main topographic features are underlined.

The slope deposits are of unknown age. Their morphologic expression is masked by the relief. The cities of Zacatecas and Guadalupe are currently growing directly on these deposits (**Figure 5**). Regarding the alluvial fans, there are a few of them in the eastern side of the Sierra de Zacatecas because most of the valleys are used for agricultural activities. The fluvial deposits are common in the western side of this sierra; however, the arroyos are currently modified due to city growth.

The buildings constructed on moderate to loosely consolidated sediments and lithological units are more vulnerable than those with well consolidated materials. In the sedimentary deposits shown in **Figure 5** the erosion of loose materials generate hazardous zones if the terrain slope is over 20°.

1.3. Land use and vegetation

The study area is located in a semi-desertic area with an average annual precipitation of less than ~500 mm/yr so most of the water supply for any all human activities is taken from underground wells. In these conditions the vegetation varieties are limited (**Figure 6**). The valleys are mostly used for agriculture. The thorn scrub and nopal (*Opuntia*) occupy the gentle hills of the sierras de Zacatecas and Tolosa. This type of vegetation requires little water and their roots extend laterally. The bushes are either natural vegetation outside the city limit or reforested areas inside those limits.

Since the original data was generated during the decade of 1970 [18-19], the situation now is different. The area shown as "Natural bushes" (**Figure 6**) is currently being substituted by thorn scrub and nopal. However, there is no up-to-date cartography.

The land use shown in **Figure 7** is taken from [18-19]; as well as in **Figure 6**, the valleys are used for agriculture. Cattle use is now greater, while the forestry area has decreased. Though the information on the criteria used to define "forestry use" is unclear, informally, these areas were considered for land conservation. However, since there are no written rules, the land use changes and modifications are based on unknown criteria. What has been seen is a continuous modification of landscape for urban purposes next to the city limit.

The land use map (**Figure 7**) was created from the INEGI (Instituto Nacional de Estadística, Geografía e Informática), institution depository of most of the cartographic information in the country. These maps are the basis for all the projects that require cartography. This being said, the conditions of the maps vary. The older ones were made different than the ones made today.

1.4. Edaphology

The soil classification used here is that of the United Nations Educational, Scientific and Cultural Organization [20]. In the Sierra de Zacatecas, where most municipalities are located, the Lithosol Eutric is dominant (**Figure 8**). It is composed by local, scarcely transported, rock clasts whose size varies from coarse sand to gravel; usually it is poor in organic matter. Its thickness is less than 15 cm.

The fluvisols are found along the arroyos; they show well-developed bedding, with normal gradation and variable amounts of organic matter. Their thickness is unknown since they are filling the valleys. The change in their texture and composition allows us to define the subclass.

The xerosols are dominant in the valleys. They have variable amounts of organic matter. A whitish layer at the top is characteristic of this soil, and it is usually due to the carbonate or sulfate accumulation.

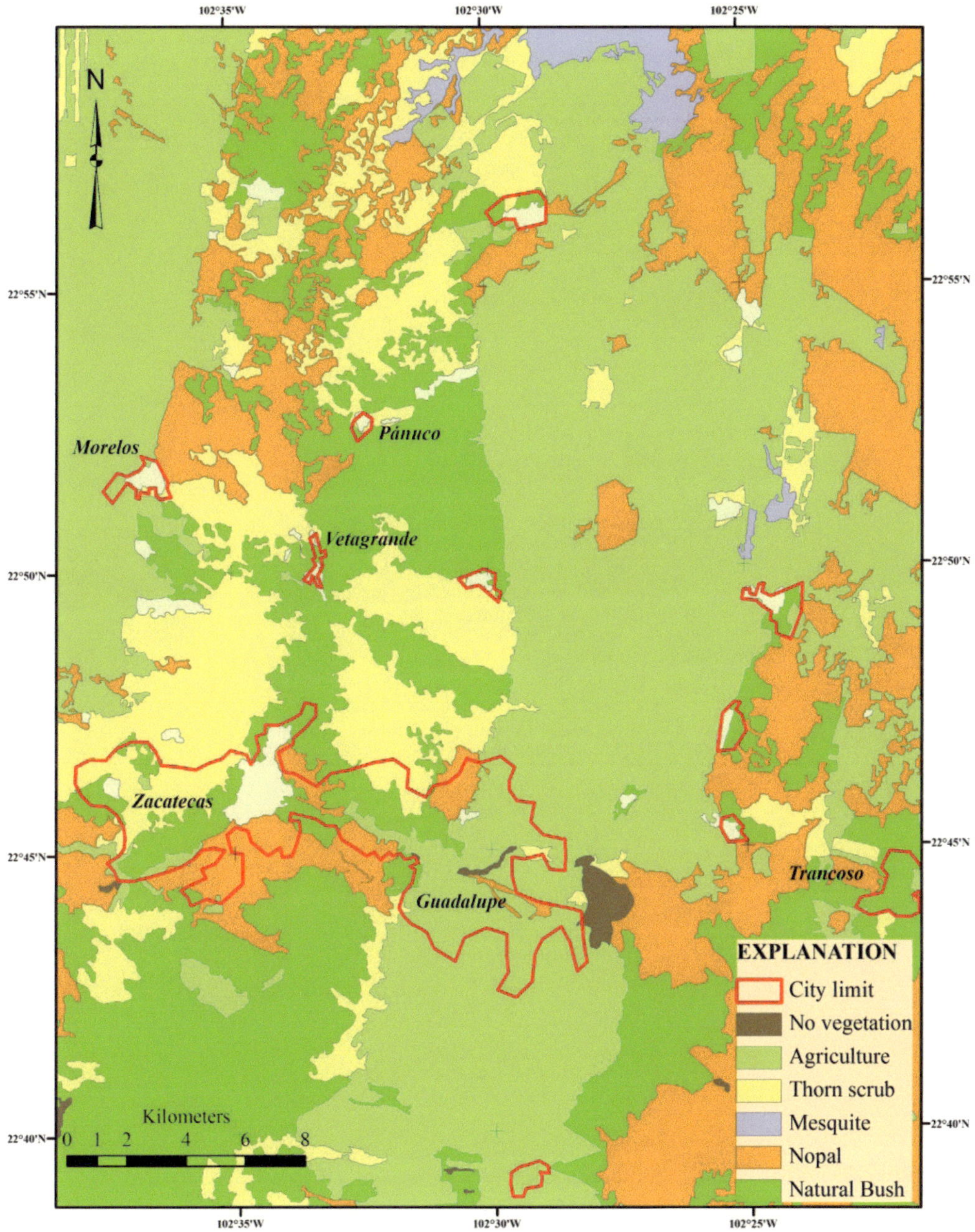

Figure 6. Vegetation of the study area, modified from [18-19]. The city limit corresponds to 2010. The names in bold, italics are the municipalities. The places with no vegetation are artificial dams. The original data are from the decade of 1970 [18-19].

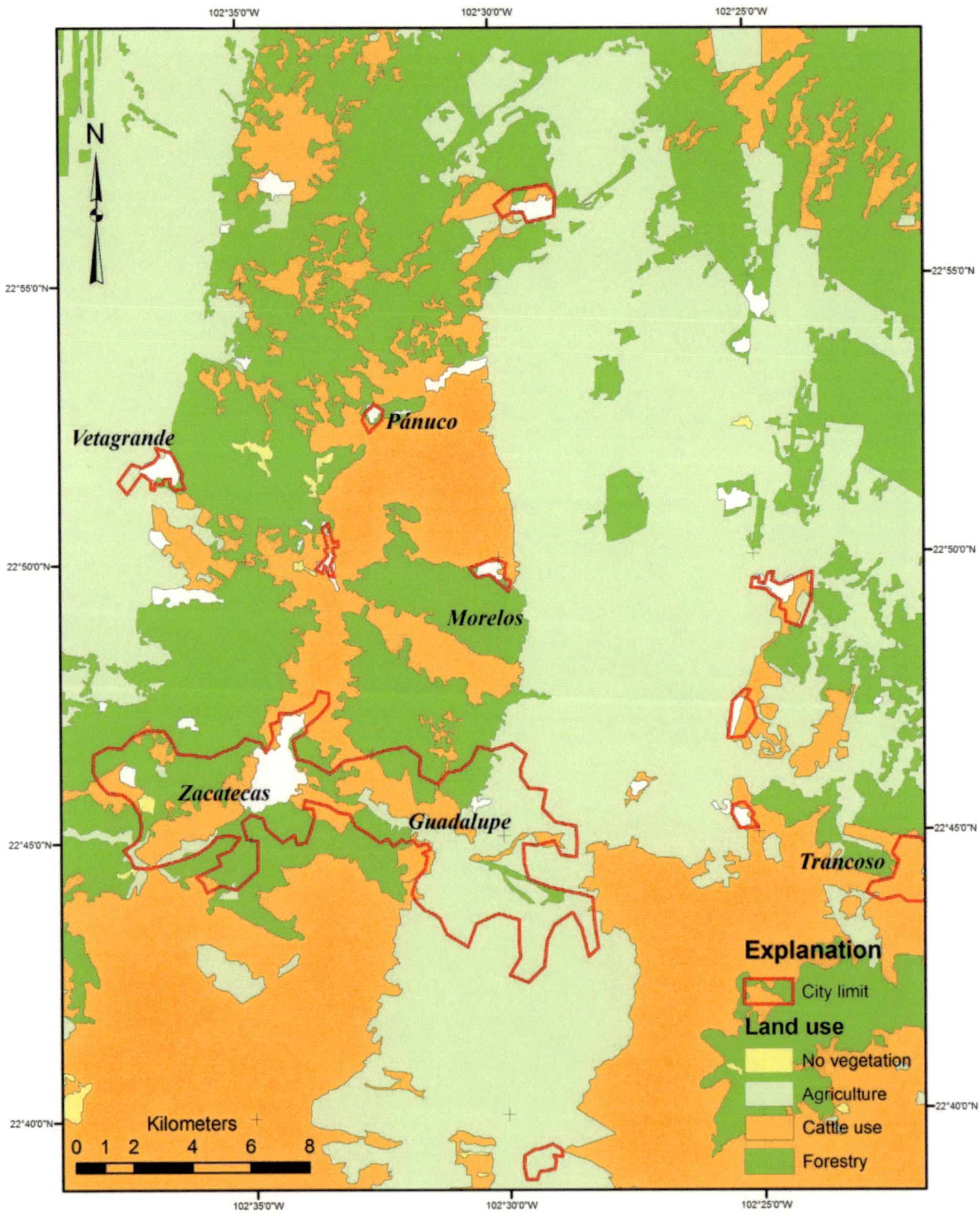

Figure 7. Land use proposed to the study area from [18-19]. The city limit is from 2010. The names in bold-italics are the municipalities. The places with no vegetation are the city limits in 1971. The original land use data are from the decade of 1970 [18-19].

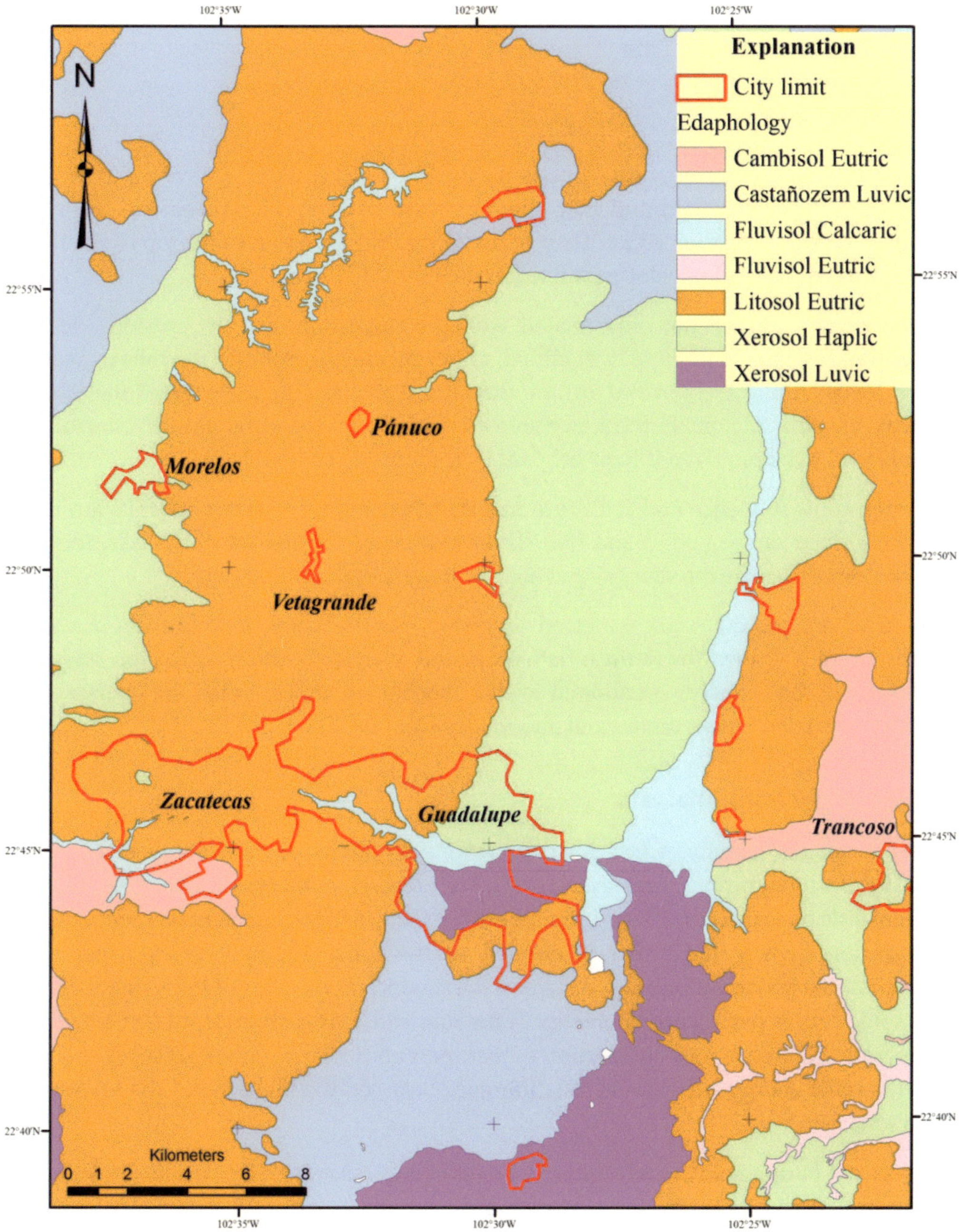

Figure 8. A simplified edaphologic map of the study area. Modified from [21-22].

2. Methodology

This chapter describes the elements used in the geologic and geomorphologic analysis. It starts with the field mapping of the elements where the hazards are occurring. These parameters were the basis for a geomorphologic study and finally all the information was integrated and analyzed in a GIS.

2.1. Field work

The risk areas are associated with slope deposits. The field work was directed to recognize the risk elements in the field and map them at scale 1:10,000. These elements are ancient landslide deposits and fallen rocks. They occur along arroyos, close to the top of the highest hills, and can be traced downslope until they reach the city limit.

The hazardous areas are not only related with the slope deposits; the erosion effects are accelerated where the landscape is modified, either by cutting through the hills or by filling in the arroyos. Along the arroyos are the fluvial terraces and at the edge of the Sierra de Zacatecas, the alluvial fans. In the study area all unconsolidated or moderately consolidated materials tend to be removed if their original conditions are changed.

The stratigraphic sequence and faults are mainly taken from [3, 13]. The rock falls are related with faults, steep slopes (> 20°) and rain. The rocks usually move less than 15 meters away from the source because the steep slopes are normally less than 10 m high.

In places where the cities are expected to grow, the landscape modification starts with vegetation removal, then the surface flattening, and eventually the construction phase. The cartography of the elements mentioned in this subchapter, unless when the landscape was modified artificially, where digitalized and managed in ArcGIS v. 9.3.1.

2.2. Geomorphologic analysis

The geomorphologic analysis was made according to the procedure described by [23]. The method uses a topographic chart scale 1:50,000 which is divided into squares; for this study the length side of each square was 1 km. In each square, four parameters were measured: 1) the dissection density (DD) that is defined as the total length of arroyos per square kilometer; 2) the general dissection density (GDD) which is the sum of the lengths of all the topographic curves per square kilometer; 3) the maximum dissection depth (MDD) which is the elevation difference measured from a creek perpendicular to the nearest highest point; and 4) the relief energy (RE) that is the difference between the highest and the lowest point in each square.

The measured parameters in each square were stored in a database in ArcGIS software. Each value is considered to be in the center of the square [23]. For each parameter, a Kriging interpolation procedure was used to define a raster image showing the spatial distribution of the variable. Since each parameter is in GIS Image format, the values can be managed for classification.

Based on field mapping, the slope deposits originated where DD> 10 km/km^2, GDD> 25 km/km^2, MDD>130 m and RE > 160. The areas defined this way were called "high erosion zones". The function used is a mathematical logical union of variables. If the variables change from 8.5 to 10 km/km^2, 20 to 25 km/km^2, 100 to 130 m and 130 to 160, respectively; they belong with the main body of the slope deposit, so they are "medium erosion zones". Whereas values ranging from 7 to 8.5 km/km^2, 15 to 20 km/km^2, 70 to 100 m, and 100 to 130, respectively, are located at the tip of the slope deposit and they are called "low erosion zones". Lower values are considered as "very low erosion zones".

The erosion areas defined this way do not take the slope into account. Therefore, due to the interpolation method used, "high erosion zones" can be located in a flat area, as well as an area with slope.

2.3. GIS analysis

The digital slope model (DSM) was obtained from the digital terrain model. The slopes were divided, according with field observations, in: 1) 0° to 5° semi-plain, 2) 5.01° to 10° gentle slope, 3) 10.01° to 20° hillside, 4) 20.01° to 30° ramps, and 5) >30.01° scarp or cliff. Since the DSM and the geomorphologic data are in the same coordinates system (NAD27; UTM-13N) the maps can overlay for spatial analysis. We used an intersection function of the selected DSM data with the geomorphologic analysis. This way, the DSM has redefined the hazardous areas by combining the slope ranges with the erosion zones. The intense erosion zones are located only in the scarps; while the medium erosion zones occur in the ramps and scarps. The low erosion zones are located in hillsides and ramps.

Once a result was obtained from the combination mentioned in the above paragraph, the next step was to combine the edaphology and land use information. All these areas are in the same kind of soil: Lithosol Eutric. This soil is less than 15 cm thick, rich in gravel and sand with variable amounts of organic matter. The vegetal coverage consists of bushes, nopals and grazing vegetation. When using the edaphology and land use data, the obtained results did not modify the previous outcome.

The analysis for rock falling was making a buffer of 15 meters length of the "Geology map". The structures were the result of the combination of the buffer on the fault and the slope range and direction.

All the information layers were spatially analyzed in ArcGis software, Ver. 9.3.1, and the results were verified in the field. For proof, we selected places where there were two or more erosion zones. Thus, the changes by erosion promoted by the lithology, sedimentary deposits, geologic structures, vegetation, or soil types could be observed.

3. Results

In this chapter all the mapping and digital analysis is integrated to get the definition of erosion zones and their relationship with hazardous zones. The obtained results were checked in the field to verify that our model is a reliable tool for urban development planning.

The basis for the analysis was the Digital Slope Model (DSM) (**Figure 9**). The cell size used was 20 m since, after testing larger and smaller sizes, that was the best dimension that defines the landscape.

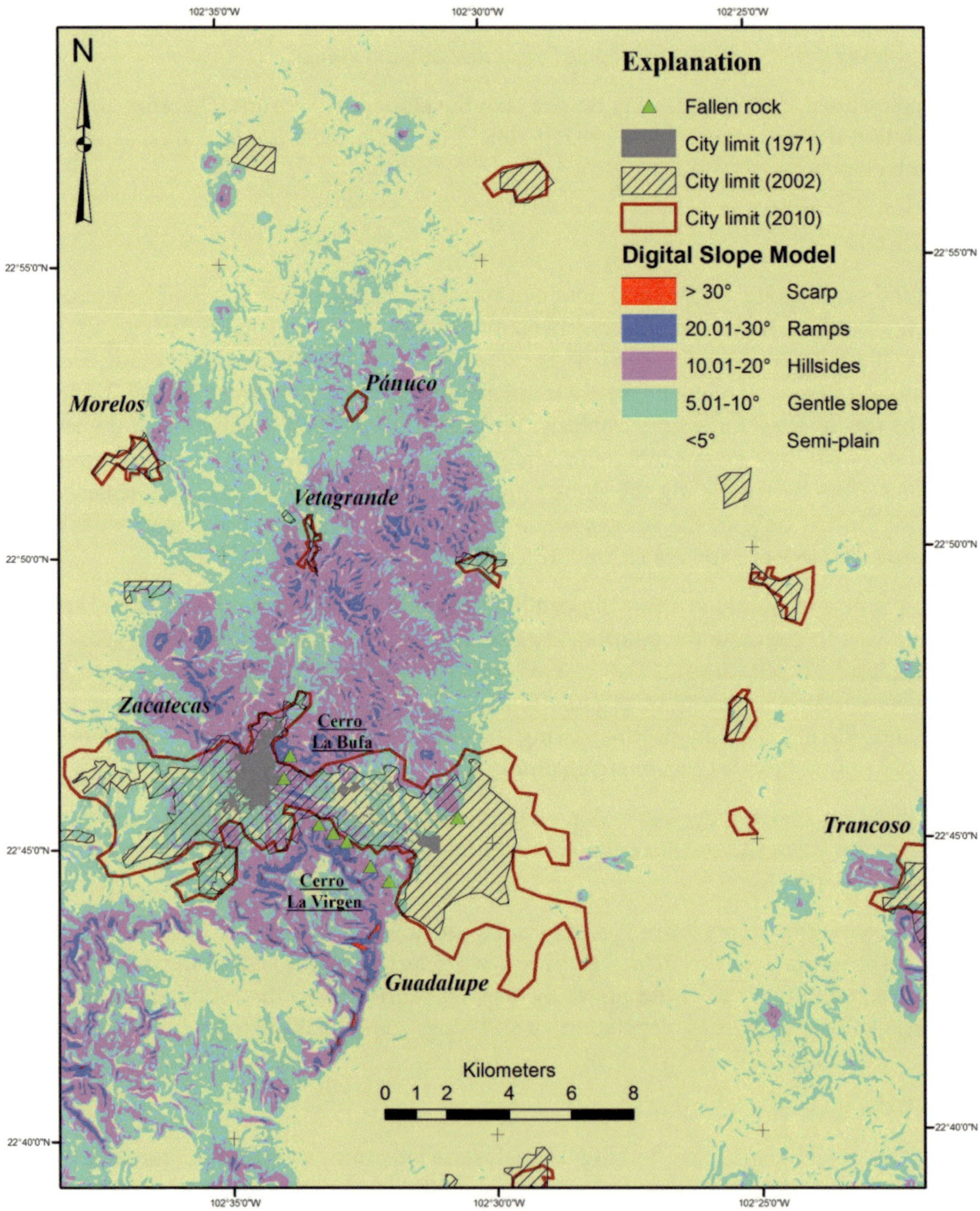

Figure 9. A digital slope model of the study area. The model was built using the contour lines each 10 m. The fallen rocks are those whose diameter is larger than 3.5 m. They are mostly related with ramps and scarps on Cerro La Virgen and Cerro La Bufa.

The **Figure 9** depicts the Zacatecas and Guadalupe cities growing in the gentle slope and semi-plain areas up to 2002; the use of hillsides and ramps was restricted to a few areas in Guadalupe. Due to population growth and continuous pressure on the territory, after 2002, the cities advanced toward the hillside areas. Moreover, due to the pressure on the territory, the arroyos became urbanized. The modification of the original conditions of the territory on hillsides and ramps favors the erosion on unconsolidated to moderately consolidated materials, both natural and artificial.

After 2002 the city grew in the direction of higher slopes, thus, aiding the territory to be affected by erosion. Furthermore, the city limit now is closer to scarps or cliffs where the fallen rocks become a common feature. The **Figure 9** shows the location of rocks whose diameter is over 3.5 m; however, the common ones are those of ~ 1 m that can be found upslope.

3.1. Geologic analysis

Rock falling is a common phenomenon in the study area. It is related with rock type and its fabric. Those rocks that are massive and deformed, if cut by faults, are most likely to be suitable for the development of this process. Additionally, the slope plays an important role in the location of hazardous areas associated with rock falling.

The analysis was made according to the observations made in modern and ancient fall rock. At present time the fallen rocks (~ 50 cm in diameter or less) observed are less than 15 m away from the fault, if there are scarps and ramps. If there is a combination of flowing water after rain with the slope, the displacement could be as far as ~ 100 m away from the source. Whereas the largest distances seen are ~ 1,000 m away from the faults (**Figure 10**).

The spatial analysis was made considering the following parameters:

1. The location of the mapped faults.
2. The lithologic units whose fabric is massive or deformed; they are: Zacatecas Formation, Deformed andesite and Undifferentiated tuff.
3. The slopes: scarp, ramp and hillslope.
4. The travel distance recorded by the fallen rocks.

The analysis is based on the location of the faults and making buffers at 15 and 100 m. The areas defined by the buffers, if they intersect ramps and scarps, are classified as hazardous zones. Due to the steep slopes no rock type was included.

The historical record indicates that a rock can travel as far as 1,000 m. For the faults mapped, this distance is in the hillslope area. During analysis, the first step was to define the buffers at 1,000 m. The next step was to perform a logical operation. If the buffer intersects a hillslope, and the lithology is considered, then the intersection identifies the area which has the possibility to have fallen rock. In **Figure 10** the areas for each distance are irregular polygons with voids inside of them. The voids are areas where the above mentioned conditions are not satisfied.

The results shown in **Figure 10** indicate the areas susceptible to have, or have the potential to be affected by fallen rocks. The field verification of these results indicates that our model defines vulnerable areas for rock falls. However, the results do not indicate the recurrence time or periodicity of the phenomena, It is merely starting where it could happen.

The hazardous areas are mostly located outside of the city limits. Additionally, **Figure 10** can be used for planning urban growth and, if necessary, make the proper preventative arrangements to avoid possible damages to the population and/or infrastructure prior to the landscape modification.

3.2. Geomorphologic analysis

In the Zacatecas and Guadalupe area the geomorphic agents, in relevance order, are: rainwater, gravity, wind and ice[3]. Since water is the main geomorphic agent, considering the semiarid climatic conditions, the effects of the erosion are most evident during the rainy season in the summer. In the winter, the wind and ice can increase their erosive effects in the loose materials.

Figure 11 shows the distribution of each geomorphologic parameters. It can be noticed that during the last 40 years the cities grew on medium to low erosion zones. However, this tendency has recently changed. Nowadays, the growth is close to the limits of Sierra de Zacatecas, getting closer to the "high erosion zones". The Dissection Density is the only parameter that barely has "High erosion" values. This is because the slope deposits start where the creeks do too; so the length of the "high erosion" is short.

The geomorphologic parameters are joined using a logical expression: If two or more "high erosion" areas intersect and the slopes and ramps are scarp, then they define an "Intense erosion zone" (**Figure 12**). The "Medium erosion zones" are defined if two or more "medium erosion" intersects with hillslope. The "Low erosion" areas are defined if two or more "low erosion" parameters intersect the gentle slope and semi-plain.

The erosion zones are defined on the basis of geomorphologic analysis and the DSM (**Figure 12**). The effects on the different lithological units, as well as the land coverage, soil type, and sedimentary deposits are verified in the field. If the natural vegetation and soil are preserved, the erosion processes do not play an important role independently of where they are located. If this is case, the surface creeping is the only geomorphic process. The vegetation and soil removal, along with the modification of arroyos and basin modifications and/or filling with unconsolidated materials, are the starting point for the erosion processes that affects the landscape.

When the original conditions are changed, the zonation that is defined here then applies. The effects are in the unconsolidated to moderately consolidated materials. There is a slow, but continuous, removal of sediments; mainly sand. The erosion occurs mainly during the rainy season in the summer. However, due to the low precipitation (~ 500 mm/yr) the monthly amount of rain could be so low that its effect as erosive agent could be minor. During the rainy season, the clay content in the sediments promotes the hydration and dehydration that, together with scarps and ramps, favors the formation of gullies.

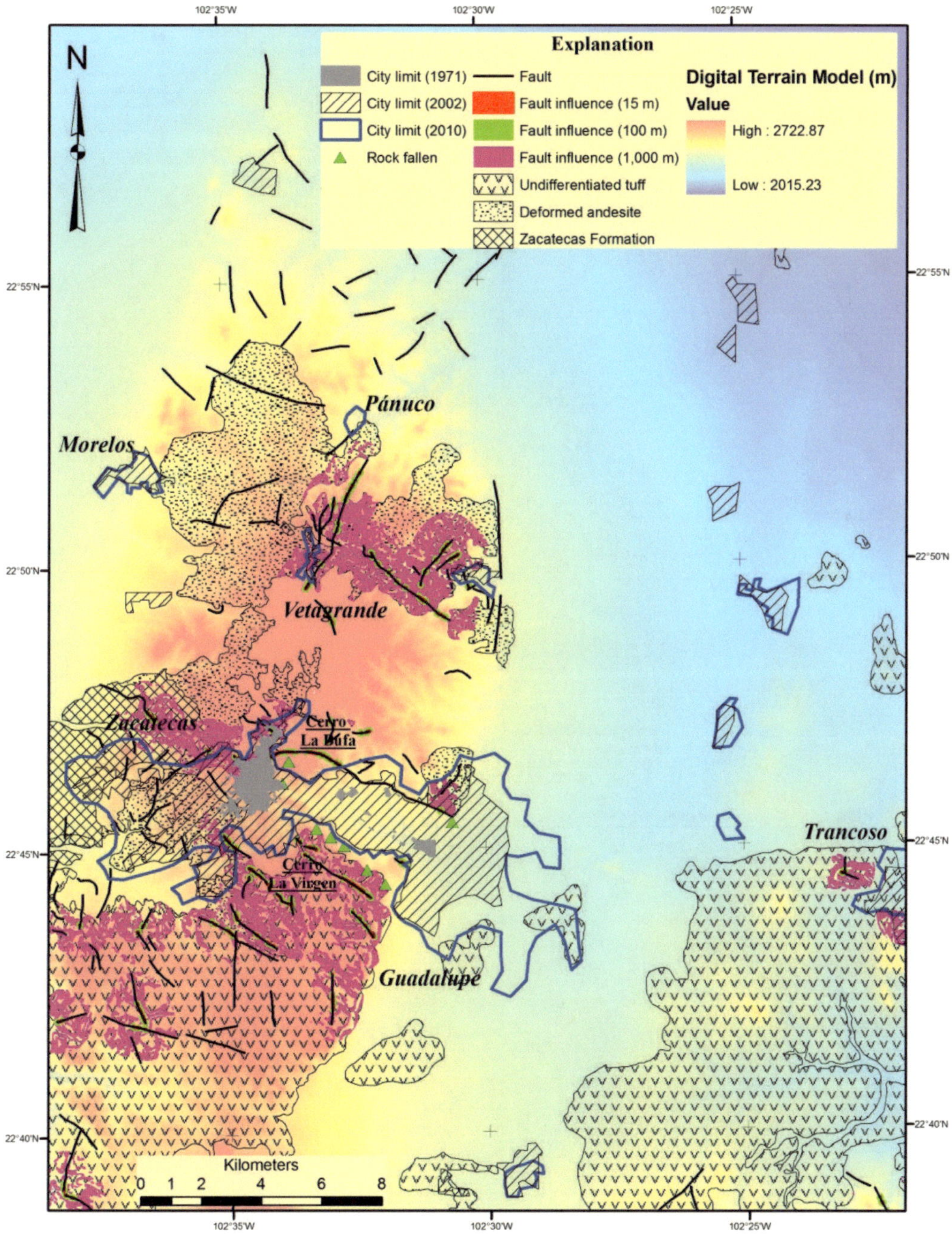

Figure 10. A digital terrain model showing the hazards associated with rock falling in scarps and ramps (red), hillslope close to the faults (green) and hillslope away from faults (magenta). It can be noticed that there are some faults not associated with erosion areas. This is because they are in gentle slopes or semi-plain areas.

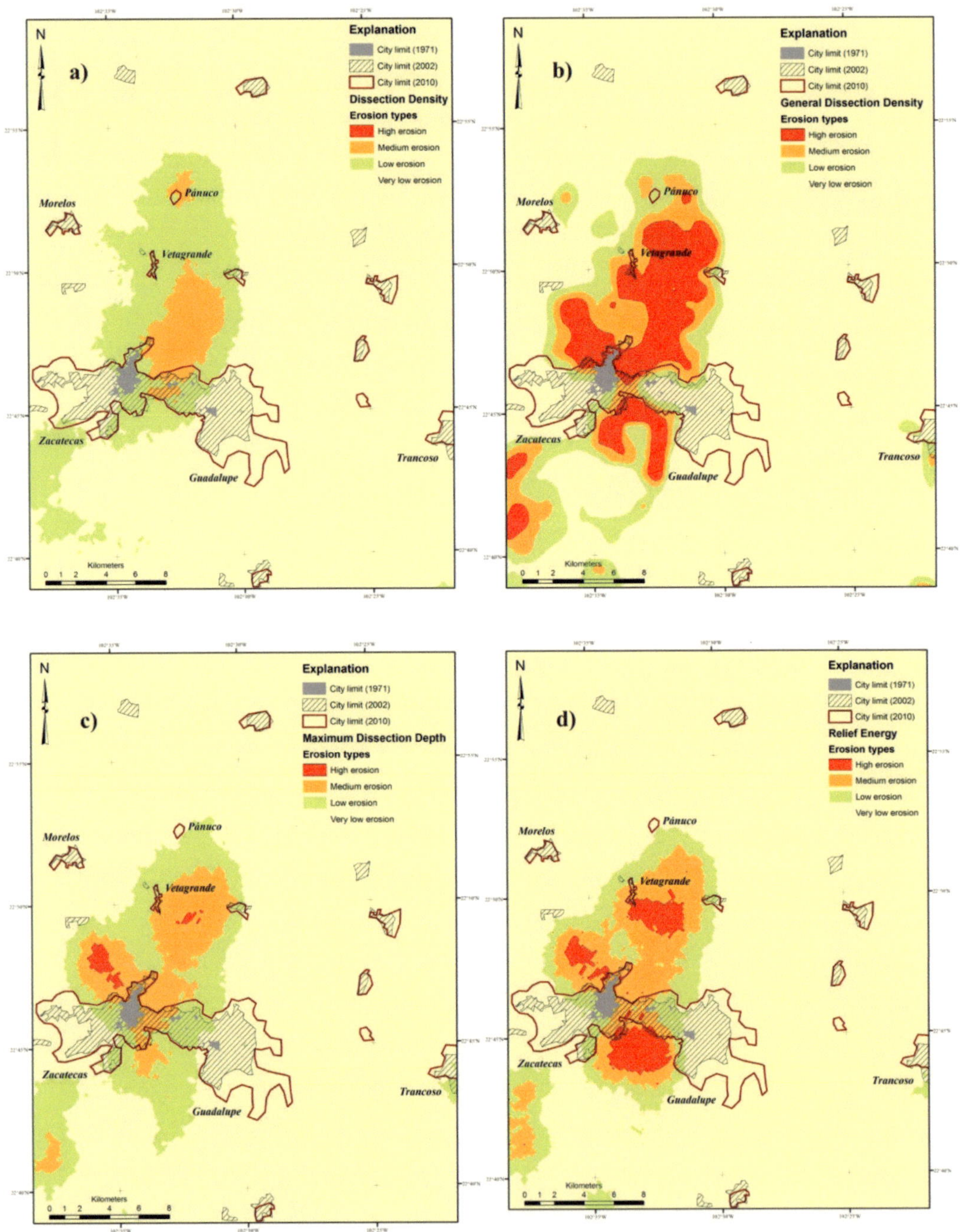

Figure 11. Maps showing the distribution of the geomorphologic parameters: a) Dissection density, b) General Dissection Density; c) Maximum Dissection Depth; d) Relief Energy. The erosion type values were defined according to section 2.2. Up to now, the city has grown in moderate to low erosion zones, although the "High erosion zone" is getting closer.

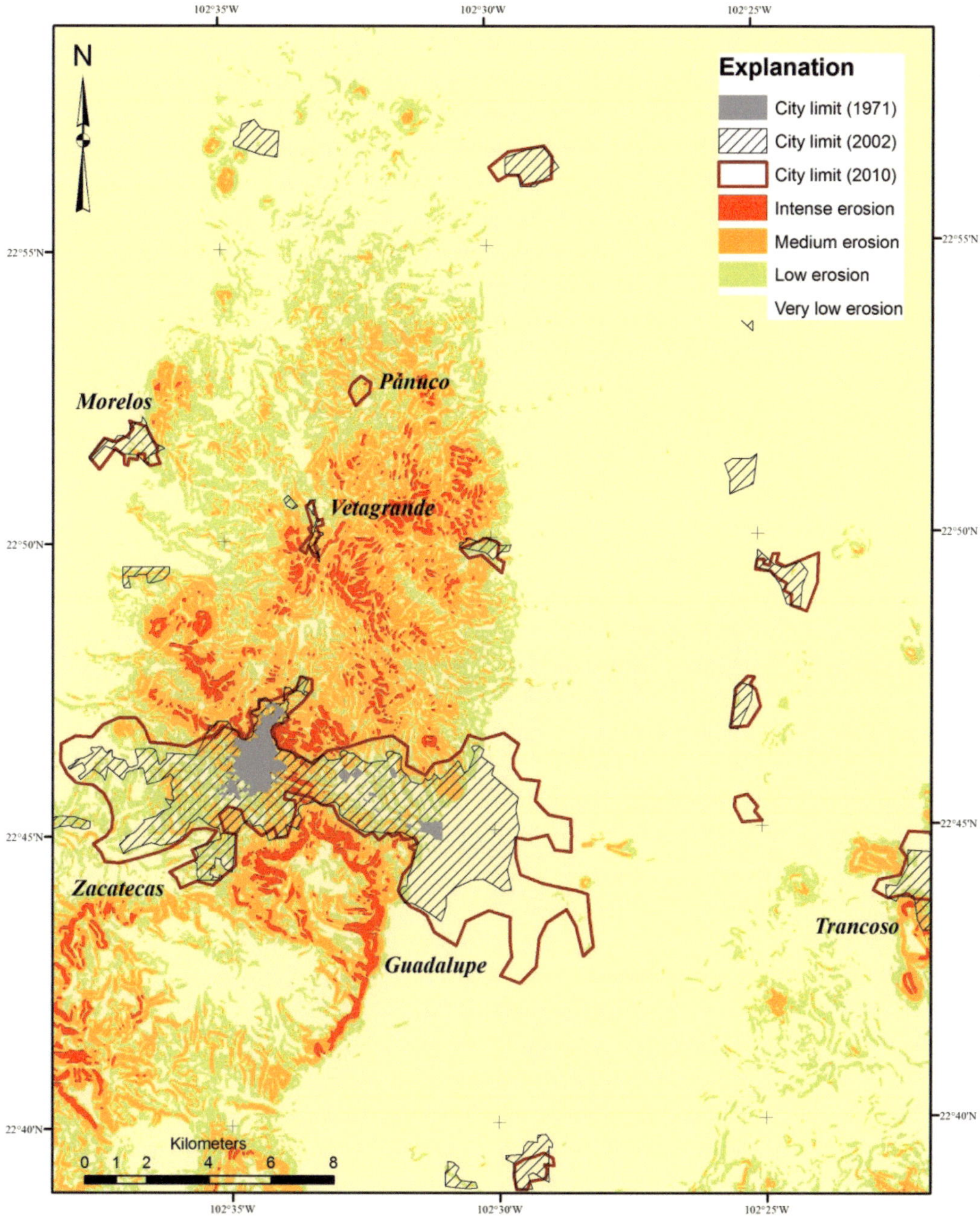

Figure 12. Erosion zones defined for the Zacatecas and Guadalupe quadrangles based on the geomorphologic and the digital slope model analyses. The "Intense erosion" areas indicate that if the natural conditions are changed, either by removing or adding new materials, they can be easily eroded according to the rain intensity. In the "Medium erosion" zones occurs the same event as in the previous one, but in a slower way. The "Low erosion" zones are located in the semi plain and gentle slope areas; their effects depend on the amount of running water after a rain.

The consequences of erosion in the rocks from the Mesozoic sequence and the Tertiary Volcanic rocks (both lithologic units well consolidated) are related with the faulting and fracturing patterns associated with scarps and ramps in intense and medium erosion zones. In the low erosion zones the effects are negligible. Most of the Zacatecas and Guadalupe cities are built on the Zacatecas Conglomerate whose facies composition and structure define differential erosion zones. Clearly, the sand and clay-rich facies can be more easily affected by the erosion processes than those well consolidated. The differential erosion removes the sandy rich strata, leaving unstable the conglomeratic ones or large rock fragments. With time, the unbalanced materials fall down.

The results shown in **Figure 12** were verified in the field and they are presented in **Figure 13**. The erosion removes unconsolidated materials if the original conditions are changed; otherwise the process is very slow. The velocity at which erosion acts depends on the zone they are in. The model here defined should be taken into account for the urban development planning. This model locates areas potentially affected by erosion if the landscape is modified.

Figure 13. Field verification of the proposed model for hazardous zones associated with erosion. a) Moderate erosion zone; in the filling deposit can be seen gullies formed after the rainy season; whereas the conglomerate is affected by differential erosion. b) Intense erosion zone. An excavation was made

for unknown reasons. The effects after the rainy season were that the wall and sidewalk fall. c) A Low erosion zone in a semi plain area; more than one year after the road was built, large gullies developed. d) An intense to moderate erosion zone. The undisturbed area is in the intense erosion where no visible effects are seen; while the moderate one is where the streets are traced.

4. Conclusions

If the landscape is not modified, the only erosion process acting is the surface creeping. This is independent of the location of the erosion zone. In the intense to medium erosion zones the mobility of loose materials is mainly achieved by rainwater when: 1) the landscape is modified, 2) the road cuts have a high angle or high angle slopes, 3) in the moderately consolidated facies of the Zacatecas Conglomerate, and 4) in the sedimentary deposits. The vegetation, land use and edaphology seem not to have any significant outcome in the definition of the erosion zones.

In the well consolidated rocks, the effects of the high erosion zones are associated with faults and fractures. In the historical record, rocks of less than 1 m in diameter can travel as far as 1 km away from the source rock in scarps to hillslopes. The model here presented detect the areas with strong possibilities of having fall rocks fallen.

The low erosion areas only have an effect in loose materials; while in the very low erosion zones the effects are along the arroyos.

5. Further research

After this study several questions were answered, but new ones arise for further research such as:

1. Until now no geophysics method has been used to define the extent and depth of the slope deposits, fluvial terraces, alluvial fans and artificially filled places. A method that could be used is surface waves, in this way the elastic parameters of the unconsolidated sediments could be obtained and may be used in construction regulations. The surface waves could also be used to locate buried slope deposits.
2. The slope deposits are of unknown age; they could be dated by looking for fossils, using U-Th-He, cosmogenic isotopes and/or paleomagnetism. The knowledge of their age and recurrence could be useful in defining the hazards´ recurrence.
3. The geomorphologic parameters used allowed us to define the erosion zones if the topography is abrupt; however they are not designed to evaluate the almost flat surfaces of the valleys. A further work is to look for the parameters that could be used to evaluate erosion in the valleys.
4. In the places damaged by the erosion processes, it is necessary to develop a mitigation plan. This should be made by an interdisciplinary professional team.
5. It is necessary to define the unconsolidated material loss; either from soil, fillings, sedimentary deposits or unconsolidated materials. In this way it could be possible to

make a precise evaluation of the sediment removal and the location of the more likely places where it will occur.

Author details

Felipe de Jesús Escalona-Alcázar*, Bianney Escobedo-Arellano, Brenda Castillo-Félix, Perla García-Sandoval, Luz Leticia Gurrola-Menchaca, Carlos Carrillo-Castillo, Ernesto-Patricio Núñez-Peña, Jorge Bluhm-Gutiérrez and Alicia Esparza-Martínez
Unidad Académica de Ciencias de la Tierra,
Universidad Autónoma de Zacatecas Calzada de la Universidad, Fracc. Progreso, Zacatecas, México

Acknowledgement

This research was partly financed by the project PROMEP No. UAZ-PTC-133 granted to Felipe Escalona. Partial support was also given by the Unidad Académica de Ciencias de la Tierra (UACT) of the Universidad Autónoma de Zacatecas, México. The authors acknowledge José de Jesús Fernández-Ávalos, Director of the UACT, for their support. Comments from Juan Carlos García y Barragán helped to improve this manuscript. The authors acknowledge the Copyediting/Proofreading by Charlotte Presley.

6. References

[1] Código Urbano para el Estado de Zacatecas (1996) Official Journal of the Zacatecas State Government. Ordenance No. 81, Published on September 11th, 1996, 122 pp.

[2] Acuerdo de la Declaratoria de Reservas de Suelos Derivada del Programa de Desarrollo Urbano de la Conurbación Zacatecas-Guadalupe (2004) Official Journal of the Zacatecas State Government. Agreement signed on August 6th, 2004, 273 pp.

[3] Escalona-Alcázar FJ, Suárez-Plascencia C, Pérez-Román AM, Ortiz-Acevedo O., Bañuelos-Álvarez C (2003) La secuencia volcánica Terciaria del Cerro La Virgen y los procesos geomorfológicos que generan riesgo en la zona conurbada Zacatecas-Guadalupe. GEOS 23(1): 2-16

[4] Atlas de Riesgos de la Ciudad de Zacatecas (2007) Ministry of Social Development. 119 pp. (Unpublished)

[5] Enciso-de la Vega S (1994) Crecimiento urbano de la Ciudad de Zacatecas y sus asentamientos humanos en zonas mineralizadas polimetálicas. Revista Mexicana de Ciencias Geológicas 11(1): 106-112.

[6] Escalona-Alcázar FJ (2010) Evaluación preliminar de los riesgos debidos a la geomorfología de la zona urbana Zacatecas-Guadalupe y sus alrededores. GEOS 29(2): 252-256.

* Corresponding Author

[7] Escalona-Alcázar FJ, Delgado-Argote LA, Rivera-Salinas AF (2010) Assessment of land subsidence associated with intense erosion zones in the Zacatecas and Guadalupe quadrangles, Mexico. In: Carreón-Freyre D, Cerca M, Galloway D, editors. Land Subsidence, Associated Hazards and the Role of Natural Resources Development; October 17-22, Queretaro, Mexico. IAHS Publication 339: 210-212

[8] XI General Census of Population (1990) National Institute of Statistics, Geography and Informatics.

[9] XII General Census of Population (2000) National Institute of Statistics, Geography and Informatics.

[10] XIII General Census of Population (2010) National Institute of Statistics, Geography and Informatics.

[11] Zacatecas Topographic Map scale 1:50,000 (F13B58) (1973) General Direction of Studies of the National Territory.

[12] Guadalupe Topographic Map scale 1:50,000 (F13B68) (1973) General Direction of Studies of the National Territory.

[13] Escalona-Alcázar FJ, Delgado-Argote LA, Weber B, Núñez-Peña EP, Valencia VA, Ortiz-Acevedo O (2009) Kinematics and U-Pb dating of detrital zircons from the Sierra de Zacatecas, Mexico. Revista Mexicana de Ciencias Geológicas 26(1): 48-64

[14] Zacatecas Geologic Map scale 1:50,000 (F13B58) (1998) National Institute of Statistics, Geography and Informatics.

[15] Guadalupe Geologic Map scale 1:50,000 (F13B68) (1998) National Institute of Statistics, Geography and Informatics.

[16] Aranda-Gómez JJ, Henry HD, Luhr JF (2000) Evolución tectonomagmática post-paleocénica de la Sierra Madre Occidental y la parte meridional de la provincia tectónica de Cuencas y Sierras. Boletín de la Sociedad Geológica Mexicana LIII: 59-71

[17] Ponce BS, Clark KF (1988) The Zacatecas Mining District: A Tertiary caldera complex associated with precious and base metal mineralization. Economic Geology 83: 1668-1682

[18] Zacatecas Land Use and Vegetation Map scale 1:50,000 (1977) National Commission on National Territory Studies.

[19] Guadalupe Land Use and Vegetation Map scale 1:50,000 (1977) National Commission on National Territory Studies.

[20] FAO/UNESCO Soil map of the World 1:5,000,000 Paris (1974) United Nations Educational, Scientific and Cultural Organization

[21] Zacatecas Edaphologic Map scale 1:50,000 (1971) National Commission on National Territory Studies.

[22] Guadalupe Edaphologic Map scale 1:50,000 (1971) National Commission on National Territory Studies.

[23] Lugo-Hubp J (1988) Elementos de geomorfología aplicada. 1st Edition. Mexico: UNAM. 128 pp.

Chapter 19

Evaluation of Soil Quality Parameters Development in Terms of Sustainable Land Use

Danica Fazekašová

Additional information is available at the end of the chapter

http://dx.doi.org/10.5772/48686

1. Introduction

Soil is a vital natural source and, at the same time, has an economic and eco-social potential. It allows the production of food and raw materials, recycles waste, creates forest-agricultural land, filters and retains water, allows the usage and valorisation of sun energy, ensures the cycle and balance of substances in nature, maintains diversity of plant and animal species. It primarily shapes the quality of the environment; it is the resource and cultural heritage of the Earth; it ensures the life and social being of the population. Agricultural activities realised in landscape affect natural resources. A rational usage of renewable and non-renewable resources which are not retrieved in real time is an essential precondition.

The farming system is the most widespread environmental technology with its positive and negative consequences. It utilises essential natural resources and, at the same time, influences other natural environments. Therefore, ecologisation of farming is a priority of farmers as well as environmentalists. Respecting the principles of soil sustainability and other components of environment is the basic precondition for life sustainability.

A United Nations [UN] conference on environment and development (Rio de Janeiro, 1992) prioritised sustainable development, which is presented in a global development programme for the end of the 20th century and especially for the 21st century (Agenda 21). The concept of sustainable development of agriculture includes such practices in farming which respect ecological aspects in growing plants and ethology of livestock in rearing, do not enhance damage in ecological land stability, respect environmental protection, including surface and underground water and monitor the quality of agricultural produce.

Sustainable agriculture is based on the principle of agriculture being a biological process which, in practice, should imitate key characteristics of the natural ecosystem. It strives to bring diversity into agro-ecosystem, recycle nutrients efficiently and maintain the priority of sunlight as a source of energy for agro-ecosystems.

Specific manifestations of soil require different approaches. In soil protection, these must be ecological (biological) approaches, as this is the only way to achieve sustainable development of ground cover and the resulting economical and social development and environmental balance in society.

Sustainable use of soil takes soil-ecological conditions into consideration and is realised in such a way and in such intensity, which gives rise to neither negative changes in soil, nor establishes trends for the development of negative characteristics in soil. The essential principle of sustainable farming system is its protection from any degradation by natural or man-induced influences. Sustainable development of soil use also encompasses the protection of the soil acreage to such an extent which ensures that all soil functions are employed.

In a number of European countries, sustainable use of soil is realised according to the principles of International Federation of Organic Agriculture Movements [IFOAM] and is referred to as ecological soil management. When introducing ecological systems of soil management, the main criterion is the application of knowledge in the functioning of natural ecosystems, which are typical of plant and animal variety and sunlight is the exclusive source of energy. In cultural (artificial) agro-ecosystems, the structure is disrupted by man drawing the production past the limit of the agro-ecosystem. The ecological system focuses on theoretical elaboration of farming arrangement in sensitive areas (the protection of underground and surface water zones, polluted zones, national parks, protected natural areas and soils heavily endangered by erosion). Continued protection of nature and natural resources is at the forefront; therefore, significant intensifying constituents of conventional agriculture (high dosage of fertilisers, full usage of pesticides, annual subsoil ploughing, major hunts, high ratio of grain crops, intensive breeding, heavy automation) are replaced by technologies with strong economical and ecological components (tillage minimisation, anti-erosive crop rotation, monitoring of plant nutrition, integrated a biological protection of plants, minimal automation, free-range breeding).

According to Organisation for Economic Co-operation and Development [OECD], an indicator is a parameter or a value derived from several parameters. It provides information about a particular observed phenomenon from the viewpoint of its quantitative or qualitative characteristics, present in a give time and area, in the environment as a whole, or its individual components by the qualitative parameters of these components influencing the health condition of the population, as well as the structure and function of the ecosystem in the area in question. From the above stated, it results that there are a number of horizontal and vertical causal links between individual environmental indicators. The "sustainability indicator" can, thus, be defined as a measurable factor, whose imbalance negatively influences the long-term performance of the whole production system. Stable agriculture has a time and space dimension. The time scale depends on the adaptability of the system (usually 5 to 10 years, or more); space can be given by the borderlines of soil-climatic units or areas. Stability indicators should be applicable to the evaluation of the main components of sustainable agriculture. Attention is mainly paid to the level of farming and its productivity regarding the ecological soil potential, maintaining diversity of plant species as well as the protection of natural resources, social-economic viability related to the regional and world economy.

From the viewpoint of agricultural practice, the stability indicators regarding productivity of agricultural production and ecological aspects of farming systems have been explored in most detail. The guaranteed yield on the level of the ecological potential of location (without further input increase), the ability of the system to return to the initial performance in a short period of time after a natural disaster, achieving a relatively high efficiency of water and plant nutrition utilisation, maintaining the soil quality environment (organic mass, soil organisms, nutrients), reliability of the methods used in integrated plant protection, ensuring the quality of water resources, maintaining the level of underground water without major fluctuations, and protecting natural resources are considered quantifiable biophysical indicators of sustainable productivity (Klír, 1997). With regard to the evaluation of ecological sustainability, the most significant indicators are maintenance and improvement of biodiversity in managed, as well as adjacent natural ecosystems, maintaining the environmental quality and avoiding pollution limit excess (Virmani & Singh, 1997 as cited in Fazekašová, 2003).

It is impossible to select universal soil parameters with regard to their suitability for sustainable soil and is subject to specialised discussions. A significant role in the selection of parameters is played by their variability in time, related to parameter stability. The following soil parameters can be distinguished: stable (such as soil depth or granularity), relatively stable (the salt content, the content of organic mass in soil, heavy metal contamination), relatively dynamic (pH, the content of nutrients), and dynamic (soil humidity and temperature, microbial activity, etc.). Stable and relatively stable parameters dominantly influence soil quality, while relatively dynamic and dynamic characteristics are more connected to its short-term changes.

Soil parameters indicate the state of soil ecosystem characteristics, which especially reflect production, buffering, filter and other soil functions. From this view, the structure of soil profile (the soil class), soil type, soil depth, skeletal nature, the content and quality of humus substances, accessible nutrient supply, soil reaction, the content of foreign substances in soil, and soil edaphon seem to be of highest importance.

Soil quality cannot be judged directly; it must be determined from the changes of its parameters. It is more accurate to evaluate the range of appropriate indicators rather than to use a single one. Soil quality is significantly affected by physical, chemical, biological and biochemical properties sensitive to changes in the environment and land management. With regard to physical properties, there are bulk density, porosity, water retention capacity, soil temperature, etc. In the group of chemical characteristics, total carbon and nitrogen content, soil reaction and content of available nutrients are observed. Evaluation of biological parameters focuses on microbial biomass and its activity, soil respiration, potentially mineralised nitrogen, the activity of soil enzymes, etc. Soil enzymatic activity can be used as a microbial indicator of soil quality, since the activity of soil enzymes is closely related to essential soil characteristics. It indicates changes sooner than other soil characteristics and can be an integrating soil-biological index reflecting soil use (Javoreková et al., 2008; Šarapatka, 2002). Wick et al. (2002) considers selected enzymatic activities as suitable indicators for long-term soil monitoring and quality assessment (Miralles, 2007; Geisseler, 2009). A decrease in soil quality is obvious from the values of critical load of risk substances. When evaluating the content of heavy metals in soil, attention must also be paid to their bio-accessibility (Bujnovský & Juráni, 1999).

The chapter deals with a synthetic and comparative analysis of scientific findings regarding the development of soil quality parameters in the conditions of a sustainable farming system. Based on the research carried out between 1997 and 2010 on a model area situated in a marginal region of north-eastern Slovakia (48° 57' N; 20° 05' E), the development of soil indicators are evaluated, focusing on physical (bulk density and soil porosity), chemical (soil pH, inorganic nitrogen, available phosphorus, potassium, magnesium and organic carbon content) and biological parameters (activity of acid and alkaline phosphatase and urease), as well as the presence of risk substances in the soil ecosystem (heavy metal content - Cd, Ni and Pb).

2. Evolution of soil parameters

At present, there is little knowledge with regard to soil development in the conditions of sustainable farming systems whose principles lie in soil maintenance. There is a major effort to increase its natural productivity by as closed a cycle of nutrients as possible with the highest possible reduction of external, mainly energetic and chemical, inputs (Lacko-Bartošová et al., 2005; Fazekašová, 2003). The present findings can hardly be compared to other research due to the different soil-ecological conditions in which they were obtained. The issue of universal methods for all soil types remains a universal problem within the research of soil development. Unless this area is unified, objective comparison will remain on a regional level. Soil parameters are usually determined only in relation to specific topsoil. Certain physical and chemical parameters in subsoil cannot be neglected, since they guarantee soil functions (Fazekašová, 2003).

2.1. Methods

The research project was carried out between 1997 and 2000 and 2008 and 2010 under production conditions in the investigated area situated in a marginal region of north-eastern Slovakia (48° 57' N; 20° 05' E). Here, the ecological farming system has been applied since 1996. The area is situated in the Low Tatras National Park at an altitude ranging from 846 to 1492 m above sea level. In terms of geomorphological division, it is a part of sub-assemblies of the Kráľovohoľské Mountains (Michaeli & Ivanová, 2005). The whole area is situated in a mild zone with a sum of average daily temperatures above 10 ºC ranging from 1600 to 2000 and average precipitation of 700-1200 mm (Fig. 1).

The soil conditions are relatively homogeneous, the largest area being represented by Cambisols, mostly moderate and strongly skeletal, mainly in the subsoil, medium-weight and heavy in granularity (loamy sand, loam, clayey loam). Cambisols are the most common soil type occurring in Slovakia. From an ecological viewpoint, Cambisols are valuable for their irreplaceable ability to retain and accumulate atmospheric fallout and also for their filtration attributes. From the relief viewpoint, the majority of the land is situated on slopes, soil is often eroded and, thus, surface water resources are threatened. With regard to pollution, there is an assumption that heavy metals are transported to crops (due to the acidity of these soils). In the current crop structure, cereal acreage represents 33.3 %, potatoes 16 % to 18 % and fodder crops 49.8%. Crops are rotated as follows: perennial fodder (clover mixture) → perennial fodder (clover mixture) → winter crops (winter wheat, winter rye, triticale and winter barley) → root crops (potatoes) → spring crops (spring barley, oats) → annual mixture (oats pea, peas, ryegrass). Arable land is fertilised with manure dosage of approximately 30 t ha^{-1} once in two years. The permitted phosphorous and potassium mineral fertilisers have not been added in the past five years. The permanent grassland and arable land were fertilised with liquid organic fertiliser in the spring season, 3 000 l ha^{-1} (minimum nutrients content: total nitrogen expressed as N in dry mass at least 15 %, total phosphorus as P_2O_5 in dry mass less than 0.2 %, total potassium as K_2O in dry mass less than 0.4 %, total sulphur as S in dry mass at least 16.5 %).

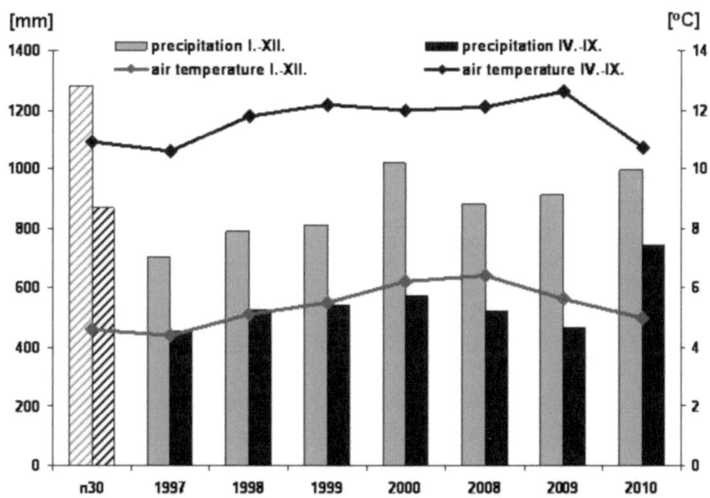

Figure 1. The course of average air temperatures (°C) and sum of precipitation (mm) during the observed period in the observed area situated in a marginal region of north-eastern Slovakia

Soil samples for physical, chemical and biological soil properties and heavy metal content determination were obtained in spring time in a connected stand on five permanent research sites, from the depth of 0.05 m to 0.15 m. Part of the soil samples were air-dried, sieved (sieve with 2 mm size opening), homogenised prior to the analysis and used for measurements of chemical and biological soil characteristics and heavy metal content. From the physical soil properties, soil bulk density and soil porosity were studied and evaluated in a Kopecky physical cylinder with a capacity of 100 cm^3 (Fiala et al., 1999). From the chemical soil characteristics, soil pH in 1M $CaCl_2$ solution was monitored and evaluated, as well as inorganic nitrogen, available phosphorus, potassium, and magnesium with Mehlich III and organic carbon content (Fiala et al., 1999). The available heavy metal content (Cd, Ni and Pb) of the samples was determined in 2M HNO_3 solution using atomic absorption spectrophotometer (Matúšková & Vojtáš, 2005). The following biological soil characteristics were monitored: activity of acid and alkaline phosphatase (Grejtovský, 1991) and urease (Chaziev, 1976). The obtained data were tested by mathematical-statistical methods from which analysis of variance and regression analysis were used (the Statgraphics software package).

2.2. Evolution of physical soil parameters

The changes in physical characteristics of soil not only result from meteorological factors, yearly farming plan, or from the course of vegetation, but also depend on the employed farming system. Larson and Pierce (1991) confirmed that soil quality can be evaluated and the sustainability of a system assessed on the basis of essential physical indicators.

Soil granularity, and especially the ratio of clay particles, primarily influences physical, hydro-physical and chemical characteristics. The soils in the monitored localities according to the content of clay particles based on Novák's classification (Fulajtár, 2006) are of loamy-sandy, loamy and clay-loamy category (Tab. 1.).

Diameter of particles [%]	Studied locality				
	I.	II.	III.	IV.	V.
> 0.25	31.3	14.5	11.5	32.5	16.0
0.25–0.05	21.6	15.5	18.9	13.9	14.9
0.05–0.01	27.8	32.4	24.3	22.2	31.3
0.01–0.001	15.5	29.3	3.3	24.0	2.6
< 0.001	3.8	8.3	11.0	7.4	8.2
I. Category	19.3	37.6	45.3	31.4	37.8
Soil	loamy sand	loam	clayey loam	loam	loam

Table 1. Particle grain-size composition of soil [%] in the monitored area situated in a marginal region of north-eastern Slovakia in depth 0.05–0.15 m.

Bulk density as an integral value of soil granularity, humus content and anthropogenic impacts on soil should not exceed the limits given for individual soil types (Tab. 2.).

Soil texture	Sandy	Loamy sand	Sandy loam	Loam	Clayey loam and clay	Clay
Bulk density	≥ 1.70	≥ 1.60	≥ 1.55	≥ 1.45	≥ 1.40	≥ 1.35
Porosity	≤ 38	≤ 40	≤ 42	≤ 45	≤ 47	≤ 48

Table 2. Critical values of bulk density soil [$t.m^{-3}$] and porosity [%] for different of soil texture (Líška et al., 2008)

Long-term research has shown that ecological soil farming regulates bulk density of soil. The measured values of bulk density were in the range of 0.94 to 1.35 $t.m^{-3}$ (Fig. 2.), in 1997 to 2009, there was a moderate decrease and values comparable to average figures for the given soil type and category according to Líška et al. (2008) were achieved (Tab. 2.), with the exception of 2010, when a mild increase in bulk density was measured. At the same time, this parameter proved to change under the influence of the water content and meteorological exposure (Kotorová, Šoltýsová & Mati, 2010). In 2010, in comparison to the previous years, precipitation reached higher values (Fig. 1.).

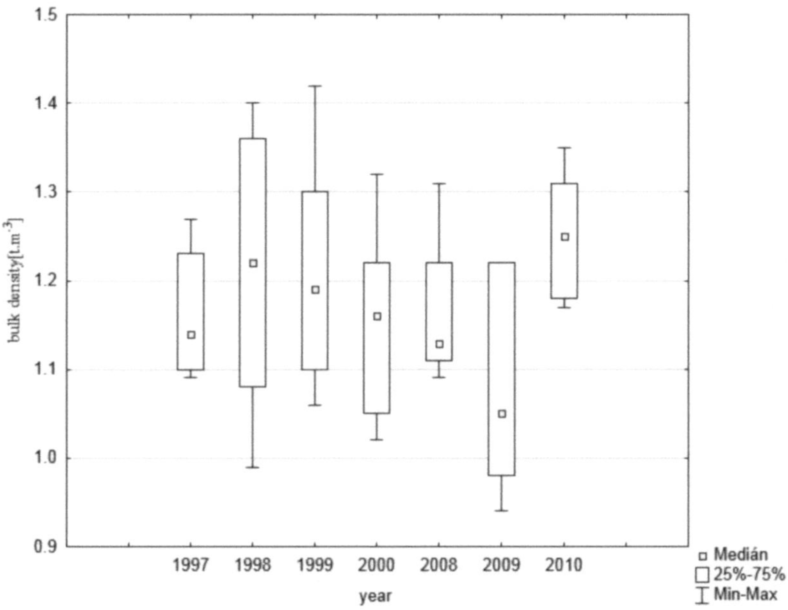

Figure 2. Bulk density of soil in the monitored area situated in a marginal region of north-eastern Slovakia expressed by descriptive statistics

General porosity is closely related to bulk density. From the total pore volume, which should not fall below 38 % for sandy soil and below 48 % for clay-loam soil (Líška et al., 2008), the share of non-capillary pores rapidly releasing gravitational water and allowing good air exchange between soil and climate should be sufficient. The share of non-capillary pores (Pn) in comparison to capillary pores (Pk) should be higher in heavy soils.

As can be seen from Fig. 3., the values show that, in the observed timeframe, porosity levels ranged between 46.43 and 64.49 %. Considering this parameter, optimum conditions were created for the growth of most arable crops, which are given by general porosity between 55 and 65 % and 20 and 25 % soil air content (Rode, 1969).

A statistically significant effect in the monitored year and locality on all observed soil physical parameters was confirmed by an analysis of variance (Tab. 3.).

Parameter	Min.	Max.	Mean	Standard error	Source of variability	d. f.	F-Ratio	P
bulk density [t.m-3]	1.04	1.35	1.18	0.016347	year	6	8.33	++
					locality	4	28.39	++
porosity [%]	49.15	60.56	55.52	0.617864	year	6	8.24	++
					locality	4	28.41	++

Table 3. Analysis of variance of soil physical parameters in the monitored area situated in a marginal region of north-eastern Slovakia
++P< 0,01 +P< 0,05

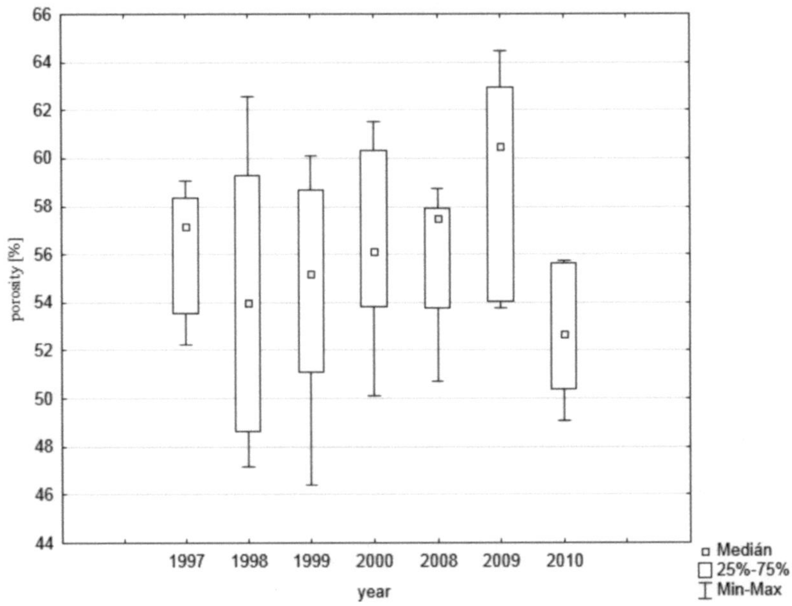

Figure 3. Porosity of soil in the monitored area situated in a marginal region of north-eastern Slovakia expressed by descriptive statistics

2.3. Evolution of chemical soil parameters

Chemical parameters are considered relatively dynamic (pH, nutrient content) and, in terms of plant growth and development, vital. Their deficiency is reflected in crop production. At the same time, they serve as indicators of additional inputs in the form of fertilisers. Sustainable farming systems exclude, or reduce, the use of artificial fertilisers; therefore, it is necessary to pay attention to the dynamics of chemical soil parameters in order to prevent one-way draining of nutrients, particularly phosphorus and potassium.

The soil pH is an important factor for soil fertility despite the fact that its values change dynamically, depending on so-called internal and external factors. It influences the buffering and filtering capacities, the quality of organic substances, nutrient accessibility for plants and the production of biomass in most crops grown. A majority of arable crops suit the range of slightly acidic to slightly alkaline soil pH – 6 to 7.5 (Krnáčová, Račko & Bedrna,

1997). A pH value lower than 5.5 is undesirable and requires ameliorative lime treatment. Similarly, from the viewpoint of productivity, alkaline soils (pH>8.4) are limiting and require appropriate measures.

In the course of monitoring the model area, the values of soil pH ranged between 5.1 and 7.2. The average values of soil pH increased moderately and were in the category of slightly acidic to neutral soil pH (6.3 – 6.9) (Fig. 4.).

This can be assigned to the ecological farming system, as physiologically acidic mineral fertilisers were not applied. On the contrary, organic fertilisers (manure at the dosage 30 t ha^{-1} and liquid organic fertilisers at the dosage 3000 l ha^{-1}) were applied. The organic matter positively influences the buffering capacity of soil, which is why the soil reaction was stabilised. Nevertheless, it is necessary to pay continuous attention to soil reaction, since soil is naturally acidified through acid atmospheric fallout as well as calcium intake by plants.

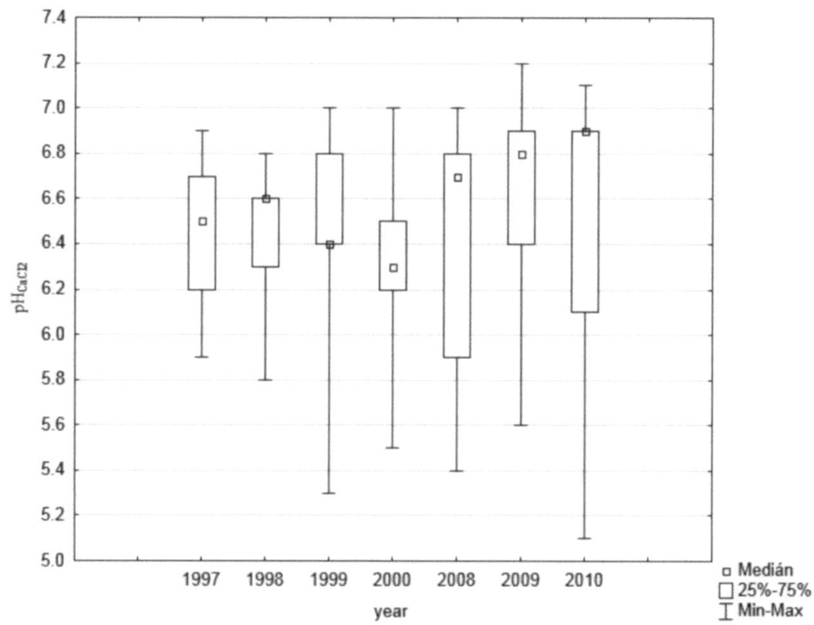

Figure 4. Soil reaction ($pH/CaCl_2$) in the monitored area situated in a marginal region of north-eastern Slovakia expressed by descriptive statistics

Nitrogen, phosphorus and potassium are the most important nutrients. Their supply to soil can be realised in various ways; fertilisation being of most importance. A lack of essential nutrients is rapidly reflected in the level of plant production. Nitrogen in soil is restored as part of its natural cycle. Its additional supply is necessary for intensified harvest, when its natural supply is not sufficient in order to achieve the targeted harvest. The supply of phosphorus and potassium by fertilisation is related to their supply in soil. Their supply in soil is not exhaustless; moreover, when constantly utilised, they are not naturally renewable.

According to Bielek (1998), there is a small probability that an increase in the total nitrogen content has a positive effect on soil fertility. This only applies to productive and highly productive soils. For soils with low production capacity, a reciprocal ratio between the total nitrogen content and soil fertility is typical. From the total nitrogen in soil, 95 % to 98 % is bound in organic forms; fertility functions determine mechanisms of its accessibility to plants. It is mainly organic nitrogen mineralisation, or, more specifically, that part of mineralisation which prevails over carbon immobilisation related to fertility. Inorganic nitrogen only represents a small part of total nitrogen and its content in the season is subject to frequent and fast changes, resulting from natural and anthropic factors. The concentrations of the main forms of mineral nitrogen (ammonia, nitrates) result from pure mineralisation and frequent nitrification of nitrogen in soil. In our research carried out in natural conditions, medium to highly favourable content of inorganic nitrogen in topsoil has been observed (Fig. 5.) in spite of the fact that in the soil-ecological conditions of the investigated area (a mild zone with a sum of average daily temperatures above 10 ºC ranging from 1600 ºC to 2000 ºC and average precipitation of 700-1200 mm), the nitrogen mineralisation is less intensive (the optimum temperature for an intensive process is 28-30 ºC); therefore, even with a high total content of nitrogen, the content of mineral (i.e. immediately available) nitrogen may not be high. The assumption is that by adding high doses of organic fertiliser, the total nitrogen content will increase. However, including legumes in the crop rotation can increase the content of immediately available nitrogen. These crops leave high amounts of nitrogen in soil (more than 100 kg ha^{-1} N), which are later available for the crops grown in the following period (Jurčová & Torma, 1998; Kováčik, 2001).

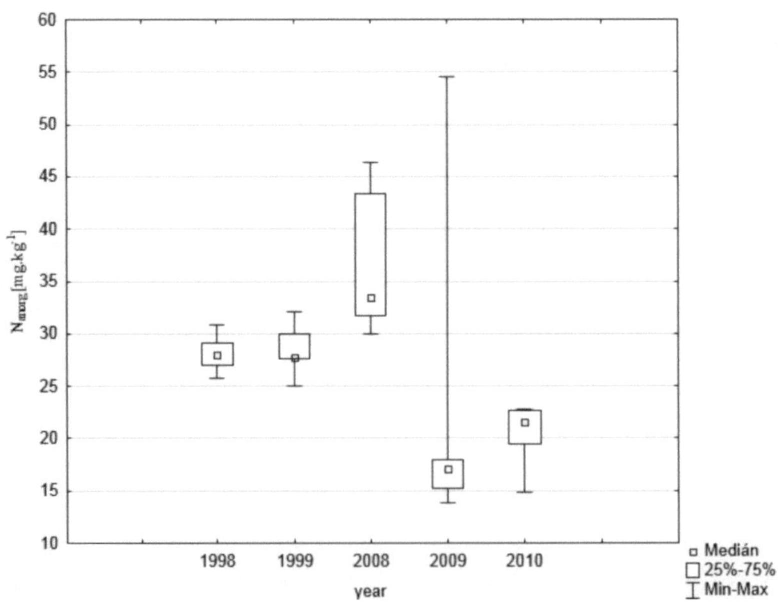

Figure 5. N_{anorg} content of soil in the monitored area situated in a marginal region of north-eastern Slovakia expressed by descriptive statistics

Phosphorus is firmly fixed in soil and its proportion is relatively stable and dependent on soil reaction values. Between 1997 and 2010, the value of soil pH did not significantly change in the investigated area. With regard to the above, the proportion of available phosphorus changed only minimally (Fig. 6.).

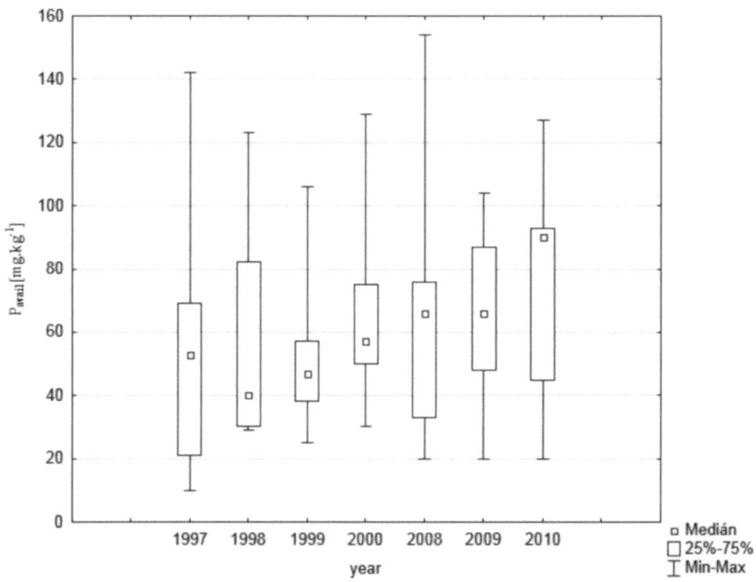

Figure 6. P_{avail} content of soil in the monitored area situated in a marginal region of north-eastern Slovakia expressed by descriptive statistics

In the observed period, the proportion of potassium and magnesium was relatively stable (Fig. 7. and Fig. 8). Due to the grain structure of the soils (medium and heavy soils), these nutrients are bound to soil particles and are not prone to soil washing in spite of high precipitation throughout the year.

Organic mass determines the soils quality, as it binds soil particles, stabilises soil (by which the risk of erosion decreases), increases water retention and cationic exchange capacity and reduces the negative impact of pesticides, heavy metals and other pollutants. A high proportion of organic carbon alone cannot guarantee a high yield; however, the influence of soil carbon on productivity increases when the levels of carbon decrease below 1%. With a content C_{org} 1.0 to 1.5, productivity decreases by 15 % and with content C_{org} under 1.0 %, it decreases by as much as 25 %. The most significant parameter is the ratio of humin acids and fulvene acids. This ratio is considered highly favourable, if it is higher than 2, satisfactory in the range between 1 and 2, and unfavourable if lower than 1. With regard to biological activity of soil, so-called non-specific humus substances play a significant role. These are a source of nutrients for soil microorganisms, participating in important cyclic biochemical processes (Hraško & Bedrna, 1988).

The content of humus in soil is a parameter prone to significant changes in the long-term. The application of high amounts of organic fertilisers and incorporating perennial fodder crops in the crop rotation influenced the preservation of humus content. The measured C_{ox} values ranged from 2.16 to 3.92 (Fig. 9.), which, when conversed to humus (conversion coefficient 1.724), are medium to good humic soils (Vilček et al., 2005).

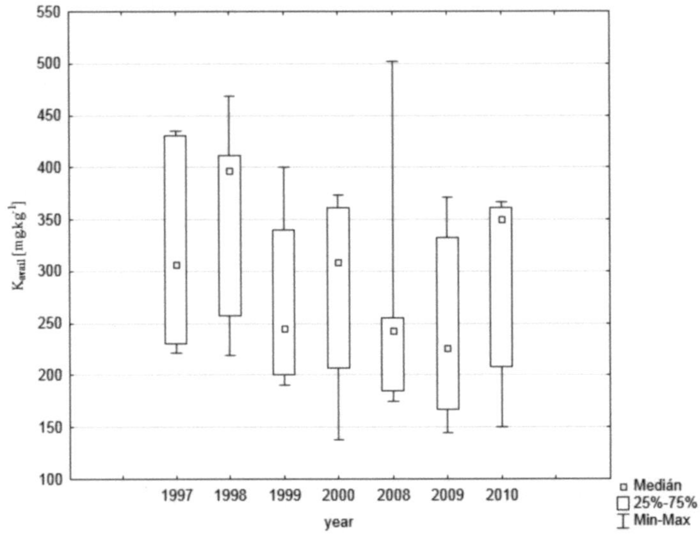

Figure 7. K_{avail} content of soil in the monitored area situated in a marginal region of north-eastern Slovakia expressed by descriptive statistics

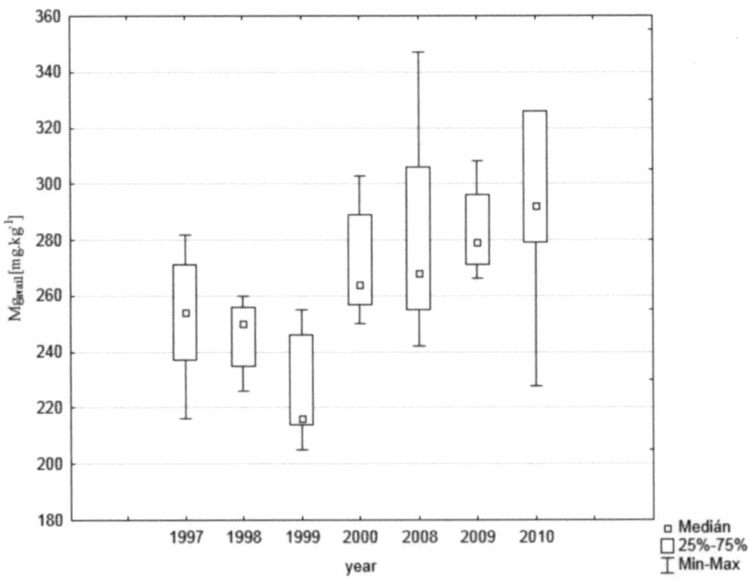

Figure 8. Mg_{avail} content of soil in the monitored area situated in a marginal region of north-eastern Slovakia expressed by descriptive statistics

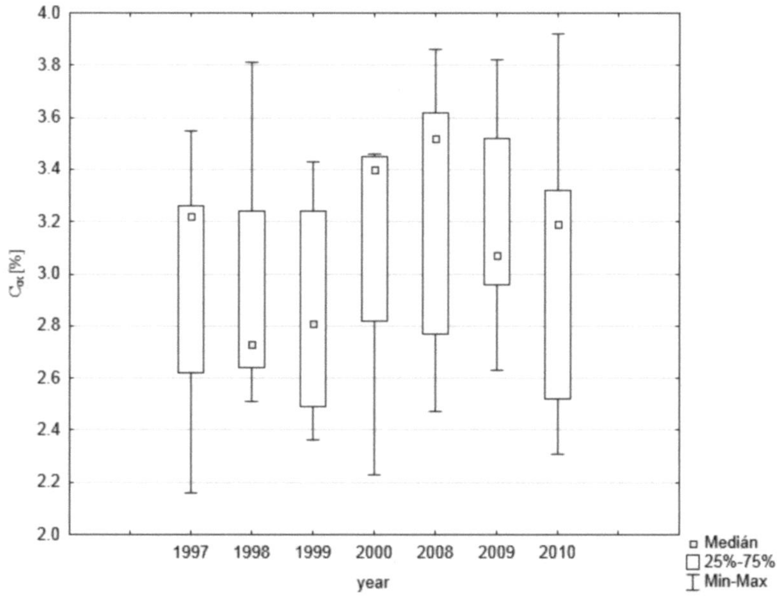

Figure 9. C_{ox} content of soil in the monitored area situated in a marginal region of north-eastern Slovakia expressed by descriptive statistics

A statistically significant effect of the monitored year on all observed soil chemical parameters was confirmed by an analysis of variance (Tab. 4.). The influence of the monitored locality on soil chemical parameters was also statistically significant, with the exception of N_{anorg}.

Parameter	Min.	Max.	Mean	Standard error	Source of variability	d. f.	F-Ratio	P
pH/$CaCl_2$	5.77	7.13	6.41	0.083124	year locality	6 4	0.82 24.51	- ++
C_{ox} [%]	2.25	3.61	3.03	0.084802	year locality	6 4	1.46 23.99	- ++
N_{anorg} [mg.kg^{-1}]	16.76	40.50	27.52	1.698623	year locality	6 4	12.77 2.22	++ -
P_{avail} [mg.kg^{-1}]	19.97	127.88	64.63	3.494827	year locality	6 4	3.02 121.06	++ ++
K_{avail} [mg.kg^{-1}]	168.59	427.98	290.91	12.5772	year locality	6 4	5.87 43.46	++ ++
Mg_{avail} [mg.kg^{-1}]	215.98	301.43	265.0	4.918103	year locality	6 4	17.52 3.73	++ ++

Table 4. Analysis of variance of soil chemical parameters in the monitored area situated in a marginal region of north-eastern Slovakia

++P< 0,01 +P< 0,05

2.4. Evolution of biological soil parameters

The information on biological soil parameters is not as plentiful as it is in the case of physical and chemical parameters, despite the fact that the effect of edaphon on biochemical processes in soil, nutrients balance, soil structure, etc. is proven in general.

There are a great number of enzymes in soil, depending on the diversity of soil organisms and the conditions of organic substances transformation.

Soil enzymes regulate the functioning of the ecosystem and play key biochemical functions in the overall process of organic matter decomposition in the soil system (Sinsabaugh et al., 1997). They are important in catalysing several important reactions necessary for the life processes of micro-organisms in soils and the stabilisation of soil structure, the decomposition of organic wastes, organic matter formation and nutrient cycling (Dick et al., 1994 cited in Makoi & Ndakidemi, 2008).

Enzymes are present in the cells of living organisms in soil (bacteria, fungi, algae, and soil fauna) and plant roots. Micro-organisms are the major source of enzymes in soil. The amount and quality of enzymes in soil is dependent on their characteristics, volumes and forms of organic matter and the activity of micro-flora. Enzymatic soil activity is higher in fertile soils with plenitudes of organic matter. The highest proportion of various enzymes can be found in the humus soil horizon (Pejve, 1966). The activity of soil enzymes can be enhanced by using organic fertilisers (Burns, 1978; Iovieno et al., 2009; Chander et al., 1997). The urease enzyme belongs to the hydrolases group of enzymes and is responsible for the hydrolysis of urea fertiliser applied to the soil into NH_3 and CO_2 with the concomitant rise in soil pH. This, in turn, results in a rapid N loss to the atmosphere through NH_3 volatilisation. Due to this role, urease activities in soils have received a lot of attention since it was first reported, a process considered vital in the regulation of N supply to plants after urea fertilisation (Makoi & Ndakidemi, 2008).

Soil urease originates mainly from plants and micro-organisms. It can be found as a free enzyme in soil solution, and yet more often firmly bound to soil organic mass or minerals, as well as inside living cells (Klose & Tabatabai; 2000; Alef & Nannipieri, 1995). Its activity depends on soil humidity (Baligar et al., 2005), pH, humus proportion and quality (Tabatabai & Acosta-Martínez, 2000) and the total nitrogen content (Nourbakhsh & Monreal, 2004). At the same time, an increased sensitivity to excess content of heavy metals (Kromka & Bedrna, 2000) and a negative effect of triazine herbicides on the activity of enzymes (Belińska & Prangal, 2007) was shown.

Phosphatases are a broad group of enzymes that are capable of catalysing hydrolysis of esters and anhydrides of phosphoric acid. In soil ecosystems, these enzymes are believed to play critical roles in P cycles (Speir et al., 2003) as evidence shows that they are correlated to P stress and plant growth. Apart from being good indicators of soil fertility, phosphatase enzymes play key roles in the soil system (Dick et al., 2000 cited in Makoi & Ndakidemi, 2008).

Soil phosphatase has certain typical characteristics. It depends on the substratum and its concentration. Two optimums levels, acidic and alkaline, are often present (Burns, 1978). An optimum pH of soil phosphatase is influenced by a great number of factors.

Soil pH differs from the pH optimal for phosphatase activity. Soil phosphatase can be inactive if the differences between soil pH and optimum enzyme pH are too great (Chaziev, 1976). The activity of soil phosphatase is higher in soils with high humidity in comparison to dry soils or soils with normal humidity. Phosphatase activity declines with an increasing soil depth, which is caused mainly by lower biological activity in lower soil profiles. Inorganic phosphate, copper, mercury and vanadium also have a considerable inhibitory effect on soil phosphatase activity (Burns, 1978; Speir et al., 2003).

There was minimum fluctuation in the measured values of soil enzyme activity in the observed period. The urease values ranged from 0.43 to 0.67 mg $NH4^+$ - $N.g^{-1}.24$ $hour^{-1}$, and the values of acidic and alkaline phosphatase between 236.8 and 336.5 μg $P.g^{-1}.3$ $hour^{-1}$ (Fig. 10., Fig. 11. and Fig. 12.). These are values typical for sparse-vegetation soils (Burns, 1978).

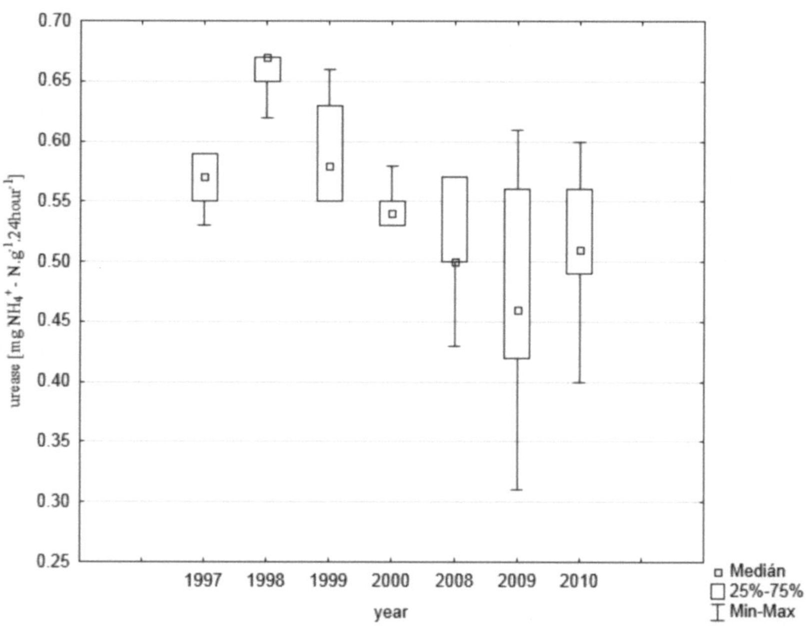

Figure 10. Urease activity of soil in the monitored area situated in a marginal region of north-eastern Slovakia expressed by descriptive statistics

At the same time, a higher activity of soil enzymes in lower temperatures was confirmed (the area is situated in a mild district with a sum of average daily temperatures above 10 °C ranging from 1600 to 2000 and average precipitation between 700 and 1200 mm) and organic fertilisers and soil organic mass stimulate the activity of soil phosphatase and significantly enhance the protection of natural soil urease (Chaziev, 1976; Bremner & Mulvaney, 1978).

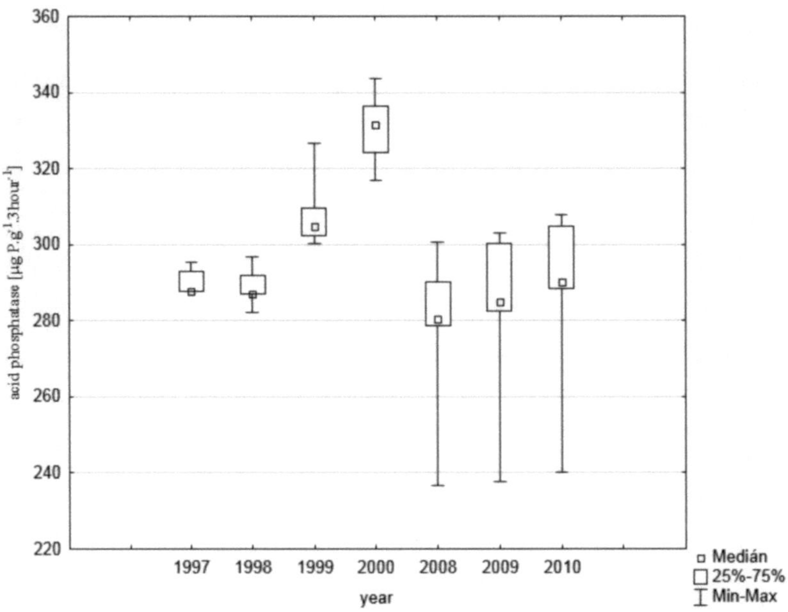

Figure 11. Acid phosphatase activity of soil in the monitored area situated in a marginal region of north-eastern Slovakia expressed by descriptive statistics

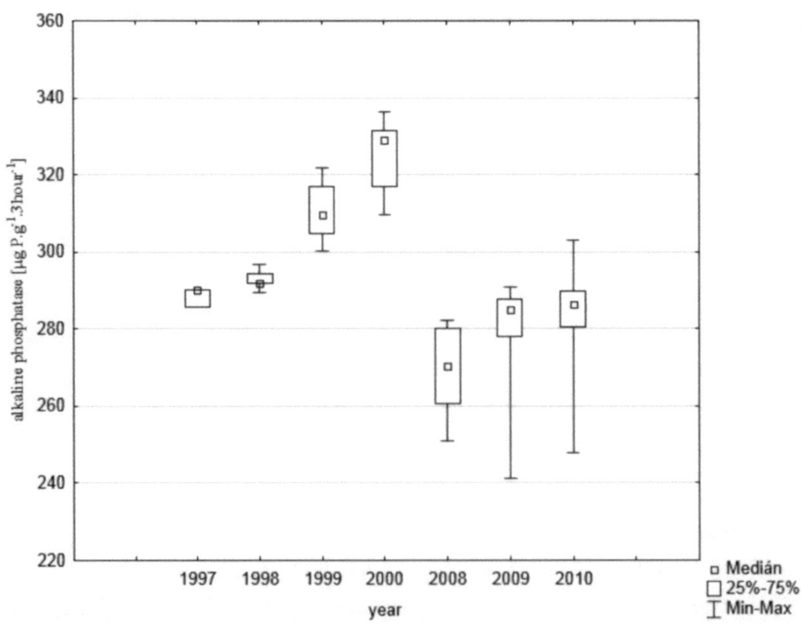

Figure 12. Alkaline phosphatase activity of soil in the monitored area situated in a marginal region of north-eastern Slovakia expressed by descriptive statistics

A statistically significant effect of the monitored year and locality on all observed soil biological parameters was confirmed by an analysis of variance (Tab. 5.).

Parameter	Min.	Max.	Mean	Standard error	Source of variability	d. f.	F-Ratio	P
urease [mg NH_4^+ - N.g^{-1} .24$hour^{-1}$]	0.454	0.674	0.551	0.007954	year locality	6 4	44.07 33.12	++ ++
acid phosphatase [μg P.g^{-1} .3$hour^{-1}$]	271.23	306.77	294.91	2.65001	year locality	6 4	36.16 15.85	++ ++
alkaline phosphatase [μg P.g^{-1} .3$hour^{-1}$]	264.35	329.33	291.97	1.96567	year locality	6 4	75.50 16.39	++ ++

Table 5. Analysis of variance of soil biological parameters in the monitored area situated in a marginal region of north-eastern Slovakia
++P< 0,01 +P< 0,05

2.5. Concentration of heavy metals in soil

An increase in inputs employed in the farming system has gradually brought about the need for studying and evaluating their potential negative influence on soil environment and production quality. Monitoring soil contamination with various degrees of biotoxicity is an important area. Fertilisers, especially industrially produced, are considered (including rock slackening, atmospheric decline and waste stock) a significant source of risk elements in soil (Beneš, Benešová, 1993). The system of farming, including the use of fertilisers, can also indirectly affect the acceptability of risk elements for plants (Beneš, 1993).

Loading agricultural soil with harmful substances is serious, since soil is not only the key to agricultural production but also has filtration and buffering capacities. Soil considerably influences the composition and quality of underground water and provides a living environment for soil micro-organisms (Tischer, 2008; Gulser, 2008). It could be assumed that accumulating higher concentrations of heavy metals in soil is a potentially serious danger to the food chain (Torma et al., 1997). It is especially toxic elements and organically highly-persistent substances that are among harmful substances entering soil.

Heavy metals as a large group of polluters are a serious problem in all components of the environment, including soil. As a great number of these have considerable toxic effects, their highest allowed concentrations are defined for the soil system, similarly to those for air and water. It is extremely difficult to define limit concentrations of heavy metals for soil, since, in contrast to air and water, soil is an extremely heterogeneous system and mobility of inorganic contaminants, closely related to the intake by plants, depends on several soil factors. The approaches towards the determination of metal concentration limits in soil vary significantly in individual countries. In some countries, the definition of limits for heavy metals concentrations is based on soil use (these are defined as so-called trigger and action values), or, possibly, on eco-toxicological data in so-called standard soil and limit values for the total and dissolvable concentration of heavy metals in soil (Barančíková, 1998; Makovníková et al., 2006).

Toxicity of heavy metals varies; it decreases in the following line Hg>Cd>Ni>Pb>Cr and their influence is enhanced by their non-degradability. Soil is only presented as a passive acceptor of heavy metals; it becomes the source of polluting other components of the environment and the food chain. Changes in soil properties are responsible for the mobilisation of metals, especially pH, humus content and quality and the proportion of clay fraction (Barančíková, 1998).

With regard to the above findings, the content of the following risk elements was observed in the conditions of sustainable use of soil: lead, cadmium and nickel (in the leachate 2M HNO_3) (Fig. 13., Fig. 14. and Fig. 15.). The evaluation showed that the content of dangerous elements in soil did not reach maximum permitted values for the Slovak Republic (Act No. 220/2004 Coll.) and the measured values corresponded with natural contents of the observed elements in soil and base rocks (Makovníková et al., 2006). At the same time, in ecological systems, no anthropogenic pollution by applying chemical substances and sediments in soil is present.

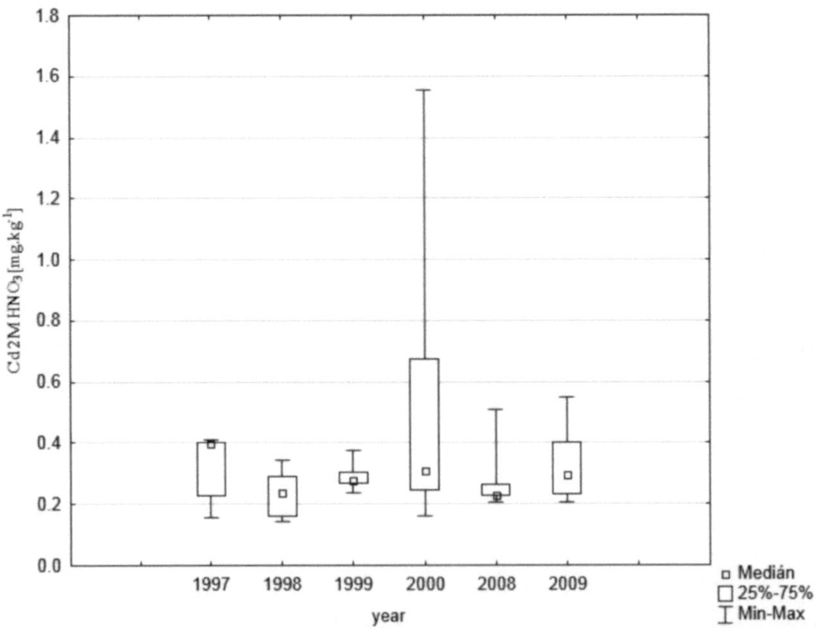

Figure 13. Cd content of soil in the monitored area situated in a marginal region of north-eastern Slovakia expressed by descriptive statistics

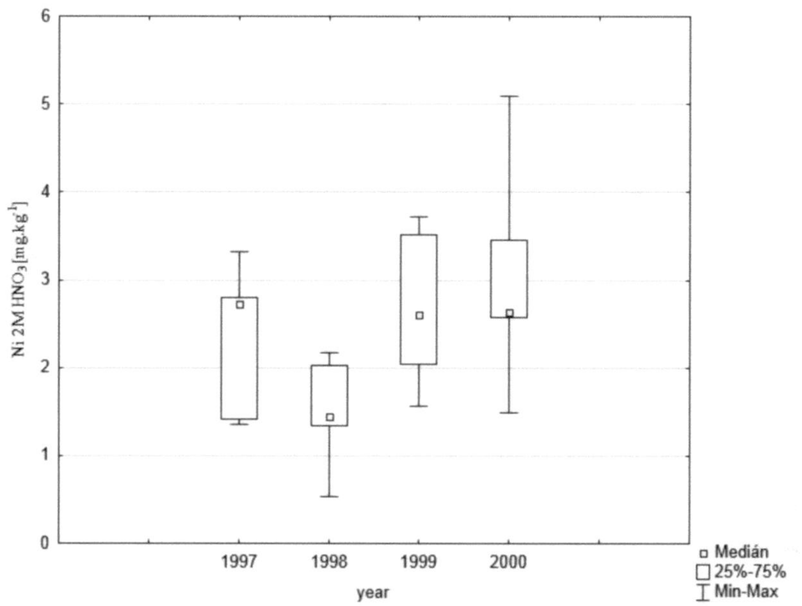

Figure 14. Ni content of soil in the monitored area situated in a marginal region of north-eastern Slovakia expressed by descriptive statistics

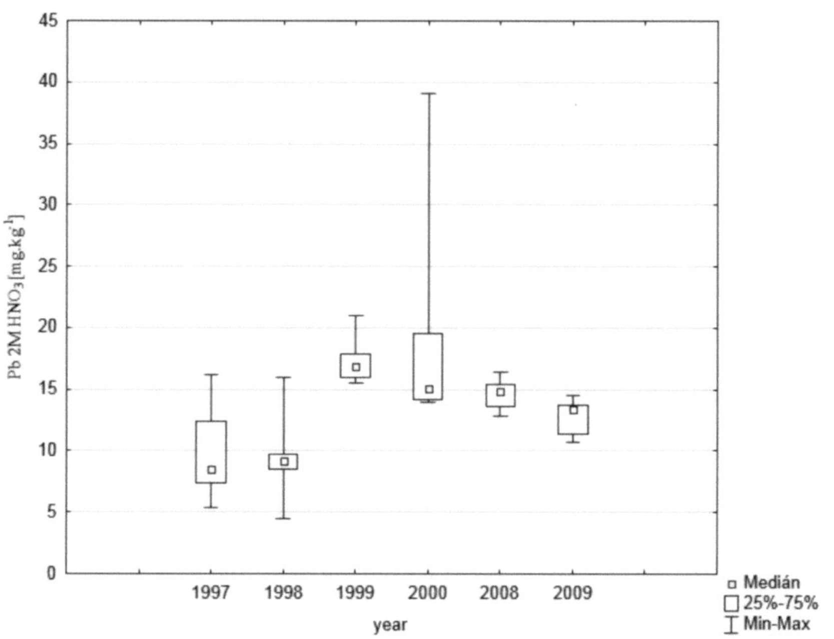

Figure 15. Pb content of soil in the monitored area situated in a marginal region of north-eastern Slovakia expressed by descriptive statistics

A statistically significant effect of the monitored year and locality on observed heavy metal content of the soil was confirmed by an analysis of variance (Tab. 6.).

Parameter	Min.	Max.	Mean	Standard error	Source of variability	d. f.	F-Ratio	P
Pb 2M HNO_3 [$mg.kg^{-1}$]	7.77	22.18	14.11	0.823555	year locality	6 4	12.30 22.01	++ ++
Cd 2M HNO_3 [$mg.kg^{-1}$]	0.129	0.697	0.343	0.037233	year locality	6 4	9.37 19.69	++ ++
Ni 2M HNO_3 [$mg.kg^{-1}$]	0.934	3.436	2.392	0.140375	year locality	6 4	24.08 25.20	++ ++

Table 6. Analysis of variance of the heavy metal content of the soil in the monitored area situated in a marginal region of north-eastern Slovakia
++P< 0,01 +P< 0,05

3. Conclusion

The farming system is the most widespread environmental technology with its positive and negative consequences. It utilises essential natural resources and, at the same time, influences other natural environments. Therefore, ecologisation of farming is a priority of farmers as well as environmentalists. Respecting the principles of soil sustainability and other components of the environment is a basic precondition for life sustainability. Sustainable agriculture is based on the principle of agriculture being a biological process which, in practice, should imitate key characteristics of the natural ecosystem. It strives to bring diversity into agro-ecosystems, recycle nutrients efficiently and maintain the priority of sunlight as a source of energy for agro-ecosystems. Sustainable use of soil takes soil-ecological conditions into consideration and is realised in such a way and in such intensity, which gives rise to neither negative changes in soil, nor establishes trends for the development of negative characteristics in soil. It is impossible to select universal soil parameters for sustainable soil, which is why the area is subject to specialised discussions. A significant role in the selection of parameters is played by their variability in time, related to parameter stability. Stable (such as soil depth or granularity), relatively stable (the salt content, the content of organic mass in soil, heavy metal contamination), relatively dynamic (pH, the content of nutrients) and dynamic (soil humidity and temperature, microbial activity, etc.) parameters are more connected to its short-term changes. Soil quality cannot be judged directly; it must be determined from the changes of its parameters. It is more accurate to evaluate the range of appropriate indicators rather than to use a single one. Soil quality is significantly affected by physical, chemical, biological and biochemical properties sensitive to changes in the environment and land management. At present, there is little knowledge with regard to soil development in the conditions of sustainable farming systems. The present findings can hardly be compared to other research due to the different soil-ecological conditions in which they were obtained. The issue of universal methods for all soil types remains a universal problem within the research of soil development. Unless this area is unified, objective comparison will remain on a regional level. Soil parameters are usually determined only in relation to specific topsoil. Certain physical and chemical parameters in subsoil cannot be neglected, since they guarantee soil functions.

The present results showed development of selected soil parameters during long-term monitoring on a model area situated in a marginal region of north-eastern Slovakia where an ecological farming system was applied. Soil physical properties change not only under the influence of weather conditions, crop year, vegetation pass, but also under the influence of applied management systems. During the year and growing season, bulk density value also varies depending on water availability in the soil, weather and farming methods. The research showed that soil physical properties get adjusted after long-term application of an ecological farming system and the measured values were stabilised, reaching levels comparable with the average values for the soil type. Agrochemical soil characteristics did not change significantly during the research period. High doses of organic fertilisers had a positive effect on soil productivity, and, thus, indirectly on maintaining soil pH, the available nutrient content and retention of humus in soil. In spite of this, it is necessary to continuously pay attention to soil reaction, because soil is naturally acidified through acid atmospheric fallout as well as calcium intake by plants. Values of selected heavy metals in the monitored period did not exceed the limit values published in Act No. 220/2004 Coll. The values of activity of phosphatase and urease changed minimally during the research period and they refer to values typical for soils with sparse vegetation. At the same time, it was proven that increasing the content of soil organic matter promotes natural protection of soil enzymes. Analysis of variance confirmed a statistically significant effect of the monitored year on all observed soil parameters. The effect of the observed locality, with the exception of pH/$CaCl_2$, C_{ox} and N_{anorg}, on other soil parameters was also statistically significant.

Author details

Danica Fazekašová
University of Prešov, Slovakia

Acknowledgement

The study was supported by VEGA 1/0601/08 *Effect of biotic and abiotic factors on ecosystem sustainability*, VEGA 1/0627/12 *Diversity, resiliency and health of ecosystems in different farming system and polluted territories in anthropogenic land* and KEGA 012PU-4/2012 *Preparation and realization of the research focused on creating teaching aids for education of environmental subject*.

4. References

Act. No. 220/2004 on Soil Protection and Agricultural Soil Using

Agenda 21 and Indicators of Sustainable Development, MŽ SR, Bratislava, Slovak Republic, ISBN 8088833035

Alef, K. & Nannipieri, P. (1995). Soil Sampling, Handling, Storage and Analysis, In: *Methods in Applied Soil Microbiology and Biochemistry*, K. Alef, P. Nannipieri, (Eds.), 49–115, Academic Press, ISBN 978-0-12-513840-6 London, Great Britain

Baligar, V. C. ; Wright, R. J. & Hern, J. L. (2005). Enzyme Activities in Soil Influenced by Levels of Applied Sulfur and Phosphorus, *Communications in Soil Science and Plant Analysis*, Vol.36, No.13-14, pp 1725-1735, ISSN 00103624

Barančíková, G. (1998). The Proposal of Special Classification of Agricultural Soils of Slovakia from Viewpoint of their Sensibility to Contamination by Heavy Metals, *Plant Production*, Vol.44, No.3, pp. 117 – 122, ISSN 0370-663X

Belińska, E. J. & Prangal, J. (2007). Enzymatic Activity of Soil Contaminated with Triazine Herbicides, *Polish Journal of Environmental Studies*, Vol.16, No.2, pp 295-300, ISSN 1230-1485

Beneš, Š. & Benešová, J. (1993). Balance of Risk Elements in the spheres of the Environment, *Plant Production*, Vol.39, No.10, pp. 941-958, ISSN 0370-663X

Beneš, Š. (1993). *The Contents and Balance of Elements in the Spheres of the Environment*, 1. part, MZ, Praha, Czech Republic

Bielek, P. (1998). *Nitrogen in Agricultural Soils of Slovakia*, Soil Science and Conservation Research Institute, ISBN 80-85361-44-2, Bratislava, Slovak Republic

Bremner, J. M & Mulvaney, R. L. (1978). Urease Activity in Soils, In: *Soil Enzymes*, R.G.Burns, (Ed.), 149 – 196, Academic Press, New York, USA

Bujnovský, R. & Juráni, B. (1995). *The Subsoil*, Soil Science and Conservation Research Institute, ISBN 80-85361-49-3, Bratislava, Slovak Republic

Burns, R. G. (1978). *Soil Enzymes*, Academic Press, ISBN 0-12-145850-4, New York, USA

Caldwell, B. A. (2005). Enzyme Activities as a Component of Soil Biodiversity, *Pedobiologia*, Vol. 49, pp. 637 – 644, ISSN 0031-4056

Chander, K.; Goyal, S.; Mundra, M. C. & Kapoor, K. K. (1997). Organic Matter, Microbial Biomass and Enzyme Activity of Soils under Different Crop Rotations in the Tropics, *Biol Fertil Soils*, Vol.24, No.3, pp 306-310, ISSN 01782762

Chazijev, F. CH. (1976). *Soil Enzyme Activity*, Nauka, Moskva, Russia

Fazekašová, D. (2003). *Sustainable Use of Soil – Definition and Evaluation of Indicators and Parameters of Soil Development*, FHPV PU, ISBN 80-8068-228-3, Prešov, Slovak Republic

Fazekašová, D.; Kotorová, D.; Balázs, P.; Baranová, B. & Bobuľská, L. (2011). Spatial Variability of Physical Soil Properties in Conditons of Ecological Farming in Protected Area, *Ekológia (Bratislava)*, International Journal for Ecological problems of the Biosphere,Vol.30, No.1, p. 1-11 ISSN 1335-342X

Fiala, K.; Barančíková, G.; Brečková, V.; Búrik, V.; Houšková, B.; Chomaničová, A.; Kobra, J.; Litavec, T.; Makovníková, L.;Pechová, B. & Váradiová, D. (1999). *Partial Monitoring System – Soil*, Binding methods, 1st ed.; Soil Science and Conservation Research Institute, ISBN 80-85361-55-8, Bratislava, Slovak Republic

Fulajtár, E. (2006). *Physical Parameters of Soil*, Soil Science and Conservation Research Institute, ISBN Bratislava, Slovak Republic

Geisseler, D. & Horwath, W. R. (2009). Short-term Dynamics of Soil Carbon, Microbial Biomass, and Soil Enzyme Activities as Compared to Longer-term Effects of Tillage in Irrigated Row Crops, *Biol Fertil Soils*, No.46, pp. 65-72, ISSN 1432-0789

Grejtovský, A. (1991). Influence of Soil Improvers on Enzymatic Activity of Heavy Alluvial Soil, *Rostl. Výr*, 37, pp 289–295, ISSN 1214-1178

Gulser, F. & Erdogan, E. (2008). The Effects of Heavy Metal Pollution on Enzyme Activities and Basal Soil Respiration of Roadside Soils, *Environ Monit Assess*, Vol.145, No1-3, pp. 127-133, ISSN 1573-2959

Hraško, J. & Bedrna, Z. (1988). *Applied Pedology*, Príroda, Bratislava, Slovak Republic

Iovieno, P. ; Morra, L. ; Leone, A. ; Pagano, L. & Alfani, A. (2009). Effect of Organic and Mineral Fertilizers on Soil Respiration and Enzyme Activities of two Mediterranean Horticultural Soils. *Biol Fertil Soils*, Vol.45, No.5, pp. 555-561, ISSN 01782762

Javoreková, S.; Králiková, A.; Labuda, R.; Labudová, S. & Maková, J. (2008). *Soil Biology in Agroecosystems*, Slovak University of Agriculture in Nitra, ISBN 978-80-552-0007-1, Nitra, Slovak Republic

Jordan, D. ; Kremer, R. J. ; Bergfield, W. A. ; Kim, K. Y. & Cacnio, V. N. (1995). Evaluation of Microbial Methods as Potential Indicators of Soil Quality in Historical Agricultural Fields. *Biol Fertil Soils*, Vol.19, No.4, pp 297-302, ISSN 01782762

Jurčová, O. & Torma, S. (1998). *Methodology for Quantifying Potential Nutrient Plant Residues*, Soil Science and Conservation Research Institute, ISBN 8085361892, Bratislava, Slovak Republic

Klír, J. (1997). *Sustainable Agriculture*, Department of Agricultural and Food Information, ISBN 80-86153-18-5, Praha, Czech Republic

Klose, S. & Tabatabai, M. A. (2000). Urease Activity of Microbial Biomass in Soils as Affected by Cropping Systems. *Biol Fertil Soils*, Vol.31, No.3-4, pp 191-199, ISSN 01782762

Kotorová, D. ; Šoltýsová, B. & Mati, R. (2010). *Properties of Fluvisols on the East Slovak Lowland at their Different Tillage*, 1. ed. CVRV, ISBN 978-80-89417-25-4, Piešťany, Slovak Republic

Kováčik, P. (2001). *Methodology of Nutrient Balance in Ecological Farming System*, Slovak University of Agriculture in Nitra, ISBN 80-7137-957-3, Nitra, Slovak Republic

Krnáčová, Z.; Račko, J. & Bedrna, Z. (1997). Indicators of Sustainable Development Use of Agricultural Soils in Slovakia. *Acta Environmentalica Universitatis Comenianae*, Supplement, pp. 165-172, ISSN 1335-0285, Bratislava, Slovak Republic

Kromka, M. & Bedrna, Z. (2000). *Soil Hygiene*, Comenius University in Bratislava, ISBN 80-223-1602-4, Bratislava, Slovak Republic

Lacko-Bartošová, M. ; Cagáň, Ľ.; Čuboň, J. ; Kováč, K. ; Kováčik, P. ; Macák, M. ; Moudrý, M. & Sabo P. (2005). *Sustainable and Ecological Agriculture*, 1st ed. Slovak University of Agriculture in Nitra, ISBN 80-8069-556-3, Nitra, Slovak Republic

Larson, W.E. & Pierce, F.J. (1991). Conservation and Enhancement of Soil Quality. *Evaluation for Sustainable Land Management in the Developing World*, Vol.2, No.12, pp. 175-203, Technical Papers. IBSRAM Proc.

Líška, E.; Bajla, J.; Candráková, E.; Frančák, J.; Hrubý, D.; Illeš, L.; Korenko, M.; Nozdrovický, L.; Pospišil, R.; Špánik, F. & Žember, J.; (2008). *General Crop Production*, Slovak University of Agriculture in Nitra, ISBN 978-80-552-0016-3, Nitra, Slovak Republic

Makoi, J. H. J. R. & Ndakidemi, P. A. (2008). Selected Soil Enzymes: Examples of their Potential Roles in the Ecosystem, *African Journal of Biotechnology*, Vol. 7, No.3, pp. 181-191, ISSN 1684–5315

Makovníková, J.; Barančíková, G.; Dlapa, P. & Dercová, K. (2006). Inorganic Contaminants in Soil Ecosystems, *Chem. Listy,* Vol.100, No.6, pp. 424 – 432, ISSN 0009-2770

Matúšková, L. & Vojtáš, J. (2005). *Guideline for Detecting Soil Contamination,* Soil Science and Conservation Research Institute, Bratislava, Slovak Republic

Michaeli, E. & Ivanová, M. (2005). Regional Geoecological Structure of Landscape and Primary Development Potential of the Prešov Self-governing Region. *Folia Geographica.* Vol. 18, No.8, pp. 116-142, ISSN 1336-6157

Miralles, I.; Ortega, M.; Sánches-Marañón, M.; Leirós, M. C.; Trasar-Cepeda, C. & Gil-Sotres, F. (2007). Biochemical Properties of Range and Forest Soils in Mediterranean Mountain Environments, *Biol Fertil Soils,* No.43, pp. 721-729, ISSN 1432-0789

Nourbakhsh, F. & Monreal, C. M. (2004). Effects of Soil Properties and Trace Metals on Urease Activities of Calcareous Soils. *Biol Fertil Soils,* Vol.40, No.5, pp. 359-362, ISSN 01782762

Pejve, J. (1966). *Soil Biochemistry,* SVLP, Bratislava, Slovak Republic

Rode, A. A. (1969). *Basic Science of Soil Moisture,* Tom II, Gidrometeorologič. izd., Leningrad, Russia

Šarapatka, B. (2002). Possibilities of Using Enzyme Activities as Indicators of Productivity and Systems Quality, *Biological indicators of soil quality,* pp. 26-31, ISBN 80-7157-64-25, MZLU, Brno, Czech Republic

Sinsabaugh, R. L.; Findlay, S.; Franchini, P. & Fischer, D. (1997). Enzymatic Analysis of Riverine Bacterioplankton Production, *Limnol. Ocecmogr.,* Vol. 48, No.1, (1997), pp. 29 – 38, ISSN 0024-3590

Speir, T. W. ; Van Schaik, A. P. & Lloyd-Jones, A. R. (2003). Temporal Response of Soil Biochemical Properties in a Pastoral Soil after Cultivation Following High Application Rates of Undigested Sewage Sludge. *Biol Fertil Soils,* Vol.38, No.6, pp. 377-385, ISSN 01782762

Tabatabaj, M. A. & Acosta-Martínez, V. (2000). Enzyme Activities in Limed Agricultural Soil. *Biol Fertil Soils,* Vol.31, No.1, pp. 85-91, ISSN 01782762

Tischer, S.; Tanneberg, H. & Guggenberger, G. (2008). Microbial Parameters of Soils Contaminated with Heavy Assessment for Ecotoxicological Monitoring, *Polish Journal of Ecology,* Vol.56, No.3, (2008), pp. 471 – 479, ISSN 15052249

Torma, S. (1999). *Potassium - an Important Nutrient in the Soil and Plant, Pedo disertationes,* Soil Science and Conservation Research Institute, ISBN 80-85361-51-5, Bratislava, Slovak Republic

Trasar-Cepeda, C. ; Leirós, C. ; Gil-Sotres, & F. Seoane S. (1998). Towards a Biochemical Quality Index for Soils: An Expression Relating Several Biological and Biochemical Properties. *Biol Fertil Soils,* Vol.26, No.2, pp. 100-106, ISSN 01782762

Vilček, J. ; Hronec, O. & Bedrna, Z. (2005). *Environmental Pedology,* Soil Science and Conservation Research Institute, ISBN 80-8069-501-6, Bratislava, Slovak Republic

Wick, B.; Kűhne, R. F. & Vielhauer, K. (2002). Temporal Variability of Selected Soil Microbiological and Biochemical Indicators under Different Soil Quality Conditions in South-western Nigeria, *Biol Fertil Soils,* No.35, pp. 115-121, ISSN 1432-0789

Chapter 20

The Compost of Olive Mill Pomace: From a Waste to a Resource – Environmental Benefits of Its Application in Olive Oil Groves

Beatriz Gómez-Muñoz, David J. Hatch, Roland Bol and Roberto García-Ruiz

Additional information is available at the end of the chapter

http://dx.doi.org/10.5772/48244

1. Introduction

Olive oil farming is a significant feature of land use in Mediterranean regions, covering over five million hectares in the EU Member States. The main areas of olive oil production are in Spain (2.4 million ha), followed by Italy (1.4 million ha), Greece (1 million ha) and Portugal (0.5 million ha) [1]. Whilst olive plantations are found over most of the Mediterranean region, around 65% of the Spanish olive oil area is located in Andalusia (southern Spain), representing 30% of the total EU olive oil production and about 35% of the utilized agricultural area. Therefore, olive oil farming in Andalusia is of great social, economic and environmental significance and any change in the olive oil industry in terms of management practices and post-processing options could be of wide importance, at least at the regional scale.

The olive oil industry generates large quantities of by-products. Almost all of the olive mills in Spain use the two–phase centrifugation system for oil extraction to reduce wastewater generation and lower the contaminant load, compared with the three–phase centrifugation system [2] which is currently used by other Mediterranean countries. The main by–product of the two–phase extraction system is olive mill pomace (OMP, hereafter), which in Mediterranean areas is produced during a short period over the winter, from November to February [3], the amount generated varying between 7 and 30 million m^3 per year [4-6]. Typically, OMP is a semi–solid to semi–liquid by-product resulting from the mix of "alpechin", the main by–products resulting from the older three–phase extraction procedure, and "orujo". This by-product is made mainly with water, seed and pulp and is a potentially harmful by–product for the environment, because of the phytotoxic and antimicrobial properties, low pH, relatively high salinity and organic load, and the phenolic and lipid constituents [7-11]. Direct application to rivers or soil is not allowed under most of the national regulations of the producer countries. The main physico–chemical characterization of OMP can be found in other reports [10,12,13]. According to these studies, OMP is acidic, with a very high content of organic matter and carbon, rich in potassium (K), poor in phosphorus (P), with intermediate levels of nitrogen (N) and may also contain phenolic and lipid compounds.

Some economic (due to costs associated with disposal) and environmental problems arise from the disposal of OMP. The various options for the fate of the large amounts of OMP which are produced annually in Andalusia can be very diverse: a general description of these can be found in [14]. Briefly, one alternative for the disposal of this large amount of OMP could be in evaporative ponds, but large areas would be required for this option which might also pose several potential environmental problems such as bad odour, leaching and insect proliferation. Another major preferred option would include the generation of renewable energy taking advantage of the relatively high calorific value of OMP. Other important uses include the transformation of OMP into an organic fertiliser and soil conditioner through composting. The process of composting OMP consists of mixing it with a blend of natural organic residues (e.g. olive leaves and twigs collected after cleaning the olive fruit in the mill, and/or straw, or manures), which is then allowed to decompose in aerated piles for 7 to 9 months. This means of re-utilization can help to improve soil fertility in olive oil farms which are characterized by low organic matter, reduce the cost of inorganic fertilisers or, for a commercial enterprise, can provide an additional source of revenue for the olive oil mill economy. The main reasons for composting are that OMP has a semi–solid consistency which makes it difficult to manage, and to eliminate any phytotoxical effects by composting for at least 18 weeks [15]. Composting OMP enables it to be sanitized; the mass and volume of the product are reduced and stabilized prior to land spreading. As already mentioned, before composting, OMP is mixed with bulking agents such as olive tree leaves, which are gathered along with the olive fruit, twigs and small branches, straw [16], cotton waste [17] and manure to increase the nutrient content, or any other materials of animal or plant origin which are available locally.

In Andalusia, composted olive mill pomace (COMP, hereafter) production has increased exponentially during the last seven years from 1000 tonnes in 2003 to 70000 in 2011 [18] and there are about 14 olive mills which are producing COMP in this region. Despite this rapid increase, there are no published studies on the main agrochemical properties and the effects of OMP application to soil. Indeed, there are few studies on the main physico-chemical changes in OMP during composting [19-21], or on the chemical characterisation of the final OMP composted product [22]. Moreover, these studies have been undertaken using only a limited number of OMP composts, which were produced in only low experimental quantities and at a small experimental scale. The main aims of this chapter are to review the information from other studies and our own on the agrochemical characterization of COMP currently produced in Andalusia and on the short- and long-term effects of its application on the soil physico-chemical and biological properties.

2. Agrochemical characterization of composted olive mill pomace

The relatively high diversity of the bulking agents used (such as olive mill pomace, olive leafy material (OLM), manure or straw) as well as the variable proportion in which they are mixed is responsible for the highly heterogeneous nature and variability of the quality of the COMP.

2.1. Composition of commercially produced composted olive mill pomace

As far as we know, there are no studies on the characterization of currently commercially produced COMP. In one of our own study, seven ready to apply OMP composts were collected from different olive mills located in several provinces of Andalusia. COMP samples differed in composition and in the proportions of primary materials such as OMP, OLM, manure and straw, as shown in Table 1 and Picture 1.

Composted olive mill pomace	OMP	OLM	Manure	Straw
COMP1	80%	7%	Sheep M 13%	
COMP2	75%	12%	Poultry M 13%	
COMP3	60%		Sheep M 40%	
COMP4	70%	8%	Sheep M 12%	10%
COMP5	85%	5%		10%
COMP6	80%	20%		
COMP7	80%	13%	Poultry M 7%	

Table 1. Composition of different composted olive mill pomaces described in this chapter. OMP and OLM refer to olive mill pomace and olive leafy materials, respectively.

Picture 1. Visual appearance of the composted olive mill pomace used in this study.

COMP pH ranged 7.45 to 8.34 and were adequate for most agricultural purposes. Those COMPs made of manure tended to have a higher pH. Mean COMP pH was 8.03, a value similar to those reported by [20,23] for other experimentally–produced composts made of OMP, and is within the pH range considered as optimal for the activity of microorganisms and plant growth [24,25]. In all cases, electric conductivity was lower than the 10 dS m^{-1}, threshold established as indicator of possible phytotoxic/phyto–inhibitory effects on plants or in soil [26]. COMP has a high content of organic matter (60.5%, on average) and carbon (30.7%, on average) (Table 2). These values are higher than those reported for cow, sheep and poultry manures and similar to those found for horse, pig and rabbit [27]. The highest values were found in those composts which included OLM and straw with little or no manure (COMP4 through to COMP7) and may be due to incomplete organic matter degradation of the larger particle sizes and the higher lignin content of OLM. A low organic matter degradation rate during composting has been reported [20], mainly because the high lignin content of the OMP and the high moisture of the initial mixture which limits sufficient aeration. For these COMPs, application to soil could be a good strategy to increase the organic matter content of soils of olive oil farms which in the Mediterranean basin are usually depleted in organic matter and are exposed to progressive degradation processes. Furthermore, the content of labile carbon in the COMPs are relatively low, indicating that respired carbon derived from compost after application to soil would be expected to be low. Therefore, COMP application to soil could increase C storage due to the high total carbon content and expected low rate of C mineralisation (see below).

Total nitrogen (TN) averaged 1.5%, and was higher for those COMP which included relatively high amounts of manure during composting (Table 2). This values is within the range of that found for compost made from plant residues [28,29] and similar to that reported for compost produced experimentally with OMP plus rabbit or sheep manure, and rice straw or almond shells [22].

The C:N ratio ranged from 27.2 to 35.8 for COMP4, COMP5 and COMP6, which did not include, or contained only limited amounts of manure, while the ratio was much lower (at 10.5 – 18.7) for COMP1, COMP2, COMP3 and COMP7, which included sheep or poultry manures (Table 2).

The total K (TK) in the composts averaged 1.7% and was highest for COMP2, which included a high proportion of poultry manure and lowest for COMP4 and COMP5, both of which included straw (Table 2). These values were within the range of values reported by other authors [20,23] for experimentally–produced compost and also similar to those of OMP [13]. This indicates that K was not lost through leaching during the composting process. The total P of the composts averaged 0.41% and was highest for those amended with manure (COMP1, COMP2 and COMP3) and lowest for those which included only OLM or straw (Table 2). TP values were lower than those of municipal solid wastes and sewage sludge, although similar to other vegetable wastes and manures [30].

Those COMPs with a high percentage of manure had lower lignin contents, but lignin content tended to increase in those COMPs with a higher proportion of OLM and straw. Polyphenol contents of the COMPs were less than 2%, even though OMP is usually characterized by high levels of polyphenols (Table 2). Polyphenols from olive oil mill waste water have been found to be toxic for some soil microorganisms [31], and this is one of the main reasons why direct application to soil cannot be recommended. The low polyphenol contents in the composts agree with values found by other researchers [32], who showed that the polyphenol content decreases during composting. In all cases, the polyphenol contents were lower than 4%: the limit which has been established where there is a shift between net N mineralization and immobilization during decomposition [33]. Those composts which are currently produced in Andalusia and which included manure showed typical lignin and polyphenol contents lower that 20% and 2%, respectively and therefore are suitable as organic fertilisers. Lignin and polyphenol contents of residues have been shown to be robust indices for the prediction of N mineralisation from residue–N after incorporation in soil [34,35], with typical thresholds for immediate net N mineralisation being < 15% lignin and < 3 – 4% total extractable polyphenol contents [36-38].

COMPs phytotoxicities (Zuconni test) were typically higher than 50%, except for COMP3 and COMP5, suggesting that a relatively high percentage of the currently produced COMPs are mature enough to be applied in the field.

	COMP commercially produced (min –mean- max)	**Others authors[1]**
Organic matter (LOI) (g kg^{-1})	272 - 605 – 879	465 - 621
Total C (g kg^{-1})	184 - 307 - 390	301 - 491
Total N (g kg^{-1})	10.7 - 15.0 - 20.0	14.0 - 27.8
C:N	10.5 - 21.9 - 35.8	14.0 - 22.7
Total P (%)	0.19 – 0.41 -1.19	0.5 - 1.5
Total K (%)	1.06 – 1.73 -2.39	20.6 - 39.5
Lignin (g kg^{-1})	76.0 - 218 - 313	410 - 426
Polyphenols (%)	0.94 – 1.33 - 2.1	-
Labile organic C (g kg^{-1})	4.6 - 16.5 - 25.0	7.3 - 10.2

Table 2. Main physico-chemical properties of seven commercially produced composted olive mill pomaces (Table 1) and results from bibliographic review of different authors for composted olive mill pomace. Values are mean of four replicates.

[1]Other authors [19,20,21,22,23] for experimentally produced COMPs.

2.2. Nutrient distribution in different particle size fractions of composted olive mill pomace

The separation and application of different COMP particle sizes could provide for better optimization of COMP management, because a fairly clear relationship between particle–size distribution of an exogenous source of organic matter and the C and N dynamics in soils for sludge compost [39] (among others) has been demonstrated. Similarly, C mineralisation and turnover was seen to differ according to the particle–size fraction in a cattle slurry compost [40,41], also from a sludge–straw mixture [42] and in an aerobically digested sewage sludge composted along with screened green waste compost, stored yard trimmings and crushed wood pallets [39]. Moreover, the nitrogen availability in compost has also been shown to be related to particle size, increasing as particle size decreased in sludge compost [43]. Generally, N mineralisation is greater in the fine and water–soluble fractions than in coarser fractions, at least for manure and city refuse composts [44] and composted sewage sludge [39]. These results suggest that the size of fractions in compost contribute substantially to the total C and N dynamics of mineralisation after compost is incorporated into soil. This is of particular interest, because depending on the range in particle sizes, composts could therefore provide a means for storage of C in soil (i.e. from the larger sized compost fractions), or a source of available N (i.e. from the finer compost fractions). The diversity and variable amount of raw materials used to obtain COMP make the final products very heterogeneous and therefore, it is expected that the nutrients and the main physico-chemical properties also differ in the particle size fractions of COMP. However, no studies existed to corroborate this, therefore in a separate study we examined whether or not this expectation about COMP was correct.

To estimate the nutrient distribution in particle size fraction of COMP, seven commercially produced COMPs were studied (Table 1) by sieving successively through to obtain 3 different sized fractions: <0.212, 0.212 – 1.0 and 1.0 – 60 mm. About 52% of COMP particles were larger than 1mm for those COMPs with a high content of manure, whereas, in contrast for COMPs without manure and with straw the consisted for 80% of these larger particle sizes. . The percentage of fine particles (<0.212 mm) doubled for COMPs with manure. These COMPs showed similar percentages of particle size fractions reported for duck manure [45], suggesting that during composting, manures generate a relatively high content of finer fractions. In general, the content of different compounds rich in organic C, such as organic matter, total carbon and lignin increased with particle size, being significantly higher for fractions between 1 to 6 mm. In sewage sludge compost made with screened refuse, yard trimmings and pallets [39], and for two types of dairy slurries others found [41], as in this study, an increase in the total C content with larger particles, although these differences were less distinct.

In contrast, nutrients contents including total N, mineral N, total P or total K were significantly higher in the fine fractions (<0.212 mm). More than 40% of the total N, P and K found in the original COMPs were in the <1 mm particle fractions in COMPs which included sheep or poultry manure during composting, whereas for COMP5, 6 and 7, the contribution of <1 mm fractions was lower than 30%. This finding agrees with those for manure and city refuse composts [44], animal slurry [41,45-47] and sludge compost [39], from workers who observed that most of the total N and various forms of N available in the short-term were mostly in the finer and water–soluble fractions.

As a result of lower N, but higher C, the C:N ratio decreased as particle size increased (Figure 1). Similar relationships between C:N ratio and particle size fractions were obtained for cattle slurry [46,47] and for sludge compost [39]. Typically, the C:N ratio for particles finer than 1 mm was lower than 18.2 in all COMPs, except those which included straw (COMP4 and COMP5), suggesting higher N availability of the organic N from these particles. During decomposition, manures with C:N ratios below 15 are likely to result in positive (net) N mineralisation after application to soil [48].

Overall, the germination index was higher than 80% in the <1 mm fractions, whereas it was lower than 50% (e.g. with some degree of phytotoxicity) for fractions 1.0 – 6.0 mm, although this was not true for COMPs made without manure (COMP5 and 6 mixtures) (Figure 1). Thus, the fractionation of the compost produced in olive mills containing <1 mm phytotoxicity–free fractions is highly recommended for commercial purposes.

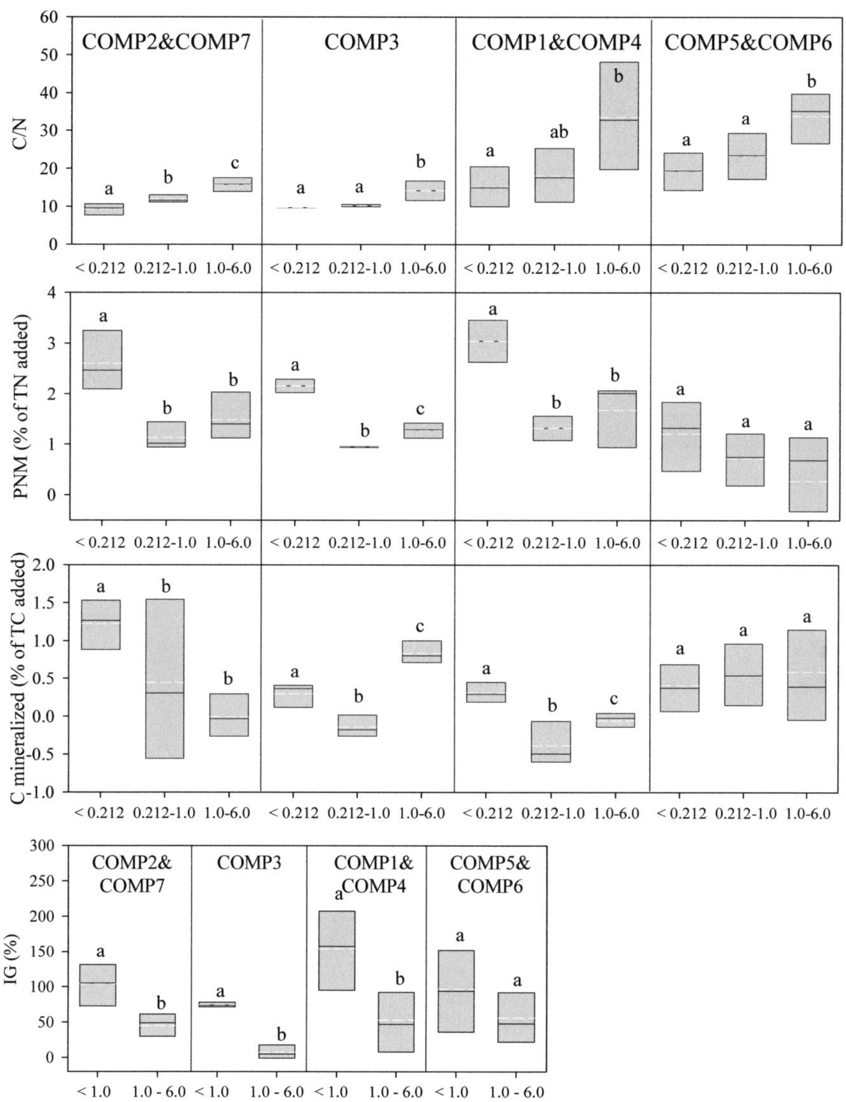

Figure 1. Box–plot of C:N, potential mineralisable N (PNM), mineralisable C and germination index of particle size fractions of combinations of COMPs according to the Principal Component Analysis. Different letters stand for significant differences (P <0.05) among particle size fractions.

The main differences among seven COMP tested according their quality was shown in the Principal Component Analysis (Figure 2). First principal component (PC1), which was negatively correlated with COMP quality indicators (e.g. total and available nutrients, C:N ratio,…), categorized the COMPs according to the pool of nutrients and organic matter and C, showing the high influence of raw materials in the quality of individual COMPs. Those COMPs made with poultry manure resulted in a high quality product, followed by COMPs made with sheep manure, whereas the quality of those COMPs made with a high proportion of OLM and straw but no manure was low. In addition, there was a trend for the quality to increase in relation to the content of finer fractions.

This highlighted the large influence of the raw materials in the final composition and quality of COMP product; a high proportion of manure increased the abundance of finer particles with high nutrients contents, whereas the addition of large amounts of OLM caused an increase in the content of larger particles with a high content of C compounds.

Overall, the separation and application of different COMP particle sizes could be useful for better optimization of COMP management, considering that the smallest particles (<1 mm) have a much higher germination index value, total N, P and K and a higher potential to mineralise N, providing a more homogeneous, high quality compost. On the other hand, larger particles (>1 mm) had higher contents of organic matter and C and tended to have the lowest C mineralisation and thus, on application to the soil in olive groves as a soil conditioner, could improve the structure and increase the organic matter of the poorer soils and increase the storage of soil organic C.

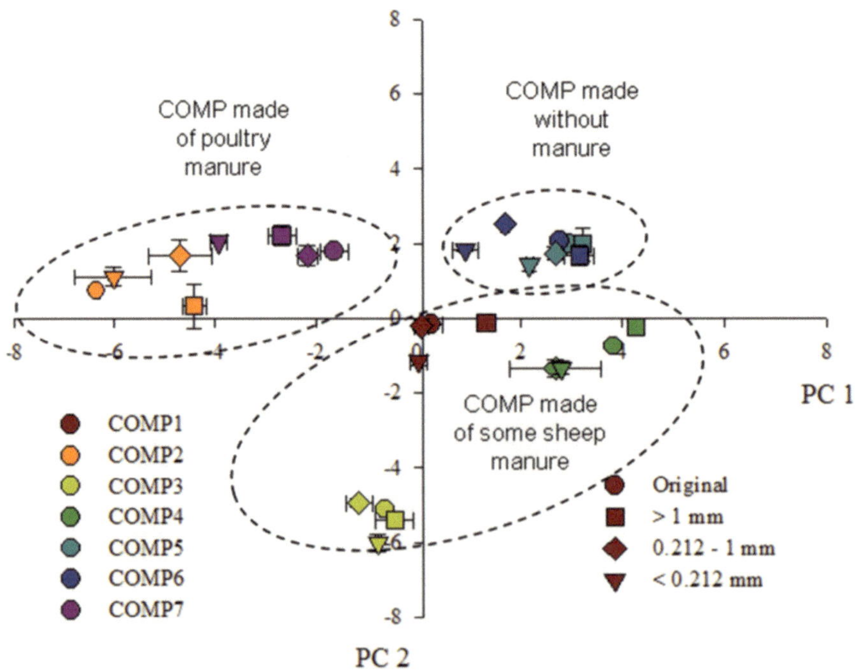

Figure 2. Scores of the whole and the particle size fractions of the COMPs in the space defined by PC1 and PC2.

3. Effects of composted olive mill pomace on nitrogen availability

Although it is expected that the application of COMP-organic N would also supply some available N for the plant, there is little information on the decomposition rate of COMP and the impact this may have on the N available for the growth and productivity of olive trees.

Previous work [49] showed a relatively low net N mineralisation in soil after application of compost obtained from olive mill wastewater taken from an evaporative pond and other agricultural by–products, suggesting that during decomposition of composted OMP, N immobilization could take place.

More detailed knowledge of the main soil N transformations and dynamics, following COMP application to soil, could be the key to regulating soil fertility through an improved capability to predict crop available N and organic matter sequestration and losses.

3.1. Nitrogen supply after the application of composted olive mill pomace

Available nitrogen supply throughout net N mineralisation (NM) and nitrification was studied for commercially produced COMPs differing in the raw materials co-composted with OMP (Table 1) in an aerobic incubation experiment under controlled conditions over two years. The highest rates of NM for the control soil and COMP samples were found during the first week of incubation. This positive net N mineralisation was probably from the N and C contained in the labile components of the COMP, which can be used directly by microbes as an energy source. After the initial stage of mineral N release, net N immobilization during COMP decomposition was found. Probably because the microorganisms began to decomposed the more recalcitrant compounds (with a higher C:N ratio) and this placed a demand on available soil N resulting in net N immobilization, which can ultimately be re–mineralised over time. Similar rates of mineralisation were found with mature composted materials [50,51]. This result was unexpected since, according to the TN contents, C:N ratios and lignin contents of the different COMPs, net N mineralisation would be expected according to the view that the addition of organic matter generally causes N mineralisation [52-54]. Raw materials co-composted with OMP had effects on net N mineralization. During the first year of incubation, N immobilization was found in all COMPs, except for COMPs made chiefly from OMP and OLM. After two years, decomposition of COMP resulted in a positive net N mineralization (i.e. net mineral N supply) for all COMPs (Figure 3), with values as high as 95% of that added as COMP-N. These results demonstrate that the organic N contained in the COMPs is not readily available over short to medium timescales, and suggests that most of the organic N may be in relatively recalcitrant forms. Indeed, during the composting process, N is immobilised in condensed aromatic compounds which might contribute to the reduced availability of N [55]. Therefore, the application of COMP may result in short–term declines in N availability in agricultural soils: it is therefore recommended that during the first year, the application of COMP should be combined with some other N-rich source of fertilizer. On the other hand, autumn COMP application could be a useful approach to lowering soil inorganic N levels, thereby reducing potential N losses by leaching.

3.2. Application of composted olive mill pomace to soil reduce nitrogen losses by leaching

Plant nitrogen use efficiency is usually lower than 50%, though highly dependent on crop type and environmental conditions. There are different ways by which the N is lost and no longer available for crop uptake, such as denitrification, ammonia volatilization and leaching, which might also be associated with environmental problems. N leaching can contaminate groundwater or surface waters through runoff. Indeed, although N contamination of groundwater arises from several sources, such as industrial waste, municipal landfills, mining, or septic systems, agricultural practices remain a major source [56-58]. About 15 – 55% of N applied to crops can be lost by leaching every year [59,60]. This is particularly important in olive oil groves as many olive crops are on areas vulnerable to nitrate pollution.

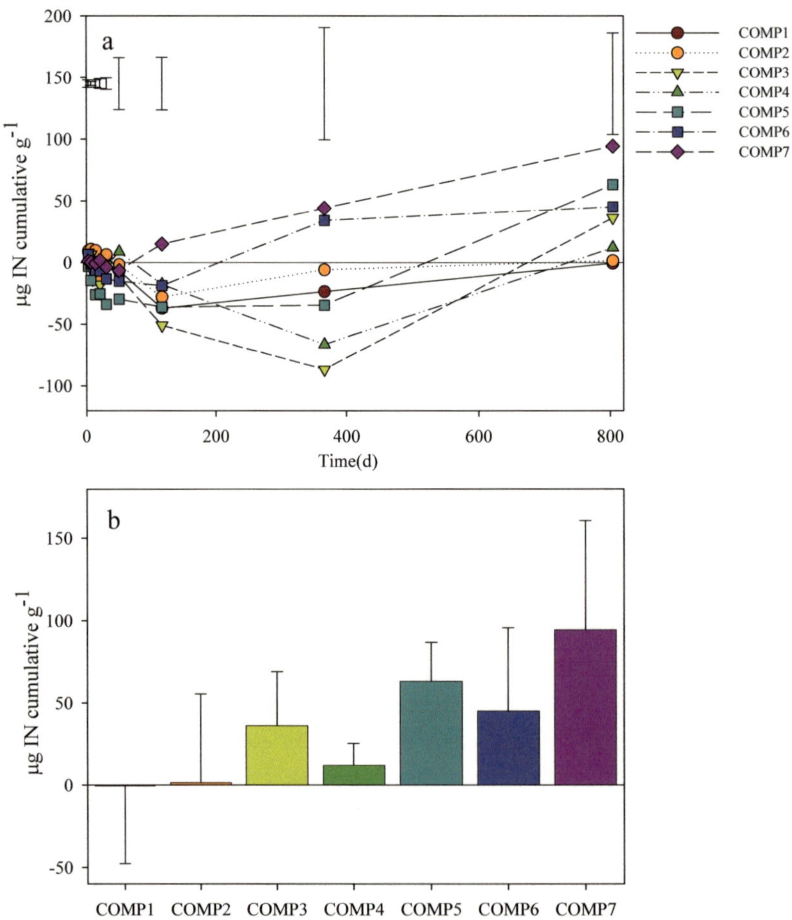

Figure 3. (a) Cumulative amounts of COMP–amended inorganic N (IN) for each sampling period and (b) at the end of incubation (365 d) for COMP amended soils. Bars for each sampling represent the mean of the standard deviations of the whole set of COMP–samples.

Taking into account that N contained in COMP is not readily available in the short-term, N lost by leaching is expected to be low after application of COMP to soils. Indeed, short-term (51 days) leaching losses of N after application of COMP were similar or even lower (between -5 to 10% of the added N) than the unamended control soils in an experiment carried out under laboratory conditions. These values were much lower than those found for commercial organic fertilizer (up to 30% of the added) and urea or sodium nitrate (up to 80% of the added) for the same amount of total N added (Figure 4).

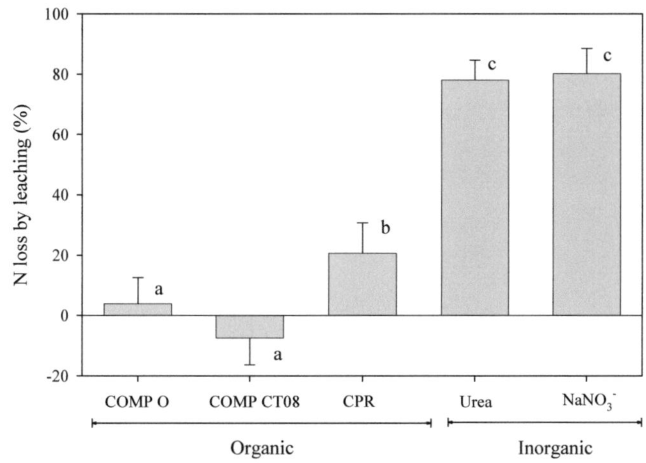

Figure 4. Percentage of N lost as nitrate by leaching of the total N added (equivalent to 100 μg N g^{-1}) after 51 d and three precipitation events (total precipitation equivalent to 80 l m^{-2}). Values are the mean of 4 replicates. Bar denotes standard deviation. Different letters denote significant differences between treatments ($P<0.05$).

Similar results were found in an experiment under outdoor conditions after one year. Nitrate leaching in soils amended with COMP was compared with soils amended with sheep manure (sheep M), commercial organic fertilizer (CPR), or inorganic fertiliser. Fertilisers were applied at two different rates (equivalent to 100 and 200 μg N g^{-1}) combined with two modes of application (on the soil surface or mixed with soil) and simulating autumn conditions. The lowest losses were for COMP amended soils (up to 7% of that applied) and the method of application had significant effects on mineral N leaching (Figure 5). In general, those soils which received COMP on the soil surface averaged negative mineral N losses (i.e. lower or similar losses to the control soil). N application rates had no effect on COMP IN leaching, regardless of the way the COMP was applied. Overall, inorganic N leaching for those treatments which received either Sheep M or CPR did not differ significantly, although leaching after surface application of CPR was higher than sheep M. Up to 37% of the fertiliser–derived N was leached for the 'double' surface application of CPR. No effects of rate or methods of application on IN leaching were found for CPR. However, for sheep M, leaching was higher at the double rate and lower for the 'surface', compared with the 'mixed in' application. The highest IN leaching reached 58% of the added N for inorganic fertiliser application (Figure 5). Results from the laboratory and outdoor experiment clearly showed the recalcitrant nature of the COMP-N (i.e. a high degree of N retention) despite the relatively low C:N ratio, confirming previous results on COMP-derived negative net N mineralisation and nitrification. Thus, application of COMP on olive oil farms distributed in the nitrate vulnerable areas might be a suitable strategy to diminish nitrate contaminated groundwater. This strategy agrees with [61] who reported that composted material can considerably lower the risk of groundwater contamination. Slow–release, controlled–release and stabilized fertilisers, such as composted OMP are recommended for several agro–environmental situations to improve nutrient use efficiency by reducing nutrient losses from the soil. Results from our studies clearly demonstrated the high efficiency of composted OMP to retain N and the beneficial reduction of nitrate leaching elicited by this compost.

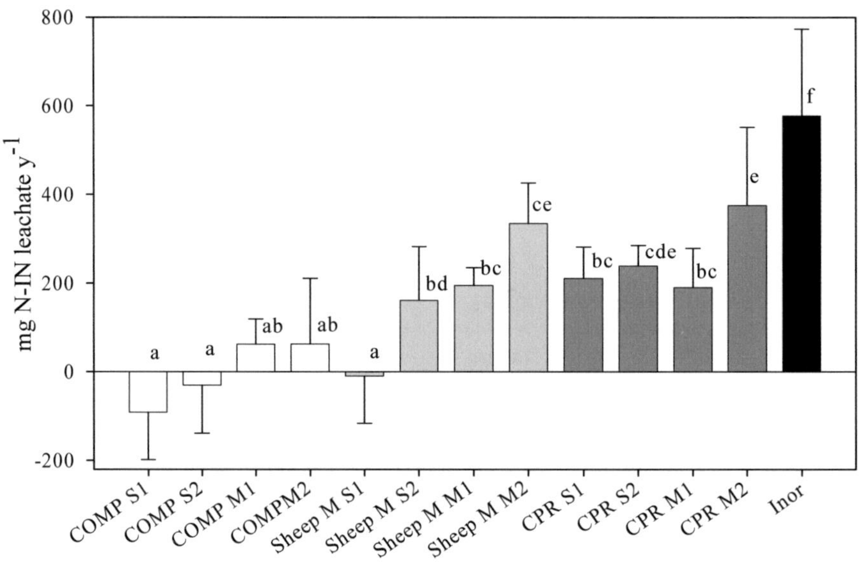

Figure 5. Cumulative fertiliser–derived mineral N (nitrate + ammonium) leaching after one year under natural rainfall and temperature in outdoor conditions for soils which received: i) composted olive mill pomace (COMP), ii) sheep manure (Sheep M), iii) commercial organic fertilisers (CPR), or iv) $NaNO_3$ (Inor) at 1 (250 μg N g^{-1}) or 2 (500 μg N g^{-1}) doses. M and S, stand for soil in which the fertilisers were mixed (M) with the soil or applied to the soil surface (S). Values are means of 4 replicates and bars denote standard deviations. Different letters denote significant differences ($P<0.05$).

3.3. N_2O emission derived of application of composted olive mill pomace

Agricultural soils are a significant source of atmospheric N_2O which is of concern because of the role this gas has in global warming [62]. Emissions of N_2O from soil have been shown to increase after the addition of plant residues [63] and organic fertilisers [64], the biochemical composition (or quality) being an important determinant of the magnitude of N_2O emissions [65].

N_2O emissions in soil amended with COMP (100 μg N g^{-1}) were determined under aerobic incubation over 51 days. Soil N_2O–N fluxes were constant and relatively low (2.71 ng N_2O–N g^{-1} d^{-1}), and did not vary significantly between COMP-amended and the control soils, except for the COMP which was made with moderate levels of sheep manure (COMP1) (Figure 6). The cumulative N-N_2O emissions after 51-days were only 0.23% of the added COMP-N. This result was not in agreement with the well documented increase in N_2O emissions after plant residue addition to soils as reported in other studies [63,65]. The very low increase in N_2O emissions after COMP application suggests that N losses via denitrification after the annual application of COMP to olive oil groves are expected to be very low, and thus application of COMP could provide organic N without increasing the emissions of N_2O

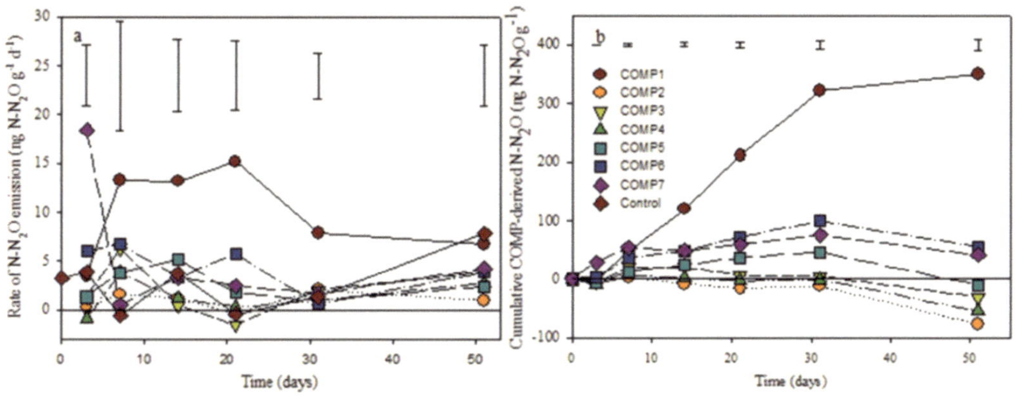

Figure 6. Time course of (a) N-N_2O emission rate and (b) cumulative N-N_2O emission in soil amended with COMP during 51 days of lab incubation under aerobic conditions. Bars denote the mean standard deviations for each treatment and sampling point.

4. COMP-carbon mineralization

As shown above, composted olive mill pomace (COMP) has a high content of total carbon, most of which of recalcitrant nature (e.g. expected low C decomposition) such as lignin or fibre. Therefore, COMP application in soil can be a good strategy to increase the organic matter content in soil, as well as the organic C. This practice could reduce CO_2 emission to the atmosphere increasing C storage in soil.

4.1. Carbon mineralization

Measurements of soil CO_2 emissions can provide useful insights into soil C cycling, and provide a basis for evaluating soil C dynamics and potential C sequestration in different agricultural systems [66], particularly in intensive production systems affected by the different cropping practices and residue management [67]. Soil C can only be assimilated, and recycled through the microbial biomass, or respired [68]. Soil respiration and mineralisation are generally thought to be related to the composition of the microbial biomass, which in soils, tends to vary both across substrate qualities [69] and in time [70]. Low C mineralisation measured as CO_2–C fluxes after mature COMP application have been shown [71]. However, the COMP used in this study was an experimentally produced COMP. The amount of COMP-C mineralized after 8 months of incubation after COMP application in eight soils differing in soil organic carbon (SOC) is shown in Figure 7. For all cases, the COMP-C mineralized (e.g. emitted as C-CO_2) was less than 10% of that added (4 ◉g COMP-C g^{-1}), and this was true, independently of soil texture or SOC contents. This result suggests that C compounds in COMP are complex and refractory, resulting in a high residence time in the soil. Moreover, the percentage of COMP-C mineralisation tended to be lower at higher rates of COMP application, which indicates a higher potential for COMP-C to accumulate in the soils at higher rates of application (Figure 8).

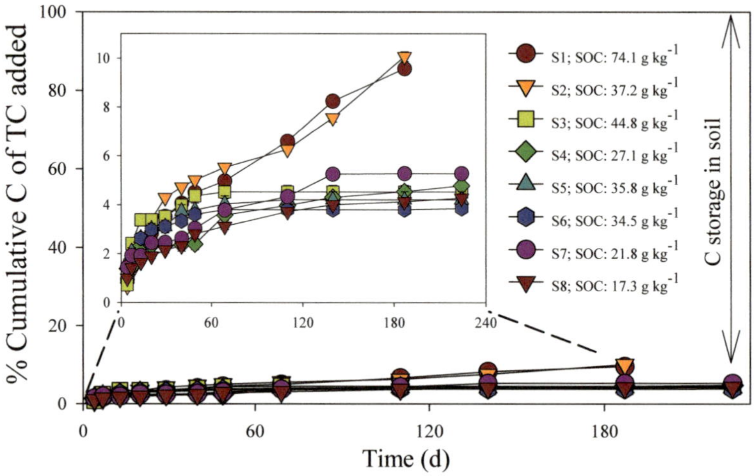

Figure 7. Percentage of C emitted as CO_2 during COMP decomposition for eight types of soil after 240 days of incubation. Bars denote standards deviations.

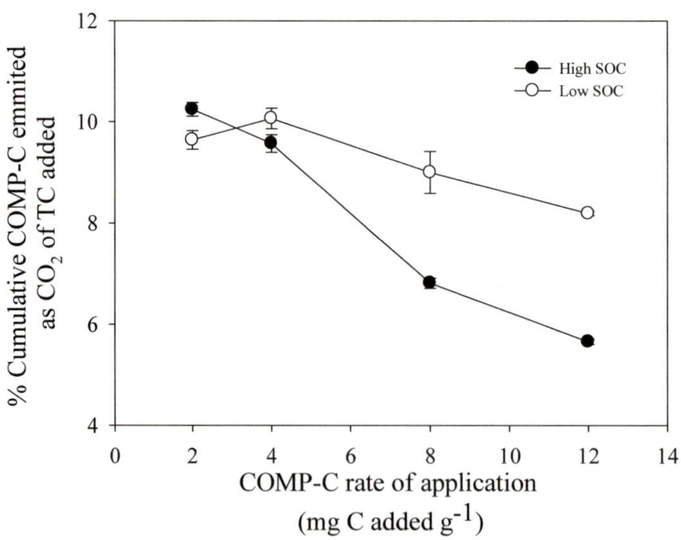

Figure 8. Effects of increasing the rate of COMP-C application to two soils differing in soil organic carbon (SOC) on cumulative COMP-C derived C-CO_2 emissions after 8 months of incubation. Bars denote standards deviations of three replicates.

5. Long-term effects of composted olive mill pomace application in olive groves

The majority of the olive grove soils of Andalusia are characterised by low levels of organic matter, and are exposed to progressive degradation processes. Thus, organic matter application is required to compensate for organic carbon deficiency and to improve soil fertility. As a consequence, the use of a transformed agro-industrial waste for agricultural use, such as composted olive mill pomace, might represent a realistic solution to overcome both the soil degradation of olive oil groves and a sustainable disposal of OMP, thus increasing the sustainability of olive groves. Regional authorities, have recognised the potential use of COMP in agriculture, and have promoted composting which has resulted in an exponential increase in the production of COMP during the last 5 years, with a production of about 70000 tonnes in the 2009-2010 harvest campaign [18]. However, the long-term effects of annual application of COMP on the soil fertility of olive oil groves should be evaluated to promote more olive mill pomace composting.

Soil samples were taken from four olive oil farms; Olvera (O), Reja (R), Tobazo (T) and Andújar (A), which annually received applications of COMP during 3, 4, 9 and 16 years, respectively (COMP olive oil farms, hereafter). Soil samples were also taken from comparable olive oil farms located in the vicinity (<20 m) of each of the COMP farms that never received composted olive mill pomace (NCOMP, hereafter). These NCOMP farms have similar environmental conditions, topography and soil texture to the COMP farms. Soil samples were characterized for physico-chemical and biological variables to determine the fertility and functionality of soil after COMP application.

In general, soil pH in the 4, 9 and 16 years COMP sites was 0.31, 0.23 and 1.47 units lower than the comparable NCOMP olive farming, respectively, whereas no differences were achieved in the site after 3 years of COMP application. Moreover, the soil water holding capacity (WHC) was significantly higher in the COMP treated soils, with increases of 2 – 4% with respect to NCOMP soils. The higher (WHC) in the soils under COMP farming was expected (Table 3), due to the relatively higher levels of organic matter in these soils: the effect of organic matter on the increased potential for soils to retain water was described previously for 77 soil profiles [72]. Moreover, soil water aggregate stability was also significantly higher after COMP application for all treated soils: up to 1.5 - 2.5 times greater for COMP compared with NCOMP treated soils (Table 3). The higher soil aggregate stability in the COMP soils agrees with the findings of [73], who showed that the application OMP to soils increased this soil property from 64 to 73% after 5 years. The cation exchange capacity of soils treated with COMP was also significantly higher than in the relative NCOMP soils, independently of the site (e.g. number of years since COMP was first applied) (Table 3).

Overall, COMP application significantly increased both organic matter and carbon contents, compared with the NCOMP soil, except at the site that received COMP for only three years. The soil organic matter content in soils treated for 4, 9 and 16 years with COMP (i.e. R, T and A sites), was from 2.1 to 8.5 times greater than the respective NCOMP treated soil (Table 3). Our data confirmed previous results which indicated that soil organic matter increased after amendment with OMP during the first weeks after application [74] or after several years [73,75]. Our results also showed that organic carbon was higher in the COMP treatment compared with the NCOMP treatment. The increase in organic carbon after COMP application indicates that the decomposition rates of COMP–C were relatively low, and lower than the rates of annual COMP–C application. These recalcitrant compounds might contribute to the reduced rate of C decomposition. Soil organic carbon was higher in the soils treated with COMP, being up to eight times higher in soil where COMP was applied during the last 16 years in comparison with NCOMP farming. When extrapolating to a hectare scale, these results show that soil organic C increased up to 30 tonnes per ha^{-1} y^{-1} in soil after COMP application. The low decomposition rate of COMP and high organic C content in COMP-amended soil indicates that application of COMP in olive oil groves is an appropriate strategy to sequester organic carbon into the soil and should be evaluated further.

Overall, the total N content was higher in the COMP than in the NCOMP soils although this depended on the site. At O and R sites, which received three and four years of COMP application respectively, differences with the comparable NCOMP soils were not significant. However, total N was 1.3 and 14.8 times higher in soil with 9 and 16 years of COMP application (sites T and A) than in the comparable NCOMP treated soils (Table 3). The higher TN in COMP soils was expected, since application of organic residues normally results in an increase in the soil content of N [49,73,76]. They showed a significant increase in the TN after two to three years of application of OMP or COMP, indicating that COMP–N is very resistant to mineralisation, and therefore is retained in the soil.

Generally, soil labile phosphorus (P) in the amended COMP soils was higher than the unamended soil. In the O, R and A sites, soil available P at the COMP soils was 13.9, 260.3 and 1607 % higher than the NCOMP soil, whereas no significant difference was found at T site (Table 3). At all sites, soil exchangeable potassium was significantly higher in the COMP treated soils than in the unamended soils.

Site		WHC (%)	SA (%)	CEC (meq100 g^{-1})	LOI (%)	TC (%)	TN (%)	Available P (µg P g^{-1})
O	COMP	26.3±0.81^a	51.8±2.0^a	22.2±1.22^a	3.95±0.95^a	2.29±0.39^a	0.25±0.04^a	10.6±0.04^a
	NCOMP	22.3±0.3^b	34.0±1.6^b	20.8±0.21^b	3.45±0.85^a	2.00±0.49^a	0.27±0.03^a	9.3±0.03^a
R	COMP	21.8±1.3^a	57.9±1.8^a	25.3±3.83^a	8.34±3.67^a	4.84±1.90^a	0.29±0.03^a	30.3±0.03^a
	NCOMP	20.5±1.7^b	37.1±0.9^b	18.6±0.27^b	3.96±1.09^b	2.30±0.57^b	0.23±0.02^a	11.5±0.02^b
T	COMP	23.6±0.5^a	56.7±0.2^a	21.1±3.46^a	6.31±2.22^a	3.66±1.15^a	0.25±0.08^a	6.9±0.01^a
	NCOMP	20.5±0.8^b	22.6±0.6^b	15.4±2.06^b	2.39±0.62^b	1.39±0.32^b	0.10±0.02^b	7.6±0.02^a
A	COMP	30.2±0.5^a	23.9±1.0^a	23.3±6.9^a	16.1±3.49^a	8.49±2.11^a	0.74±0.33^a	57.4±0.33^a
	NCOMP	28.7±1.3^a	23.7±1.5^a	10.6±1.2^b	1.88±0.41^b	1.09±0.21^b	0.05±0.02^b	3.57±0.02^b

Table 3. Water holding capacity (WHC), soil stables aggregates (SA), cation exchangeable capacity (CEC), organic matter (LOI), total C (TC), total N (TN) and available P of olive oil farming which received (COMP) or not (NCOMP) composted olive mill pomace at Olvera (O), Reja (R), Tobazo (T) and Andújar (A). Data are the mean of five replicates ± standard deviation. Different superscript letters for each site denote significant differences between COMP and NCOMP farming (one way ANOVA; $P< 0.05$).

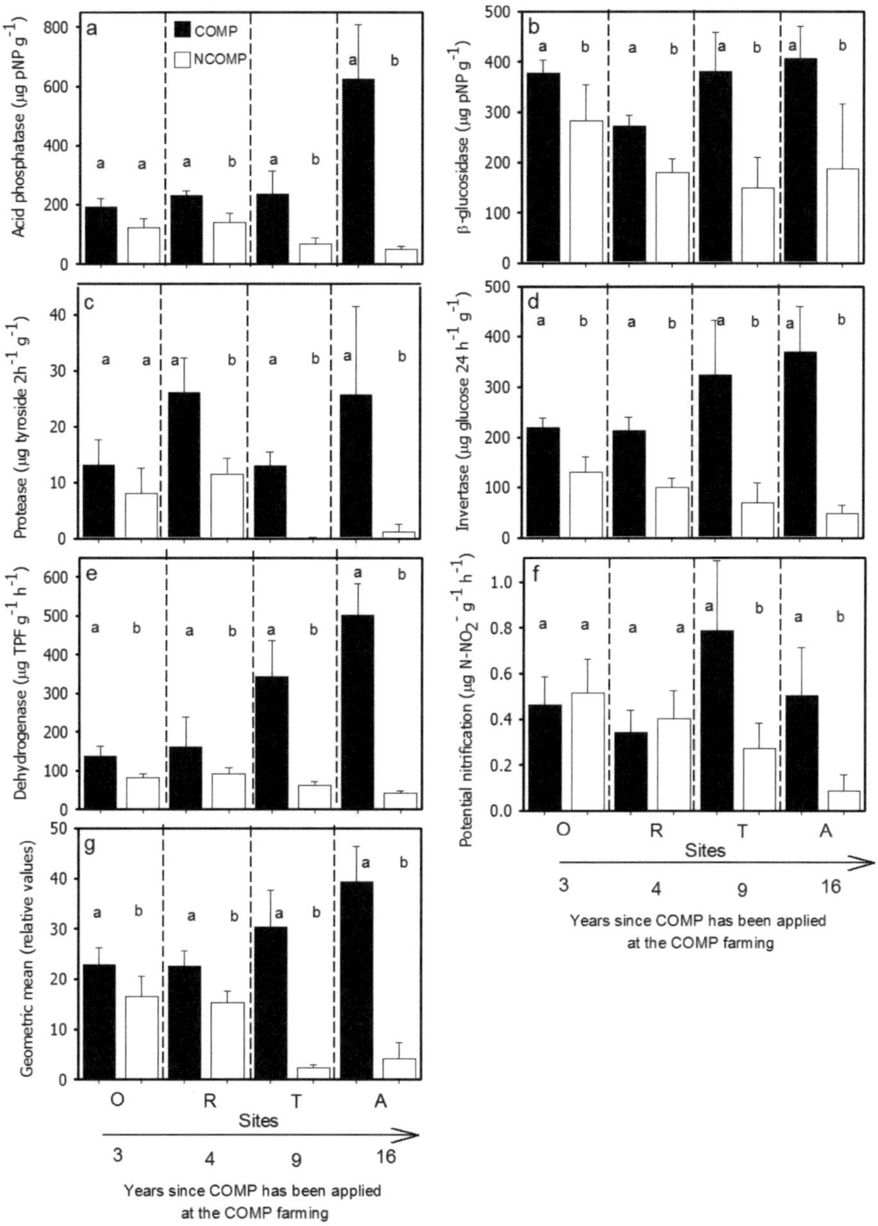

Figure 9. Soil acid phosphatase (a), β-glucosidase (b), protease (c), invertase (d), and dehydrogenase (e) activities, potential nitrification rate (f) and geometric mean of the assayed soil enzyme in olive oil farms which received (black) or not (white) COMP at Olvera (O), Reja (R), Tobazo (T) and Andújar (A). The number of years since composted olive mill pomace has been applied to COMP farming is also indicated. Different subscript letters stand for significant differences ($P < 0.05$) between COMP and NCOMP farming of each site.

Soil enzyme activities related to carbon, nitrogen and phosphorus cycling have been proposed as a tool to assess soil quality/health and functioning [77]. Enzymes activities related to the C, N and P cycles were all significantly higher in soils amended with COMP than in NCOMP soils (Figure 9). Others authors [78,79] have found similar results after application of manures of different origin.

When pooling all the analysed variables in a Principal Component Analysis it was found that the first principal component (PC1) was negatively correlated with soil organic matter, total N, available P, soil aggregate stability and soil enzyme activities and therefore, PC1 is strongly related to soil functioning. PC2, on the other hand, was positively correlated with clay content, WHC and cation exchange capacity. COMP-treated soils shifted upwards (higher WHC and CEC) and towards the left (higher soil functioning) with respect to their NCOMP soils in the PC1–PC2 space (Figure 10). Therefore PCA, which included all the analysed variables, clearly separated paired plots according to the COMP application, supporting the hypothesis that COMP application improves overall soil functioning.

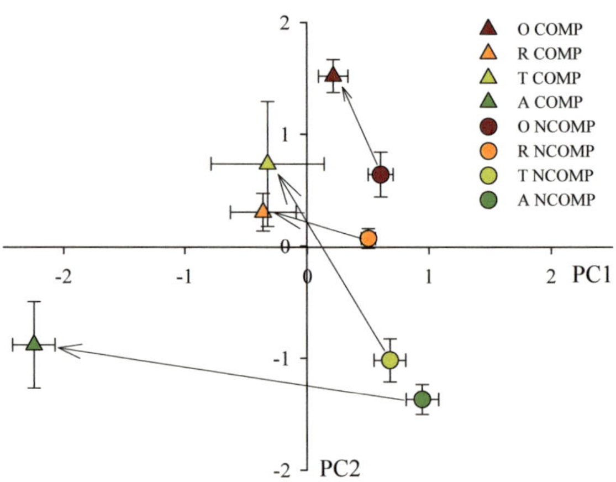

Figure 10. Ordination of the COMP and NCOMP farms at Olvera (O), Reja (R), Tobazo (T) and Andújar (A) in the space defined by the PC1 and PC2 axis resulting from the PCA analysis carried out with physico–chemical and biochemical soil properties. Coordinates are the means of t five replicates and bars represent the standard deviations of the mean. Arrows illustrate the differences in the position of the COMP and NCOMP farming at each site.

Finally, the vectorial distance between comparable COMP and NCOMP farms (e.g. differences in soil functioning) tended to increase with the period of COMP application, suggesting that there is potential for further increases in soil fertility and functioning after long-term application of COMP.

6. Ecological services associated with the use of composted olive mill pomace as organic fertiliser

There is a global trend towards developing agricultural production systems which are sustainable. This involves the more efficient utilisation of inputs and the reduction of waste products. Ideally, organic by-products should be transformed into useful products by processes such as nutrient recycling, replenishment of soil organic matter or generation of energy. In this context, the cost of waste disposal would be avoided and environmental pollution reduced. This fact has led y to a widespread increase in the value of by-products of the agricultural industry. This is also now the case for OMP.

Recycling of OMP through composting (relatively easy-to-use and low costing methodology) could be a sound strategy to provide some ecological services to olive oil groves (Figure 11). Firstly, composting OMP reduces most of the potential environmental pollution problems linked with the disposal of approximately 4 million of tonnes of OMP produced in Andalusia over a relatively short-time span (3 months). On the other hand, most of the nutrients (especially nitrogen, phosphorus and potassium) harvested with the yield, are contained in the OMP, and therefore after composting and application to olive oil groves helps to recycle these nutrients, reducing the need for chemical fertilisers. Our estimates show that between one to two-thirds of the Andalusian olive oil groves could be fertilised annually with the OMP produced in Andalusia after composting, with a subsequent reduction of about 25 – 60% in chemical fertilisers. In addition, the main beneficiary of the economic and environmental profits of composting OMP and application to olive oil groves is the farmer.

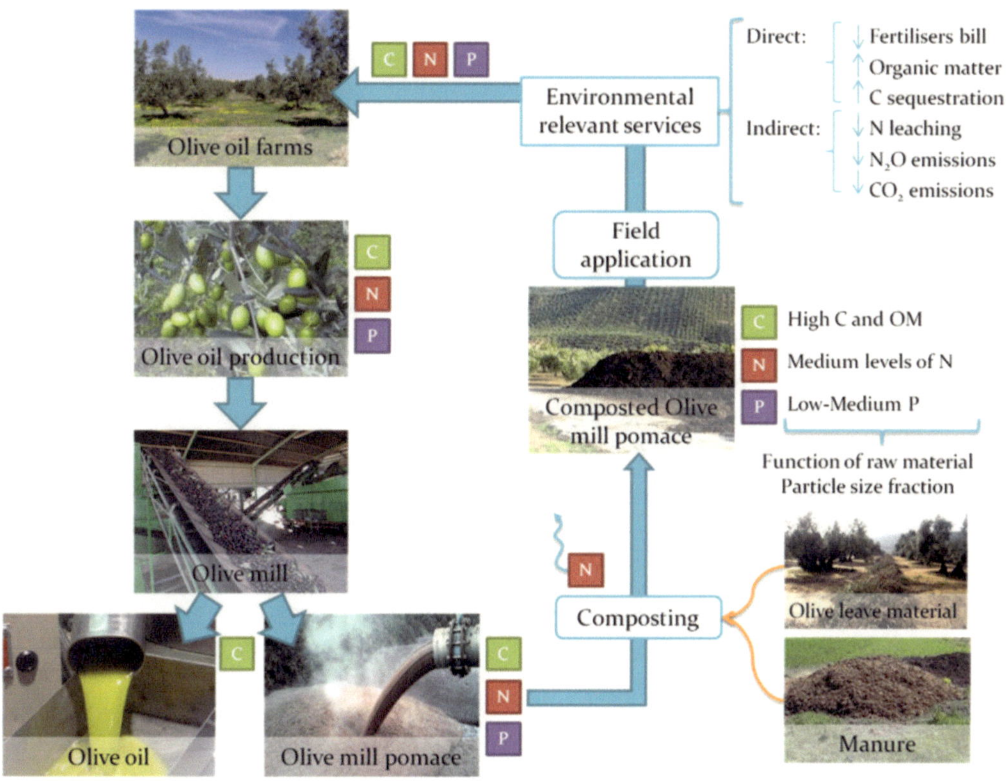

Figure 11. Fate of C, N and P of the fruit harvested in the olive oil farming when olive mill pomace (main by-product of the olive oil mill industry) is composted and applied to the olive oil groves. Some of the environmental services linked to the recycling of olive mill pomace throughout composting, are also indicated.

Some of the environmental services restored after application of COMP and shown in this chapter include: i) Increased soil organic matter and carbon. Indeed our data showed that both increased significantly after regular applications of COMP and there was a trend for increased differences in SOM and SOC between soils amended or unamended with COMP according to the number of years when COMP was applied. Other variables related directly or indirectly with the increase in SOM include i) an increase in soil microbial activity related to nutrient recycling (e.g. soil enzyme activities), cation exchange capacity, soil aggregate stability and available P and K, ii) increase in soil carbon sequestration. COMP-C derived mineralization is rather low (< 10% over one year at optimal conditions) and thus, regular applications of COMP could contribute to soil C sequestration and help alleviate the soil C-CO_2 emissions linked to olive oil cultivation, iii) reduction of the potential nitrate leaching and N_2O emissions. Our data on seven currently produced COMPs have demonstrated that during decomposition of OMP, soil mineral N is retained (i.e. net N immobilization) reducing not only nitrate leaching, but also N_2O emissions.

7. Conclusions

The recycling of nutrient and organic matter of the olive mill pomace after its composting and application to the soil of olive oil farming is a worthwhile strategy to avoid the potential environmental harm of olive mill waste disposal and could lead to increased soil fertility and functionality. The characteristics of the COMP currently produced are adequate for agricultural purposes (high organic matter and carbon, high level of potassium and from low to medium levels of N and P, and lack of phytotoxicity) and the quality was highly dependent on the proportion of manure co-composted with olive mill pomace. COMP-N is well humified and during decomposition soil mineral N can be immobilised depending on the proportion of raw materials co-composted, and therefore it is recommended to combine N rich fertiliser with COMP during the first years of COMP application. COMP-nitrate leaching (at a temporal scale of months to a year) and nitrous oxide emissions were negligible after COMP application to soils. COMP-C mineralization was very low (< 10% of that added after one year) and, therefore, COMP application to soils could enhance C sequestration in olive oil farming. Soil fertility and functioning was improved after three years of regular applications of COMP to soils of olive oil groves a there was a clear trend for a further increase over longer periods of application.

Overall, composted olive mill pomace is a worthwhile strategy to reduce the environmental problems associated with the disposal of OLM, and increases the sustainability and ecological services of olive oil cultivation.

Author details

Beatriz Gómez-Muñoz
Corresponding Author
Ecology Section, University of Jaén, Campus Las Lagunillas s/n, Jaén, Spain

David J. Hatch
Sustainable Soils and Grassland Systems Department, Rothamsted Research, North Wyke, Okehampton, Devon, UK

Roland Bol
Institute of Bio-and Geosciences, Agrosphere (IBG-3), Forschungszentrum Jülich GmbH, Jülich, Germany

Roberto García-Ruiz
Ecology Section, University of Jaén, Campus Las Lagunillas s/n, Jaén, Spain

Acknowledgement

This research was carried out in the framework of the "Olive grove project" of the General Secretary for Rural Development and organic production of the Junta de Andalucía, and with the economic help of the Minister of Science and Technology of Spain throughout the project referenced CGL200908303. We would like to thanks to Jose María Álvarez and the various olive oil farmers and olive oil mills from which soil and composted olive mill pomace samples were taken.

8. References

[1] Beaufoy G, Pienkowski M (2000) The environmental impact of olive oil production in the European Union: practical options for improving the environmental impact. European Forum on Nature Conservation and Pastoralism. European Commission, Brussels.

[2] Dios–Palomares R, Haros–Giménez D, Montes–Tubio F (2005) Estudio estructural del sector oleícola de Andalucía. Nivel de calidad y respecto medioambiental de las industrias almazaras. Comunicaciones EXPOLIVA.

[3] Moraetis D, Stamati FE, Nikolaidis NP, Kalogerakis N (2011) Olive mill wastewater irrigation of maize: impacts on soil and groundwater. Agr. Water Manage. 98: 1125-1132.

[4] Cabrera F, López R, Martinez-Bordiú A, Dupuy de Lome E, Murillo JM (1996) Land treatment of olive oil mill wastewater. Int. Biodeter. Biodegr. 38: 215-225.

[5] Kavvadias V, Doula MK, Komnitsas K, Liakopoulou N (2010) Disposal of olive oil mill wastes in evaporation ponds: effects on soil properties. J. Hazard. Mater. 182: 144-155.

[6] López-Piñeiro A, Cabrera D, Albarrán Á, Peña D (2011) Influence of two-phase olive mill waste application to soil on terbuthylazine behaviour and persistence under controlled and field conditions. J. Soil Sediment. 11: 771-782.

[7] Paredes MJ, Moreno E, Ramos–Cormenzana A, Martinez J (1987) Characterization of soil after pollution with waste waters from olive oil extraction plants. Chemosphere. 16: 1557–1564.

[8] Pérez J, de la Rubia T, Moreno J, Martinez J (1992) Phenolic content and antibacterial activity of olive mill wastewater. Environ. Toxicol. Chem. 11: 489–495.

[9] Linares A, Caba JM, Ligero F, De la Rubia T, Martínez J (2003) Detoxification of semisolid olive–mill wastes and pine–chip mixtures using Phanerochaete flavido–alba. Chemosphere. 51: 887–891.

[10] Sierra J, Marti E, Garau MA, Cruanas R (2007) Effects of agronomic use of olive oil mill wastewater – field experiment. Sci. Total Environ. 378: 90–94.

[11] Piotrowska A, Rao MA, Scotti R, Gianfreda L (2011) Changes in soil chemical and biochemical properties following amendment with crude and dephenolized olive mill waste water (OMW). Geoderma. 161: 8-17.

[12] Paredes C, Cegarra J, Roig A, Sánchez–Monedero MA, Bernal MP (1999) Characterization of olive mill wastewater (alpechin) and its sludge for agriculture purposes. Bioresource Technol. 67: 111–115.

[13] Alburquerque JA, Gonzálvez J, García D, Cegarra J (2004) Agrochemical characterisation of "alperujo", a solid by–product of the two–phase centrifugation method for olive oil extraction. Bioresource Technol. 91: 195–200.

[14] Roig A, Cayuela ML, Sánchez–Monedero MA (2006) An overview on olive mill wastes and their valorisation methods. Waste Manage. 26: 960–969.

[15] Cayuela ML, Millner P, Slovin J, Roig (2007) Duckweed (Lemna gibba) growth inhibition bioassay for evaluating the toxicity of olive mill wastes before and during composting. Chemosphere. 68: 1985–1991.

[16] Madejón E, Galli E, Tomati U (1998) Composting of wastes produced by low water consuming olive mill technology. Agrochimica. 42: 135–146.

[17] Cegarra J, Amor JB, Gonzálvez J, Bernal MP, Roig A (2000) Characteristics of a new solid olive–mill–by–product ("alperujo") and its suitability for composting. In: Warman, P.R, Taylor, B.R. (Eds.), Proceedings of the International Composting Symposium ICS_99, 1. CBA Press Inc, pp. 124–140.

[18] Álvarez de la Puente JM, Jáuregui J, García–Ruiz R (2010) Composting olive mill pomace: The Andalusian experience. Biocycle. 31–32.

[19] García–Gómez A, Roig A, Bernal MP (2003) Composting of the solid fraction of olive mill wastewater with olive leaves: organic matter degradation and biological activity. Bioresource Technol. 86: 59–64.

[20] Alburquerque JA, Gonzálvez J, García D, Cegarra J (2006) Measuring detoxification and maturity in compost made from "alperujo", the solid by–product of extracting olive oil by the two–phase centrifugation system. Chemosphere. 64: 470–477.

[21] Canet R, Pomares F, Cabot B, Chaves C, Ferrer E, Ribó M, Albiach MR (2008) Composting olive mill pomace and other residues from rural southeastern Spain. Waste Manage. 28(12): 2585–2592.

[22] Alburquerque JA, Gonzálvez J, García D, Cegarra J (2007) Effects of a compost made from the solid by–product ("alperujo") of the two–phase centrifugation system for live oil extraction and cotton gin waste on growth and nutrient content of ryegrass (Lolium perenne L.). Bioresource Technol. 98: 940–945.

[23] Hachicha S, Sellami F, Medhioub K, Hachicha R, Ammar E (2008) Quality assessment of composts prepared with olive mill wastewater and agricultural wastes. Waste Manage. 28(12): 2593–603.

[24] Mustin M (1987) Le Compost: Gestion de la Matie`re Organique. Dubusc, F. (Ed.), Paris, pp. 117–263.

[25] Das KC (2007) Co–composting of alkaline tissue digester effluent with yard trimmings. Waste Manage. 28: 1785–1790.

[26] Iannotti DA, Grebus ME, Toth BL, Madden LV, Hoitik HAJ (1994) Oxygen respirometric method to assess stability and maturity of composted municipal solid waste. J. Environ. Qual. 23: 1177–1183.

[27] Moral R, Moreno–Caselles J, Pérez–Murcia MD, Pérez–Espinosa A, Rufete B, Paredes C (2005) Characterisation of the organic matter pool in manures. Bioresource Technol. 96(2): 153–158.

[28] Sánchez–Monedero MA, Roig A, Paredes C, Bernal MP (2001) Nitrogen transformation during organic waste composting by the Rutgers system and its effects on pH, EC and maturity of the composting mixtures. Bioresource Technol. 78: 301–308.

[29] Goyal S, Dhull SK, Kapoor KK (2005) Chemical and biological changes during composting of different organic wastes and assessment of compost maturity. Bioresource Technol. 96(14): 1584–1591.

[30] Pascual JA, Ayuso M, García C, Hernández T (1997) Characterization of urban wastes according to fertility and phytotoxicity parameters. Waste Manage. Res. 15: 103–112.

[31] Capasso R, Evidente AA, Schivo L, Orru, G, Marcialis, M.A, Cristinzio, G, 1995. Antibacterial polyphenols from olive oil mill waste waters. J. Appl. Bacteriol. 79: 393–398.

[32] Ait Baddi G, Alburquerque JA, Gonzalvez J, Cegarra J, Hafidi M (2004) Chemical and spectroscopic analyses of organic matter transformations during composting of olive mill wastes. Int. Biodeter. Biodegr. 54(1): 39–44.

[33] Palm CA, Gachengo CN, Delve RJ, Cadisch G, Giller KE (2001) Organic inputs for soil fertility management in tropical agro ecosystems: application of an organic resource database. Agr. Ecosyst. Environ. 83: 27–42.

[34] Fox RH, Myers RJK, Vallis I (1990) The nitrogen mineralization rate of legume residues in soil as influenced by their polyphenol, lignin and nitrogen contents. Plant Soil. 129: 251–259.

[35] Palm CA, Sánchez PA (1991) Nitrogen release from the leaves of some tropical legumes as affected by their lignin and polyphenolic contents. Soil Biol. Biochem. 23: 83–88.

[36] Melillo JM, Aber JD, Muratore JF (1982) Nitrogen and lignin control of hardwood leaf litter decomposition dynamics. Ecology, 63: 621–626.

[37] Constantinides M, Fownes JH (1994) Nitrogen mineralization from leaves and litter of tropical plants—relationship to nitrogen, lignin and soluble polyphenol concentrations. Soil Biol. Biochem. 26: 49–55.

[38] Palm CA, Rowland AP (1997) A minimum data set for characterization of plant quality for decomposition. In: G. Cadisch and K.E. Giller, Editors, Driven by Nature: Plant Litter Quality and Decomposition, CAB International, Wallingford. 379–392.

[39] Doublet J, Francou C, Pétraud JP, Dignac MF, Poitrenaud M, Houot S (2010) Distribution of C and N mineralization of a sludge compost within particle–size fractions. Bioresource Technol. 101(4): 1254–1262.

[40] Grilo J, Bol R, Dixon L, Chadwick D, Fangueiro D (2011) Long term release of carbon from grassland soil amended with different slurry particle size fractions: a laboratory incubation study. Rapid Commun. Mass Sp. 25: 1514–1520.

[41] Fangueiro D, Chadwick D, Dixon L, Bol R (2007) Quantification of priming and CO_2 respiration sources following the application of different slurry particle size fractions to a grassland soil. Soil Biol. Biochem. 39: 2608–2620.

[42] Robin P, Ablain F, Yulipriyanto H, Pourcher AM, Morvan T, Cluzeau D, Morand P (2008) Evolution of non–dissolved particulate organic matter during composting of sludge with straw. Bioresource Technol. 99: 7636–7643.

[43] Tester CF, Sikora LJ, Taylor JM, Parr JF (1979) Decomposition of sewage sludge compost in soil: III. Carbon, nitrogen, and phosphorus transformations in different sized fractions, J. Environ. Qual. 8: 79–82.

[44] Aoyama M (1985) Properties of fine and water–soluble fractions of several composts. Micromorphology, elemental composition and nitrogen mineralization of fractions. Soil Sci. Plant Nutr. 31: 189–198.

[45] Fangueiro D, Gusmão M, Grilo J, Porfírio G, Vasconcelos E, Cabral F (2010) Proportion, composition and potential N mineralisation of particle size fractions obtained by mechanical separation of animal slurry. Biosyst. Eng. 106: 333–337.

[46] Fangueiro D, Bol R, Chadwick D (2008a) Assessment of the potential N mineralization of six particle size fractions of two different cattle slurries. J. Plant Nutr. Soil Sc. 171: 313–315.

[47] Fangueiro D, Pereira J, Chadwick D, Coutinho J, Moreira N, Trindade H (2008b) Laboratory assessment of the effect of cattle slurry pre–treatment on organic N degradation after soil application and N_2O and N_2 emissions. Nutr. Cycl. Agroecosys. 80: 107–120.

[48] Beauchamp EG, Paul JW (1989) A simple model to predict manure N availability to crops in the field. In: J.A. Hansen and K. Henriksen, Editors, Nitrogen in Organic Wastes Applied to Soils, Academic Press, London.

[49] Cabrera F, Martín–Olmedo P, López R, Murillo JA (2005) Nitrogen mineralization in soils amended with composted olive mill sludge. Nutr. Cycl. Agroecosys. 71: 249–258.

[50] Parker CF, Sommer LE (1983) Mineralization of nitrogen in sewage sludges. J. Environ. Qual. 12: 150–156.

[51] Hérbert M, Karam A, Parent LE (1991) Mineralization of nitrogen and carbon in soils amended with composted manure. Biology, Agriculture and Horticulture, 7: 349–361.

[52] Chae YM, Tabatabai MA (1986) Mineralization in an aridisol mixed with various organic materials. J. Environ. Qual. 15: 193–198.

[53] Bernal MP, Kirchmann H (1992) Carbon and nitrogen mineralization and ammonia volatilization: from fresh, aerobically and anaerobically treated pig manure during incubation with soil. Biol. Fert. Soils, 13: 135–140.

[54] Hadas A, Portnoy R (1994) Nitrogen and carbon mineralization rates of composted manures incubated in soil. J. Environ. Qual. 23: 1184–1189.

[55] Senesi N, Plaza C, Brunetti G, Polo A (2007) A comparative survey of recent results on humic–like fractions in organic amendments and effects on native soil humic substances. Soil Biol. Biochem. 39: 1244–1252.
[56] Ferguson RB, Shapiro CA, Hergert GW, Kranz WL, Klocke NL, Krull DH (1991) Nitrogen and irrigation management practices to minimize nitrate leaching from irrigated corn. J. Prod. Agric. 4: 186–192.
[57] Hallberg GR (1985) Agricultural chemicals and groundwater quality in Iowa: Status Report, Cooperative Extension Service, CE–2158q, Iowa State Univ, Ames, IA, p. 11.
[58] Stanley EH, Short RA, Harrison JW, Hall R, Wiedenfeld RC (1990) Variation in a nutrient limitation of lotic and lentic algal communities in a Texas (USA) river. Hydrobiologia. 206: 61–71.
[59] Hallberg GR (1989) Nitrate in ground water in the United States. In R.F. Follett (ed.) Nitrogen management and ground water protection. Elsevier, New York, p. 35–74.
[60] Yadav SN (1997) Formulation and estimation of nitrate–nitrogen leaching from corn cultivation. J. Environ. Qual. 26: 808–814.
[61] Insam H, Merschak P (1997) Nitrogen leaching from forest soil cores after amending organic recycling products and fertilizers. Waste Management Research, 15: 277–292.
[62] IPCC (2007) Changes in Atmospheric Constituents and in Radiative Forcing, Cambridge University Press, UK and New York USA.
[63] García–Ruiz R, Baggs E (2007) $^{15}N-N_2O$ emissions from decomposition of plant residues in olive oil orchard. En: Towards a better efficiency in N use, pp. 250–253.
[64] Baggs EM, Stevenson M, Pihlatie M, Regar A, Cook H, Cadisch G (2003) Nitrous oxide emissions following application of residues and fertiliser under zero and conventional tillage. Plant Soil. 254: 361.
[65] Baggs EM, Rees RM, Smith KA, Vinten AJA (2000) Nitrous oxide emission from soils after incorporating crop residues. Soil Use Manage. 16: 82–87.
[66] Duiker SW, Lal R (2000) Carbon budget study using CO_2 flux measurements from a no till system in central Ohio. Soil Till. Res. 54: 21–30.
[67] Adiku SGK, Narh S, Jones JW, Laryea KB, Dowuona GN (2008) Short–term effects of crop rotation, residue management, and soil water on carbon mineralization in a tropical cropping system. Plant Soil, 311: 29–38.
[68] Manzoni S, Proporato A (2009) Soil carbon and nitrogen mineralization: Theory and models across scales. Soil Biol. Biochem. 41, 1355–1379.
[69] Cleveland CC, Liptzin D (2007) C:N: P stoichiometry in soil: is there a "Redfield ratio" for the microbial biomass? Biogeochemistry. 85: 235–252.
[70] Jensen LS, Mueller T, Magid J, Nielsen NE (1997) Temporal variation of C and N mineralization, microbial biomass and extractable organic pools in soil after oilseed rape straw incorporation in the field. Soil Biol. Biochem. 29: 1043–1055.
[71] Bernal MP, Sánchez–Monedero MA, Paredes C, Roig A (1998) Carbon mineralization from organic wastes at different composting stages during their incubation with soil. Agr. Ecosyst. Environ. 69: 175–189.

[72] Hollis JM, Jones RJA, Palmer RC (1977) The effect of organic matter and particle size on the water retention properties of some soils in the West Midlands of England. Geoderma. 17, 225– 238.

[73] López–Piñeiro A, Albarrán A, Nunes JM, Barreto C (2008) Short and medium–term effects of two–phase olive mill waste application on olive grove production and soil properties under semiarid Mediterranean conditions. Bioresource Technol. 99: 7982–7987.

[74] Piotrowska A, Iamarino G, Rao MA, Gianfred L (2006) Short–term effects of olive mill waste water (OMW) on chemical and biochemical properties of a semiarid Mediterranean soil. Soil Biol. Biochem. 38: 600–610.

[75] Montemurro F, Convertini G, Ferri D (2004) Mill wastewater and olive pomace compost as amendments for rye–grass. Agronomie. 24, 481–486.

[76] Benítez C, Tejada M, González JL (2003) Kinetics of mineralization of nitrogen in a pig slurry compost applied to soils. Compost Sci. Util. 11: 72–80.

[77] Dick RP (1994) Influence of long–term tillage and crop rotation combinations on soil enzyme activities. Soil Sci. Soc. Am. J. 56: 783–788.

[78] Mäder P, Fliessbacth A, Dubois D, Gunst L, Fried P, Niggli U (2002) Soil fertility and biodiversity in organic farming. Science. 296: 1694–1697.

[79] Melero S, Porras JC, Herencia JF, Madejón E (2006) Chemical and biochemical properties in a silty loam soil under conventional and organic management. Soil Till. Res. 90: 162–170.

Chapter 21

Social-Ecological Resilience and Maize Farming in Chiapas, Mexico

Nils Max McCune, Francisco Guevara-Hernández, Jose Nahed-Toral, Paula Mendoza-Nazar, Jesus Ovando-Cruz, Benigno Ruiz-Sesma and Leopoldo Medina-Sanson

Additional information is available at the end of the chapter

http://dx.doi.org/10.5772/45898

1. Introduction

Chiapas is considered one of the areas of origin for maize, and indigenous production systems remain a major element of maize farming and the food system of Mexico's southernmost state. These traditional systems often include a complex, long-term relationship between maize farming and the larger landscape (Medellín and Equihua, 1998). However, what often appear to be static, steady states of small farmer land use in fact are highly dynamic systems that are in the midst of major adaptations to new climatic and economic conditions. The globalization of the agribusiness model that came to dominate the U.S. landscape during the 20th century has put intense pressures on small farmers in Chiapas, adding to complications arising from shifting growing seasons and other effects of global climate change. However, history has shown that Mexican small farmers are never passive objects of their circumstances, and their responses to early 21st century marginalization take both centralized and local forms, in political and productive terms (Guevara-Hernández et al. 2011a). The balance of forces—capitalist agriculture, external input dependence, neoliberal national food policy on one hand; traditional knowledge, agroecological transitions, social movements on the other—in Mexican agriculture tends to distinguish two contending models for food systems, even as climate change, resource scarcity, and economic crises limit humanity's options.

Shifting cultivation and other traditional systems of food production in Chiapas are being highly influenced by the slow-motion arrival of industrial agriculture, which is based on maximizing short-term productivity through the use of synthetic inputs such as fertilizers and pesticides, as well as commercial seeds (Garcia-Barrios et al. 2010). Most farm inputs are supplied by Mexican subsidiaries of foreign-owned multinational corporations such as

Monsanto, Syngenta, and Dow. At the same time, an industrial food system means that large amounts of capital are invested in food sales, creating giant monopolies in processing and distribution, which significantly reduce the portion that farmers receive of the price paid by consumers. In other words, control over the food system—the set of activities that are built around capital flow and labor command in food production, shipping, transformation, consumption—is highly concentrated in the hands of input manufacturers and food processing, trading, and retail corporations, while the riskiest part of agriculture—the actual farming process—is still in the hands of hundreds of thousands of small farmers (Magdoff et al. 2000).

In Chiapas, these farmers live in hills and valleys, in forested land and former forests, in dry shrub lands and in lush jungles. Their communities and farmlands compose a peasant landscape, in which patches of forest are interspersed in a complex mosaic of farms, backyards, homes, schools, rivers, roads, and towns. Depending on the type of agriculture practiced, soil may be highly degraded or intact. Chiapas is a center for biodiversity; many endemic species live in and around forest patches and agroecosystems (Ramírez-Marcial et al. 2001). In this setting, the demands of rural social movements such as *¡Sin Maíz No Hay País!* (Without Corn There is No Country!) increasingly refer to the goal of food sovereignty. Food sovereignty means a fundamental emphasis on local and domestic production, based on land access for small farmers and ecological production practices. It rejects food as a commodity to be included in free trade agreements or dumping schemes meant to undermine countries' domestic production capacity. As a political proposal, food sovereignty implies a radical democratization and decentralization of the agriculture-food system, including the destruction of corporate power over food. On a more cultural level, food sovereignty is an affirmation of rural community, local knowledge, and gender equality.

Both the agribusiness model and the food sovereignty model are highly complex, integrated systems that involve the relationship of society and nature. The stability of each system depends on distinct factors that combine social and ecological drivers, while the capacity of each to respond to shock or disturbance will depend on unique intrinsic qualities. In the case of industrial or export-focused agriculture, we have a system that has been shown to be destructive to peoples and ecosystems, due to its focus on short-term profits through maximizing monoculture productivity. Here we draw from theoretical contributions from restoration ecology that use models of alternative stable states to study change in complex systems (Suding et al. 2004). We argue that monoculture/capitalist agribusiness represents one pull of attraction, or alternative state, indeed a food system unique to late-stage global capitalism. On the other hand, we propose that the agroecology/food sovereignty framework may in fact represent another alternative state, far more promising for building resilient food systems in the 21st century. In order to develop this line of inquiry, we start by examining more closely the resilience paradigm and the two proposed stable states at an abstract/global level. Then we describe historical, agroecological, and political elements of food system resilience in the case of a maize-growing community in Chiapas.

2. The resilience paradigm

Research into sustainable agriculture has increasing come to embrace the conceptual approach of food systems, as these reflect the interface of alimentation, human activities, public policies, cultural norms and social well-being, along with land, farms, ecosystems, and economies. The complex interactions between these processes at distinct scales, and involving various institutional and economic actors, may produce the outcome of food security. The food system approach may be useful for developing cohesive strategies across policy sectors, including agrarian and land access sectors, natural resource and environmental management, agriculture, trade, economy, industry, science and technology, health, and education, among others (Ericksen et al. 2010). Efforts to achieve food system sustainability in the midst of global environmental and economic changes are beginning to coalesce around certain concepts that help determine the most significant problems in food systems and identify management strategies at several levels of analysis (farm, community, national, international) to increase social, ecological, and economic sustainability (Pretty et al. 2011).

The management of complex, adaptive systems has become a dynamic field of new trans-disciplinary theory, especially with regard to life supporting systems of human activities, such as agriculture, in sensitive ecological contexts. Social-ecological systems (Berkes and Folke, 1998), or coupled human and natural systems (Liu et al. 2007), have become a central concept to allow greater understanding of the interdependencies and feedbacks between social and ecological systems. The contributions of systems ecology are applied in order to understand the complex internal dynamics and adaptability of these coupled systems. Many of the concepts that inform such studies of systems originate from ecology, for two reasons: one, its emphasis on qualities that emerge from a set of relationships between elements, rather than the reductionist focus on elements in isolation; and two, the growing academic and popular concern for the relationship between humanity and the biosphere that tenderly exists on the surface of the Earth's crust (Lang, 2009).

In the efforts to understand the intrinsic qualities of social-ecological systems, researchers from several different disciplinary backgrounds have approached the concept of resilience (Shattuck, 2012). The resilience principle stems from systems ecology theory (Hooper, 1973) that suggested that instead of static, unchanging climax communities, natural ecosystems could evolve between several alternative stable states, with biotic and abiotic feedback mechanisms accelerating or preventing system change. Disturbances began to be seen as an integral part of ecosystem function, and resilience as an emergent system capacity to absorb a certain magnitude of shock and maintain key system functions before reaching a critical threshold and switching to an alternative stable equilibrium with new system properties (Holling, 1973; Noy-Meir, 1975). Noy-Meir (1975) used the analogy of a mechanical ball-in-container (figure 1) to describe alternative steady-states. The original steady-state is stable to fluctuations within a certain range, but too hard a push in one direction will send it over the turning point and toward a new steady-state. The major concern in light of global environmental change is that ecosystems will be pushed beyond their limits, into new steady-states that provide less ecological services (Walker et al. 2004).

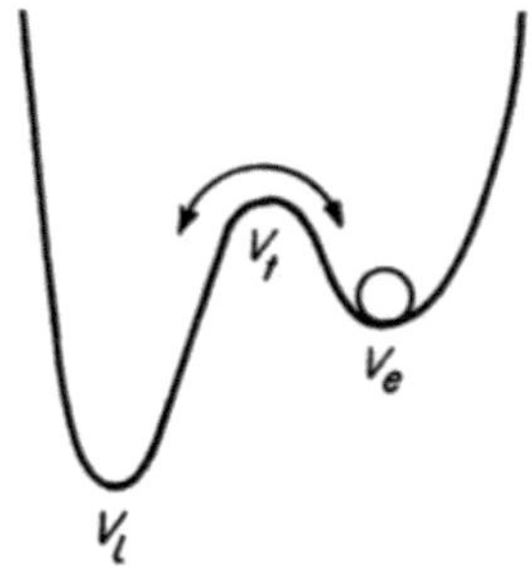

Figure 1. A physical model of the two-steady-states situation (from Noy-Meir, 1975).

The two-steady-state model is a very simple illustration of a key concept in resilience studies: the threshold. The high point of the center curve in figure one is the threshold, or point of no return, for the original system. Resilience systems may absorb strong shocks below this point; one tiny push above it will result in what could be irreversible and accelerated change. One of the major objectives of resilience research is to identify and characterize system thresholds, in order to understand what makes systems able to absorb some shocks without changing overall function, what makes some changes temporary and others permanent, and how to shift thresholds through system adaptations. Social and ecological dynamics in industrial food states can be very different from dynamics in traditional or sovereign food states. Efforts to promote transition to a sovereign food state need to better understand feedbacks and constraints of the industrial food state. Alternative state models, used in restoration ecology to focus on internally reinforced states and recovery thresholds (Suding et al. 2004), may help guide historic conversions to sovereign food systems.

The recent attention paid in academic literature toward the concepts of risk, robustness and resilience in social and economic systems is surely related to such troubling events as the global financial crisis of 2008; the Great Recession that is still very much limiting employment and well-being among most nations in the world; the so-called Arab Spring that has produced protest movements for systemic change in diverse countries like Egypt, Tunisia, Yemen, Libya, Syria, Spain, Greece, Portugal, France, and the United States; the global food price crises of 2007 and 2011; the impending scarcity of hydrological and energetic resources; and finally, the arrival of such damning evidence as increased incidence of extreme weather events (e.g. hurricanes, droughts, floods) and long-term changes associated with excessive greenhouse gases in the Earth's atmosphere.

The sudden growth in use of the term "resilience" has been studied elsewhere and associated with a shift toward the understanding of complex systems as being more dynamic and less tied to any one climax or stable state. The Resilience Alliance and the Stockholm Resilience Centre, high-profile scientific platforms that bring together orthodox neoliberal economists with systems ecologists (Walker and Cooper, 2011), emphasize an adaptive cycle within complex systems that includes phases of growth, decadence, self-destruction and renewal. Curiously, this self-organization that complex systems are seen to have is used to argue against inserting planning mechanisms and regulation on economic systems. Indeed, the free market is argued to be a complex system capable of self-regulation. The same analysis considers the shock of deregulation and structural adjustment programs to be a healthy opportunity for renewal. A skeptic might insert here the example of New Orleans or Iraq, both places where disaster (to be located somewhere on a gradient between "natural" and human-made) created an opportunity to rebuild in a new image and for the benefit of a distinct group, in both cases resulting in massive profits for contractors that, incidentally, worked under U.S. government contact (i.e., the planned economy). Essentially, orthodox neoliberals create conditions for huge profit by monopoly capital, rather than constructing any sort of self-organized or resilient world state. On the contrary, we argue that the resilience of a local food system is often inversely related to its integration into the world capitalist economy.

Shattuck (2012) proposes a framework for studying resilience in food systems, in order to effectively prioritize goals of human well-being and biodiversity conservation. This author combines literature on biophysical restraints, adaptive capacity, and political economy to develop a food system resilience framework that considers ecological processes and underlying macroeconomic causes of livelihood vulnerability. Here we presuppose that as an emergent property, resilience of a given food system-state is always subject to influence by resilience of systems that operate at larger or smaller scales. We have built our analysis around a resilience assessment in a rural community of Chiapas, but have sought to contribute to a much-needed debate on the global scale. In the study community, we evaluated aspects of economic, social and ecological resilience, based on interviews, surveys, and data sampling that were carried out over the course of two years of fieldwork in the Fraylesca region of Chiapas during 2010-12.

3. Resilience of the industrial food system

Allenby and Fink (2005) define resilience as the "capability of a system to maintain its functions and structure in the face of internal and external change and to degrade gracefully when it must." Accepting this definition, we should define the functions of the capitalist industrial food system and describe its structure, before looking for the critical thresholds that it may not cross without converting into a qualitatively different system. As with all activities driven by capital, the paramount function of the capitalist food system is to reproduce capital in greater quantities. This function has lead to a rationalization of economies surrounding food, based on the principle of maximizing the difference between costs and revenues in the application of capital to production, processing, distribution, and sales. In the productive sphere, rationalization means the maximization of commodity production. In order to maximize revenues, capitalist food systems have developed enormous structures for food processing in order to add to commodity value, as well as advertising to boost revenues. As capital has been slow to penetrate the actual farming process itself (Levins, 2007), due to its risky nature and bio-physical limitations, it has instead reduced the on-farm value added to commodities (i.e. cut into the farmer's income) by creating an industry of costly farm inputs, which are generally accepted as a part of the modern, monoculture form of growing food.

To think of the industrial agriculture model as a stable state on a food system continuum requires identifying its major components, defining the limits and time scale of the system, determining the values, peoples, and natural resources involved, analyzing the political economy and legal character of industrial agriculture, and recognizing the cross-scale interactions that all have an impact on system resilience (Kinzig et al. 2007). The essential components of a mature industrial agriculture model include the integration of food into free trade agreements, government support for agribusiness, market control over land and water resources, external input-intensive production models, monoculture and specialization, as well as an agricultural research establishment that focuses on developing profitable technologies. These conditions are largely met in Mexico as a whole, except for the millions of small-scale producers who continue to meet food needs at the local and national level using few external inputs and some form of communal land rights. The contradictory proliferation—and hybridization—of this model in socially and ecologically adverse circumstances deserves further attention.

The industrial food system represents the technological and organizational apex of a model first put into place in lands colonized by European powers. The monoculture is an invention of colonial economies, which treated dominated nations only as sources for cheap raw materials (including cheap food for a growing industrial working class). In Ireland, the monoculture potato production system was enforced by British colonial law that prevented the Irish from planting other crops. When a common pathogenic fungus destroyed the potato harvest, millions of Irish were killed by the ensuing famine. This is a classic example of the risk of the food system built on monoculture. Despite many such examples of spectacular failure, the industrial food system remains deeply committed to monoculture production systems around the world. The development of a global food system based on the increasing excursion of capital into farming has been a complex process, by which capital completely surrounded farming by taking over farm input and post-harvest economies while only slowly moving into the actual farming itself. Early stabs at industrial agriculture included the guano boat and phosphorus mining fertilizer industries in the late 1800s. In California, industrial agriculture and land takeovers were always linked (Walker, 2004), as wheat farming prospecting triggered a new "gold fever" and led to a bonanza period of often-falsified speculation on real estate.

A great new era began for industrial agriculture after the Second World War. Many countries were in ruins, and baby booms gave impetus to the US war materials industry to "convert swords to plowshares" and sell them to reconstruction programs. Factories that produced nitrogen-based explosives already had the entire infrastructure necessary to produce nitrogen fertilizers, tank assembly lines could easily be converted to create tractors, and many of the nastier chemicals used in war efforts were found to have satisfyingly lethal effects on insects and unwanted plants in agriculture. Even more importantly, the call for technical solutions to hunger was seen as an antidote to the more radical demands from structural change and wealth redistribution in order to combat poverty-caused hunger in the hopeful post-war and post-colonial world. Thus it was that the "green revolution," a broad program of agricultural research and technology development mostly focused on producing new high-yield varieties, was developed as production-focused solution to hunger in Asia, Africa, and Latin America.

Industrial agriculture is based on the intensive use of external inputs—such as improved seeds, fertilizers, pesticides, and irrigated water—to maximize yields. In the United States, this led to more overall production, causing the prices that farmers receive for their production to fall. As input costs rose and farm prices fell, small farms were squeezed by the low per-unit return on farming. The large farms that could produce at such a scale as to be profitable—despite the narrow margin between input costs and sales prices—started to swallow a much greater share of U.S. farm income. In 1969, the 1.2 percent of U.S. farms with the greatest annual income earned 16 percent of net farm income; by the end of the 1980s, they earned nearly 40 percent (Rosset, 1998). It is not necessarily the technologies of industrial agriculture that cause this concentration of agricultural income, but their application in societies where the advantage is already with wealthier growers and large agribusiness corporations. In such a social context, green revolution technologies tend to accelerate the concentration of food system resources, such as land and capital, in the hands of a few large players.

At a landscape level, the agribusiness model becomes best consolidated in conditions of potential ecological homogeneity and market control over economic resources (Perfecto et al. 2010). For this reason, its arrival to the Fraylesca region of Chiapas has been uneven—dominating the landscape in the large valleys, and barely felt in the most remote *ejidos* and family farms. The *ejido* system, as a form of collective property embedded in the national food system, has been a buffering element that kept local food systems viable in much of the Mexican countryside, especially in the southern states. Meanwhile, the large population of northern Mexico has created a more drastic contrast between delicate, rain-fed systems, and industrial farms built on fossil water.

This partially explains the highly disproportionate amount of private and public investment in agriculture in the northern states, while most support for farmers in southern states take the form of social welfare programs (Fox and Haight, 2010). The capitalist agribusiness model has been consolidated in northern Mexico in the generations since the end of the Second World War. In southern Mexico, it has arrived in waves—the most tidal of which was the destruction of state-owned grain warehouses and price regulations—which have yet to completely break the small farmer food system that retains a large geographic and nutritional importance, despite the dismantling of the economic structure that had been built around it since the Revolution.

Resilience in the industrial food system depends on two major objective factors: avoiding ecological destruction that would affect profit margins, and continued growth into new markets to prevent negative effects of overproduction. As global economies become more integrated than ever before and resource scarcity on a global level seems imminent, agribusiness corporations have moved into biotechnologies as a way to absorb huge sums of capital and—they hope—create vast new seed, energy, and pharmaceuticals markets. As a global system, agribusiness can easily leave behind devastated ecosystems and farm communities once degradation has reached the point that farming is no longer profitable. This has especially been the case in areas where long-term irrigation and synthetic fertilizer use have increased salt content of soils beyond thresholds of productivity, or on deforested land where original soil fertility is quickly exhausted to abysmal levels. While this kind of ecological and economic destruction is not threatening to the agribusiness model as a whole, it does threaten to create a social blowback, in the form of rural social movements and consumer groups, strong enough to threaten the future of the agribusiness model. In this sense, widespread social rejection of industrial agriculture is a subjective factor (i.e. dependent upon people's consciousness) that deeply influences the resilience of the system.

4. Resilience of the sovereign food system

Many efforts have been made to define food sovereignty (Patel, 2009). As an evolving concept, it has also been subject to growing social and academic interest, giving its full meaning an emergent quality in the historical conditions of 21st century social struggle (García-Linera, 2011). Nonetheless, we present a non-exhaustive list of components for economic, social and ecological resilience in sovereign food systems. While definitions are still being agreed upon, economic resilience among rural peoples may consist, at this particular historical moment and in many parts of the world, of several interacting components: 1) land access and unalienable rights to produce; 2) capacity to produce an abundance and variety of food necessary to meet most local food needs, essentially the potential to subsist with local production; 3) minimal dependency upon external inputs (e.g. hybrid seeds, pesticides) the availability and price of which are controlled by monopolies or foreign corporations; 4) maximum capacity to use local and renewable sources for energy and material needs (e.g. water, light, soil nutrition, farm labor); 5) use of diversified land-use and production systems, that may include extraction, agriculture, animal production, and small-scale processing in order to appropriate the value added by labor; 6) diverse income sources and form, which may include local products and off-farm employment, local and regional markets, direct contact with consumers, or in-kind payments; 7) real participation in the planning, design, and implementation of economic activities, through grassroots organizations or through governmental planning processes; and 8) the capacity to adapt and transform economic systems to better suit ecological and social necessities. Economic resilience allows systems of economic activities to withstand climatic shock, sudden scarcity or loss of markets, or long-term disturbance.

Social resilience, perhaps more difficult to define, includes at least the following components: 1) free and universal access to education and culture; 2) methods for sharing information and ideas vertically and horizontally in a way that combines theory and practice; 3) strong social organizations organized with democratic principles; 4) access to a common identity that admits and is strengthened by diversity; 5) access to universal and affordable health care; and 6) respect for social and economic human rights, such as the human right to food. This is clearly not a static situation, but rather a dynamic learning and adaptation process subject to conflict and contestation, as well as consensus. Ecological resilience could consist of: 1) ecosystems based on material cycles (e.g. hydrological, carbon, nitrogen), energy flow through trophic webs, and ecological interactions between biotic agents (e.g. competition, predation, mutualism, commensalism, parasitism); 2) productive patterns that maintain the possibility for indefinite temporal continuity without degradation to material cycles; 3) diversity of function in the ecosystem, with the greatest possible number of niches filled; and 4) redundancy of function, so that the potential loss of some species can be compensated by the activity of others.

In order to think of food sovereignty as a stable state or domain of attraction, it becomes necessary to define the "pull" that is capable of directing a food system transition, once the thresholds of the industrial food system have been reached. The direction and strength of this force of attraction almost certainly depends on which thresholds have been reached in the industrial food system-state. Sometimes in order to better understand the resilience pull effect, we can ask ourselves, in any given context, what the easiest option is. In the contradictory dominant order of a globalized, capitalist food system, where are there activities that, when certain thresholds have been passed, become easier to do by conforming to the logic of food sovereignty rather than the logic of capital accumulation? Clearly, the access to conventional farm inputs is a defining pull toward the capital-influenced agricultural model. Thus maybe one starting point for a regime shift could be the end of access to conventional, yield-intensifying chemicals and seeds. This was indeed the case in Cuba, the world's greatest example yet of a national-level transition from an industrial agriculture model to an organic, diversified, low-external input agricultural model. When conventional inputs, petroleum, and imported food all become unavailable due to the fall of the Soviet Union and the US trade embargo, Cuba's small farmers, scientific community, and government teamed up to direct a national inward-looking agricultural effort, based on organic urban gardens, agroecological small farms, and the breaking-up of unproductive state farms into cooperatives more directly controlled by workers (Rosset and Benjamin, 1994). In this case, the perturbation was an acute food crisis; the response was a rapid, dramatic regime shift toward the food sovereignty framework.

In other cases, the defining pull that defines food state transitions could be the social demand for land, as was the case in those that accompanied the Mexican Revolution. There is also the eternal drive for greater social justice and equality, which has been a major component of food sovereignty-themed social movements in countries such as the United States and Brazil. In the case study below, the driving pull toward a food sovereignty state of the food system is the concern for human health.

5. The focal system

By focusing on one maize-growing community in rural Chiapas, we set out to understand the resilience of a maize production system to external disturbances and internal contradictions that jeopardize its natural resource base and the health of its inhabitants. That is to say, we are interested in the factors of social-ecological resilience in a small farmer agroecosystem at the community/landscape level over the next couple decades. In Mexico, land reform resulted in the creation of *ejidos*, or agrarian communities of small producers with internal political structures. The *ejido* system is itself a complex adaptive system that has survived decades of neoliberal food and land policy at the national level. Within one *ejido*, our study focuses especially on the food production, distribution, and consumption surrounding what is known as the *milpa* system, or the fields where maize is grown, in 25 hectares within the limits of the *ejido*. The overall shift toward the agribusiness state in Mexican maize farming is an uneven, long-term trend which takes place over the course of decades and has a series of social and ecological feedbacks.

The critical components of the current system include farmer families, land access, synthetic fertilizers, hybrid seeds, and chemical herbicides, as well as maize purchasers. Critical components of the restored agroecosystem will include local knowledge and farmer identity, community interest in health and nutrition, soil fertility, functional agrobiodiversity, farmer-to-farmer knowledge exchanges, strong local organizations, traditional seed varieties, organic fertilizers, and crop rotation (Milestad et al. 2010). Important natural resources in the focal system include biodiversity, clean water, fertile soil, forest carbon, and knowledge in the form of traditional maize varieties. Key people include the producer families, including elders, women, men, and children, while critical values include the relative interest in short-term income versus long-term economic sustainability, the capacity to coalesce around the concept of health, and the impacts of belonging to the small farmer social class (Guevara-Hernández et al. 2011b).

The *ejido* system is a form of local governance, not only for land issues but also for other social issues such as health. Property rights reflect a mix of collective and private land-holdings, with complex informal arrangements used for producing food on the land that is closest to the community, regardless of who is the legal owner. There is a definite lack of strong rural organizations in the region, leading the *ejido* structure, the elementary school, and the local church to hold a monopoly over collective action in the community. Maize farming communities in the region, as in most of Mexico, are highly dependent on government anti-poverty programs, as these have come to replace most productive subsidies and credit mechanisms. Obviously, multinational farm input corporations are untouched by democratic institutions that might be used to control the use of toxic chemicals in the community or promote local seeds (Bakan, 2004). Indeed, scale factors deeply influence the capacity to characterize the resilience of the focal system, because its resilience is intertwined with that of the agribusiness model at the international level, as shown below in table 1.

In order to assess social-ecological resilience, at the focal scale of one maize-growing community in Chiapas during the recent past and near future (10-30 years), we will look for the cross-scale interactions between components in table 1 and describe their feedback mechanisms (Buchmann, 2010). In the next section, we give a context for understanding the local food system in the study community.

Scales of variables	*Food system components*
Spatial Scales	
Microbiological	soil ecology, nutrition assimilation, chemical exposure to organisms
Field	net primary productivity, cost of production, maize yield, planned agrobiodiversity, diversity of associated plant and insect species, soil health, use of trees, use of organic fertilizers, pest and plant disease
Community/landscape	**attitudes, knowledge, organization, experimentation, crop diversity, out-migration, conflict, human health, nutrition, economic necessity, access to social assets, access to land, landscape matrix quality**
Subregional	health indicators, land use, public policies, level of influence of agribusiness, growing seasons, farm prices, social equality, farmer organizations, political economy and actors, capacity for self-reliance in key crops
National	food policy, environmental policy, trade policy, agrarian policy, education policy, health care policy, popular participation in democracy, adaptive governance, level of influence of transnational corporations
Global	climate change, global social movements, capitalism
Temporal Scales	
Hourly, Daily or Weekly	food security, physical activity, household labors, planting dates, weed control, soil biological processes
Seasonal	production cycles, climate factors, weed and insect communities, off-farm income, farm prices, training programs, implementation of government supports
Annual	farm productivity, learning-by-experimentation, community demographics, government policies, farm prices
Decade-level	**soil erosion and compaction, landscape mosaic quality, land use changes, trade policy, market influences, climate change, crop suitability**
Longer-term	land use, population factors, national sovereignty, world economic system

Table 1. Spatial and temporal scales in food systems. Focal scale is in bold.

6. Regional characteristics and historical context of the study community

The Fraylesca region is a hot and dry tropical zone that comprises the Central Valleys of Chiapas. Its major city, Villaflores, is located about two hours' drive south of the state capitol of Tuxtla Gutierrez, but the region continues another 100 km to the southeast. The Fraylesca traces its name back to the monks who habited the zone during the early colonial period. It is one of the regions of Chiapas with the least presence of indigenous language-speaking groups, probably due to the productivity of its lands and resulting displacement of the indigenous population during the colonial period. During the 1960s and 1970s, the Fraylesca region was known as the maize equivalent of the breadbasket (*granero*) for southern Mexico, where flat, alluvial valley floors gave typical maize harvests of 5-8 metric tons per hectare. Green Revolution technology, introduced through concentrated efforts to modernize maize farming systems in the flatlands, trickled upstream into the hills as population growth and limited land access pushed families upwards. By the 1990s, the vast majority of traditional maize varieties had been lost in the region, due to adoption of hybrid varieties in government seed programs and corporate advertising. The traditional shifting *milpa* agriculture system was generally replaced by a system of permanent fields in which maize is planted in monoculture during the first rainy season from June to July. Chemical laden maize fields came to dominate the landscape, while posters advertising agricultural chemicals and hybrid seeds are pinned to trees along the highways.

Recent decades have produced change in the Fraylesca region. Increased costs of maize farm inputs, together with soil degradation and low farm prices, appears to be putting the commercial maize farming system in economic jeopardy. Cattle-ranching has been increasingly embraced by hillside farmers, who graze beef cattle on maize stalks during the dry season and set them into the forest during the rainy season. More commercial lowland cattle operations buy the chicken manure from massive chicken farms in the region and feed it to their cattle, in order to produce greater volumes of milk and beef. This practice is generally disliked by the population, but that has not prevented it from becoming the conventional practice adopted by ranchers and dairy farmers. Intertwined with the growth of cattle-raising and the uncertainty of maize cultivation, the changing climate has added to the insecurity of social-ecological systems in the Fraylesca region. Growing seasons have shifted as annual precipitation has begun to concentrate in the second rainy season of the year from September to November, increasing the risk of cob-rot fungal disease. Rainfall has become scarcer during the long dry season from December to May, leading farmers to concentrate their cropping activities between the months of June and October.

Land tenure in the Fraylesca region is subject to similar social tensions to those that have characterized Chiapas as a whole during the last 50 years. The agrarian reform of the Mexican Revolution was slow in arriving to Chiapas, and the *finca* system of large landlord estates remained intact into the 1920s. The first *ejidos*, or agricultural communities created and protected by Mexico's agrarian reform laws, were created in the region as a result of social struggle in the 1920s in valley floors, and currently resemble small towns of paved streets, parks, and residential neighborhoods. As population pressure increased, peasant families have challenged landlord estates across the landscape, building makeshift communities in remote hills and asking for government recognition. Until the constitutional counter-reforms of 1992, Mexican land policy included a legal process for recognizing land claims through the Secretariat of Agrarian Reform (SRA) and reimbursing landowners for the forfeiture of unused land to new agrarian communities, which in turn could become *ejidos*. These second- and third-generation settlements differ in many respects from the older *ejidos*, in that they have much less access to health and education services, markets, transportation, and government supports. In newer *ejidos*, maize farming and cattle grazing take place on slopes that are much more vulnerable to erosion than the alluvial flatlands of the valley floors. As the Zapatista rebellion and federal military occupation of much of the Los Altos region took place in 1994-95, indigenous communities displaced by the violence began to look for land far into the hills of the Fraylesca region. These communities, fleeing from bloodshed, occupied land belonging to large and small landholders alike and sought federal recognition. While many such communities have obtained a certain level of land security by gaining *ejido* status, others remain in situations of precarious land tenure despite more than a decade of waiting for governmental recognition, even while cultivating the landscape and constructing homes.

7. Characterizing small farmer maize production in 24 de Febrero

The *ejido* of 24 de Febrero is about 45 minutes' drive south of Villaflores, in the municipality of Villa Corzo. The community is situated at 16°06′30′′ north and 93°22′33′′ west, at an altitude of 900 meters above sea level (Figure 2). The climate is considered subhumid tropics, with an annual precipitation of 1,248mm, concentrated in five months from June to October, and a mean temperature of 24° Celsius. Of the 1,240 hectares that belong to the *ejido*, 650 hectares are considered forestland and areas of important habitat for rare species of mammals, birds, reptiles, and amphibians. No more than 40 hectares are dedicated to maize cultivation in any given year, and often on fields with between 5 and 20 years of continuous maize monoculture production. Meanwhile, the area dedicated to cattle production shifts across the landscape, from pastures to former maize fields to forestland.

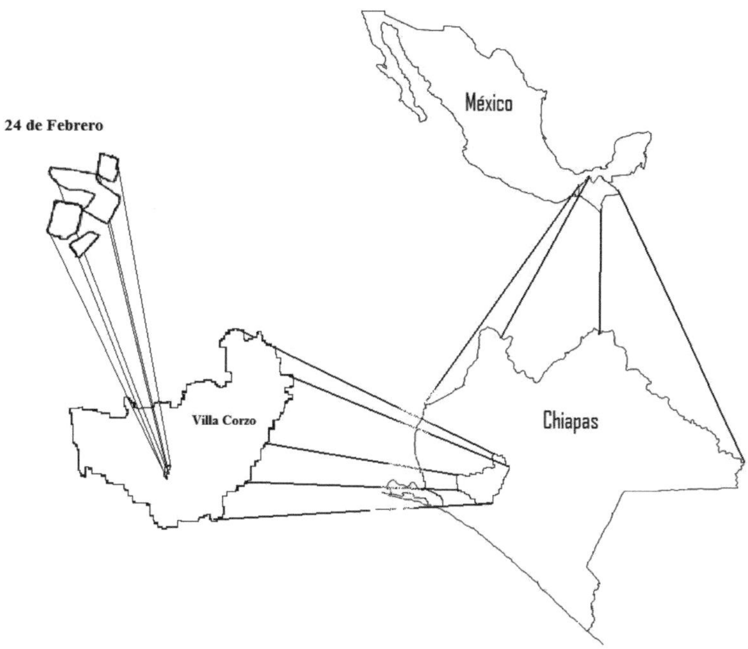

Figure 2. Geographic location of the *ejido* 24 de Febrero, in Villa Corzo municipality, state of Chiapas. Source: Rural Development Studies Network (2011).

The community of 24 de Febrero is primarily made up of one extended family of blood relatives and in-laws. None of the residents of the community speak an indigenous language. In contrast to many *ejidos*, the founders of the settlement were small farmers in nearby lands between the current community and the valley floor. In the mid 1980s, a group of peasants from a different part of the Fraylesca region organized with a lawyer to occupy lands belonging to one of these small farmers. In response, a large part of the extended family organized to create an *ejido* on the contested lands, and thus avoided a land conflict between peasant groups. So it was that the *ejido* 24 de Febrero was founded in 1986 with 5 homes. In subsequent years, the settlement has grown to include over 50 homes, with most new construction by sons and daughters of the community's founders. In the entire community, there are only 38 maize farmers, due to cost/benefit pressures that have taken out of production several former fields that are now considered to be too far to walk from the community.

Transition from the original subsistence system

OG plus no fires, OG plus fertilizante, etc.

Risks in the degraded state: soil erosion and cob-rot disease

We need to evaluate soil erosion in the 18 fields, as well as damage due to the cob-rot disease.

Farmers in 24 de Febrero plant both purchased maize seed and several native or mixed varieties that have been used by local farmers for generations. Planting usually takes place in the months of June and July, when torrential rains soften land. Maize fields are typically about 15 to 45 minutes' walk from the population center of the community, and are generally on sloped hillsides that are deeply eroded. Some farmers mix their seeds with chemical pesticides in order to limit damage from ants, while others use local herbs such as *epozote* to the same effect. Planting is carried out using hollow gourds to contain seeds, and a wooden stick with a metal tip to open up small holes in the untilled soil. Two or three seeds are tossed into each hole, which are made every 40 cm in rows of 80 cm width. Some farmers still follow the traditional practice of mixing squash seeds in with their maize seeds, in order to plant squash every 3m or so throughout the field.

Generally farmers clear fields for planting by using a systemic herbicide such as glyphosate, applied soon after the first rains in May. A few farmers still burn fields before planting maize, although only in areas that haven't been planted with maize in several years. Around 4 days after planting, most farmers apply a contact herbicide, such as 2,4D amina or paraquat. At 15-20 days, farmers apply a dose of nitrogen or phosphate fertilizer. Another herbicide treatment is carried out at 40 days with 2,4D amina and paraquat. At 45-50 days, a second fertilization is carried out. Maize plants are bent over below the ears only in fields where beans are planted in between rows, during the months of September and October. Sweet maize is harvested in September and October for family consumption, while the vast majority of maize ears are left to dry in the fields and harvested from December to March (table 2).

Jan	Feb	Mar	Apr	May	June	July	Aug	Sep	Oct	Nov	Dec
Maize harvest		Home gardens, sugar cane and off-farm activities			• Plant maize • Plant squash	40 days of heat without rain		• Plant beans and harvest sweet maize		•Bean and maize harvest begins	
Dry season					First rainy season			Second rainy season		Dry season	

Table 2. A seasonal calendar for cropping activities in 24 de Febrero.

During harvest, farmers remove only the corn ear, and leave all crop residues in the field. Soon afterwards, cattle are generally moved into harvested fields to graze upon the maize stalks. Cattle manure and pulverized maize stalks encompass the major sources of soil organic matter. Crop rotation is a very rare practice in the community, as maize is the only commercial crop and additional food crops such as sugar cane, banana, chili peppers and yucca are grown in small, separate batches. The practice of leaving fields in fallow, or several years of woody bush and chaparral tree regrowth in between cycles of maize production, was largely abandoned with the adoption of synthetic fertilizers by the community. Crop association, however, remains as a traditional practice in several fields, where farmers plant squash seeds along with their maize and beans in the latter rainy season.

7.1. Conventional and alternative practices

Figure 2 shows the ecological quality of several chemical practices in the community. A value scale from 0 to 3 was applied to four indicators of chemical usage, following a method for quantifying ecological quality of management practices (McCune et al. 2011). A value of 0 denotes practices with no benefit and with harmful impacts to ecological processes, 1 represents practices with no benefit but with a minimum of harmful effects, 2 denotes practices with minimal or insufficient benefits to ecological processes, and 3 represents practices with broad ecosystem benefits and that are applied with ecological criteria.

For example, in the case of chemical inputs as shown in Figure 2, the value attributed to each product is essentially the inverse of usage intensification; e.g. the higher value for 2,4D amina shows that this product is less widely used than paraquat. The indicator practices were chosen to be sensitive to changes from the conventional practices found in the community, in order to indicate where processes of innovation may be entering into maize farming. Combinations of organic with conventional fertilization practices remain very rare in the community, as do responses that indicate that farmers believe that it is possible to produce without conventional inputs. Incipient processes of innovation were found in the use of composts for fertilization, and in substitution for paraquat, a contact herbicide that is also a respiratory toxin. In 2011, four farmers were experimenting with liquid mixes made from the leaves of two common trees (*Ficus* spp. and *Byrsonima crassifolia*), along with fine salt and one-tenth the normal dosage of paraquat. Results were encouraging, although the next steps for expanding the usage of these homemade liquids are unclear. Despite the farmer experimentation taking place, chemical fertilizers and paraquat-based herbicides were most intensively used of the four indicators of agrochemical use among the 18 farmers surveyed.

The information in Figure 3 can be useful for determining the kinds of dependence produced within small farmer communities in Chiapas. The extreme dependence on synthetic fertilizers is an indicator of the level of soil erosion present in agricultural fields. Aside from highly unusual, small-scale efforts at growing organic maize for specialty markets or home use, the only maize-growing systems in Chiapas without this dependency

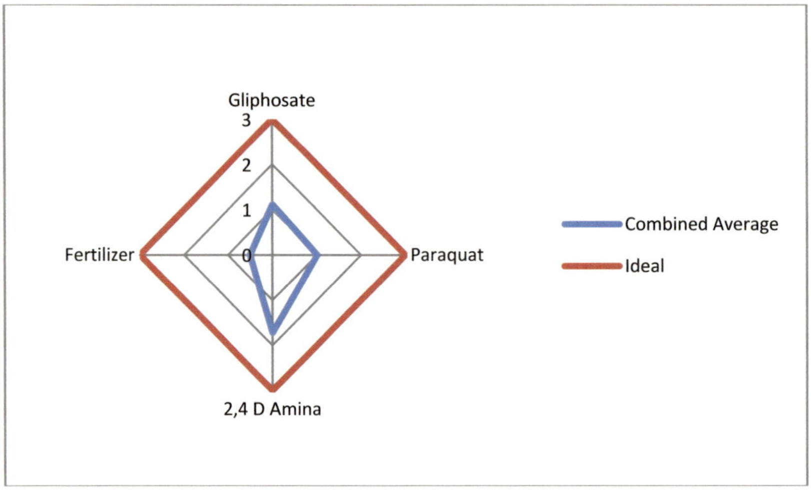

Figure 3. Ecological quality of chemical input use for maize production in the ejido 24 de Febrero, using an indicator system on a scale of 0 to 3. The value of zero represents maximum chemical application, while the value of three represents no usage.

on chemical fertilizers are the traditional shifting agriculture systems in which the forest regrowth of long-fallowed fields is cut and burned, and seeds planted into the rich layer of ash.

It is useful to identify the constraints and feedbacks of soil fertility management in Chiapas maize production. Fertilizers represent the greatest cost of maize farmers, with typical costs in 24 de Febrero reaching well over $1,000 US per hectare. Yet substitution of synthetic fertilizers with organic soil amendments is difficult, because hillside soils are so badly eroded that existing soil has almost no nutritional content, and up to 40 tons of organic matter per hectare would need to be applied in order to satisfy nutritional requirements of maize. The production and transportation of this volume of organic fertilizers would require large inputs of labor, difficult for farmers to provide as livelihood diversification strategies have left less time for maize-production activities than before. Thus the availability of time and labor (or cash in the case of purchased organic fertilizers) is a limiting factor for the efforts to break the dependency on synthetic fertilizers. The sloped maize fields represent an additional restraint, in that erosion is likely to undo most soil amendment applications until a massive labor effort goes into erosion-reduction practices, such as stone or stick terracing.

Feedbacks between management and ecological factors also complicate efforts to reduce fertilizer dependency. For example, populations of soil organisms that could improve soil structure and nutrition over time are likely to be negatively affected by the application of chemical fertilizers. In addition, cattle grazing in maize fields during the dry season can exacerbate erosion and also cause soil compaction. Compared to other crops, maize is hardy to degraded soil structure as long as sufficient nutrients are present. Indeed, farmers feel that its capacity to adjust to poor soils is an aspect of maize's centrality as "the" subsistence crop. Thus the economic need to produce every year is combined with the fact that maize is the only crop that can be produced under such marginal conditions, to create a system of monoculture maize production year after year in the same fields. Legal prohibitions on burning are meant to limit landscape degradation, but they also effectively end shifting agriculture, since fire is the typical way to open sloped fields to agriculture. Without shifting fields or rotations, soil degradation is accelerated to alarming levels.

The use of herbicide cocktails is a characteristic of maize farming in the Fraylesca region. The major restraints on reducing herbicide dependence are related to the labor opportunity cost, as manual weed removal takes a great deal of time that can be otherwise used productively by households. New ideas such as the organic weed retardant may represent the most promising directions for reducing herbicide dependence. Management feedbacks also exist with regard to herbicide use. After herbicide disturbance to fields, pioneer species such as aggressive weeds are the first plants to take advantage of nutrients and light in the newly opened spaces. The practice of herbicide application tends to increase the relative abundance of plants species that establish competitive relationships with maize crops. Fields with frequent use of herbicides tend to have major problems with a few key weeds, whereas fields without herbicide use have a greater diversity of associated plant diversity, including beneficial and non-competitive species.

A final interesting aspect of farmers' chemical use is the high dependence on herbicides and very infrequent of insecticides in the maize cropping system. In general, insect pests are not considered to be more than a nuisance, while farmers identify over 10 beneficial insects, mostly insect predators and parasitoid wasps (see below).

Over the course of two or three years, the community has progressively given more importance to ecological considerations, partly as a result of a training course in environmental education offered by the National Forestry Commission and carried out in the *ejido* by a local NGO in 2010. As a result of this course and follow-up activities by a participatory research team from the Autonomous University of Chiapas, several farmers have engaged in communication and experimentation with the purpose of substituting organic and traditional inputs for chemical inputs in agricultural activities. These and other alternative activities, such as a promotion of herbal medicine, are being adopted explicitly out of concerns for human health that emerged in monthly *ejido* assemblies, where the residents attributed poor health to chemical usage and the diminishing quality of diets. A first practice of seed saving using such traditional materials as ash, lime, and several kinds of herbs was carried out with the participation of 18 farmers. Of these, five attempted to grow fields of organic maize in 2011 using compost and organic fumigants. Figure 4 shows indicators for the appropriation of agroecological practices by the same 18 maize producers in the *ejido*. A similar value scale as with chemical use was applied to four indicators of alternative productive activities: nutrient cycling, crop rotation, crop association (intercropping), and seed management.

Our results showed a much greater appropriation of nutrient recycling practices than alternative seed management and crop association activities, basically due to the local customs of leaving crop residues in the field, moving cattle into former crop areas to eat the maize stalks, and the total absence of plowing. Crop rotation was shown to be a major

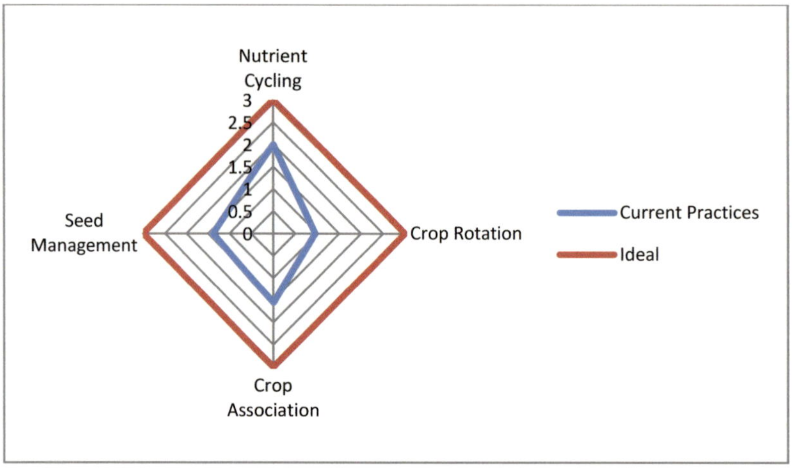

Figure 4. Ecological quality of alternative practices for maize production in the ejido 24 de Febrero, using an indicator system on a scale of 0 to 3. The value of zero represents total absence of alternative practices, while the value of three represents widespread use of ecological practices for reasons that farmers understand.

problem in the community, as annual crops of maize dominate the agricultural landscape during the single growing season. Maize is the preferred crop due to its importance as a food crop, its durability as a commercial crop, and its response to fertilizers even in highly eroded soils, providing a lightly positive cost-benefit balance to farmers for years even as soil quality declines. Cultural preference for maize makes diversification of the productive landscape a complex and sensitive process.

7.2. Biological interactions in maize production systems in 24 de Febrero

Weed and insect communities within the maize fields show that even under existing conditions and technological patterns, the small farmer landscape is capable of supporting a rich diversity of species and functional groups. This is an especially important finding, given that agriculture and biodiversity conservation are often considered in neoliberal theory to be mutually exclusive, even competitive uses for land in the tropics (Grau and Aide, 2008). The idea of contradictory agricultural and conservation goals, and the necessary segregation of the two, has led neoliberal resource economists to support wilderness reserves in some parts of the rural tropical landscape and industrial agriculture in the rest (Aide and Grau, 2004). The problem, as pointed out by rural organizations, is that small farmers are essentially excluded from both parts of the landscape, and conservation policy then becomes a tool for the dispossession of family farmers and rural communities. In addition to small farmer objections, the neoliberal model of biodiversity conservation has been challenged on ecological grounds, as recent decades of theory on metapopulations has shown the importance of migration between habitat patches for species survival. According to this conservation paradigm, also known as the convergent model of mixed land-use (Miki et al. date unknown), agricultural systems that retain elements of the original ecosystem can promote successful migration between patches of wilderness. This may make the ecological quality of agroecosystems even more important than the conservation of habitat patches to biodiversity conservation in the tropics.

In 18 fields of 24 de Febrero, weeds and crops were measured monthly for the percentage of area they covered within three 50cm x 50cm quadrants placed using a random block design in each field, for a total of 54 quadrants, during the six-month maize growing season. In the total studied area of 13.5 square meters, over 30 species of weeds were found, reflecting high overall richness despite the use of herbicides. Tree cover and type varied among fields, but crucially, trees were present in all fields. The uses for weeds and trees were various; eight weeds were considered to be edible, and 11 were identified as medicinal plants. Among trees, several were nitrogen-fixing legumes and others were fruit-bearing, with the remainder having social use as building material, medicine, or firewood.

We identified 29 families of insect herbivores, seven families of secondary consumers or predators, 12 families of parasitoids, and two families of pollinators in five samplings across the eighteen study fields during the growing season of 2011. The insect community within maize fields reflects a high level of biodiversity and is likely to have the net result of stabilizing yields, creating an "ecological homeostasis" through complex networks of trophic, life-cycle, and density-dependent interactions (Vandermeer et al. 2010). The critical interactions within such an autonomous agroecological service such as pest control may be highly complex and occur on various spatial and temporal scales.

In 24 de Febrero, maize fields also bear beans, squash, tomato, edible herbs, several medicines, and several use categories of trees. This multi-use aspect of agricultural fields may lend itself to system resilience, since any detrimental impact on maize production is partially offset by the other functions of the same land. Land-use diversity is an element of system resilience that is pronouncedly strong in small farmer settings (Altieri, 2010). While the use of the farm landscape in 24 de Febrero retains an important level of diversity, it is also useful to ask why it doesn't have even more, especially given the supposition that small farmers maintain diverse productive systems. To understand the drivers of land-use change and agricultural intensification, it becomes necessary to examine the social and economic vulnerability of small farmers in Chiapas.

7.3. Characterizing social resilience

Livelihood is an important concept for understanding risks in social-ecological systems. Both vulnerability and livelihood trace their conceptual roots in the search by Sen for adequate measures of well-being (1993). Livelihood has to do with the relationship between households and the conditions of their production and reproduction as an economic unit, including housing, employment, income, access to basic necessities and to consumer goods, transportation, health, and education. It is generally used to define baseline measures of human well-being, and as such applied to small scale rural producers and the rural or urban classes without property. While livelihood studies generally examine immediate aspects of economic life at a household level, vulnerability studies tend to focus on structural factors, such as legal, political, cultural, ecological or economic factors that threaten livelihoods.

In order to make an initial characterization of livelihood resilience factors in the community of 24 de Febrero, an indicator system was created to include among its variables: food access, health care access, access to credit, access to public programs, access to markets, access to alternative technologies, and access to education (Figure 5). These access indicators are significant to everyday life under normal circumstances, but they are also indicative of social risks that could become urgent under changing conditions.

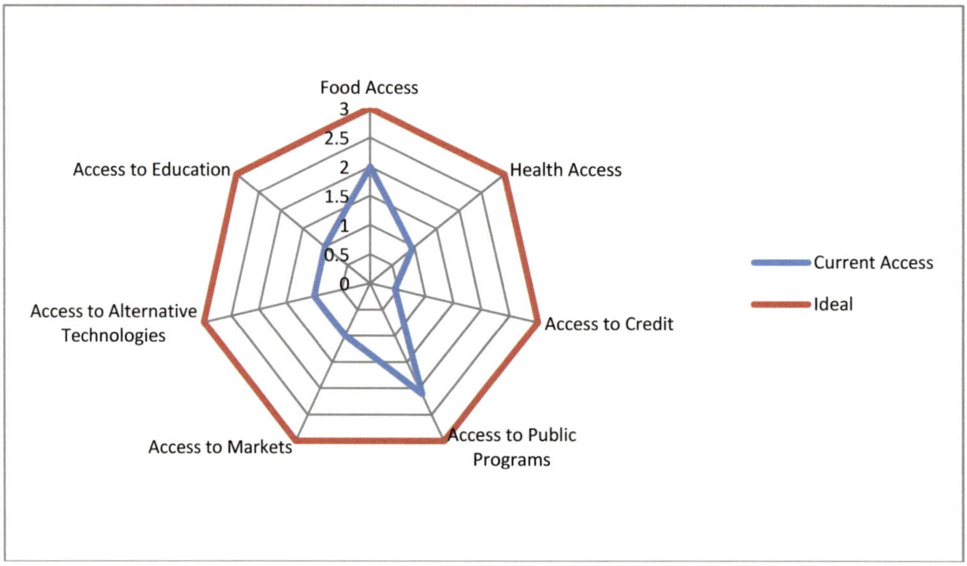

Figure 5. Factors of livelihood resilience in the ejido of 24 de Febrero, using an indicator system on a scale of 0 to 3. The value of zero represents abandoned or systematically denied access rights, while the value of three represents free and universal access as well as participation within planning or implementation processes.

Results confirm the existence of several types of social vulnerability in the community. Indicator values for access to food and public programs were substantially greater than those for other variables. Values were notably low for access to education and access to credit, two indicators related to opportunity. With regard to education, the attained value of one means that the average response to interview questions was that beyond elementary school, monetary costs associated with education made it inaccessible. The very little access to credit for farmers in the maize growing regions of Chiapas is a matter of considerable importance for the community of 24 de Febrero, and contributes to migration by young people to the United States in pursuit of sufficient cash to construct homes or purchase fertilizers. At the same time, the conservation of certain traditional practices is often attributed to the lack of farm credits, which would enable farmers to pursue a more technified production strategy.

8. Food sovereignty and the nation-state

While we have characterized many aspects of the maize production system in 24 de Febrero, the question remains: where is the community located on a gradient between agribusiness and the food sovereignty model? Clearly, the food system of the community is somewhere in between the two states that we have described. Its combinations of traditional and conventional technologies, cash- and subsistence-oriented agriculture, monoculture and multi-use systems, indeed, environmental stewardship and degradation, give the local food system a character highly compatible with the sovereignty system state. However, the community is within a nation that has been subjected to the full formula of capitalist agribusiness.

Here we come to a fundamental issue of scale: the local food system in 24 de Febrero can only be understood in its larger context, as part of Mexico's food system. On one hand, it is an adaptation on the traditional *milpa* system of maize production on collectively held lands that has characterized Mexican food systems for millennia. On the other hand, it is the result of the extension of capital logic and conventional technologies to far corners of the Mexican countryside, bringing junk food and chemical input dependencies to the rural household. The ambiguity is a signal of the importance of scale, and it may well be possible that systems within the gravitational pull of one steady state could also exist within another. For example, the Procede land certification policy enacted during the 1990s was thought to be the end of collective landholding in Mexico, as it partitioned private titles for *ejido* lands and legalized land sales (De Ita, 2000). However, the internal resilience of the *ejido* system, based on social and political feedback mechanisms, was strong enough that certification did not have the same effect that it has in other parts of the world, such as Africa and the Middle East.

At a larger historical scale, collective resource-use regimes such as the *ejido* system may be momentarily compatible with both capital-driven and socially-planned economies. In this sense, a valid comparison can be made between *ejidos* of Mexico and agricultural production cooperatives in Cuba, both of which are based on profound land reform and collective agrarian property governed by local assembly. Such institutions can exist within countries dominated by the industrial food model, but the dynamic of the overall food system will determine how long they last and how they change. The *ejido* was created as a compromise between the radicalized peasantry of the Mexican Revolution and conservative groups of power that were interested in limiting resource redistribution. By giving *ejidos* to peasants, militant rural organizations could be demobilized and wages could be kept low during industrialization, since industrial workers' wages were supplemented by their access to productive land. Essentially, the *ejido* was a major tool for the consolidation of a new bourgeois regime after the Mexican Revolution. It was the eventual reorientation of the national economy toward global capital and away from nation-building in the late 1970s that brought the *ejido* system into conflict with the emerging neoliberal resource management regime. The temporary compatibility of agrarian systems that have a food sovereignty character, such as the Mexican *ejido*, within industrial food systems that are in the process of consolidation, is a matter of scale and historical contingency. Components embedded in one kind of food system can reflect a distinct qualitative character, as long as they are so limited in scale and impact as not to push the larger system to a threshold.

The dependence of the maize production system in 24 de Febrero on foreign multinational corporations and the chemicals they sell is a sign of its integration into the international capitalist development model that dominates Mexico from outside. As maize sellers, they are limited to the white maize varieties sought by middlemen buyers en route to large-scale tortilla production (mixed first with maize from the state of Sinaloa or, increasingly, the United States) or exportation to Guatamala. Surrounded by an agricultural landscape of maize fields with hybrid seeds, the integrity of their traditional varieties is at risk of contamination. It would seem that in neoliberal Mexico, the community is within the outer reach of the industrial food model and as such, subject to its contradictions. Nonetheless, on a household level, there remains a level of resistance to the industrial food system, which takes the form of self-sufficiency in basic grains, conservation of landrace maize and bean varieties, use of home gardens, traditional labor-sharing arrangements and artisan food processing. These are all components of the local food system consistent with a food sovereignty framework, but they are gradually disappearing from the landscape. What is the role of the Mexican State in this transformation?

8.1. Devolutionary governance

Since the Mexican Revolution of 1910, national policy toward agriculture has reflected a struggle between peasant groups that have fought for land access and favorable policies toward small farmers, and a combination of business and political elite from within and outside of Mexico that have sought to develop a capitalist, export agriculture model. Since the dawn of the twentieth century, capital investments began creating two basic tracks for Mexican agriculture: the capital-intensive irrigated, specialized farms in Central and Northern valleys and plains, and subsistence agriculture in most other non-urban land in Mexico. This split in land use reinforced the nation's conception of monoculture as "modern," and diversified, low-input farming as "backward."

The agricultural research and technology program that came to be known as the Green Revolution was largely a US Cold War-era effort to resolve issues of hunger and poverty in Mexico with technical solutions, rather than new social and economic policy (Perfecto et al. 2010). The proliferation of new, "modern" seed varieties, as well as irrigation infrastructure, synthetic fertilizers and pesticides during the 1960s, 1970s, and 1980s, was a highly uneven process that reflected the compromise between corporatist governmental policy beneficial to large agricultural interests, and the commitment to

Since the 1980s, Mexico's government has opted for a free trade strategy in all productive spheres, as a result of changes in the economic ideology of the ruling party, as well as external pressure from international lenders and the United States. For Mexican industry, this meant the final and unequivocal abandonment of the import-substitution industrialization strategy. In commerce, it eventually led to the signing of the North

American Free Trade Agreement in 1993, which opened up Mexican markets to a flood of cheap products from the United States. In agriculture, the adoption of the free trade model meant a shift in strategy from the goal of self-sufficiency that had characterized agricultural and land policy since the revolution. Guaranteed farm prices for basic grains such as maize disappeared, as the state reduced its presence in the countryside and international private actors stepped into the void. Cheap grain from subsidized farmers in the United States began to flood Mexican markets, adding to the economic insecurity of millions of Mexican maize farmers. Meanwhile, the price of tortilla, the basic and essential form of maize in the Mexican diet, has more than tripled for consumers since NAFTA was signed, as a result of concentration of the maize storage and processing sectors by several transnational corporations.

One of the most controversial issues in contemporary Mexico is the entrance of genetically modified maize into the country, almost universally from the United States, as seed, feed, or food. In 2001, Mexican and U.S. researchers accidently found traces of genetically modified maize in landrace varieties of rural Oaxaca (Quist and Chapela, 2001, Nature 414), and subsequent studies have confirmed the contamination of maize landraces by modified genes across Mexico. Given the extraordinary cultural and alimentary importance of maize in Mexico, the loss of traditional agrobiodiversity in this crop represents a loss of national patrimony and sovereignty. In 2007, President Felipe Calderon created by decree a federal program to support *in situ* conservation of landrace maize varieties by farmers. However, the Secretary for Agriculture, Livestock, Fish and Food (SAGARPA for its initials in Spanish) was cold to the proposal, as it went against the productivity focus of its programs. Thus it fell to the Secretary for Natural Resources and the Environment (SEMARNAT) to take on the maize biodiversity program. SEMARNAT, in turn, sent the new law to its National Commission for Natural Protected Areas (CONANP), which began to apply the program, its implementation having now been reduced to agricultural areas within nature reserves.

CONANP's Program for *in situ* Conservation of Landrace Maize, or PROMAC (*Programa de Maíz Criollo*) as it is more commonly known, is still a new fish in a very complex pond of federal and state programs that combine agriculture and natural resource conservation. It pays about $100 US per year to farmers who have been growing landrace maize varieties in nature reserves to continue growing them, and advocates the conservation of the traditional *milpa* productive system (maize in association with squash, beans, and other edible plants). PROMAC funds are used based on the discretion of each nature reserve, and can be used to hold seed exchange fairs, conduct capacity-building trainings for farmers, build seed banks and even create maize-based cultural centers. While this program clearly has the potential to strengthen the peasant maize production system, its capacity to help small farmers and protect landrace maize varieties depends on how it is implemented in each nature reserve. In interviews, many nature reserve officials compare PROMAC to PROCAMPO: a program created during the administration of Carlos Salinas (1988-1994) to buffer his free trade economic shocks and which pays annual subsidies to all citizens who show documents proving that they grow crops or raise livestock. Despite its populist appeal, PROCAMPO is a notoriously inefficient program, as its implementation provides ample opportunity for fraudulent payment claims and requires no participation in training programs or production plans. Many Mexican politicians at the national level oppose support of any kind for maize production, because they see it as a marginal subsistence activity that is outside of the free market agricultural strategy of specialized crop exports and basic grains imports. Thus from its origins, PROMAC has been born into a hostile and disjointed institutional atmosphere, in which some nature reserves have ignored the program while others have encouraged farmers to enter into it.

While PROMAC is among the more important federal programs for maize farmers in Chiapas, due to the significant amount of farmland within protected areas, on the state level there is a program called Solidarity Maize that is closely connected with the governor's office. This program ostensibly gives payments in the form of agricultural inputs to all maize farmers in the state. Given that fertilizers represent the greatest expense in maize farming in many regions of Chiapas, small farmers are generally in favor of Solidarity Maize and eager to participate. Unfortunately, the program reaches a relatively small portion of the actual maize farmers in Chiapas, while creating a massive informal market for sacks of fertilizers that are often exchanged for political allegiance long before reaching farmers.

In 2008, a group of farmers and advocates formed the Landrace Maize Network (*Red de Maíz Criollo*) in Chiapas in order to stem the loss of traditional peasant varieties of maize and defend the *milpa* production system. This group protested the fact that the supports from the state government through the Solidarity Maize program set small farmers on a course toward conventional, chemical-laden agricultural practices. The Landrace Maize Network achieved a commitment by the state government to offer organic fertilizers to those producers who request them, setting a new precedent for governmental support for alternative agriculture. Unfortunately, to date very few farmers know that they have the option to request organic farm inputs.

The farmers of 24 de Febrero have yet to receive support from PROMAC, despite their long-term commitment to growing traditional maize varieties, and they have not received support from the Solidarity Maize program either. In fact, PROCAMPO is the only government support that they receive. Despite being relatively close to population centers, and following the requirements to be considered in state and federal programs, they have been left out of the little support for small-scale agriculture that exists in Mexico.

9. Conclusions

The future is unknown, social and ecological drivers of change are linked, and periodic, qualitative change is part of life. That is one view of the world, carefully developed in resilience theory since Holling's (1973) seminal essay. Here we have posited two contrasting (but not exhaustive) food system possibilities, in part to demonstrate the openness of history. We do not see evidence for necessary evolution toward stable equilibrium in either social or ecological systems. Rather, the last several hundred years have shown that history is full of surprises, and theory of stages of development is often more hindering than helpful. In its insistence on the existence of the unknown, orientation toward emergent properties, and focus on feedback mechanisms, resilience theory is of extraordinary usefulness in social science.

Despite such impressive strides in systems thinking, we find the brave new world embraced by resilience science to itself be exclusionary in terms of possible outcomes for humanity. Far too often, the capital system is naturalized into the feedbacks of the social-ecological framework, rather than understood as a historical system that is liable to the same phases of growth, decadence, collapse and renewal as something qualitatively different. Indeed, in the present conditions of global financial crisis and historical highs of economic inequality, it would be quite blind not to accept the collapse of global capitalism as a historical possibility. With natural resource exhaustion and an exploitative human-nature relationship increasingly understood as inevitable contradictions of this economic system (O'Conner, 2001), the alternative stable states model is highly relevant for the testing of alternatives at distinct scales.

In *ejido* assemblies, the residents of 24 de Febrero have identified the directions that they would like the community to take. The local vision is of organic agriculture as a response to what appear to be increasing health problems, such as high blood pressure, diabetes, and cancer. The community has made collective decisions to begin a process of experimentation to combine what people remember of the practices used by past generations with technical advice from a research team of the Faculty of Agronomic Sciences at the Autonomous University of Chiapas. The role of local innovation is being filled (Milestad et al. 2010). But is this enough to pull the system toward a food sovereignty stable state?

The resilience literature identifies the need for adaptive governance (Allen and Holling, 2010; Folk et al. 2005). This can be interpreted in several ways, but the creation of public-private partnerships is often suggested, albeit in terms of "bridging organizations" or multi-stakeholder groups. In sum, governance for resilience is understood to take place in what Walker and colleagues (2002) call with refreshing honesty "more-or-less democratic, pluralistic, capitalist" societies. The resilience principle, as applied to social-ecological systems, has been applauded and feared because it normalizes—and absorbs all critique to—the neoliberal development model (Walker and Cooper, 2011). When adaptive governance is understood to mean increased private influence over formerly public spheres, especially natural resource management, then the objectives of its research agenda may well include building resilience to "shocks and disturbances" like market crashes, critical social movements, and dissent.

Unfortunately, history may show that the degraded system is global, and that what is "too big to fail" in the international economic system has indeed already failed, in terms of its social and ecological impacts. The destruction of global agrobiodiversity—a fundamental component of food system resilience—that took place during the last half-century has not been accompanied by a solution to world hunger, as more than one billion undernourished people make plain. If resilience is the "capability of a system to maintain its functions and structure in the face of internal and external change and to degrade gracefully when it must,

(Allenby and Fink, 2005)" then perhaps the most prudent option would be to begin a wide debate about the graceful exit of the agribusiness model. The development of sovereign food systems at the local level requires cross-scalar interactions that should not be limited to farmer innovations, but include farmer-to-farmer networks, strong rural organizations, and redistributive public policy. If these requirements cannot be met within the neoliberal development model, then the next step is to ask what kind of transformation is necessary at the global level, beyond the narrow market valuation of Nature.

Author details

Nils McCune
Ghent University, Belgium
Red de Estudios para el Desarrollo Rural AC, Mexico

Francisco Guevara-Hernandez
Facultad de Ciencias Agronomicas, Universidad Autonoma de Chiapas, Mexico

Jose Nahed-Toral,
El Colegio de la Frontera Sur, San Cristobal de las Casas, Mexico

Paula Mendoza-Nazar
Facultad de Medicina, Veterinaria y Zootecnia, Universidad Autonoma de Chiapas, Mexico

Jesus Ovando-Cruz
Red de Estudios para el Desarrollo Rural A.C., Mexico

Benigno Ruiz-Sesma
Facultad de Medicina, Veterinaria y Zootecnia, Universidad Autonoma de Chiapas, Mexico

Leopoldo Medina-Sanson
Idem

10. References

Allen, Craig, and C.S. Holling. 2010. Novelty, adaptive capacity, and resilience. *Ecology and Society* 15(3): 24-39.

Allenby, Brad, and Jonathan Fink. 2005. Toward inherently secure and resilient societies. *Science* 309: 1034-1036.

Bakan, Joel. 2004. *The Corporation: the pathological pursuit of profit and power.* New York: Free Press.

Buchmann, Christine. 2010. Farming System Dynamics: The quest for a methodology to measure social-ecological resilience in subsistence agriculture. Presentation in: *Adapative management in subsistence agriculture*. 9th European IFSA symposium. 4-7 July 2010, Vienna (Austria).

García-Linera, Álvaro. 2011. *La Potencia Plebeya: acción colectiva e identidades indígenas, obreras y populares en Bolivia*. Havana: Fondo Editorial Casa de las Américas.

Ericksen, Polly, Beth Stewart, Jane Dixon, David Barling, Philip Loring, Molly Anderson and John Ingram. 2010. The Value of a Food System Approach. In: *Food Security and Global Environmental Change*. London: Earthscan.

Folk, Carl, Thomas Hahn, Per Olsson, Jon Norberg. 2005. Adaptive governance of social-ecological systems. *Annual Review of Environmental Resources* 30: 441-473.

Fox, Jonathan and Libby Haight. 2010. Mexican agricultural policy: Multiple goals and conflicting interests. In: *Subsidizing Inequality: Mexican Corn Policy Since NAFTA*. Woodrow Wilson International Center for Scholars.

García-Barrios, Luis, Yankuic M. Galván-Miyoshi, Ingrid Abril Valdivieso-Pérez, Omar R. Masera, Gerardo Bocco, and John Vandermeer. 2010. Neotropical forest conservation, agricultural intensification, and rural out-migration: the Mexican experience. *BioScience* 59(10): 863-873.

González, Humberto and Alejandro Macías. 2007. Nutritional Vulnerability and Mexico's Agro-Alimentary Policy. *Descatos* 6: 14-31.

Grau, H. Ricardo and Mitchell Aide. 2008. Globalization and Land-Use Transitions in Latin America. *Ecology and Society* 13(2): 16-28.

Guevara-Hernández, Francisco, Nils McCune, Heriberto Gómez-Castro, René Pinto-Ruíz, Francisco Medina-Jonapá, Adalberto Hernández-López, and Carlos Tejada-Cruz. 2011a. Conflicting regulatory systems for natural resource management in Southern Mexico: An ethnographic case study. *International Journal of Technology and Development Studies* 2(1): 30-62.

Guevara-Hernández, Francisco, Nils McCune, Luis Rodríguez-Larramendi, and Gillian Newell. 2011b. Who's who? Power mapping, decision making and development concerns in an indigenous community of Oaxaca, Mexico. *Journal of Human Ecology* 36(2): 131-144.

Holling, C.L. 1973. Resilience and stability of ecological systems. *Annual Review of Ecological Systems* 4: 1-23.

Kinzig, Ann, Lance Gunderson, Allyson Quinlan, and Brian Walker. 2007. *Assessing and managing resilience in social-ecological systems: a practitioner's workbook.* Resilience Alliance: Stockholm.

Levins, Richard. 2007. The Maturing of Capitalist Agriculture: Farmer as Proletarian. In: Lewontin, Richard, and Richard Levins, *Biology Under the Influence: Dialectical Essays on Ecology, Agriculture, and Health*. New York: Monthly Review Press, pp. 329-341.

Lin, Brenda. 2011. Resilience in Agriculture through Crop Diversification: Adaptive Management for Environmental Change. *BioScience* 61: 183-193.

Magdoff, Fred, John Bellamy Foster, and Frederick H. Buttel. 2000. *Hungry for Profit: the agribusiness threat to farmers, food, and the environment.* New York: Monthly Review Press.

McCune, Nils, Yanetsy Ruíz González, Edith Aguila Alcántara, Osvaldo Fernández Martínez, Calixto Onelio Fundora, Niria Castillo Arzola, Pedro Cairo Cairo, Marijke D'Haese, Stefaan DeNeve, and Francisco Guevara Hernández. 2011. Global questions, local answers: soil management and sustainable intensification in diverse socio-economic contexts of Cuba. *Journal of Sustainable Agriculture* 35(6): 650-670.

Medellín, Rodrigo and Miguel Equihua. 1998. Mammal species richness and habitat use in rainforest and abandoned agricultural fields in Chiapas, Mexico. *Journal of Applied Ecology* 35: 13-23.

Milestad, Rebecka, Susanne Kummer, and Christian Vogl. 2010. Building farm resilience through farmers' experimentation. Presentation in: *Knowledge, innovations and social learning in organic farming*. 9th European IFSA symposium. 4-7 July 2010, Vienna (Austria).

Miki, Rosa, Mariana Zarazua, Adriana Aguilar, Lizzette Luna, Juan Carlos Hernández, Martín Cadena, and Mariano Torres. Date Unknown. Divergent vs. Convergent Development Models. Sustainability Science Seminars.

O'Conner, James. 1998. *Natural causes: essays in ecological Marxism.* New York: Guilford Press.

Patel, Raj. 2009. What does food sovereignty look like? *Journal of Peasant Studies* 36(3): 663-673.

Perfecto, Ivette, John Vandermeer, and Angus Wright. 2010. *Nature's Matrix.* Earthscan: London.

Ramírez-Marcial, Neptalí, Mario González-Espinosa, and Guadalupe Williams-Linera. 2001. Anthropogenic disturbance and tree diversity in Montane Rain Forests in Chiapas, Mexico. *Forest Ecology and Management* 154: 311-326.

Rosset, Peter. 1998. Do we need new technology to end hunger? In: Lappé, Francis Moore, Joseph Collins, Peter Rosset and Luis Esparza, *World Hunger: 12 Myths.* Grove Press/Earthscan.

Rosset, Peter, and Medea Benjamin. 1994. *Cuba: the greening of the Revolution*. Food First Press: Oakland.

Sen, Amartya. 1992. Capacidad y Bienestar. In: Nussbaum, Martha, and Amartya Sen (Eds.), *La Calidad de Vida.* Oxford: Oxford University Press, the United Nations University, pp. 54-80.

Shattuck, 2011. Resilience and Vulnerability in Agriculture: A Cross-Disciplinary Approach. Forthcoming.

Suding, Katharine, Katherine Gross, and Gregory Houseman. 2004. Alternative states and positive feedbacks in restoration ecology. *Trends in Ecology and Evolution* 19(1): 46-53.

Vandermeer, John, Ivette Perfecto, and Stacy Philpott. 2010. Ecological Complexity and Pest Control in Organic Coffee Production: Uncovering an Autonomous Ecosystem Service. *BioScience* 60: 527-537.

Walker, Brian, Stephan Carpenter, John Anderies, Nick Abel, Graeme Cumming, Marco Janssen, Louis Lebel, Jon Norberg, Garry Peterson, and Rusty Pritchard. 2002. Resilience management in social-ecological systems: a working hypothesis for a participatory approach. *Conservation Ecology* 6(1): 14-30.

Walker, Jeremy, and Melinda Cooper. 2011. Genealogies of Resilience: From Systems Ecology to the Political Economy of Crisis Adaptation. *Security Dialogue* 14(2): 1-28.

Walker, Richard. 2004. *The Harvest of Bread: 150 years of agribusiness in California.* Berkeley: University of California Press.

Chapter 22

Methodology for the Regional Landfill Site Selection

Boško Josimović and Igor Marić

Additional information is available at the end of the chapter

http://dx.doi.org/10.5772/45926

1. Introduction

One of the most important causes of environmental pollution is certainly an inadequate waste management. The three factors that have primarily influenced this problem area are: ever increasing amount of municipal solid waste (which causes increasingly pronounced space occupation), increasing amount and types of hazardous waste, as well as lack of awareness on the importance of acting promptly in this field. Particular problems in waste management occur in developing countries, where the awareness of the importance of environmental protection has not yet achieved a satisfactory level and where, out of economic or political reasons, professional guidelines associated with waste management are not observed. Problems emerge either due to a lack of legislation, or obsolete legislation, lack of trained personnel, inadequate waste management infrastructure, financial constraints in the establishment of a modern waste management system, population lacking the awareness about solid waste management, impossibility of selecting appropriate space for landfill development, lack of standards, etc.

Great problems in waste management in Serbia are caused by increasing amount of waste, lack of sanitary landfills built under international standards (which is inefficient and ecologically acceptable), as well as by the fact that the principle of hierarchy in waste management is not observed at all. Problems emerging in the field of environmental pollution and the manner of responding to pollution through the planning documentation, only increase the importance of this problem area.

Waste management is a complex process which implies a control of the entire waste management system (from waste generation, through collection and transportation of waste, to waste treatment and disposal), along with the support of legislation and appropriate institutional organization. The accent in the present paper is placed on spatial planning as an inevitable instrument for strategic waste management. The paper also points out the importance of spatial aspect in the waste management planning process.

2. Disposal of solid waste

The final functional element in the waste management system is waste disposal. Waste disposal is a final fate of all types of waste, either municipal solid waste, collected and transported directly to landfills, or industrial waste or other materials from waste treatment facilities which are of no use-value any longer [1].

Landfill forms the basis of every waste management plan, because there will always be waste to be disposed of.

Sanitary landfills are sites selected for waste disposal, such as natural or artificial (excavated) depressions, engineered facilities, where the waste is, through appropriate technological processes, compacted as densely as practicable to minimize its volume and covered with a layer of soil or some other material in a systematic and sanitary manner. Before proceeding with such work, a terrain to be used must be selected, surveyed and prepared [2].

Sanitary landfills are necessary in any combination, even for some other form of solid waste treatment, because there will always be waste to be disposed of. Uncontrolled dumps must be closed along with necessary sanitation. This requires knowledge of a series of notions, processes and activities which should enable proper landfill planning, design, construction, exploitation and funding, as well as control of landfill environmental impacts [3].

3. Spatial aspect of waste management

Considering that waste management system is realized in space, it is quite clear that characteristics of space greatly determine the choice of an adequate management system, i.e. its spatial organization. This primarily refers to the selection of sites having physical elements of the system, such as, primarily, sanitary landfills, transfer stations, recycling centers, etc. In this context, physical-geographical and anthropogenetic characteristics of space are of great importance. Relative to these characteristics, conceptual solutions to the waste management system are defined, and landfill site selection process is carried out for elements of the waste management system.

3.1. Requirements for the regional landfill site selection

In the waste disposal process, a controlled disposal procedure is unavoidable, either for the disposal of genuine waste or materials that remain after the treatment process, or, as necessary, if the main process cannot be carried out in certain period because of interruption, defect, overhaul, or out of other reasons. Sanitary landfills are necessary in any chosen waste management option, because there will always be waste to be disposed of on landfills. In this sense, locating potential landfill sites, as the most commonly used process through which a huge amount of collected waste is treated, should be given great attention in the waste management process, i.e. in spatial planning process. This is a very delicate and very important process from the viewpoint of the protection of key environmental factors (land, water and air), landscape values, as well as the protection of population health [4]. Out of this reason, it is also necessary to dedicate great attention to the investigation of a character, as well as potential and real landfill impact on the environment. This enables the elaboration and implementation of measures to eliminate or minimize negative effects.

Sanitary landfill planning and construction is only a part of a complex solid waste management process which encompasses the treatment of waste from its generation, through minimization of its amount, selection, recycling, collection, transport and disposal, to landfill recultivation and bringing of land to new use. However, although sanitary landfills are only a part of a wider waste management process, this activity is characterized by a very complex and long-term process which must take into account natural and anthropogenetic characteristics of space.

Sanitary landfill is available land for solid waste disposal at which engineering methods of waste disposal are used in a manner in which threats to the environment are minimized. The landfill site selection and technology of devices and equipment for sanitary waste treatment and disposal should be in the function of the protection and rational use of space.

Sanitary landfill development implies activities in several phases where certain sequence must be obeyed. The process is usually carried out in four phases:

- landfill site selection (site investigation process),
- identification of a landfill site (through the planning documentation) and elaboration of conditions for bringing it to the intended use,
- elaboration of construction (technical) documentation,
- landfill construction.

3.1.1. Preconditions for sanitary landfill construction

It is necessary to consider the following requirements and requirements for sanitary landfill construction:

- Spatial and urban planning requirements
- Spatial and regional requirements
- Landfill site selection
- Required land area
- Transportation distances
- Local site conditions
- Topography
- Climate conditions
- Hydrogeological conditions
- Geological conditions

- Geo-mechanical conditions
- Environmental protection

Landfill site selection.- In planning, landfill site selection occupies extremely important place. In the widest sense, the natural, social, political, economic and technical factors have an important role in landfill site selection, thus it follows that the selection is to be made by a multidisciplinary team of experts. Given that landfill construction is considered as a non-economic activity, a special task is to select landfill site from the aspect of the use of buildable land and its price, as well as other natural or urban values which have an important role in relation to rationality and planned landfill remediation or its rehabilitation, i.e. reconstruction [5]. From technical and technological aspects, for the planning, design, construction and exploitation needs, it is necessary that on each landfill site the following is ensured:

- Complete sanitary security for people living in the surrounding residential areas, as well as personnel working at landfills
- Protection of land, air, ground and surface water from pollution
- Rational use of land, as well as save land (increased levels of waste compaction using special machines, as well as a deposition height)
- Maximum number of machines and equipment for all types of works

3.2. Implementation of GIS tools in waste management

Information system is an arranged set of information on things and facts in surroundings, with the aim to get acquainted with a system. Right decision making in planning and space organization depends to a great extent on knowledge, i.e. the quality and importance of information available to decision makers.

The GIS is a powerful set of computer tools for collecting, storing, searching as necessary, transformation and display of real-world data for various purposes [6].

As one of the most complex information systems that cover all spatial problems, the GIS has many advantages out of which the most significant are:

- It covers all elements of geo-space and ecological elements
- It includes natural and social elements of space

The use of GIS is appropriate for:

- Spatial planning
- Mapping for various purposes
- Traffic planning
- **Waste management planning**
- Natural resource inventory and management
- Computer mapping of population and entering of census data
- Creating hazard maps and programs of procedures in such cases, etc.

Essentially, GIS contains data on:

- Air
- Earth's surface
- Water areas
- Lithosphere
- Soil
- Biocenosis
- Anthropogenetic spatial elements, etc.

The GIS, as already mentioned, consists of spatial data and descriptive data.

Spatial data refer to locations, i.e. spatial relationships between phenomena and objects. They are obtained based on literature, maps, aerial photos taken from aircrafts, etc., and they are useful only if they can be converted into maps.

Descriptive data are linked to localities, polygon line or body and are the system accompanying content.

The GIS key features are:

- Possibility of spatial search of phenomena
- Possibility of overlapping contents and combining individual contents into a new quality
- Logical operations with spatial and descriptive data

Geographic information systems are most frequently compatible with most of related systems (geodetic, agricultural, geologic, mining, water resources management, forestry, urban planning systems, etc.), but also with census databases, statistical information systems, technological databases, databases associated with health, education, science, etc. Using GIS mapped data, we carry out precisely what an information system should enable: solve a problem, make queries, reach answers, or examine possible solutions. Here, data are manipulated digitally, and not manually, because we manipulate the data on events and activities using digital cartographic objects. In other words, the points, lines and areas in this cartographic database are used for data management.

Therefore, the GIS is a general tool for problem solving. It is created for making a certain project. A successful GIS is built, not bought, and indented for analysts to draw out relevant data for forecasting and planning, as well as various pieces of information associated with a specific space, as well as the problem area which is a subject of analysis.

The role of GIS tools in waste management planning is dominant in landfill site selection process. In addition, GIS tools are also used for distribution and identification of locations for other elements of waste management system such as transfer station network, waste selection and processing centers, for defining transportation corridors, etc.

The method of multicriteria analyses and evaluation is used for identifying locations of elements of a waste management system in the GIS. This approach is inevitable in locating complex objects, such as, for example, regional municipal solid waste landfills. Its complexity is reflected both in the size and function of objects, as well as in relation to various possible spatial impacts, also in negative context.

The use of GIS in defining strategies, analyses and visualization of solutions and alternatives helps us consider and clearly represent various scenarios, as well as select the most suitable solutions through a prism of different relevant criteria (spatial, ecological, hydro-geological criteria, etc.) [7].

Therefore, in using the GIS in the selection of the most suitable landfill sites, two things of key importance are [8].:

1. Analysis of space, i.e. all of its physical-geographical and anthropogenetic characteristics. It is necessary to comprehensively consider the space on which the problem is to be solved or which can be useful for problem solving. In this process, because of social sensitivity associated with this issue, it is necessary to be impartial in considering a possible landfill site. This can only be achieved if the entire space is considered to the same level of detail and in the same manner;
2. Visualization of space and its characteristics and impacts. This is necessary so that all participants in the project could have equal chance to perceive and understand the subject problem area. This enables active participation in searching for solutions to an acceptable compromise [9]. All participants must consider the space, as well as its advantages and disadvantages for landfill site selection. This is precisely one of the most important advantages of using GIS tools in landfill site selection, as well as of choosing other elements of a waste management system.

Defining landfill site selection criteria is the main step in landfill site selection process. In the first phase, based on exclusiveness, the sites which do not satisfy these criteria are eliminated. Positive areas within which it is possible to search for the most suitable solutions are the result of this process. This phase represents an activity of microzoning. Using GIS tools, through overlapping cartographic presentations of a certain space carried out based on exclusion criteria, it is rather simple to eliminate unsuitable landfill sites.

After eliminating the unsuitable landfill sites, the attention is dedicated to the nomination of landfill sites within the remaining "conditionally suitable" zones. In this process, local governments and professional institutions can and must be of great importance, but soil investigations and collecting relevant data on physical-geographical and anthropogenetic characteristic of space are indeed of utmost importance.

Through nominating potential landfill sites, preconditions for the selection of the most suitable landfill site are created, which is followed by multicriteria analysis and evaluation of candidate sites. Site selection criteria are entered into tables and weighted for each candidate site based on the entered value scale. In this way, the evaluation process using GIS tools is carried out in an efficient manner and in a short period of time.

The role of GIS tools in the landfill site selection process is in that it enables faster singling out and clearer presentation of suitable and unsuitable sites based on previously given criteria.

In this context, it is evident that selection criteria and value scale for evaluation of candidate landfill sites are of key importance in this process, while GIS tools represent a powerful means which to a great extent facilitate and speed up the process. This refers not only to the landfill site selection process, but also to defining the spatial organization of the entire waste management system, as well as defining the transfer station network.

3.3. Landfill site selection criteria

3.3.1. Landfill site selection criteria

The most important step in this process is to define landfill site selection criteria.

There are two groups of criteria. The first group includes the so-called exclusion criteria that are used in the first phase of the landfill site selection process. Exclusion criteria are defined relative to the specific situation and they represent restriction criteria.

Some of exclusion criteria can be classified into a group of the following indicators:

- Distance from natural elements of space (watercourses, water sources, protected natural resources, etc.)
- Distance from anthropogenetic elements of space (infrastructure facilities, settlements, protected cultural structures, etc.)
- Terrain morphology
- Hydrological and geological characteristics of space
- Degradation of space
- Recommendations of local authorities in a form of intermunicipal corporation agreements, etc.

According to exclusion criteria, areas which should not be further analyzed are discarded, i.e. areas that will be analyzed and evaluated in consecutive phases singled out. In the elimination phase, a single-criterion method is mainly used.

After that, in cooperation with local institutions and experts, certain number of sites are nominated for which a multicriteria evaluation is carried out. In this context, criteria based on which each candidate site will be evaluated in the same way are defined. This is a second group of criteria.

Site evaluation criteria are mainly classified into several basic groups. Commonly, there are three basic groups of criteria whose definition varies from author to author:

- Ecological or environmental criteria,
- Socio-economic or social or spatial criteria,
- Technical and operational criteria (which usually also involve certain economic, spatial and ecological criteria).

Any variation of groups of basic indicators is possible. Regardless of the formulation of basic groups of criteria, they include approximately either the same or almost the same number of indicators and criteria that are analyzed and compared in the process of selection of the most suitable site for a landfill.

Number of landfill site selection criteria ranges from 20 to over 40. They are classified (or not classified) into groups of criteria to which they belong, which are also similar, but can be differently formulated.

A particularly sensitive and important step in landfill site selection that follows the choice of relevant criteria is to define value scales based on which each individual criteria is evaluated (valued, ranked). Each criteria is assigned its corresponding weight (value) which is determined based on expert's evaluation and evaluation of participants in the process of sanitary landfill site selection. Here, quantitative evaluation is commonly used (e.g. scores from 1 to 10, or from 1 to 5).

Qualitative/expert assessment can also be used, where criteria can be assessed as suitable, conditionally suitable or unsuitable. Qualitative assessment is today increasingly less used, because the use of new technologies enable more accurate and more sophisticated assessment under the principle of quantitative assessment. In this case, accurate and objective data are obtained that can be compared and used for making right decisions.

When a potential site is assessed according to all given criteria, it is possible to carry out the following two steps:

1. Adding up all obtained scores
2. Multiplying the obtained scores by importance values (weights).

The first step in evaluating candidate sites is the simplest one and will low requirements. The best score is obtained through adding up all obtained scores for each criterion. Evaluation of candidate site in this case does not have different scenarios that can be of great help to decision makers.

The second step is more complex as different scenarios can be used. For example, if criteria for locating candidate landfill site are classified into several basic groups, then the number of scenarios to be considered is consistent with the number of criteria groups. Criteria from one group are favored in the first scenario, the most important criteria in the second scenario are those from the second group, and so on. The final option is a situation when groups of criteria are multiplied by the same importance value, without favoring any of criteria group. By presenting the scenarios in synthesis Table, it is easy to identify which candidate sites are the most suitable in which scenarios. The PROMETHEE method [10] is an example of this approach.

The basic advantage of this procedure is in that decision makers have a clear idea of which is candidate site is the most suitable if criteria from a certain group of criteria (ecological or economic or spatial, etc.) are assessed as the most worthwhile criteria, and if basic criteria groups are dealt with equally. This greatly facilitates decision making. Regardless of which of the many methods for evaluation of potential landfill sites are used, the question of objectivity of the procedure arises taking into account that the selection of evaluation elements (criteria, weights), but also the very decision-making process, is a matter of objectivity of experts and decision makers. This can be considered as a common disadvantage of all methods for potential landfill site selection. Therefore, the subjectivity in this process must be minimized to the utmost limit, while objectivity must be maximized.

4. Case study: Regional landfill site selection in the Kolubara region in Serbia

4.1. Analysis of the present state in the Kolubara region

We have chosen the area of the Kolubara Region comprising 11 municipalities with 382,000 inhabitant as an example of theoretical knowledge presented in the first part of the present paper.

The Study on the Selection of Micro-location for the Regional Landfill with Recycling Centre and Regional Center for Municipal Solid Waste Management, Regional Plan for Solid Waste Management, as well as Strategic Environmental Assessment (SEA) for the same Plan have been elaborated for the Kolubara Region.

Municipal waste from the territory of 11 municipalities in the Region is disposed of to 10 unarranged sanitary city landfills and a certain number of illegal dumps. All existing landfills should be closed or remediated and recultivated in the shortest possible time. It is recommended to prolong the life-time of the existing landfills through the mentioned remediation projects. Recognizing the need for the final, contemporary waste disposal and management, 11 municipalities of the Region have united together in forming the regions for the development of a waste management. The initiatives that have been launched in this context have resulted in the elaboration of the "Study on the Selection of Micro-Location for the Regional Municipal Waste Landfill with Recycling Center for the Kolubara Region", based on which the location for regional landfill has been selected.

4.2. Locating the regional landfill site for Kolubara Region

The sanitary landfill construction implies carrying out activities in several phases, whereby it is necessary to observe a specific sequence. The process is mainly carried out in four phases as follows:

1. Identifying (selecting) the location
2. Determining the location (through the planning and design documentation) and creating conditions for bringing land to intended use
3. Elaborating the construction documents (technical documentation)
4. Landfill construction

The most sensitive and the most important step in making a concept of regional municipal solid waste management is a regional landfill site selection. Relative to the selected regional landfill site, other elements of a waste management system are also located and their spatial distribution carried out.

Once the selection of the most suitable landfill site is made, it is necessary to incorporate it in the planning solutions in order to create conditions for the elaboration of technical documentation, as well as for landfill construction.

4.3. Regional landfill site selection criteria

The elaboration of the Study on Landfill Site Selection represents the first step in making a concept of municipal solid waste management in the Kolubara Region.

The first step in landfill site selection is to define exclusion criteria.

Taking into account current legislation, Intermunicipal Agreement on Joint Waste Management, basic exclusion criteria used in practice, available data on the space, as well as relevant characteristics of a specific space, the following exclusion criteria have been defined, see [6]:

- Seismic activity over 9 MCS
- Distance of less than 500 meters from watercourses
- Distance of less than 500 meters, or 1.5 km from settlements, if not sheltered
- Distances of less than 500 meters from water supply sources
- Collision with the existing planning documents
- Distance of less than 500 meters from roads of the first category if the site is not sheltered
- Terrains with an inclination of over 30%
- Terrains of more than 300 meters above sea level
- Alluvial plains and karst terrains

Their corresponding areas have been identified using the GIS tools. Through overlapping the corresponding areas, the following is obtained:

1. Potentially suitable areas for landfill
2. Unsuitable areas within which it is not possible to locate the landfill.

Once the potentially suitable areas within which its is possible to search for the regional landfill site are singled out, the regional plans for waste management for the Kolubara District and Belgrade administrative area have been considered in which the area of the Kolubara lignite basin has been determined for a macro-location for the regional municipal waste landfill.

Besides, in the Intermunicipal Agreement on the Joint Waste Management it has been agreed that the landfill site will be located in the territory of the Ub municipality since Ub has agreed to accept the waste generated in the newly formed region for municipal solid waste management in its territory.

The location of Carić (within the territory of the Valjevo municipality) has been also considered taking into account that this location has been previously analyzed several times and assessed as a suitable site for a landfill.

In nominating the location, the following has been taken into account:

- Preliminary analyses of the entire area and possible central position of potential sites in the region
- Data collected during field visits
- Consultations and recommendations of relevant local institution and experts
- Guidelines set by the EU and the Waste Management Strategy of the Republic of Serbia
- Data and information from the existing planning documentation.

Based on the above mentioned, the following three potential landfill site locations have been proposed (Figure 1).

1. **Location KALENIĆ**, in the area of open pits in the Ub territory
2. **Location BOGDANOVICA ,** city landfill (dump) in Ub
3. **Location CARIĆ**, for which certain investigations have already been carried out which have indicated certain advantages for landfilling. The location is within the territory of Valjevo municipality.

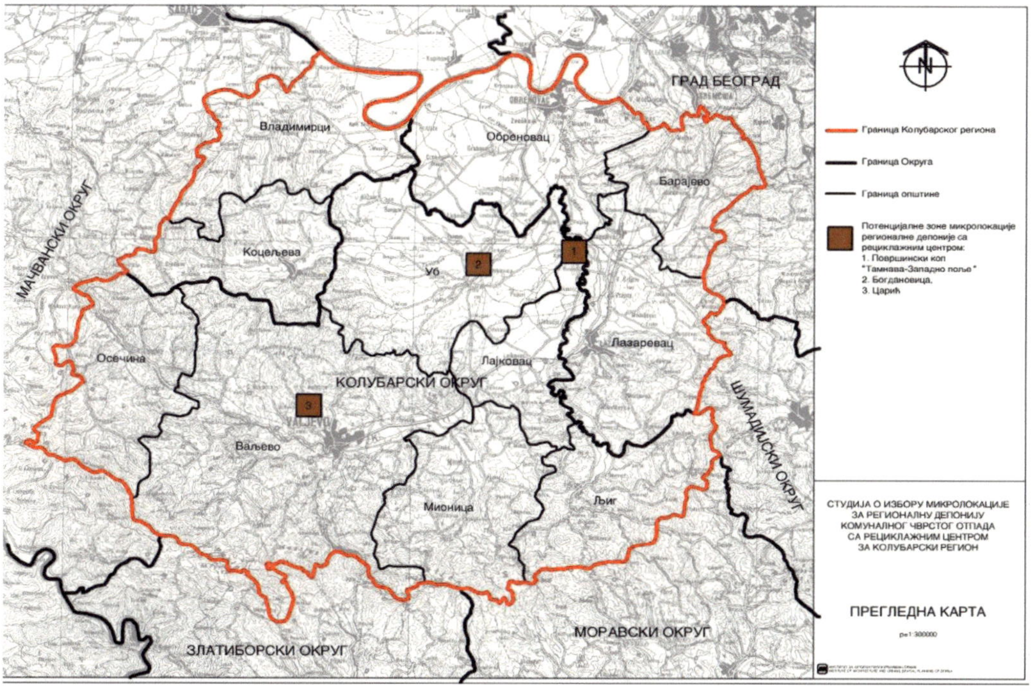

Figure 1. Position of potential landfill sites in the Kolubara Region

After the nomination of locations, the criteria for the evaluation and selection of the most suitable landfill site have been defined.

New criteria have been defined based on investigation and analysis of previous experiences of other countries and the EU guidelines, as well as on available relevant data for their evaluation. In this context, the following criteria for the selection of micro-location for regional landfill in the Kolubara Region for the municipal solid waste management have been defined:

1. **Hydrogeological characteristics**

a. Rock masses with fissure–cavernous porosity and a high water permeability (karstified rocks, limestones and dolostones)
b. Rocks with intergranular porosity, coarser grained rocks (coarse-grained gravel)
c. Rocks with low porosity (alluvial and glacial sediments)
d. Materials with a low water permeability, mainly impermeable complexes of 10-6≥ k ≥ 10-9m/s, or with a low water impermeability, but with small layer thickness of less than 1.0 m;
e. Water impermeable materials (clay, flysch) of k≤10-9m/s, the layer thickness ≥1.0 m.

2. **Ground water**

a. Aquifer is, over a brief period and at high water levels of greater frequency, above the bottom of the landfill in one part of its bottom area, while at other water levels, it is beneath the bottom; occasional flooding also occurs at the landfill site
b. Aquifer, at high water levels of small frequency, rarely rises above the landfill bottom; wetting of the landfill bottom is possible
c. Aquifer at water levels of 1 to 3 m is beneath the landfill bottom
d. Aquifer at high water levels > 3 m is beneath the landfill bottom
e. Aquifer does not exist

3. **Distance from the boundaries of zones of sanitary protection of water supply sources**

Distance from the boundary of

(a) narrower protection zone:(b) wider protection zone:

a. 0 to 0.2 kmbelt along the protection zone contours
b. 0.2 to 0.5 kmup to 0.5 km
c. 0.5 to 1.0 km0.5 to 1.0 km
d. 1.0 to 1.5 km1 to 2 km
e. more than 1.5 km> 2 km

4. **Geological-tectonic characteristics**

a. Pronounced fault zone
b. Fault carbonate rock masses with numerous surface and underground karst shapes or flat terrain
c. Flysch sediments, shales, marlsones, sandstones, etc.
d. Glacial sediments
e. Magmatic rocks

5. **Distance from the closest settlements with concentrated development or residential zones of urban settlements**

a. Distance 1.5 - 2 km, or 0.75 - 1 km if shelter
b. 2 -3 km, or 1 – 1.5 km with shelter

c. up to 4 km, or 1.5 - 2.0 km with shelter
d. up to 5 km, or 2.0 – 2.5 km with shelter
e. more than 5 km, or more than 2.5 km with shelter

6. Relief characteristics of the terrain

a. Broken relief, very uneven terrain, particularly pronounced in karst landscapes, incompact (scattered) spatial entity encompassing several valleys
b. Broken relief, uneven terrain, compact spatial entity
c. Incompact (scattered) spatial entity encompassing several valleys, naturally shaped terrain suitable for formation of valleys
d. compact waste entity, naturally shaped for locating a landfill site in a steep terrain or in natural depression
e. Mildly inclined or flat terrain, naturally shaped for locating a landfill site or possibly a landfill in excavated depressions or on earth fills

7. Available space for waste disposal and ancillary facilities

a. up to 5 yrs
b. up to 10 yrs
c. up to 15 yrs
d. up to 20 yrs
e. 20 yrs

8. Site acceptability

a. General landfill site disagreement
b. General agreement, but disagreement from local community
c. General agreement, but disagreement from certain individuals form local community
d. General agreement and somewhat moderate disagreement from local community
e. General acceptance of a landfill site

9. Engineering-geological characteristics

a. Incoherent rock masses, unstable slopes, slides and falls, active landslides
b. Complex of incoherent and semi—coherent rock masses (deluvial sediments), possible occurrence of landslides due to undercutting the foot of an existing slope
c. Semi-coherent rocks, possible occurrence of landslides due heavy falls
d. Coherent rocks, slightly stoned rock, stable slopes
e. Solid rocks, stable slopes even those of greater inclinations

10. Current land use

a. Cultivated agricultural land (ploughland, orchards), individual houses and other residential buildings within holdings, sportsgrounds, etc.
b. Quality tall forests;
c. Meadows
d. Pastures, shrub woods
e. Uncultivated land, thickets, barren land, excavations, quarries

11. Distance from individual water supply (wells)

a. 100 - 200 m, downstream of the landfill or approximately on landfill level

b. up to 500 m, downstream of the landfill or on the same level as the landfill

c. 500 to 1000 m, downstream or on the same level as the landfill

d. downstream of the landfill at the distance up to 200 m, downstream of the landfill at the distance of 1-1.5km;

e. downstream of the landfill at the distance of more than 200 m, downstream of the landfill at the distance of more than 1.5 km.

12. Landscape characteristics

a. Highly disturbed and completely changed natural ambience during landfill exploitation and after its closure

b. Highly disturbed natural ambience during landfill exploitation, and partly after the landfill closure

c. Natural ambience disturbed during landfill exploitation, and to a less extent after its closure

d. Natural ambience slightly disturbed during landfill exploitation, and undisturbed after its closure

e. Ambience not disturbed either during landfill exploitation or after its closure.

13. Linear distance from roads and railroads

more important roads | other roads

without shield | with shield | without shield | with shield

a. 500 m | 300 m | 300 m | 200 m

b. 600 m | 400 m | 400 m | 250 m

c. 800 m | 500 m | 500 m | 300 m

d. 1000 m | 600 m | 600 m | 400 m

e. >1000 m | >600 m | >600 m | >400 m

14. Distance to sacral structures, monuments of culture or protected natural resources

a. Distance 1.0 – 1.25 km, or 0.5 – 0.75 km where there is a shield

b. 1.25 -1.50 km, or 0.75 - 1,0 km with shield;

c. 1.5 – 2.0 km, or 1.0 - 1.25 km with shield;

d. 2 – 2.5 km, or 1.25 – 1.5 km with shield;

e. more than 2.5 km, or more than 1.5 km with shield

15. Seismic Activity

a. 9-8 MCS

b. 7 MCS

c. 6 MCS

d. 5 MCS

e. < 5 MCS

16. Existing site infrastructure

a. Absence of any infrastructure
b. Poor infrastructure
c. Only one infrastructure segment (access road, water supply line, electricity);
d. Several infrastructure segments
e. All or most of the infrastructure segments

17. Distance from surface watercourses

a. Permanent rivers or standing waters at the distance of 500 to 1000 m, there is a risk of flooding during high waters, defense measures against high waters required
b. Small watercourses, permanent or periodic ones (brooks, torrents), there is a flood risk, it is necessary to displace or channel these waters
c. Heavy inflow of rain waters from immediate catchments, defense against these waters requires more complex facilities; there is no flooding
d. Permanent watercourses at the distance greater than 1 km, no risk from flooding; defense standard solutions applicable
e. Great distance from watercourses, no risk from flooding, very low inflow of rain waters, simple protection against these waters possible

18. Terrain preparation

a. Very complex terrain leveling works, including intensive blasting on the greatest part of the site
b. Complex terrain leveling works, blasting required only in some parts of the landfill site
c. Terrain leveling works on the greatest part of the landfill site using machines
d. Terrain leveling on the smaller part of landfill site using machines
e. Simple terrain leveling works on the smaller part of the landfill site

19. Earth for covering the disposed waste – distance from the borrow site

a. greater than 5 km,
b. 2-5 km,
c. 1-2 km,
d. up to 1 km,
e. on site.

20. Position of the site in the Region

a. Completely dislocated relative to the central position in the Region; at the edge of the Region
b. Within the radius of 20 km relative to the central point in the Region,
c. Within the radius of 10 km relative to the central point in the Region,
d. Centrally positioned relative to the Region,
e. Within the radius of 10 km relative to the central point in the Region, but closer to the municipalities with the largest amounts of municipal solid waste.

21. Ownership of land

a. 100 % of land under private ownership, greater number of smaller plots
b. 100 % of land under private ownership, greater plots

c. About 75 % of land under private ownership, about 25 % of land under state ownership
d. About 50 % of land under private ownership and about the same amount of land under private ownership
e. 100 % of land under state ownership

22. Precipitations

a. 1500 mm
b. 1000 to 1500 mm
c. 600 to 1000 mm
d. 300 to 600 mm
e. < 300 mm

23. Air temperature

a. < 6° C
b. 6-9° C
c. 9-12° C
d. 2-15° C
e. > 15° C

24. Air flow

a. Very frequent high intensity winds, with prevailing wind direction towards settlements and other localities where people stay and work
b. Less frequent lower intensity winds with prevailing wind direction towards relevant facilities
c. Prevailing winds of changeable direction towards relevant facilities
d. Dominant winds blowing in the opposite direction, from settlements and other places where people stay and work, as well as low intensity winds blowing in direction towards the settlements
e. Most of the winds blowing in opposite direction, from settlements and other places where people stay and work

25. Distance to individual houses outside settlements

a. < 250 m
b. 500 m
c. 1000 m
d. 1500 m
e. 500 m

26. Site shelterness

a. Visible from all distances and all angles
b. Locality sheltered to a smaller extent
c. Locality sheltered to a greater extent
d. The glimpse of the locality can be caught in the great distance
e. Not at all visible, except when you come in the locality itself

27. Access road – reconstruction, or construction of a new road

New roadRoad reconstruction

a. > 1000 m,>1500 m
b. 500-1000 m800 – 1500 m
c. 200-500 m300 – 800 m
d. <200 m< 300 m
e. There is an access road of satisfactory characteristics

28. Providing electricity supply via the distribution network at the distance of:

a. > 2 km
b. 1 - 2 km
c. 0. 5 - 1 km
d. 300 - 500 m
e. < 300 m

29. Water supply in the locality

a. From the public water supply system via connection longer than 4 km, or from a local water supply via a connection longer than 3km
b. From the public water supply system via connection 2 to 4 km long, or from a local water supply via a connection up to 3 km long
c. From the public water supply system via connection from 1 to 2 km long, or from a local water supply via connection up to 1 km long
d. From the public water supply system via connection from 0.5 to 1 km long, or from a local water supply via connection up to 500 m long
e. From the public water supply system via connection up to 500 m long

30. Distance to agricultural land

a. < 100 m
b. 100 - 300 m
c. 300 - 500 m
d. 500 - 1000 m
e. > 1000 m

31. Distance from the main transmission line, gas pipeline, crude oil pipeline, drinking water pipeline

a. up to 100 m
b. 100 - 200 m
c. 200 - 300 m
d. 300 - 500 m
e. 500 m

32. Possibility of construction in phases and extension

a. No possibility of construction in phases or of extension
b. Limited possibility of construction in phases, but not of extension
c. Possibility of construction in phases, but not ofextension

d. Possibility of construction in phases and of limited extension
e. Possibility of construction in phases and of unlimited extension

Criteria are presented under the principle of exclusion criteria. More precisely, no detailed guidelines for evaluation have been given for criteria save for exclusion criteria which define requirements which a potential site MUST meet in locating the municipal solid waste landfill site.

4.4. Implementation of multicriteria evaluation method in landfill site selection

Potential micro-location for regional landfill in the Kolubara Region has been determined through multicriteria analysis and evaluation. Chosen criteria have been evaluated by assigning scores from 1 to 5 for each candidate site.

At the same time, depending on their importance in evaluating the locality quality, criteria have been classified into 3 pondering categories (PC). Each weight category has its specific value – weight, which is multiplied by the score of corresponding criteria. In this way, a final score is obtained for each criterion. Values by pondering categories are:

- PC1 = 1
- PC2 = 1.5
- PC3 = 3

The relation between pondering categories (PC) is: $K_{i+1} = K_i/1.5$

PC 3	PC 2	PC 1
	Landfill site selection criteria	
1 - 8	**9 - 20**	**21 - 32**

Table 1. Grouping the criteria by pondering categories (PC)

Table 2 indicates that after assigning a score to each criterion, the Kalenić location has been singled out as the most suitable one. The other two locations (Bogdanovica and Carić) have been assigned much poorer scores compared to the Kalenić location. However, in cases when the difference in ranks between candidate locations is extremely small at the end of evaluation process, it is difficult to make a final decision on which site is the most suitable. In this case, it necessary to carry out an additional evaluation which implies the evaluation of candidate sites by different scenarios. The chosen site selection criteria are then grouped into basic groups, while in the "additional" valuation process, the criteria from one of the basic groups are favored in each scenario, see [6].

There are so many scenarios as groups, plus one for the scenario according to which each basic criteria group is evaluated equally (for the last scenario, the data taken from basic evaluation or criteria are multiplied by weight value). In this way, decision makers are given opportunity to choose the option based on their policy, and thus select the most suitable site.

In regional landfill site selection for Kolubara Region, no "additional" evaluation has been required due to evident advantages of location Kalenić.

Criteria	PC	Kalenić	Bogdanovica	Carić
1. Hydrogeological characteristics	**PC 3**	12	9	9
2. Groundwater	**PC 3**	15	6	12
3. Distance from the boundaries of zones of sanitary protection of water supply sources	**PC 3**	12	15	15
4. Geological-tectonic characteristics	**PC 3**	12	9	9
5. Distance from the nearest settlements with concentrated development or residential zones of urban settlements	**PC 3**	12	3	12
6. Relief characteristics of the terrain	**PC 3**	15	15	12
7. Available space for waste disposal and ancillary facilities	**PC 3**	15	3	6
8. Site acceptability	**PC 3**	15	6	6
9. Engineering-geological characteristics	**PC 2**	3	3	3
10. Current land use	**PC 2**	7.5	7.5	1.5
11. Distance from individual water supply (wells)	**PC 2**	7.5	7.5	7.5
12. Landscape characteristics	**PC 2**	7.5	4.5	1.5
13. Linear distance from roads and railroads	**PC 2**	7.5	1.5	7.5
14. Distance to sacral structures, monuments of culture or protected natural resources	**PC 2**	7.5	7.5	7.5
15. Seismic Activity MCS	**PC 2**	3	3	3
16. Existing site infrastructure	**PC 2**	7.5	3	1.5
17. Distance from surface watercourses	**PC 2**	6	1.5	7.5
18. Terrain preparation	**PC 2**	4.5	4.5	4.5
19. Earth for covering the disposed waste – distance from the borrow site	**PC 2**	7.5	3	1.5
20. Position of location in the Region	**PC 2**	7.5	6	3
21. Ownership of land	**PC 1**	5	4	1
22. Precipitation	**PC 1**	3	3	3
23. Air temperature	**PC 1**	2	2	2
24. Air flow	**PC 1**	4	4	3
25. Distance to individual houses outside the settlement	**PC 1**	3	2	1
26. Location shelterness	**PC 1**	5	3	4
27. Access road - reconstruction or construction of new road	**PC 1**	3	3	1
28. Providing electricity supply via the distribution network at the distance of	**PC 1**	3	3	2
29. Water supply in the locality	**PC 1**	4	3	1
30. Distance to agricultural land	**PC 1**	5	4	1
31. Distance from main transmission line, gas pipeline, crude oil pipeline, drinking water pipeline	**PC 1**	5	2	5
32. Possibility of construction in phases and extension	**PC 1**	5	3	2
Total sum of criteria scores		**231.5**	**154.5**	**154**

Table 2. Evaluation of potential site by chosen criteria, see [2].

Had the final results of evaluation for all candidate locations been equal, the "additional" evaluation process would have been carried out in a manner as described in the text that follows. Namely, chosen criteria for additional evaluation would be classified into three basic groups (Table 3).

ECOLOGICAL	SPATIAL	SOCIO-ECONOMIC
Landfill site selection criteria		
1, 2, 3, 4, 6, 9, 10, 12, 15, 22, 23, 24, 26	5, 7, 11, 13, 14, 17, 19, 20, 25, 30, 31	8, 16, 18, 21, 27, 28, 29, 32

Table 3. Classification of chosen criteria into basic criteria groups

The scores of each criteria obtained in the basic evaluation process would then be multiplied by weight values for criteria groups according to different scenarios (Table 3). Weight values would actually be the percentage values whose sum is 100%.

Scores of each criteria from the basic evaluation process would be then multiplied by weight values for criteria groups according to different scenarios (Table 4).

Weight values would actually be the percentage values whose sum is 100%.

Scenario / Basic criteria group	SC 1	SC 2	SC 3	SC 4
EKOLOGICAL	**0.50**	0.25	0.25	**0.33**
SPATIAL	0.25	**0.50**	0.25	**0.33**
SOCIO-ECONOMIC	0.25	0.25	**0.50**	**0.33**

Table 4. Criteria weight values according to different scenarios

After multiplying the criteria values from basic evaluation by weight criteria according to different scenarios and their sum for each candidate site, the ranking of candidate sites according to different scenarios would be obtained (Table 5).

Scenario	SC 1	SC 2	SC 3	SC 4
	Site ranks			
Candidate site	Kalenić (83.12)	Kalenić (77.50)	Kalenić (73.32)	Kalenić (76.23)
	Bogdanovica (63.12)	Carić (53.25)	Bogdanovica (47.40)	Bogdanovica (50.98)
	Carić (58.62)	Bogdanovica (48.87)	Carić (45.07)	Carić (50.82)

Table 5. Ranking of candidate sites according to different scenarios

Through multicriteria evaluation according to different scenarios, several options and different arguments for the selection of the most suitable site are made available to decision makers. Implementation of different scenarios is based of the PROMETHEE method.

In this case, it has been shown that the location Kalenić has the best values in all four scenarios, while it is evident that the remaining two locations differ depending on scenario. The location Carić is better valued for the scenario 2, while location Bogdanovica is better valued in other three scenarios.

4.5. Multicriteria analysis and evaluation using GIS tools

In selecting the landfill site in the Kolubara Region for municipal solid waste management, the GIS tools have been implemented in singling out areas to be eliminated. The areas to be eliminated have been singled out based on defined exclusion criteria. Each of exclusion criteria has been presented graphically (cartographic presentation), and corresponding areas have been identified using GIS technology. Through overlapping maps of each exclusion criteria, negative areas have been singled out that should not be further analyzed in the landfill site selection process for the municipal solid waste management. Negative areas are shown in the synthesis map (Figure 2).

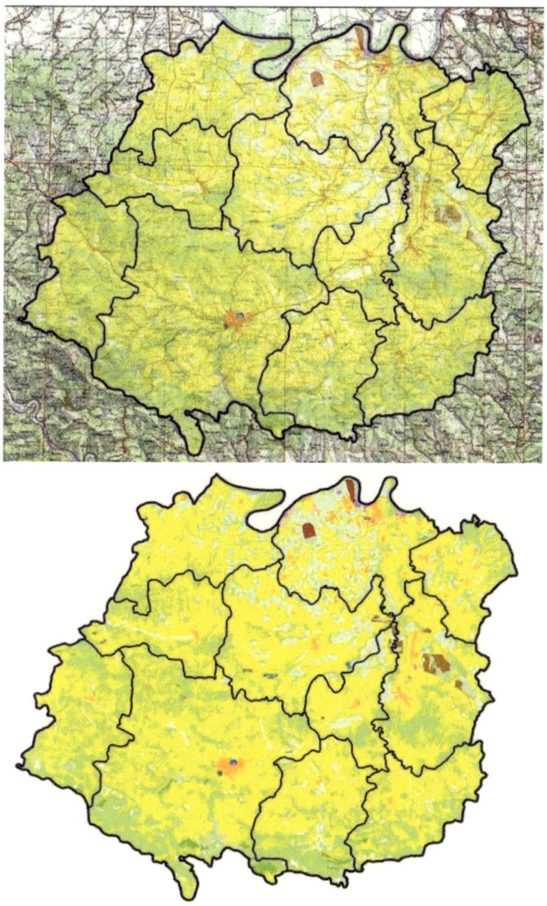

Figure 2. Analsys of land use structure based on CORINE information system [11].

Figure 2. depicts the structure of land use in Kolubara Region based on the CORINE (Coordination of Information on the Environment). The CORINE program is the European information base as support to the sustainable development policy of the European Union. The database contains data on: urban areas, crop yield, meadows, forests and natural vegetation, waters,. as well as other dynamic processes in the environment. All mentioned data are cartographically presented, which enables a more simple analysis of the subject area. The CORINE program was initiated in 1985. At the beginning, the program was developed and tested on 10 regions of the European Union by demonstrating the feasibility of the approach. Satellite photographs on which the CORINE database is based have been geometrically and radiometrically supplemented and with abundance of data which are in the CORINE Land Cover organized hierarchically in three levels classified in 44 classes (correspondingly presented spatial features and data). After showing positive results, in 1994 the European Environment Agency based in Copenhagen undertook the maintenance and use of the CORINE Land Cover database. Since then, the CORINE Land Cover (CLC) has been affirmed which is reflected in the fact that an increasing number of European countries are involved in the CLC project which has provided them with an opportunity to more efficiently pursue their environmental protection and sustainable development policies, as well as to carry out analyses for various needs and development strategies. Today, 64 European countries are involved in the CORINE Land Cover 2000 Project, with clearly defined and synchronized methodology for collection, processing, as well as presentation of data, in the function of the elaboration of environmental management plans [12].

In the elaboration of the "Study on the Selection of Micro-Location for the Regional Municipal Waste Landfill with Recycling Center for the Kolubara Region", the CORINE information base has not been available for Serbia, thereby for the Kolubara Region either. However, once the information on the environment from the CORINE program have become available for users in Serbia, all results from the elimination phase of landfill site selection contained in the Study have been checked and, what is even more important, confirmed. By using the CORINE program in accordance with the defined eliminating criteria for Kolubara Region, the selection of “negative” areas has been much easier and faster. The CORINE Information base to a great extent meets the needs of elimination phase in landfill site selection, thus this phase should be used as much as possible.

On the synthesis map (Figure 3), which is a final phase in the process of elimination of “negative” areas, the areas which do not satisfy basic conditions relative to the established exclusion criteria are denoted by red color. These are mainly corridors along watercourses, first category roads, distances to settlements, areas at over 300 meters above sea level, water supply sources, etc. Thus, it is the matter of exclusion criteria represented by minimum required distance of the future landfill site relative to them [13].

In the elimination phase, it is also possible to use some other criteria such as, for example, central position of a landfill relative to the Region. This means that, because of the cost-effectiveness of the waste management system, i.e. transportation costs, it is necessary to position a landfill within the radius of 20 or 30 km relative to the central point of the Region. However, in such case, a great number of areas that merit further analysis by their characteristics can be excluded, while the problem of central positioning of a landfill can be overcome through a good organization of transfer station network in the Region. In this context, it is important to emphasize that it is not necessary to introduce a great number of exclusion criteria, but to limit the choice of exclusion criteria to the most relevant ones, as shown on the example of the Kolubara Region.

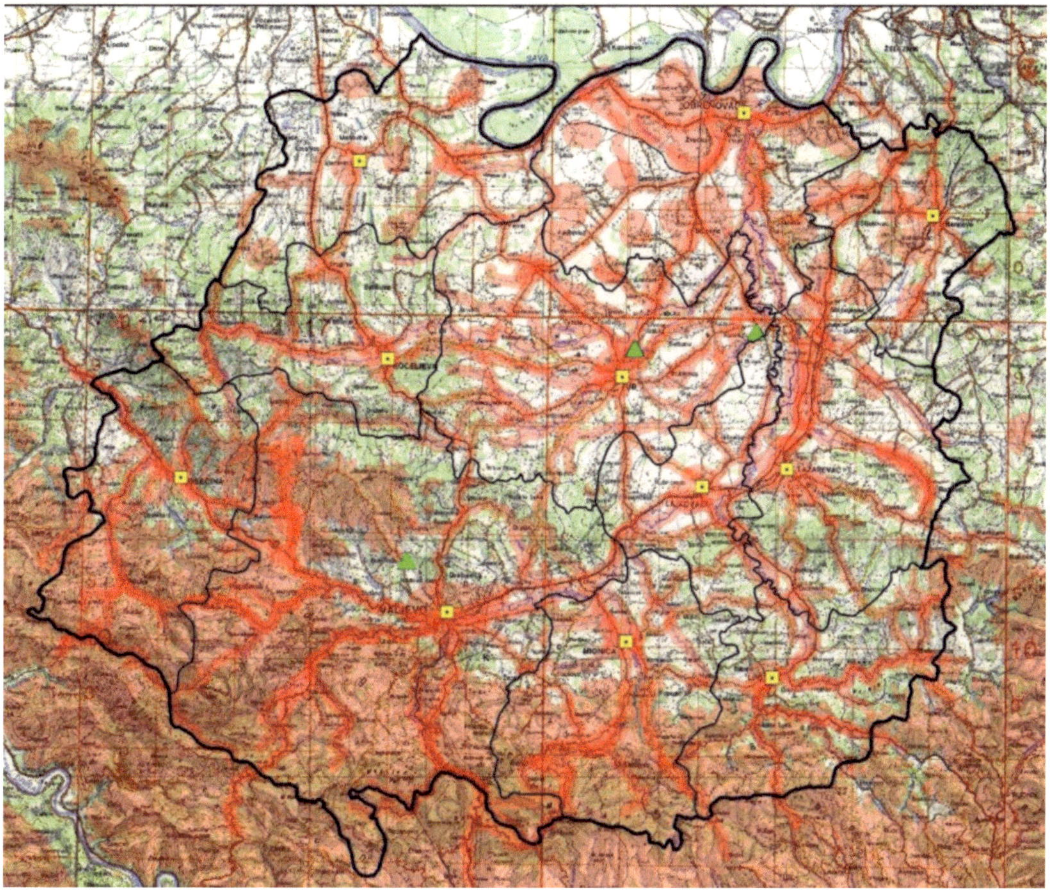

Figure 3. Suitability/elimination map

Once the consultations with relevant entities have been carried out, as well as preliminary analysis of areas that have not been eliminated in the first phase of the landfill site selection process, three sites have been singled which have been included in the process of detailed analysis and multicriteria evaluation.

The use of GIS information base in this phase of landfill site selection has considerably accelerated the process of evaluation according to 32 given criteria. Once the location Kalenić has been assessed as the most suitable one and singled out as the most acceptable one, the landfill site selection process has been completed. However, this is not where the use of GIS tools ends. Their role is also in implementation of a uniform information system for waste management which consolidates data on landfills, transfer stations, waste generation, waste flows, as well as other data important for an efficient waste management.

In this sense, GIS tools represent an information support in the functioning of the waste management system.

5. Conclusions

Landfill site selection is the most sensitive task placed before the participants in the process of planning spatial organization of a waste management system, particularly in countries in which there is insufficient awareness and lack of information in the population, and, consequently, there is a resistance to plans to locate a landfill in their area, known as NIMBY (not in my backyard) syndrome. Out of these reasons, this problem is overcome through defining the elimination and basic criteria for landfill site selection based on which an multicriteria evaluation is carried out, along with mandatory inclusion of all relevant stakeholders in the process of selecting the most suitable landfill site. In the present paper, such exclusion criteria have been chosen that are appropriate to the specific space which has been the subject of our investigation, as well as according to available spatial data. In this context, the paper emphases the fact that the choice of exclusion criteria is conditioned by a specific physical properties of space. After the phase of elimination of "negative" areas, a multicriteria analysis of sites that have been nominated based on a set of basic criteria has been carried out. Altogether 32 criteria have been defined that are based on efficient functioning of a landfill, as well as on efficient environmental protection at specific landfill site and its surroundings. A multicriteria evaluation model has been offered and value scale for evaluation of each criteria defined. The multicriteria evaluation model has been also used for different scenarios. In this context, basic criteria for landfill site selection have been grouped into several basic groups, while in the evaluation process for each criteria only one of the basic groups has been evaluated. Such approach enables decision makers to choose the most suitable option and to make best decision according to their policy.

Comprehensive consideration of the problem associated with landfill site selection for physical elements of waste management system implies the use of GIS tools, thus providing a more sophisticated process of spatial analysis and searching for better options, as well as accelerating and visually enriching the process. Advantage of using GIS tools is in that it enables faster singling out and clearer presentation of suitable and unsuitable landfill sites based of previously given criteria. The paper shows the example of advantages and disadvantages, as well as possibilities of implementing GIG in regional landfill site selection for municipal solid waste management in the Kolubara Region. The GIS applications are particularly suitable for elimination phase where, based on the given exclusion criteria and spatial data, the "negative" areas within which potential landfill sites are not be searched are very quickly and easily eliminated. The entire process is presented cartographically. The possibility of implementing the CORINE Program – a uniform European information base on the environment and space usage, which is particularly suitable for elimination phase in landfill site selection as it provides abundance of geospatial data, has been highlighted. Furthermore, the possibility of efficient waste management using database in the GIS is highlighted. The system supported by such data enables quality and fast waste management, monitoring, waste data updating, as well as the best basis for planning waste management strategy at regional level.

Author details

Boško Josimović and Igor Marić
Institute of Architecture and Urban & Spatial Planning of Serbia, Serbia

Acknowledgement

This work has resulted from research within the scientific project: "Sustainable spatial development of Danube area in Serbia " (TR 036036), which was financed within the program Technological development by the Ministry of Education and Science of the Republic of Rerbia from 2011 to 2014.

6. References

[1] Ilić M (2005) Waste Management in the Focus of Controversial Interests. In: Lechner P, editor. Waste Management. Vienna: Boku University. Pp. 95-96.
[2] Boško Josimović, Marina Ilić, Dejan Filipović (2009) Planning of Municipal Waste Management. Belgrade: Institute of Architecture and Urban & Spatial Planning of Serbia. 157 p.
[3] Waste Management Strategy of the Republic of Serbia 2010-2019.
[4] Boško Josimović (2003) Implementation of Environmental Management System in spatial planning – master's thesis. Belgrade: Faculty of Geography. 146 p.
[5] UK Department for the Environment (1995) Landfill Design, Construction and Operation Practice.
[6] Josimović B, Marić I, Manić B (2011) Methodological approach to the determination of landfill location for municipal solid waste - Case study: Regional landfill in Kolubara region. Architecture and Urbanism. 32: 55-64.
[7] Catalano A, Zhang M, Rice J (2006) The Use of GIS to Manage, Analyze, and Visualize Data Collected During an Investigation of a Proposed Landfill. Available: http://gis.esri.com/library/userconf/proc98/.
[8] Margeta J, Prskalo G (2007) Sanitary Landfill Site Selection, Građevinar. 58. No. 12.
[9] Higgs G (2006) Integrating multi-criteria techniques with geographical information systems in waste facility location to enhance public participation. Waste Management & Research: pp 105-117.
[10] Brans J.P, Vincke Ph (1985) Preference Ranking Organisation Method for Enrichment Evaluations – The Promitee Method for Multiple Criteria Decision Making. Management Science. 31: 647-656.
[11] Josimović B, Crnčević T (2010) Implementation of Strategic Environmental Assessment in Serbia with special reference to the Regional Plan of Waste Management. In: Santosh Kumar Sarkar, editor. Environmental Management. Rijeka: SCIYO. pp. 95-113.

[12] CORINE Land Cover 2000, http://terrestrial.eionet.europa.eu/CLC2000.
[13] Josimović B, Krunić N (2008) Implementation of GIS in the selection of locations for regional landfill in the Kolubara Region. SPATIUM. 17: 72-77.

Chapter 23

Land Degradation, Community Perceptions and Environmental Management Implications in the Drylands of Central Tanzania

Richard Y.M. Kangalawe

Additional information is available at the end of the chapter

http://dx.doi.org/10.5772/45897

1. Introduction

Land degradation particularly through soil erosion is an important concern in many parts of the world including semiarid areas of central Tanzania. One of the issues that have gained importance is the concern on implications of local perceptions in resource management. The Irangi Hills (Figure 1) are severely affected by soil degradation; hence they provide a vivid example for studying environmental degradation, local perceptions, and land-management strategies that the local population use to cope with the degradation problems [1]. The Irangi Hills have been influenced by various land management interventions, for example, the implementation of various soil-conservation measures that have contributed to considerable changes in the spatial and temporal land-use patterns during the 20th century, and particularly over the last four decades. The Irangi Hills, located in Kondoa District in semiarid, central Tanzania (Figure 1), constitute about 10% (c. 1256 km^2) of the District area that has been particularly affected by sheet and gully soil erosion [2, 3 ,4, 5, 6,7]. In many places soil erosion has reduced the agricultural potential of the land by the physical removal of topsoil, sand deposition on lower slopes and valley floors, and gullying and incipient badlands development [7]. The sub-humid, north-eastern parts of the Irangi Hills are more severely degraded than the semiarid south and south-west.

A combination of factors makes soil erosion a particularly serious problem in these semiarid areas of Tanzania. The problem has often been associated with local mismanagement of the land resources through among others, overgrazing, over-cultivation, burning of grasslands and woodlands, resulting in over-exploitation and consequent soil erosion [8]. However, it has been shown in recent studies in the Irangi Hills that the problems and causes may be more complex than was earlier presumed, including factors such as tectonic activities and historical changes in climatic conditions [9, 10, 11, 12].

In southern Africa, where many researchers are engaged in land management and conservation issues, many have included the aspect of perception and their influence on resource management by various groups of the society [13, 14, 15]. Local perceptions can be established by interviewing local people about how they viewed various resource conditions [16]. The importance of acknowledging the socio-economic environment of the various decision making groups involved is discussed in [17]. Also Dahlberg [18] accorded considerable importance on local views in the study of environmental change and degradation in Botswana. Understanding the local people's perceptions on environmental issues is thus a prerequisite in making successful and sustainable resource management strategies.

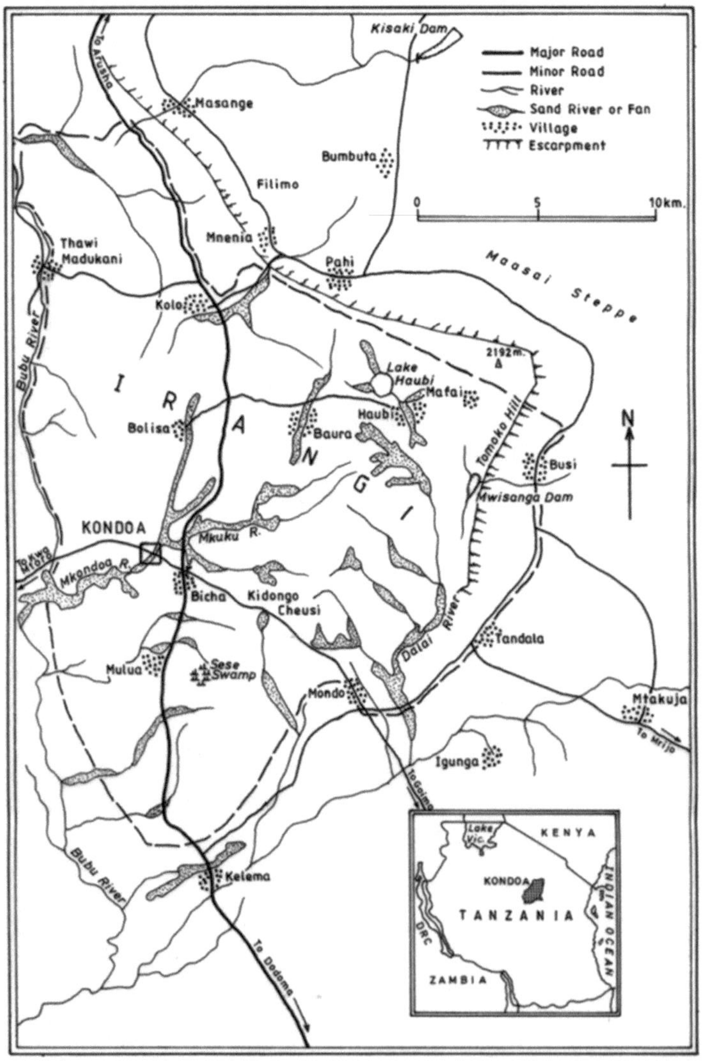

Figure 1. Map of the Irangi Hills showing the Kondoa Eroded Area and the surveyed villages of Mafai, Baura and Bolisa. The insert is a map of Tanzania showing the location of Kondoa.

Local people's perceptions of environmental issues can be looked at from three perspectives. First, people will perceive land degradation on the basis of their socio-economic interests. In this case, farmers will be more aware and concerned about environmental changes and damages that affect agricultural productivity such as soil erosion. Secondly, when these people understand that their physical environment is deteriorating they will attempt to control some of their activities leading to degradation [19], thereby be more willing to support land management programmes if they are aware that their actions are harmful to the environment [20]. Various social, economic, technological and ecological often exist, changing degrading practices especially where communities are aware of the negative impacts of current land management practices. Communities in the Irangi Hills have demonstrated their willingness to participate in conservation initiatives as indicated by their adoption of various conservation measures reported later in the chapter.

The third perspective from attitude survey shows that a large majority of farmers are concerned about soil and/or land degradation as a general community problem, disregarding the fact that their own holdings are likely to be also at risk [21]. Under such circumstances then no actions may be taken although such people hold positive attitudes towards conservation. However, it is believed that when the landowners themselves have been involved in fact-finding on their own land they become instrumental in implementing planned courses of action [22]. Thus basing on the local people's perceptions of environmental resources and knowledge then it is possible to develop methods which can allow the people themselves to provide the solutions to their environmental problems [19, 22, 23]. Generally, planning for and implementation of effective soil and water conservation measures in a site require, among other things, a detailed understanding of the extent, risk and spatial distribution of the problem [24, 25], including local concerns.

While sustainable management of resources leads to sustainable development, the deep-rooted poverty leads to overdependence on natural resources for livelihood which in some instances has undermined the capacity of the communities to manage their resources sustainably. This problem is more critical in developing countries where rapid population growth leads to the invasion of marginal lands and unsustainable land use practices that in turn encourage environmental degradation and perpetuate poverty.

This chapter is based on a study undertaken in the Irangi Hills to examine the farming community perceptions and awareness on environmental degradation. The main objective was to examine on methods used in the area to improve agricultural productivity and control soil degradation. The specific objectives were: to assess farmers' awareness on soil degradation and conservation, and to study the farming system and resource use in the area. It also examined the mechanisms that are taken by the local communities to address land resource management issues, including approaches used in soil conservation and how such approaches help to improve agricultural productivity and local livelihoods in general. It further examined the factors limiting effective community involvement in land/soil conservation initiatives.

2. The study area

The Irangi Hills (Figure 1) are located in the severely eroded area of Kondoa District known as the Kondoa Eroded Area (KEA). Kondoa District is located in the northern part of Dodoma Region at latitudes 4º10' - 5º44' South and longitudes 34º54' - 36º28' East. The land area of the Kondoa District is approximately 13,210 km^2, out of which the Kondoa Eroded Area covers 1256 km^2. The Irangi Hills forms the largest part of KEA [5]. The altitude of the Irangi Hills ranges from 1200 to 2000m above sea level. The climate of Kondoa District is semiarid, characterised by an average annual rainfall of between 600 and 800 mm, with a long-term average of 640 mm per year [5]. However, in the more elevated parts of the Irangi Hills up to 900 mm of rain have been recorded [26]. The rainfall season is usually between November and April/May, with a dry spell in February. The period between May and October is usually dry. The rainfall pattern in this area is highly variable and most of the precipitation occurs in short duration storms. The district is characterised by high evapotranspiration rates that double the amount of precipitation [27].

The majority of soils in the semiarid areas of central Tanzania originate from granitic, gneissic and schistic parent material. These soils are of low fertility, base-exchange capacity, bulk density and water-retention capacity [6]. These soils also have low organic matter content, a condition that makes them extremely erodible [7]. The Irangi Hills are severely affected by soil erosion. Studies on soil erosion in the area indicate rates of between 27 and 37t ha^{-1} yr^{-1} [28], in the sub-humid and semiarid parts of the hills respectively. The soils in Irangi Hills are generalised as coarse loamy to sandy loams in texture, being sandiest in the surface horizon. This implies the need for proper management in order to sustain agricultural productivity. Different strategies used by farmers in KEA, for instance, in coping with poor soil fertility have been described by Kangalawe [1, 29].

Soil conservation initiatives started in Kondoa district since the colonial administration in the 1930s. During the 1940s to 1950s soil conservation involved measures such as reduction of livestock numbers, ridge cultivation, contour bunding of uncultivated land, rotational grazing and gully erosion control [30]. Farmers were also required to plant sisal around farmlands to save the arable land from further destruction.

Soil conservation measures during that period were associated with colonial force, where some of the activities were assigned to people as punishment for disobedience of local rulers and tax aversion. They were thus considered as an interference with local traditions and became quite unpopular [4]. In 1973 the government of Tanzania started a state-run soil conservation project in Dodoma region, popularly known as HADO (Hifadhi Ardhi Dodoma). This was a deliberate attempt to come to grips with the menace of soil erosion and degradation in the region [5]. Three kinds of approaches were undertaken to enhance soil conservation, including mechanical, biological and administrative measures [4]. *Mechanical measures* involved barriers such as terraces and earth banks across the slopes that were built to slow down surface runoff. *Biological measures* involved earth binding with plantations of different kinds, such as grass strips across the slope, planting grass in sandy rivers and in gully bottoms, rotation of crops, and the spread of residue on the fields. The *administrative measures* encompassed tree planting demonstration plots, organising grazing and farming techniques such that the land is protected as much as possible. These measures were complemented by evacuation of all domestic animals in 1979 [5]. Expulsion of livestock from the Irangi Hills in 1979 as part of soil conservation approaches instituted in the KEA resulted in ecological transformation from the heavily browsed shrubs and scattered trees to impressive regeneration of herbaceous, woodland and woody shrubs, as well as grassland vegetation cover. Economically, however, this agro-pastoral society was no longer able to depend on livestock as a form of capital, or as a security against the harsh climatic conditions of the region.

3. Methodology

3.1. Sampling procedure

The Kondoa Eroded Area (KEA) covers twenty-eight villages. A list of these villages was made and three of them were selected at random, namely Mafai, Baura and Bolisa. The selected villages had 370, 350, and 750 households respectively. The selected villages are subdivided into sub-villages for administrative purposes. To allow for adequate representation, 10% of households from each sub-village was randomly selected for inclusion in the sample. A total sample of 147 households was selected for interviews. A random sampling procedure was employed in selecting the sample households from lists of household heads that were made for each of these villages. Sampling is a common practice in research. The random sample of 10% of the villages and households selected for this study is considered to be representative enough for statistical analysis [31]. Under certain circumstances, such as resource constraints, even a smaller sample of 5% is regarded as being representative enough [32].

3.2. Data collection and analysis

Household interviews, using a standard questionnaire, were the major means used to collect both qualitative and quantitative information. The questionnaire survey was complemented by informal surveys that involved discussions with key informants, including village leaders, extension workers, district agricultural officials and HADO staff. These informal surveys were conducted in order to get some general overview on soil degradation, community perceptions and agricultural performance in general. These surveys also provided a means and direction in crosschecking the responses from formal interviews. The key informants were found in the respective villages and/or at district level. Information from key informant interviews was analysed by triangulation with all other sources. To determine the level of awareness of soil degradation three exploratory questions were asked. Firstly, whether the study community perceived land/soil degradation as a problem in their villages. Secondly, what criteria are used by this community to determine the quality of land/soil in general. Thirdly, whether they associated land/soil degradation with crop cultivation or livestock management systems of the area. These aspects are addressed in the following sections.

The surveys were complemented by field observations in farmers' fields. Field visits involved observations of various land degradation features, such as soil erosion and sedimentation, surface runoff, sandiness of soils, crop vigour, presence of indicator-plant species; and agricultural practices, including among others, types of crops grown, cropping patterns and on-farm soil conservation measures. Field observations also included sampling soils from selected transects for subsequent laboratory analysis of soil nutrients. Three farmer-led transect walks were undertaken with small groups of farmers in each village and soil samples were collected from representative sites of major soil groups as identified by farmers. Soil profile pits were dug to a depth of one meter and samples taken at 20 cm intervals [33]. A total of eighty samples were collected for analysis. Nutrient content of the soil was determined to provide an estimate of the inherent soil fertility status and for soil nutrient balance assessment. The data was analysed using statistical measures of central tendency (means), and frequency distribution (percentages) [34, 35]. The frequency distribution data was cross-tabulated into contingency tables.

4. Results and discussion

4.1. Community' perceptions on land/soil degradation

4.1.1. Local perceptions of soil degradation

Response to the inquiry on whether the study community perceived soil degradation as a problem in their villages have shown that 58% of the respondents considered soil degradation as being a serious problem in their vicinities. This perceptions may be influenced by differences in socio-economic characteristics inherent among the local people. Socio-economic characteristics such as endowment of livelihood assets by households determine the ability of a household to use, for example, agricultural inputs like fertilisers or manure as a way of improving soil productivity. In the Irangi hills, for instance, wealthy farmers who could afford using fertilisers and/or manure did not perceive soil fertility as a major issue. Those who perceived soil degradation as a problem mentioned the generally low but declining soil fertility of the Kondoa soils, soil erosion and runoff, sandiness of soils and sedimentation as key indicators of soil degradation in their villages. Figure 2 presents the proportions of responses on indicators of farmers' awareness of soil degradation processes.

The small percentage of respondents mentioning soil erosion in Mafai village could be attributed to that this village is surrounded by a protected catchment forest and is generally less degraded compared to the other two villages. The presence of these indicators seem to show that rural people are aware of their environment and its related problems, and particularly so with those which affect the farm productivity and/or those that resulted into more visible landscape changes such as soil erosion. However, the fact that less than half of the respondents indicated that soils are inherently infertile suggests that productivity has declined significantly within living memory and that people were unaware that their yields were probably rather low from the outset. It is explained by Tosi et al. [36] that the low inherent soil

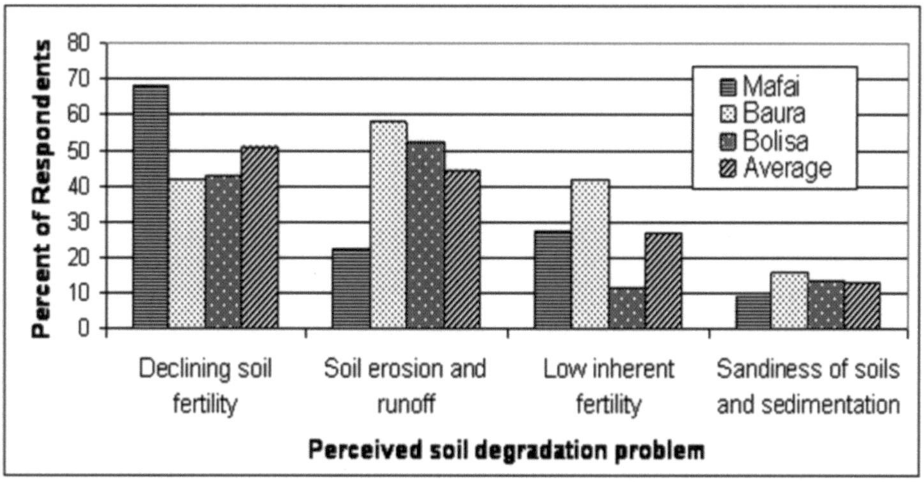

Figure 2. Percent responses on community awareness of soil degradation problems.

fertility is attributed to parent material factors. Soils in this area are reported to have formed from metamorphic and igneous rocks which are poor in plant nutrients. The dry climatic condition of the study area also limits high productivity of organic matter, resulting in poor surface cover and low incorporation into the soil as binding agent and for fertility enhancement [37].

Declining soil fertility was perceived as the major indicator of soil degradation in the studied villages. A majority of the farmers (80%) attributed such decline to continuous cultivation without resting the fields, whereas 20% ascribed it to inadequate application of manure and/or fertilisers. One explanation to continuous cultivation was the increasing land shortage that has led to intensified crop cultivation and short or no fallow periods. Studies conducted in neighbouring villages and in other parts of the Irangi Hills also revealed that most farms are cultivated every season without fallow and are thus subjected to continuous loss of soil fertility [1, 38].

The general assessments of soil fertility in Kondoa District that soils in the area are of low fertility [1, 7, 39, 40]. Results from laboratory analyses for soil nutrients (Table 1) confirm the low levels of soil nutrients in all the three villages studied. These results also indicate that Mafai soils had more advantage in terms of nitrogen content than both Baura and Bolisa. These soils had about four times as much total nitrogen as Baura and Bolisa, whereas Baura had soils richer in available phosphorus than the other two villages. The variations in nutrient contents may be attributed to, among other factors, the severity of soil degradation that characterises the studied villages and different management practices of the farms. Many soil conservation structures like contour bunds and planted trees were also observed in Mafai compared to the other two villages. Soils in Mafia village are less eroded, hence have more nutrients in the sampled surface layer attributable to presence of higher organic matter content. A large part of the village is covered by a protected catchment forest and generally less severely degraded compared to Baura and Bolisa.

Soil characteristic	Mafai	Baura	Bolisa
pH (H_2O)	4.80	5.20	5.40
Total Nitrogen (%)	0.29	0.07	0.08
Available P (ppm)	4.20	7.20	3.20
CEC me/100g	17.12	3.84	13.89
Exchangeable K (me/100g)	0.61	0.61	0.52
Exchangeable Ca (me/100g)	4.39	1.83	2.74
Exchangeable Mg (me/100g)	1.61	0.40	1.20
Exchangeable Na (me/100g)	1.86	0.92	0.80
Organic carbon (%)	2.62	0.31	0.54
Particle size			
% Sand	54.83	80.27	69.20
% Silt	14.57	10.60	17.63
% Clay	30.60	9.13	13.17

Table 1. Some characteristics of soils from the study area (measured at 0-20 cm depth)

Soil erosion and surface runoff featured as indicators of soil degradation as indicated by about 44% of respondent farmers (Figure 2). Awareness of soil erosion as a soil degrading process featured more prominently among Baura and Bolisa respondents. Visual observation of the landscape in these villages confirms the local people's response. Both Baura and Bolisa have landscapes dissected by more pronounced gullies (Figure 3) compared to Mafai village (Figure 4). Discussions with key informants in these villages indicated that historically the two villages had large numbers of livestock prior to destocking in 1979 that rendered many places devoid of vegetation because of overgrazing. This situation exposed the land surface to agents of soil erosion, such as runoff. The extensive gullies seen today in these and many other villages in the Irangi Hills are said to have formed along former cattle tracks aligned down the slope [6].

Figure 3. Gully erosion in the studied villages. This is a common feature of the landscape and in many parts of the Kondoa Eroded Area

Figure 4. An uneroded part of the landscape wihin the Kondoa Eroded Area.

Sedimentation and sandiness of the soil was perceived as a problem by only a few farmers (see Figure 2). This response was particularly obtained from farmers whose fields laid in stabilising sandfans that have soils with very low organic matter levels, low moisture holding capacity and poor fertility status. Such soil characteristics are also common in other parts of the Irangi Hills, such as in Haubi and Mulua [1, 7]. Sedimentation was reported to take place in depositional footslopes and valley bottoms where the eroded materials from hillslopes accumulate. In many places sedimentation of sandy materials buried the former fertile clayey topsoil [1, 7]. One would expect this indicator to be mentioned by most respondents, however since farmers have had their settlements and fields in that kind of environment for generations they do not often mention it as a major concern. The explanation to this situation would be that soil erosion and sedimentation in this area dates long in history to the extent that very few benchmark areas remain that could show the earlier landforms not affected by sedimentation [7].

4.1.2. Assessing land and soil quality

Findings from this study showed that there are several traditional ways communities use to evaluate and to explain the quality of the land and the soils they are cultivating. Three categories of responses appeared to be most prominent, namely crop vigour and crop yields, presence of indicator-plant species and density of vegetation under fallow [29].

A healthy and vigorous crop growth, reflected by a good crop stand in the field, was used as an important indicator that the soil is fertile enough, if moisture and other factors are not limiting [29]. Under such circumstances, even if the weather conditions worsen during the growing season such that final yields are poor, the farmer would have realised the potential fertility of a certain piece of land. A stunted crop with less vigorous growth in the field when other factors such as moisture are considered not limiting was locally perceived to indicate a high probability that soils on which the crop is growing are of low quality and infertile.

Majority of respondents (95%) considered crop yields as the best measure to comprehend land/soil quality. It was noted that low or declining crop productivity could be a clear indicator of declining soil fertility, and hence soil degradation. The use of this indicator by the local farmers in evaluating land/soil quality is also appreciated by experts in land degradation, where crop output decline is regarded as a proxy indicator of soil degradation in farmlands [1, 40, 41]. It is particularly important because it affects people directly in terms of food availability and security. However, this factor alone is not sufficient to establish that degradation is taking place since cropping conditions vary considerably between years and between individual farmers. The influence of other factors such as crop pests and diseases and climate variability may affect crop yields [42, 43]. In the Irangi Hills most of the respondents indicated also that low crop yields could be due to low and/or erratic rainfall. This aspect needs to be investigated further to establish detailed linkages between climatic patterns and crop yield trends in the area. Nevertheless crop yields are an important indicator of proximate soil conditions if other factors are not constraining.

4.2. Locally perceived association between farming systems and soil degradation

Table 2 presents the locally perceived relationships that were cited by the respondents as being the contribution of the farming practises to the observed land/soil degradation in the study areas. About 52% of the respondents associated soil degradation to continuous cropping while 23% considered inadequate manure application to be responsible for the diminishing soil quality. The overuse of the soil in continuous tillage without fertiliser supplementation, coupled by grazing on plant residues, weeds and crop stubble, has deprived the soils of both nutrients and organic matter [1, 29].

Perceived Relationship	Mafai (n=28)	Baura (n=26)	Bolisa (n=42)	Total (n=96)
Continuous cropping	57.1	30.7	61.9	52.1
Lack or inadequate availability of manure	10.7	53.8	11.9	22.9
Overgrazing in the past decades	14.3	11.5	11.9	12.5
Cultivation on steep slopes	14.3	4.0	4.8	7.3
Lack of on-farm conservation measures	3.6	0	9.5	5.2

Table 2. Perceived relationships between cropping/livestock management systems and soil degradation in the Irangi Hills (%)

Overgrazing was pointed out to be one of the processes that facilitate degradation by 12.5% of respondents. The low figure is explained by the fact that to the Rangi community, which is traditionally agro-pastoralist, having big herds of livestock is just part of their culture thus locally overgrazing is rarely seen as a major problem. A similar explanation regarding perceptions on overgrazing has been reported recently among communities in the Sukumaland and in the Iramba area [42, 44]. Overgrazing of livestock had similar effects on the soils of steep slopes and on shallow and stony soils, where continuous cultivation has not been practised [36]. Lack of on-farm conservation measures, especially before HADO started its activities, significantly contributed to the degradation features witnessed in the present days [1, 7]. Thus elaborate extension services are probably needed regarding various mechanisms that may contribute to sustainable farm production, such as on-farm erosion control, agroforestry practices and proper residue management. Proper farmer education would inculcate the culture of conservation among communities. Other associations presented in Table 2 did not feature as important concerns among farmers, but because of their role in soil degradation they are worth some attention.

While incorporation of crop residue and manure were meant to improve organic matter content and replenish soil fertility in the farm, contour ridges were constructed to check runoff and control erosion, and as such prevent further loss of soil fertility through nutrients washed away in eroded soil material. Only limited quantities of manure were however applied per unit area, especially since 1979 when livestock were evicted from the KEA. The limited supplies of manure and the high fertiliser prices are responsible for their low usage as adaptive mechanisms in fighting against soil degradation at farm level [29]. The only reliable way of replenishing soil fertility has been through crop residue incorporation into the soil. However, during dry seasons the residue provides valuable feed for livestock, whereas considerable proportions are burnt when preparing the land for a succeeding crop hence not much residue is left for incorporation into the soil. This may have negative consequences on subsequent crop productivity. Similar experiences are reported for other parts of the Irangi Hills, for example in Haubi and Mulua villages [1].

4.3. Community involvement in land resource management

4.3.1. Community participation in soil conservation

The pressure on land has practically increased all over the country particularly during the 20th century as a result of population growth. This has, in many instances resulted in unsustainable cultivation techniques including shortened fallow periods [38, 45] that consequently impoverish the soil. Enhanced long-term productivity and sustainability of the land resource thus require sound soil conservation measures in the farming systems that enhance maintenance and/or improvement of soil and land quality in general. This is an important consideration as it influences agricultural productivity and local livelihoods.

In many instances environmental degradation has stimulated a variety of responses and adaptation mechanisms by local communities. This study made an enquiry on whether farmers had undertaken any deliberate efforts to protect their land holdings from soil degradation. Majority of respondents (95%) indicated to have used one or more conservation techniques in their farms as a means of adjusting and adapting to soil degradation processes. Table 3 presents the various soil conservation approaches as mentioned by the interviewed farmers. The first three combinations of approaches, that is, contour ridges; tree, sisal and grass planting; manure application and incorporation of crop residue in soils, appeared to be the most prominent conservation strategies adopted by majority of farmers who were practising conservation, accounting for 97% of farmers in Mafai, 85% in Baura and 81% in Bolisa village. The generally sloping terrain in Mafai village partly explains the reported increase in the use of contour ridges, trees, sisal and grass planting to protect the soil from erosion.

Measures taken	Mafai (n = 35)	Baura (n = 34)	Bolisa (n = 70)	Total (N = 139)
Contour ridges; tree, sisal and grass planting; Manure application and Incorporation of crop residues in soils	80	55.9	68.5	68.1
Contour ridges and tree planting	17.1	29.4	12.9	19.8
Tree planting only	0	5.9	12.9	6.3
Stall-feeding cattle	2.9	5.9	4.3	4.4
Crop rotation	0	2.9	1.4	1.4
Total	100	100	100	100

Table 3. Soil conservation measures undertaken by respondent farmers (in %)

A majority of farmers have also planted trees as one of the soil conservation practices advocated by HADO. Table 4 presents a list of tree species that are planted in the study area and associated uses. As for crop rotation, the low response reported in Table 3 was mainly attributed to the small farm holdings that necessitate farmers to practice continuous cultivation of same fields.

Name	Scientific name	Purpose[1]	Mafai (n = 31)	Baura (n = 31)	Bolisa (n = 67)	Total (N = 129)
Silk oak	*Grevillea robusta*	Ct, Fw, Fi	87.1	90.3	95.5	91.0
Guava	*Psidium guajava*	Fr	51.6	51.6	49.3	50.8
Pawpaw	*Carica papaya*	Fr	29.0	29.0	76.1	44.7
Lemon	*Citrus limon*	Fr	19.4	22.6	17.9	20.0
Orange	*Citrus sinensis*	Fr	25.8	0	26.9	17.6
Eucalyptus	*Eucalyptus sp.*	Fw, Ct	35.5	3.2	1.5	13.4
Mango	*Mangifera indica*	Fr, Sh	0	16.1	14.9	10.3
Pomegranate	*Punicum granatum*	Fr	3.2	3.2	11.9	6.1
Leucaena	*Leucaena leucocephala*	Fd, Fi	3.2	3.2	1.5	2.6
Cypress	*Cuppressus sp.*	Fi, Ct	3.2	3.2	0	2.1
Iron wood	*Senna siamea*	Ct, Fw	0	3.2	1.5	1.6

[1]Ct = construction material/timber, Fr = fruits, Fd = fodder, Fi = soil fertility improvement, Fw = fuelwood, Sh = shade.

Table 4. Tree species planted in the study area, ranked according to preferences and percent of respondent farmers that have planted them

Findings from this study are in agreement with studies elsewhere that farmers often attempt to adjust to environmental degradation by using various measures and strategies [19, 46]. The measures taken may be different depending on the natural environment and socio-cultural backgrounds of the area concerned. According to Nsiah-Gyabaah [19], farmers' adjustment to their environment can generally be effective when they are able to predict the short term inter-annual variability patterns (e.g. in soils and weather). Under such situation they may be able to successfully use available innovations and local expertise to maximise the benefits of both soils and weather. However, where changes are unpredictable farmers may face difficulties in establishing adaptation mechanisms.

4.3.2. Challenges to community participation in soil conservation practices

Community participation in conservation practices is of great importance as it seeks to guarantee access and control over resources by the communities living in them, but who depend on these resources to satisfy their various needs (ecological, economic, social, cultural and spiritual needs). Community participation ensures more commitment in ensuring that resources are more sustainably managed, where apart from communities depending on these resources for a living and conserving them, they at the same time become their guardians [47, 48]. The active participation of various stakeholders in decision-making is crucial for ensuring the long term sustainability of community-based resource management initiatives. In several occasions however, soil conservation has not received the expected involvement of local communities. Some of the reasons that have influenced the local people's attitudes towards land/soil conservation efforts in the Irangi Hills are discussed here.

One of the reasons put forward was the tendency to underestimate the seriousness of the soil erosion problems by many people in the area. Where the tenure system is not elaborate individual farmers may not be concerned with problems of land degradation regardless of their holdings being at risk as such land degradation is considered as a general community problem. Such attitudes may result in no action being taken against land degradation even when there are no clear hindrances. The implication of the foregoing is that effective conservation is likely to be achieved when land tenure systems are properly articulated [1]. Thus efforts are needed to ensure integrated community-level planning that could promote individual farmers efforts without undermining community interests. Recent experiences from studies in Kishapu and Kahama Districts in Shinyanga Region, and in Nyarugusu area in Geita District Mwanza region indicated that many people practice soil and water conservation mainly in their own lands/farms [42], including setting aside private conservation areas such as Ngitili [49]. Adoption and/or practicing certain conservation measures are much influenced by the farmer's economic situation, including resource endowments. For instance, farmers with sufficient land holdings can afford to conserve some of it under the Ngitiri system, while land constrained farmers may not. Similar experiences would be the case for other conservation measures that require heavy investment by the farmer, for example making of soil erosion control structures that may need additional labour, and using fertilisers and/or manure.

Apart from the underestimation of the land degradation problem and inarticulate tenure system, it is also evident that the Rangi people themselves are traditionally not conservationists. When the pressure on land increases and harvests dwindle, some villagers will leave to take up new land elsewhere while others remain. The Rangi as a group can be said to practice an expansionist permanent agriculture [2]. Their principal solution to problems of soil erosion, for example, has thus been to move temporarily or permanently to areas with better conditions, while also retaining a foot in the Rangi core areas [2, 50, 51].

Experiences from the Irangi Hills and elsewhere in East Africa indicate that resource management is closely influenced by the prevailing socio-political environment [52]. It was reported, for instance, that some people in the Irangi Hills were of the opinion that the low adoption of soil conservation measures was based on negative attitudes that were inculcated among people during the colonial era, when such activities were basically coercive [1, 29]. During the colonial period soil conservation was seen as a form of oppression [50, 53]. In Kenya, for example, soil conservation was made compulsory under colonial administration and forced labour was used for community conservation programmes [22]. Although some of the soil conservation techniques employed were effective (e.g. in Machakos district), the coercive methods used were very unpopular [4, 22, 53] resulting into some local communities withholding their own conservation initiatives. This was also experienced in the Irangi Hills and among other communities in Kondoa District [50].

In the Kondoa case, the opposition to conservation during the 1940s was closely linked to a wide discontent within the Rangi community with the way the colonial government appointed local leaders. Thus a more explicit political discourse (struggle against colonialism) became linked to soil conservation policies [50]. Consequently, during the nationalist struggle in the late 1950s it was no longer possible to enforce communal turnout and most conservation work came to a halt [54]. The Kondoa Chief of the time came to side with the Rangi underground opposition against regulations on land-use. It is not surprising then that even independent governments found it difficult to reverse the previous nationalist attitudes and support conservation measures [22, 53]. Land degradation was thus left to continue. These experiences reflect to the need for local community participation in developing and implementing resource management strategies.

4.4. Land-use dynamics and environmental degradation

4.4.1. Land availability over time

In general, there have been changes in land-use use types in the study area due to several factors. Such changes can be considered to be adaptive mechanisms to population growth, economic development and to changing climate. Population growth is usually associated with increased exploitation of natural resources. According to local knowledge in the study area, prior to Tanzania's independency in 1961, arable land for agriculture and livestock was relatively plenty and used for different purposes, including cultivation and livestock grazing. The increase in population has led to expansion of agricultural and livestock activities. This has contributed significantly to changes in land-use intensity and cover types.

An assessment was made on perceptions regarding land availability over three time periods, from post independence, villagisation and current period. Experiences from the selected villages indicated that during post independence and villagisation period agricultural land was fairly easily available at the areas close to homesteads. It was reported that, land availability is currently very difficult than ever before, the main cause being increase in population pressure, which has led to increased demand for land by villagers. Eventually this has created land shortage in most of the villages. The pattern of land availability over years is also experienced in other parts of Kondoa District [55]. Farmers reported that currently the places with easily available land are those located far from homesteads, which was considered to be a major limitation for people who are not able to manage distant farms.

4.4.2. Decreasing farm sizes

For arable land use, there has been a general decline in the farm sizes. The local people attributed the decline in the farm sizes to several other factors (see Table 5). Thus there is a concern that land is increasingly being insufficient. The big proportion of the people reporting to experience land shortage reflects to that the problem of land shortage is much bigger than currently envisaged, and may worsen particularly with the fast growing population.

Perceived cause declining farm sizes	Percent
High population growth/increase	80.3
Expansion of settlements	6.9
Soil erosion and gullies	4.0
Expanding livestock keeping (more livestock)	2.9
Expansion of family sizes	2.0
Low soil fertility - land is tired/exhausted	2.0
Tree conservation	0.9
Intensive rainstorms (more erosion)	0.9
Total	100

Table 5. Different factors considered by the local community to cause declining farm sizes

The decreasing farm size is one of the causes of household food insecurity in the area. The small farm sizes limit the possibilities to practice fallow rotation. Consequently continuous cultivation culminates into declining soil fertility and reduced crop productivity. The problem may as well be aggravated by loss of land as a result of soil erosion, particularly gully erosion.

4.4.3. Chancing land use patterns and local adaptive mechanisms

Experiences on land use and the way rural land users interact with environmental resources show that communities have increasingly interacted with their local environment by faltering land use practices. However, continued land degradation has had severe environmental and social-economic consequences resulting in poor agricultural productivity, perpetuating food insecurity and poverty among the concerned communities. This necessitates the analysis of land-use dynamics, land degradation and their inter-linkages with livelihoods and poverty alleviation strategies adopted by the respective communities. Generally, there have been changes in land use patterns, which in many instances has involved increasing land use intensity as adaptive mechanisms to increased population growth, economic development and to changing climate [25]. Increase in population has led to expansion of agricultural and livestock activities, contributing significantly to changes in land use intensity, decline in farm sizes, land fragmentation and land degradation. Sustainable management of natural resources can thus be achieved by having integrated land use practices, including developing elaborate village land use plans.

Livelihood diversification into non-agricultural enterprises was also reported to have been a means of adapting to the changing environments and for poverty alleviation. Diversification as applied in the rural context is a process whereby rural households construct an increasingly diverse portfolio of activities and assets in order to survive [56]. Diversification can be viewed from different angles. It may imply a matter of survival associated with harsh local environmental conditions, or alternatively, it may be considered a matter of opportunity involving pro-active household strategies employed to improve living standards [57]. Generally, however, a diversity of livelihood sources is considered to be one of the ways through which households may develop security against agrarian environments, particularly in semiarid areas.

5. Conclusions

This chapter has examined how perceptions and knowledge facilitate and/or act as barriers to sustainable land management. It has been realised that farmers in the Irangi Hills are aware of land/soil degradation and its various processes, with levels of perception varying between villages and among respondents depending on the severity of the land degradation problem. Declining soil fertility and soil erosion, demonstrated by existence of spectacular gullies and extensive depositional sand fans, have been found to contribute significantly to the general understanding of land/soil degradation problems.

Awareness of land degradation was also reflected by the various criteria that the local people use in assessing the potentials and constraints that farmlands and the landscape in general are facing. Low crop productivity has been identified as one of the important constraints; attributed mainly to declining soil fertility, unreliable rainfall, and to a lesser extent, soil erosion. Farmers seemed to be quite aware of the association between cropping and/or livestock management systems and land/soil degradation. An extension service focusing on the various mechanisms that may contribute to sustainable farm production, such as on-farm erosion control, agroforestry practices and proper residue management is necessary. This is particularly important in a situation where continuous cultivation has become the norm because of increasing land shortage. Thus the various adaptations that farmers are already exposed to with regard to improving agricultural productivity in the erosion-stricken farmlands need to be promoted and developed further.

Limiting factors to local participation in soil conservation initiatives would probably be successfully corrected through proper farmer education that inculcates the culture of conservation among communities. It is important therefore that steps are taken to address the diversity of the intricate attitudes and socio-political environments among rural communities. It is recommended that further comparative studies be conducted so as to come up with sound strategies that will motivate the local people to participate more in issues related to sustainable land resource management. Where relevant such strategies and approaches to extension work need to be tailored to individual communities rather than using large homogenous programmes.

Increase in population has led to expansion of agricultural and livestock activities, contributing significantly to changes in land use intensity, decline in farm sizes, land fragmentation and land degradation. Since agricultural production (crops and livestock) is the major means of poverty alleviation in these semiarid areas, it calls for enhanced environmental conservation so that agricultural productivity can be sustained for the betterment of the community livelihoods. This is particularly important now given the additional factor of climate change. Thus, sustainable management of land and other natural resources can only be achieved by having integrated land use practices, including developing elaborate land use plans. Poverty alleviation strategies and other policies also need to ensure sustainable development with minimal impacts on the environment. The diversification of livelihood activities (farm and non-farm) recorded in the area indicates that local communities are struggling towards poverty alleviation and as a way of adaptation and coping to environmental degradation.

Author details

Richard Y.M. Kangalawe
Institute of Resource Assessment, University of Dar es Salaam, Dar es Salaam, Tanzania

Acknowledgement

This study was undertaken with a financial support from NORAD through its Management of Natural Resources and Sustainable Agriculture fellowship programme. Many thanks are extended to the farmers in Bolisa, Baura and Mafai villages who enthusiastically participated in this study and for their inspirations that paved a way towards completion of this work.

6. References

[1] Kangalawe RYM. (2001) *Changing land-use patterns in the Irangi Hills, central Tanzania: A study of soil degradation and adaptive farming strategies*. PhD Dissertation No. 22.

Department of Physical Geography and Quaternary Geology, Stockholm University: Stockholm.

[2] Östberg W (1986) *The Kondoa Transformation: Coming into Grips with Soil Erosion*. Research Report No. 76. Scandinavian Institute of African Studies: Uppsala.

[3] Christiansson C, Kikula IS, [51] W (1991) Man-land interrelations in semiarid Tanzania: A multidisciplinary research programme. Ambio 20 (8): 357-361

[4] Christiansson C, Mbegu A, Yrgård A (1993). The Hand of Man: Soil conservation in Kondoa Eroded area, Tanzania. Nairobi: Regional Soil Conservation Unit.

[5] Mbegu AC (1988). The HADO Project: What, Where, Why, How? Forestry and Beekeeping Division, Ministry of Lands, Natural Resources and Tourism: Dar es Salaam.

[6] Payton RW, Christiansson C, Shishira EK, Yanda P, Eriksson MG (1992) Landform, Soils and Erosion in the north-eastern Irangi Hills, Kondoa, Tanzania. Geografiska Annaler 74 (A 241): 65-79

[7] Payton RW, Shishira EK (1994) Effects of soil erosion and sedimentation on land quality: Defining pedogenetic baselines in the Kondoa District of Tanzania. In: Syers JK, Rimmers DL. Editors. Soil Science and sustainable land management in the tropics. England: Wallingford CAB International 88-119.

[8] Darkoh MB (1986) Experiences of arid land development in Tanzania: In: Arid land development and the combat against desertification: An integrated Approach. UNEP: Moscow.

[9] Eriksson MG (1998) Landscape and soil erosion history in central Tanzania. A study based on lacustrine, colluvial and alluvial deposits. Dissertation No.12. Department of Physical Geography, Stockholm University: Stockholm.

[10] Eriksson MG, Olley JM, Payton RW (1999) Late Pleistocene colluvial deposits in central Tanzania, erosional response to climate change. GFF 121:198-201

[11] Eriksson MG, Olley JM, Payton RW, (2000).. Soil erosion history in central Tanzania based on OSL dating of colluvial and alluvial hillslope deposits. Geomorphology 36: 107-128

[12] Ngana JO (1996) Climate and hydrology of the Kondoa Eroded Area. In: Christiansson C., Kikula IS. Editors. Changing Environments: Research on Man-Land Interrelations in Semiarid Tanzania. Report No. 13. Nairobi: Regional Soil Conservation Unit:.

[13] Whyte AVT (1977) *Guidelines for field studies in environmental perception*. MAN Technical Notes No 5. Paris: UNESCO.

[14] Hackel JD (1990) Conservation attitudes in southern Africa: A comparison between KwaZulu and Swaziland. Human Ecology 18 (2): 203-209

[15] Hunter Jr ML, Hitchcock RK, Wyckoff-Baird B [1990] Women and wildlife in the southern Africa. Conservation Biology 4(4): 448-451

[16] Showers KB, Malahleha GM (1992) Historical environmental impact assessment: A tool of for analysis of past interventions in landscapes. Working Paper No. 8. The Project of

Agriculture of the Joint Committee on African Studies, ISAS: National University of Lesotho.

[17] Rydgren B (1993) Environmental impacts of soil erosion and soil conservation. A Lesotho case study. PhD Thesis, UNGI Report No 85. Upsalla: Uppsala University.

[18] Dahlberg AC (1996). Interpretation of environmental change and diversity: A study from North-eastern Botswana. Dissertation Series No. 7. Department of Physical Geography, Stockholm University: Stockholm.

[19] Nsiah-Gyabaah K (1994) *Environmental degradation and desertification in Ghana. A study of the Upper West Region*. Aldershol & Brookfield: Avebury Studies in Green Research..

[20] Herberlein TA (1972) The land ethic realised: Some social psychological explanation for changing environmental attitudes. Journal of Social Issues 28: 78-82

[21] Pitt MW, Yapp TP (1992) Perceptions of land degradation and awareness of conservation programmes in North-eastern New South Wales. In: Haskins PG, Murphy BM. Editors. People protecting their land, Proceedings of the 7th ISCO conference: Sydney. pp. 115-124

[22] Critchley W (1991) Looking after our land. Soil and water conservation in dryland Africa. Oxfam.

[23] Toulmin C, Chambers R (1990) Farmer-First: Achieving sustainable dryland development in Africa. IIED Paper No. 19. London: IIED.

[24] Bewket W, Teferi E (2009) Assessment of soil erosion hazard and prioritization for treatment at the watershed level: Case study in the Chemoga Watershed, Blue Nile Basin, Ethiopia. Land Degradation & Development 20: 609–622, DOI: 10.1002/ldr.944

[25] Kangalawe RYM (2009) Land use /cover changes and their implications on rural livelihoods in the degraded environments of central Tanzania. African Journal of Ecology 47 (Suppl. 1): 135–141

[26] Ngana JO (1990). *Modelling for periodic features in seasonal rainfall and its implications to water resources and agricultural planning*. Research Report No. 27. Dar es Salaam: Institute of Resource Assessment, University of Dar es Salaam.

[27] Christiansson C (1981) Soil erosion and sedimentation in semi-arid Tanzania. Studies on environmental changes and ecological imbalance. Uppsala: Scandinavian Institute of African Studies.

[28] Eriksson MG (1999) Influence of crustal movements on landforms, erosion and sediment deposition in the Irangi Hills, central Tanzania. In: Smith BJ, Whalley WB, Warke PA. Editors. *Uplift, erosion and stability: Perspectives on long-term landscape development*. London: Geological Society of London, pp.157-168

[29] Kangalawe RYM (1995) *Fighting soil degradation in the Kondoa eroded Area, Kondoa District, Tanzania: Socio-economic attributes, farmer perceptions and nutrient balance assessment.* M.Sc. Thesis. Aas: Agricultural University of Norway.

[30] Mbegu AC, Mlenge WC (1983) Ten Years of HADO. Dar es Salaam: Forest Division, Ministry of Natural Resources and Tourism.

[31] Clarke R (1986) The handbook of ecological monitoring. Oxford: Clarendon Press. 298p

[32] Boyd HK, Westfall R, Stasch F (1981). Marketing Research: Tests and Cases. Illinois: Richard D. Inc.

[33] Hodgson JM (1985) Soil survey field handbook: Describing and sampling soil profiles. Technical Monograph No. 5. Harpenden Herts: Soil Survey, Rothamsted Experimental Station. 99 pp

[34] Johnson RA, Bhattacharyya GK (1992). Statistics: Principles and methods. Second edition. John Wiley & Sons, Inc.

[35] Ryan BF, Joiner BL, Ryan Jr TA (1992) Minitab Handbook. Second edition. : Belmont California: Duxhury Press.

[36] Tosi JA, Hartshorn GS, Quesada CA (1982) *HADO Project Development Study and Status of the Catchment Forestry*. A Report to the Ministry of Natural Resources and Tourism, Tanzania. San Jones: Tropical Science Centre.

[37] Smith JL, Elliott LF (1990). Tillage and residue management effects on soil organic matter dynamics in semi-arid regions. Advances in Soil Science 13: 69-88

[38] Mohamed SA (1996) Farming systems and land tenure in Haubi Village of the Kondoa Eroded area. In: Christiansson C, Kikula IS. Editors. Changing environments: Research on Man-Land Interrelations in Semi-Arid Tanzania. Report No. 13. Nairobi; RSCU/SIDA. pp.77-83

[39] Mowo JG, Floor J, Kaihura FBS, Magoggo JP (1993) Review of fertiliser recommendations in Tanzania. Part 2 - Revised fertiliser recommendations for Tanzania. Soil fertility report No. F6. ARI Mlingano-Tanga: National Soil Service.

[40] Dejene A, Shishira EK, Yanda PZ, Johnsen FH (1997) Land degradation in Tanzania: Perceptions from the village. World Bank Technical Paper NO. 370. Washington DC: World Bank.

[41] Kikula IS (1989) Possible implications of the similarities and differences in perception of land degradation between planners and the planned. Research Paper No. 19. Dar es Salaam: Institute of Resource Assessment, University of Dar es Salaam.

[42] Kangalawe, RYM, Liwenga, ET, Majule, AE (2007) The Dynamics of Poverty Alleviation Strategies in the Changing Environments of the Semiarid Areas of Sukumaland, Tanzania. Research Report submitted to REPOA, Dar es Salaam.

[43] Kangalawe RYM, Lyimo JG (2009). Climate Change and Rural Livelihoods in Semiarid Tanzania. Proceedings of the 10th WaterNet/WARFSA/GWP-SA Symposium: "Integrated Water Resources Management within the context of Environmental Sustainability, Climate Change and Livelihoods". Entebbe: 28 – 30 October 2009.

[44] Kangalawe RYM., Majule AE, Shishira EK (2005) Land-use dynamics and land degradation in Iramba District, central Tanzania. Addis Ababa: Organisation for Social Science Research in Eastern and Southern Africa.

[45] Madulu NF, Mbonile MJ, Kiwia HDY (1993) Environmental impacts of migration in rural Tanzania. In: *Population, Environment and Development*. New York: United Nations, pp.73-91

[46] Cummingham OR, Jenkins QAL (1982) Natural disasters and farmers: A neglected area of research by rural sociologists. The Rural Sociologist 2(5): 325-330

[47] Nyega N (2008) Assessment of the beach management units strategy on the Lake Victoria fisheries resources in Ilemela District, Tanzania. MSc Dissertation, University of Dar es Salaam.

[48] Shadrack S (2009). Effectiveness of community based forest management approach in transforming local use of forest resources in Rufiji District. MSc Dissertation, University of Dar es Salaam.

[49] Mlenge W (2004) Ngitili: An indigenous natural resources management system in Shinyanga. Nairobi: Arid Lands Information Network - East Africa.

[50] Mung'ong'o CG (1995) *Social processes and ecology in the Kondoa Irangi Hills, Central Tanzania*. Meddelanden Series B 93. Stockholm: Department of Human Geography, Stockholm University.

[51] Östberg W (1995) *Land is Coming Up: The Burunge of Central Tanzania and their environments*. Stockholm Studies in Social Anthropology, 34. Stockholm: Department of Social Anthropology, Stockholm University.

[52] Mung'ong'o CG, Kikula IS, Mwalyosi RBB (2004). Geophysical and socio-political dynamics of environmental conservation in Kondoa District. Dar es Salaam: Dar es Salaam University Press.

[53] Kauzeni AS, Kikula IS, Shishira EK (1987) Developments in Soil Conservation in Tanzania. In *History of Soil Conservation in the SADCC region*. Research Report No. 8. SADCC Soil and Water Conservation and Land Utilisation Programme, Co-ordinating Unit: Maseru.

[54] Berry L, Townshend J (1973) Soil Erosion Conservation Policies in the Semi-Arid regions of Tanzania, A Historical Perspective. In: Rapp A, Berry L, Temple P. Editors. *Studies on Soil Erosion and Sedimentation in Tanzania,* BRALUP Research Monograph No. 1. BRALUP: Dar es Salaam; 241-253.

[55] Liwenga ET, Kangalawe RYM, Masao, CA (2007) The Implications of Agricultural Commercialization on Agro-diversity Management and Food Security in the Drylands of Central Tanzania. Research Report submitted to the Research Programme on Sustainable Use of Dryland Biodiversity (RPSUD), University of Dar es Salaam.

[56] Scoones, I. (1998). Sustainable rural livelihoods: A framework for analysis. IDS Working Paper No. 72.

[57] Ellis, F. (1998). Household strategies and rural livelihood diversification. Journal of Development Studies 35 (1): 1-38.

Chapter 24

Developing a South-European Eco-Quarter Design and Assessment Tool Based on the Concept of Territorial Capital

Stella Kyvelou, Maria Sinou, Isabelle Baer and Toni Papadopoulos

Additional information is available at the end of the chapter

http://dx.doi.org/10.5772/45885

1. Introduction

Many studies have been undertaken with regard to eco-neighbourhoods in Europe. However, most of the projects that have been completed and are being analysed in relevant studies are located in Northern Europe, i.e. the BedZed in England, Hammarby (Stockholm) and BO01 (Malmö) in Sweden, Kronsberg (Hannover) in Germany, Vesterbro (Copenhagen) and Kolding in Denmark, Vauban in Freiburg and others. Findings from these projects permit nowadays to speak about a Northern European eco-neighborhood model (Souami, 2009). However, it would be interesting to investigate eco-neighborhood projects in Southern Europe that are either already realised or still in the design phase. The questions that rise are on one hand the sustainability approach that was followed for these projects and on the other hand the specific criteria involved in each case. To this end, an investigation is undertaken regarding the tools that are being used in terms of the environmental principles-criteria which are taken into account by each of them and how easy these tools are to use. Finally, a comparative analysis follows regarding the different Southern European projects and the environmental criteria involved in their implementation.

The present chapter can be summarised in three fundamental objectives: a) the investigation of contemporary tools and methods of planning and assessment of eco-neighborhoods aiming at identifying similarities and differences but also issues that can lead to an efficient Mediterranean methodology, b) the study of examples of Mediterranean eco-neighborhoods in order to create a good and bad practice guide and g) the proposal of a new assessment tool for the Mediterranean eco-neighborhood, based on the concept of territorial capital (OECD, 2001).

The methodology that was followed regarding the first objective focuses in the parametric analysis of basic criteria of existing tools seeking common ground and differences. As for the second objective the environmental criteria of examples under consideration are investigated with the use of the One Planet Living framework, while the new "SDMed eco-neighborhood tool" was based on:

1. the research and parametric analysis among the tools that concern the development of eco neighborhoods;
2. the SDMed building performance assessment tool (Sinou & Kyvelou, 2006);
3. the concept of territorial capital (OECD, 2001) and its expoitation at local level and the approach of territorial cohesion (both internal and external) that should govern an agglomeration or urban development, even at the scale of a neighborhood.

2. Definition of eco-neighbourhoods

Different approaches and perspectives can be identified regarding the definition of the term "neighborhood" and therefore "eco-neighborhood". One of the most common is to do with density and population. The link between the levels of density and land take in a typical neighborhood of 7500 people. The message is clear: the lower the density, the larger the amount of area that is occupied by buildings, roads and open space. Density per se is not an indicator of urban quality. An interesting definition of eco-neighborhood is given by Barton where he categorises according to different spatial scales. The smallest scale is the building scale; the next one is the home place scale, then the neighborhood scale, the small town scale and finally the city scale. The key sustainability and health issues identified by Barton are: plan for local facilities with attractive walking routes, local hubs to support healthy lifestyles and development of local food, waste, water and energy capture systems (Barton, 2010).

Neighborhood is defined as a residential or mixed use area around which people can conveniently walk. Its scale is geared to pedestrian access and it is essentially a spatial construct, a place. It may or not have clear edges. It is not necessarily centred on local facilities, but it does have an identity, which local people recognize and value (Barton, 2000). Moreover, it is interesting to note the three different facets of neighborhood that Barton distinguishes. Firstly, the neighborhood perceived as the base for home life, education, leisure and employment activities. Secondly, seen as a place, as an aesthetic experience and thirdly as the locus for community (Barton, 2000). All these three facets can be identified in the tools that will be presented in section three; however each tool seems to have a slightly different orientation according to whether it will be used from town planners, or urban designers, or even at an earlier stage from stakeholders during decision making. Moreover, it is significant to note at this point that the most coherent and complete approach is the one that takes into consideration in planning not only issues of energy, transport and resources but also social and economic parameters.

About the model of eco-neighborhood or eco-district, we should mention that a meaningful typology has been drafted by Souami (Souami,2009). He distinguishes three phases of eco neighborhoods' creation, in less than two decades. According to Souami, each phase is corresponding to one of the three different types of eco-neighborhood and it is the second type that permits more performances compared with the other two types, thus representing in a certain way 'the very model of the so-called 'eco-neighbourhood'.

2.1. The initial eco-neighbourhood type of the '80s

According to Souami, the initial eco-neighborhood type was most often a small pool of buildings located in the periphery of cities or in rural areas. The initiators of such projects were usually professionals and experts, politically active, enrolled in so-called alternative movements. Convinced about the importance of 'green' development and construction, the founders of eco-neighborhoods adapted both the idea and the process before choosing the site to accommodate and implement their ideas which was chosen later on. During the 1980s, we meet neighborhoods of this type in Austria, in the Netherlands and in Germany. They are, in fact, eco-villages transformed into neighborhoods and the organisation in a community or associative form of development is often used to group the inhabitants, in order to organise the public areas and implement the project.

2.2. The 'prototype' of eco-neighbourhood of the '90s

In this case, some communities have taken advantage of exceptional urban events to initiate sustainable districts on their territory: World's Fair in Hanover, B01 exhibition in Malmö, London Olympics, Olympic Games bid in Paris, Zaragoza's candidature for EXPO 2008 that initiated the 'Ecociudad Valdespartera' etc. These events have been all opportunities to initiate positive processes that go beyond conventional practices, showing ambitious environmental goals.

The projects are accompanied by an important work of communication, especially internationally. They are developed as exemplary neighborhoods, particularly successful demonstration projects addressing both to technicians and local politicians. Nevertheless, some of them constitute events by themselves. As far as governance and investment schemes are concerned, we should note that:

- partnerships involve communities, private and public developers, social housing agencies, several operators of urban services and facilities (energy, water, etc.),groups of experts and many contractors
- funds are cumulative and come from various sources: local (municipalities, public and private developers), national (sectoral programmes, exceptional ministerial grants or subsidies) and international (various European programmes).

Leaders of local projects show innovative and mainly broadly applied technical solutions (systematic recycling of rainwater, deployment of extended solar panels, photovoltaic panels, etc.) and technicians and policy-makers have the opportunity to test, validate and correct certain choices. These projects are also considered as places that promote a learning procedure for stakeholders and citizens (Kyvelou, 2010).

2.3. From the mid-90s a new type of eco-neighbourhood appears based on environmental quality criteria

In the third category, eco-neighborhood projects are initiated in a conventional manner since they mobilise ordinary tools of development and construction but they integrate environmental quality objectives. In other words, these districts adopt common and unexceptional production methods in order to integrate sustainable development perspectives. Some of them clearly refer to the achievements of the 'prototype' sustainable neighborhoods (the so-called European 'vedettes'). These projects, often modest in size, are being planned in a long-term period and they are sometimes considered as resulting from the dissemination procedure of the proto-neighborhoods and the 'prototype' ones. Nevertheless, Souami argues that, according to his investigations and research, there is no systematic genealogy and explicit influences between these neighborhoods (Souami, 2009).

Furthermore, history shows that eco-neighborhood projects concern issues arising from territorial, urban, social and economic aspirations that go beyond environmental considerations. These socioeconomic and urban contexts are different in each case: economic prosperity for some of them, acute socioeconomic crisis for others, reshaping of political and institutional context, poverty, etc. This diversity proves that there is no common profile to serve as a basis for the creation of an eco-neighborhood. In all cases, eco-neighborhoods are implemented in order to enhance image of the city and local identity. The objective is to protect the sites concerned and therefore the cities where they belong, from their prior image. This image is part of the elaboration of public policies across the whole city. The classification of a district based on its environmental performance depends on a long-term work on the construction of place identity and the deepening of the feeling of belonging to this neighborhood. The place would no longer be identified by its history, its people, its animation or attendance. It would initially live through the image of environmental performance that circulates outside. The existence of prior approaches and actions on environmental issues is not always an asset for the development of operational projects aimed at sustainable development. In some cases, certain environmental policies have applications in connection with sustainable urban projects. In other cases, affected communities have not been able to mobilise their achievements of past policies to initiate and carry out operational projects of environmental quality. Sometimes, sustainable neighbourhoods are not preceded by pre-existing environmental policies. Summing up, we should note that:

- there is not a prerequisite for environmental policy or a prerequisite for sustainable development to achieve eco-neighbourhoods;
- eco-neighborhoods are often conducted in parallel with the establishment of local policies for sustainable development and these projects may contribute to developing local sustainable development policies;
- eco-neighborhoods are not the operational implementation of existing policies for sustainable development which prove and demonstrate their effectiveness.

3. Eco-neighborhood tools

There are numerous tools developed so far, to assess performance of eco-neighborhoods and provide guidance for their planning and design. The following are the most popular ones.

3.1. To BioRegional one planet living framework

This framework consists of ten principles which should govern sustainable communities, namely, zero carbon, zero waste, sustainable transport, sustainable materials, local and sustainable food, sustainable water, land use and wildlife, culture and heritage, equity and local economy and lastly health and happiness. This approach is a very simple one and can be easily used to help individuals and local stakeholders to examine the sustainability challenges and develop appropriate solutions (Table 1).

Zero carbon	Making buildings more energy efficient and delivering all energy with renewable technologies.
Zero waste	Reducing waste, reusing where possible, and ultimately sending zero waste to landfill.
Sustainable transport	Encouraging low carbon modes of transport to reduce emissions, reducing the need to travel.
Sustainable materials	Using sustainable healthy products, with low embodied energy, sourced locally, made from renewable or waste resources.
Local and sustainable food	Choosing low impact, local, seasonal and organic diets and reducing food waste.
Sustainable water	Using water more efficiently in buildings and in the products we buy; tackling local flooding and water course pollution.
Land use and wildlife	Protecting and restoring biodiversity and natural habitats through appropriate land use and integration into the built environment.
Culture and heritage	Reviving local identity and wisdom; supporting and participating in the arts.
Equity and local economy	Creating bioregional economies that support fair employment, inclusive communities and international fair trade.
Health and happiness	Encouraging active, sociable, meaningful lives to promote good health and well being.

Table 1. Principles of BioRegional One Planet Living framework

The One Planet Communities programme uses a set of Common International Targets against each of the 10 One Planet principles to ensure that international partners' projects are guided towards a shared end-point by 2020 and to determine what level of performance is required for a development to be endorsed. Behind the One Planet initiative there are three overarching environmental drivers:

- sustainable ecological footprint;
- sustainable carbon footprint; and
- clean (non-polluting) activities.

3.2. The Eco Town framework by the Cambridge quality charter of growth

The Eco-Town framework focuses on state of the art green building, energy and transport technologies and materials to be used in an urban development context. The task is to ensure zero-carbon housing and that energy efficiencies are achieved through waste reduction, energy conservation technologies and use of more sustainable sources of energy.

The Eco Town approach refers to new settlements with a minimum of 5000 homes where the developments should reach zero carbon standards, should provide good range of facilities and affordable housing. The framework consists of four fields, the four Cs, namely, climate, connectivity, community and character. Each one of the four is subdivided in several criteria (Table 2).

UK Eco-Towns	
Climate	Energy
	Water
	Environment
	Planning for low carbon
	Low-environmental
Character	Place-making
	New design and High Design Standards
	Attractiveness and desirability
	Investment
	Locally-based facilities
Connectivity	Employment opportunities
	Transport
	Services
Community	Social mix
	Sustainable community principles
	Governance
	Delivery organisation

Table 2. Principles of Eco-Town framework

3.3. The DPL approach

DPL is an approach for sustainable urban planning that attempts to quantify and measure sustainability of urban areas (districts) based on 25 environmental, social and economic indicators (Planet, People, Profit) (Table 3). Planet indicators are subdivided into two categories, namely, stocks and local environment. People indicators are subdivided into four categories namely, safety, services, green space water and quality. Profit indicators are subdivided into three categories namely, economic vitality, sustainable businesses and capacity change. DPL was developed by IVAM in cooperation with TNO Environment and Geosciences and with financial support from the Dutch Ministry of Housing Spatial Planning and the Environment (VROM).

DPL – sustainability profile of a district		
PLANET	**PEOPLE**	**PROFIT**
stocks	safety	economic vitality
1. Material use	10. Social safety	20. Local employment
2. Energy use	11. Traffic safety	21. Local economic activity
3. Land use	12. External safety	sustainable businesses
local environment	services	22. sustainable businesses
4. Water management	13. Quality of services	capacity to change
5. Soil contamination	14. Access to services capacity to change	23. Flexibility
6. Waste management	green space and water	24. Mixed use
7. Air pollution green space and water	15. Local green space	25. ICT infrastructure
8. Noise	16. Local water	
9.Smells	quality	
	17. Quality of the district	
	18. Quality of the dwellings	
	19. Social cohesion	

Table 3. The indicators of sustainability assessment according to DPL

3.4. The Med Eco-Quartier approach

The objective of the Med Eco-Quartiers Project was to define precisely the criteria and tools for creating eco-neighborhoods in the Mediterranean region, by studying different cultures, procedural approaches and environments. As part of the project four working tools were developed. They range from the phase of project design to the final realisation phase. The four tools are *Med Eco-urbanisme, Med Eco-constructibilité, Med Eco-gouvernance* and *Med Eco-sensibilisation* (Tables 4 & 5)

Criteria of MED ECO-urbanism
1. Government and organisation
2. Economic and social growth
3. Natural and cultural/archaeological heritage
4. Restriction of urban sprawl
5. Local services
6. Mediterranean natural and climatic characteristics

Table 4. The six criteria for the development of Mediterranean eco-neighborhoods by the MED Eco-urbanism

Preoccupations for the Mediterranean eco-neighborhoods
Preserving resources
Reducing pollution
Reducing waste
Managing natural and technological risks
Improving comfort
Preserving health
Culture and heritage
Integration of the public into the project
Maintenance evolutions
mastering cost excess
Local networks
Low nuisances building sites

Table 5. Preoccupations for the Mediterranean eco-neighborhoods, source : "MED-Ecoquartiers"

The Med Eco-planning tool consists of a grid of decision support and audience piloting development project. The overall approach, crossing the various themes of sustainable urban design leads to the fundamental objectives of the project. The study of the Med Eco-planning tool recommendations allows the specification of the characteristics of the new district. Even if the Med Eco-planning tool provides a framework for reflection and action, it remains a deliverable that has not proved its replication value since it has not been linked with implementation of eco-neighborhoods in the Mediterranean (Kyvelou & Papadopoulos, 2011).

3.5. The LEED for Neighborhood development project scorecard

The tool is subdivided into five categories. Each of them has several analysis criteria, which are either prerequisite or are being given a score (Table 6). The main categories are Smart Location and Linkage, Neighbourhood pattern and Design, Green Infrastructure and

Buildings, Innovation and Design Process and Regional Priority Credit. The tool similarly to LEED for buildings calculates a certification estimate and gives five total scores, namely certified, silver, gold and platinum.

LEED for Neighborhood Development	
Smart Location and Linkage	27 points
Neighborhood Pattern and Design	44 points
Green Infrastructure and Buildings	29 points
Innovation and Design Process	6 points
Regional Priority Credit	4 points

Table 6. The basic categories of LEED for Neighborhood tool

3.6. The HQE aménagement

Formalised in March 2010, the "HQE aménagement" has been subject of a Guide issued under the auspices of the HQE Association. It is, primarily, a pragmatic and ambitious methodology, mostly based on the feedback from concrete operational projects of development. It is based on business and professional logic, which is the one of the developers. The "HQE aménagement" also aims at equipping every stakeholder involved in the development with a reference framework and a common vocabulary for conducting eco-districts and improving professional practices.

HQE Aménagement	
Objectives	**Themes**
Ensure integration and cohesion in relation to the urban tissue and other territorial scales	1. Wider territorial unity and local frame 2. Density 3. Mobility and accessibility 4. Cultural heritage, landscape and identity 5. Adaptability and evolutivity
Preserve natural resources and promote environmental and health quality	6. Water 7. Energy and Climate 8. Materials and equipment 9. Waste 10. Ecosystem and biodiversity 11. Natural and technological risks 12. Health
Promote social life and support dynamics of local economy	13. Economy of the project 14. Mixture of uses and land uses 15. Atmosphere and public spaces 16. Integration and training 17. Local economy dynamics

Table 7. The basic topics of HQE Aménagement

It is a thematic approach that describes the objectives that are sought within the sustainable operation of development. Organised in 17 themes, it allows the direct choice of sustainable actions for the implementation of the following characteristics and subjects of interest (Table 7). The 17 themes are divided into three major sets: territorial analysis - technical and environmental analysis - socio-economic analysis.

- Territorial Analysis: Ensure integration and consistency of the eco-district with the urban area and other territorial levels.
- Technical and Environmental Analysis: Preserve natural resources and promote environmental quality and health.
- Socio-Economic analysis: Promote social life and strengthen the dynamics of local economy.

4. Is there a South-European model of eco-neighbourhood?

The north-european eco-neighborhood model is mainly described by its technical and environmental performance in terms of energy, saving water or recycling materials. They seem to be the main mechanism to move from principles to the effective implementation of sustainable urban development. The urban planning and design and the implementation of networks are resulting from this approach. This model governed by the environmental approach and its performance is mainly used by communities as a powerful tool of communication, promoting the region and even as leverage to reverse social and economic depreciation. However, these social and economic aspects are not sufficiently highlighted and are not explicitly included within the agreed content of the model. On the contrary, Southern European countries and especially the Mediterranean seem to prefix social, economic and governance issues and less attention is paid to environmental performances at least from the point of view of their initial definition and specification.

4.1. The case of France: Relative delay, centrally directed movement, focus on societal issues

In France, despite a relative delay, many cities have been engaged recently in the process of sustainable neighborhoods. Most of the projects are being actually studied and implemented, so their status does not allow to fully evaluate the results and present a meaningful analysis. In addition, the famous "Grenelle de l' Environnement" has proposed to initiate a plan of voluntary eco-neighborhoods driven by local governments: at least one eco-neighborhood in every municipality that intends to realise programmes of housing development until 2012 (in continuity with the existing urban texture and integrated in the city master plans) as well as fifteen large-scale projects of energy, architectural and social innovation, while the release of growth is planned to occur by 2012 through the creation of ten "Ecopolises", that is cities of at least 50,000 inhabitants integrating environmental quality and new information and communication technologies ("Attali" commission).

What is important in France is the strong political will and the consequent centrally directed generalisation: the "Sustainable City Plan" (Plan Ville Durable), presented to the Council of

Ministers on 22 October 2008, aimed at fostering the emergence of a new way to design, build, develop and manage the city. As defined by the Ministry of Ecology, Sustainable Development, Transport and Housing (MEDDTL), the eco-neighborhood is a sustainable operation of high demonstrative and exemplary value. Considered as key measure of the Sustainable City Plan of MEDDTL, it contributes to improving quality of life, while adapting to tomorrow's challenges: "preserving resources and landscapes, while preparing the conditions for the creation of a suitable housing supply". A first call for projects for the competition entitled "EcoQuartier" was launched in October 2008 with local communities to bring together stakeholders of quality operations within an operational Club, enhance their actions and allow the dissemination of good practices. 160 projects have been submitted by communities, coming from all French regions. In 2009, the following cross-cutting issues have enabled to distinguish the winning communities through a ranking of 28 projects:

- Relevance of the urban project, governance, management and structuring of the project, mixt uses and environmental aspects of the development.
- Water, waste, biodiversity, mobility, energy efficiency and renewable energy, density and urban forms, sustainable construction.

The National Award was delivered to the ZAC of Bonne in Grenoble (38). The second call for eco-neighborhood projects was launched in January 2011. A new eco-neighborhood grid was set up in order to serve the project analysis by experts but also to provide a "framework for thought" by any community seeking to implement an eco-neighborhood. In 2011, 393 applications were submitted. A double Grand National Award was delivered to:

- Nancy, and Laxou Maxéville (54 - Urban Community of Greater Nancy) - The Hague Plateau
- Roubaix, Tourcoing and Wattrelos (59 - Urban Community of Lille) - The Union.

The success of the two calls for proposals (2009 and 2011) demonstrated the enthusiasm of local communities to develop sustainable operations, whether in cities, towns or rural communities. A committee charged to form a proposal for an "écoquartier label" was formed. This committee issued, late 2011, recommendations addressing to the Minister of Housing and Urban Development for the establishment of a national eco-district label.

These centrally directed processes in cooperation with local authorities highlight already major projects and underline the fact that the phenomenon of eco-neighborhoods is widespread in France and building resilience is ongoing for these innovative projects of urban development.

Moreover, the HQE Association, a public character platform of stakeholders dealing with sustainable building and sustainable urban development, in collaboration with SNAL (Syndicat National des Aménageurs Lotisseurs) which is a federation of over 300 private developers who contribute to the production of more than 25,000 houses annually, have produced a guide on how to integrate the HQE process for buildings in urban projects. This was the concept that led to the creation of a new HQE process entitled "HQE Aménagement" which has been subject of experimentation since 2006 in 10 pilot community projects. This approach aims at ensuring that all environmental, social and economic concerns are handled by developers and should allow private or public developers to monitor the project so that it incorporates a variety of concerns and all stakeholders likely to be involved: professionals, residents, technical services of the community. The "HQE Aménagement" is fully compatible with the future label "écoquartier" planned by the Ministry of Ecology, Sustainable Development, Transport and Housing (MEDDTL) as far as both piloting, management system and process of the development project are concerned.

To note that the design of eco-neighborhoods in France gives emphasis on the social dimension of the outputs (Lefèvre & Sabard, 2009) even by means of the rebirth of the cooperative movement, which is evident since the beginning of 2000 in France.

4.2. A mediterranean eco-neighborhood model elaborated in the framework of an interregional cooperation programme

The overall objective of the MED-Ecoquartiers project, carried out in the frame of the Medocc-INTERREG IIB and based on a European regional partnership, was to produce a common methodology in order to serve as a set of principles for the creation of new neighborhoods that are consistent with the principles of sustainability (planning, building, energy, mobility, quality of life, activities, natural resources, historic buildings and landscape) in the countries of the Western Mediterranean. It grouped the cities of Pezenas (eco-neighborhood of Saint-Christol, 29 ha for 1700 inhabitants), Dos Hermanas (Spain, eco-neighborhood Montequinto for 12,000 inhabitants), Faenza (Italy, eco-district of San Rocco, 350 dwellings for 1,000 inhabitants), Elefsis (Greece, eco-neighborhood of 88 apartments on 3 hectares).

4.3. Strategic spatial planning prevailing in an Italian eco-district project

Under the hypothesis that "there is no a unique Mediterranean city, but only many different Mediterranean cities" and looking for describing as much as possible the Mediterranean diversity, the Italian partners of the Med-Ecoquartiers Project have elaborated one of the most significant examples in St Rocco neighborhood in the City of Faenza. The San Rocco neighborhood project in Faenza addressed the two fundamental issues for the construction of new neighborhoods; the first one related to land use and consequently town planning, while the second one related to experimental aspects, implementation techniques, the use of materials and innovative technology that can improve the overall quality of the ecosystem. The San Rocco neighborhood seems to be totally oriented towards the "relationship style of planning", even though a great deal of attention has been paid to the physical product, if only to ensure environmental sustainability. The experimental planning and building of the San Rocco neighborhood bridges the gap between product focused and relationship focused planning and is strongly Mediterranean in culture. Residents' involvement was also enhanced. Another useful conclusion deriving from the Italian Project is that a Mediterranean eco-neighborhood results from strategic planning (Nonni, Laghi, 2008). The project is a strategic planning one with people and their relationships at the centre.

4.4. Greek eco-neighborhood projects linked with workers' housing projects

Eco district ideas are still in their infancy in Greece, despite a boost in green development initiated recently by the government. There is no real "eco-neighborhood" implemented in Greece and many questions need to be answered on the way to implement eco-neighborhood projects in the country. Nevertheless, Greece has participated in the Med-Ecoquartiers through a project to construct 88 housing units in the city of Elefsis. The project was carried out by the Workers' Housing Organization, a public institution that since its inception in 1954, is responsible for the construction of social or workers housing dwellings, but the Elefsis project was already in construction phase which is a fact that has hampered the implementation of the Med-Ecoquartiers tools. They only contributed to a partial modification of the original design and improved the environmental performance of the project in some areas. However, the involvement of the Organization in the Project has contributed to a broader learning process to the extent that achieved awareness and knowledge around the question of eco-neighborhoods has had a replication effect through the adoption of the criteria introduced by the programme to other projects managed by the same organisation : a new settlement in Iasmos (Rodopi) has been designed as a pilot village for implementing as much as possible the methodological tools produced by Med-Ecoquartiers and major part of the sustainable planning was the consultation between the various stakeholders, the local government and the residents in order to build the necessary resilience (Kyvelou & Papadopoulos,2011). Moreover, the Organisation has announced a European architectural competition in collaboration to the Greek Institute of Architecture (EIA) regarding the environmental design of a new social housing settlement and is currently expanding its ecological action, by participating in the ELIH-MED Project dealing with energy refurbishment of low-income housing in the Mediterranean.

4.4.1. "Green neighborhood" projects in depreciated areas of western Athens

The most recent attempts to develop eco-neighborhood projects in Athens is the one led by the Ministry of Environment, Energy and Climate change and the Centre for Renewable Energy Sources (CRES) in the depreciated and low-income area of western Athens, namely the municipalities of Aigaleon and Aghia Varvara. Nevertheless the scale of the projects is too small and no real sustainable approach can be implemented. In fact, a block of 4 social housing buildings has been chosen to serve for the pilot implementation of both zero energy buildings and an interior urban oasis as well, to improve microclimate conditions. The ministry is also attempting to implement a particular public-private partnership scheme, through voluntary agreements with small construction enterprises which can provide construction materials and building products to affordable prices. The success of this scheme is crucial for accomplishing a major objective of the project, that is to stimulate local economy dynamics. Another remark that has to be made concerns the failure of a firstly launched project in Aigaleo due to the non capacity of the local authority to fully understand the project and find the necessary institutional tools to cope with the land and buildings ownership related constraints.

As fas as the Aghia Varvara project is concerned, both on-site visits, use of interactive questionnaires and data collection from the Public Power Corporation confirmed the need for interventions in buildings in three key areas: The first one concerns exterior insulation of the building shell, replacement of old simple glazed window frames with double glazing, replacement of old blinds and use of cold paintings. The second axis is to replace the various heating and cooling systems with an energy efficient central heating and cooling system. Finally, ensuring hot water through a central solar system is the third axis of the operation.

The programme is aiming at maximising energy efficiency of the neighborhood, achieving thermal comfort for residents and improving significantly their quality of life and at the same time minimise environmental impacts.

Figure 1. Social housing apartment buildings and the urban tissue of the Aghia Varvara (Western Athens) green neighborhood project.

4.4.2. Private eco-developments in high-income suburbs, focusing on marketing

Private projects can be often met in the Mediterranean area as they concern the creation of eco-villages either of secondary residences and touristic complexes or high income level housing in prestigious suburbs. In these projects, innovation and marketing for commercial reasons are usually prevailing. An example of such a development is the "Designer Village" developed by a private construction company in Dionysos, on the foothill of the Pendeli mountain. The project concerns the development of 85 plots in which 240 dwellings are being erected. Each plot has approx. an area of 1500m2. In an attempt to use green marketing tools, nine Greek architectural cabinets of different architectural perceptions and tendencies have been invited by the construction company, to put down their inspirations. "Designer Village" is already referred to as a kind of park of exemplary Mediterranean architecture, characterized as an "excellent project" by an EC programme competition. It focuses on energy efficient techniques and improvement of thermal and visual comfort and indoor air quality. Although the environmental objectives and targets associated with the rational use of resources (energy, water etc) are ambitious, the complex is far from being characterised as eco-neighborhood since it doesn't promote neither social mixing, nor economic efficiency, accessibility and affordability or cooperation among inhabitants. In conclusion we would say that an eco-neighborhood may be regarded as such only if it is the result of social dynamics and not a simple consumer product (Kyvelou & Papadopoulos, 2011).

5. Findings of the comparative study of the existing assessment tools

The investigation with regard to the existing tools showed that they address different parameters and have different focus. It is not easy to select the most appropriate tool, since there is none of them addressing all the issues and remaining, at the same time, easy to use. It was shown that most of the environmental, social and economic parameters are being implemented in different ways for each project.

However, in many of the cases, there are parameters that were neglected. Two of them is good design and place-making. It is strongly believed by the authors that the latter should be included to an eco-neighborhood approach and addressed at an earlier stage. The aspect of good design is not only related to environmental design and its many criteria but also to issues such as attractiveness and high design standards which are addressed by the Cambridge Quality Charter of Growth. Place-making, that is making of a real coherent neighborhood with social and territorial cohesion is also often neglected.

On the other hand, since current economic and financial crisis lead to fragmentation of society, special attention has to be paid, in these projects, on social equity issues as one of the aims of sustainable territorial development. The principle of social equity focuses on the right to work and housing but also on the access to services and public goods and on the fight against poverty and social exclusion. Each citizen must have access to a job and decent housing, but also to the essential public goods. The access to housing, medical care, education and information should lead to a more stable society capable of solidarity, tolerance and generating participation. It can develop its traditions while promoting a sustainable lifestyle. Finally, 'social equity' can be defined as the addition of the merit principle of equality. This aim can be divided into four targets: ensuring everyone an adequate housing, ensuring access to efficient public services, promoting access to employment, fighting against occupational and social exclusion. To achieve the first target, the issues in a territorial project are:

- promote the social mix
- promote diversity of housing supply
- promote affordable housing policy
- integrate urban projects in a social housing policy consistent with the entire city.

Apart from social equity principles, an eco-neighbourhood initiative should be based on adaptive and flexible governance schemes (Chouvet, 2007), that is bring together community stakeholders, property developers, utilities, and the city to solidify a shared sense of purpose and partnership through the following actions:

- establish municipal policy and organisational structures to support the eco-district development;
- create an engagement and governance strategy to build community support, set priorities and actions;
- develop an assessment and management to guide project development and track ongoing performance;
- identify commercialisation opportunities for the private sector to test promising products and practices;
- implement sustainability projects through technical and economic feasibility analysis, assembly of project financing, and establishment of public-private partnerships (Kyvelou & Karaiskou, 2007).

6. Comparing Mediterranean eco-neighbourhoods

This part of the research has identified examples of eco-neighborhoods in Southern Europe, either completed or in the design phase. These neighborhoods have various scales and cover regions approximately from 10 to 250 hectares (Table 8). This is an ongoing investigation aiming to identify and classify examples, which can be used as models in future eco-neighborhood design.

Amongst the 16 presented cases, one of them is located in Portugal, three in Spain, three in France and nine in Italy. In Table 8 some general characteristics are presented for each example while in Table 9 an analysis of the different environmental, economic and social parameters that were implemented in each project is undertaken. For methodological reasons, a single tool was selected for the purposes of this classification, the "One Planet Living framework" initiated by Bioregional. The 10 principles of the tool form the corpus of the analysis which will follow.

We shoud note here that the North-European model is mainly described by its technical and environmental components. The performance in terms of energy, saving water or recycling materials are particularly highlighted. They seem to be the main mechanism to move from principles to the effective implementation of sustainable urban development. The urban planning and design and the implementation of networks are resulting from this approach. This 'model governed by the environmental approach and its performance' is mainly used by communities as a powerful tool of communication, promoting the region and even as leverage to reverse social and economic depreciation. However, these social and economic aspects are not sufficiently highlighted and are not explicitly included within the agreed content of the model of sustainable urban development.

On the contrary, Southern European countries and especially as far as the examples examined in the Mediterranean, are concerned, seem to prefix social, economic and governance issues and less attention is paid to environmental performances at least from the point of view of their initial definition and specification. In Table 8, it is also clear that sustainable transport is also a field of Mediterranean interest together with health and hapiness issues.

6.1. What would be the model of a Southern European eco-neighborhood?

The study of a series of eco-neighborhoods in Southern European countries led us to the conclusion that a Southern European eco-neighborhood model is certainly emerging. Specifically, the experience of Southern Europe shows that eco-neighborhoods are nether merely expressions of integration of sustainable development in city planning nor only products integrating new technologies and alternative energies.

The eco-neighborhoods, as developed in Europe are, in our view, important local aspects of strategic spatial planning, as this is reborn and reshaped today in Europe, thus being products of a dynamic political and social process. Eco-neighborhoods are manifestations of the change of regulatory planning (based on physical planning) to a territorial management process where "territorial marketing" has a predominant role (Kyvelou, 2010).They reaffirm, moreover, that sustainable development is an exceptional unifying element and point of recasting spatial policies. Sustainable neighborhoods are also emphasizing the issue of scale which is nowadays one of the most important theoretical discussions in spatial and urban planning. Another important conclusion is that an eco-neighborhood mainly depends on the extent to which there is a tradition of strategic spatial planning, on the cultural tradition and level of collective and community involvement (Chiotinis, 2006) and on the presence of catalytic investments and the possibility of implementing public - private partnerships towards a process of stimulating local economy dynamics (Kyvelou & Karaiskou, 2006).

The Mediterranean countries usually lagging behind as far as the above conditions are concerned, present a spontaneity and a delay of implementation. Constraints of implementation show that an eco-neighborhood initiative should be based on adaptive and flexible governance schemes, and on engagement strategies that would bring together community stakeholders, property developers, utilities, and the city to solidify a shared sense of purpose and partnership through the following actions :

a. Establish municipal policy and organisational structures to support the eco-neighborhood development;
b. Create an engagement and governance strategy to build community support, set priorities and actions;
c. Develop an assessment and management toolkit to guide project development and track ongoing performance;
d. Identify commercialization opportunities for the private sector to test promising products and practices; and
e. Implement sustainability projects through technical and economic feasibility analysis, assembly of project financing, and establishment of public-private partnerships (Kyvelou, Marava & Kokkoni, 2011).

Both lessons from the North-European examples of eco-neighborhoods and observation from different types of eco-districts in Southern Europe, either public or private projects, can provide helpful criticism and a good framework to discuss issues of efficient implementation in the near future. This framework could help to modeling work that is both to the construction of a model that can predict future trends and to the affirmation of the model, meaning that it can be used in the reproduction and replication of certain practices and related projects.

Project name	plan	area	dwellings	inhabitants	total cost
Portugal					
Mata de Sesimbra, Lisbon		5300 he+4800he natural	8000 units tourist	25.000 inhabitants	800 mil Euros
Spain					
Entrenucleus, Dos Hermanas, Seville		39,5 hectares, 7, 2 hectares settlement 160,50 km2	13500 units homes and offices	126.000 inhabitants	244 mil Euros
Ecociudad Valdespartera, Zaragoza		243,3 hectares	9.678 social housing units	31.000 inhabitants	N/A
Logroño Montecorvo, Rioja Province		56 hectares	3.000 carbon neutral housing units	130.000 inhabitants	388 mil Euros
France					
Andromède in Blagnac		210 hectares	3.700 housing	1 million	N/A
Ville de Pézenas, Quartier Saint Christol		29 hectares	90.000 sq met.	250 famillies	31 mil Euros
Saint Lys-Haute Garonne		7 hectares	N/A	7500	N/A
Italy					
St.Rocco, Faenza		8,3 hectares	390 housing units	950	9,4 mil Euros
Malizia Ecologic quarter, Siena		5 hectares	150 housing units +commercial etc	N/A	N/A
Cognento project, Modena		11,7 hectares	220 housing units	770	5,5 mil Euros
BIOPEP		14, 7 hectares	340 housing units	N/A	N/A
Sanpolino quarter, Brescia		N/A	2.000 apartments	N/A	N/A
Parco Ottavi project, Reggio Emilia		53 hectares	131.000 sq m surface,1.500 housing units	4.500	N/A
Pietrasana, Vigevano, - Pavia		N/A	220 lodgings	N/A	5,7 mil Euros
Villa Fastiggi, Pesaro		15 hectares	333 dwellings	N/A	N/A
Bologne, quartier de la Bolognina ouest		30 hectares	N/A	64.000	N/A

Table 8. Examples of Mediterranean eco-neighbourhoods

ONE PLANET LIVING PRINCIPLES	CATEGORIES IDENTIFIED FROM THE PROJECTS	Portugal	Mata de Sesimbra, Lisbon	Spain	Entrenucleus, Seville	Valdespartera, Zaragoza	Logroño Montecorvo, Rioja	France	Andromède quarter, Blagnac	Ville de Pézenas,Saint Christol	Italy	St.Rocco, Faenza	Malizia ,SienaCognento	BIOPEP, Modena	Sanpolino quarter, Brescia	Parco Ottavi , Reggio Emilia	Public Buildingsr, Pietrasana	Villa Fastiggi, Pesaro	Quartier de la Bolognina	Greece	Elefsis	Iasmos	Aghia Varvara
Zero Carbon	Solar - panels, wind control					•	•															•	•
	orientation					•									•			•				•	•
	Renewable energy		•				•		•	•												•	•
	biomass									•			•				•						
	Photovoltaic systems		•		•							•				•						•	
	Thermal stations											•		•	•	•							•
Zero Waste	Waste management		•						•	•				•		•					•	•	
Sustainable Transport	Pedestrian routes		•						•	•		•	•	•	•	•		•	•			•	
	Bicycle network		•						•	•		•		•	•	•		•	•		•		
	Green belts		•		•	•				•		•			•	•	•	•	•				
	Urban bus lines		•		•				•						•								
	Covered parking															•	•						
Sustainable Materials	Sustainable materials		•			•							•	•				•			•	•	•
Local-Sust. Food	Local and sustainable foods		•			•																	
Sustainable Water	Rainwater storing		•		•				•	•		•	•		•	•							
	Water reutilisation								•	•		•	•			•							
	Water nets						•							•	•	•							
	Water devices					•							•	•									
Land and Wildlife	Land and wildlife		•			•									•			•			•	•	
Culture and Heritage	Connection with historical centres		•							•		•							•				
	New cultural centres					•													•		•		
Equity and Local Economy	Social services		•						•	•		•	•			•		•					
	Generation mix								•	•													
	Integrating urban environment					•																	•
Health and Happiness	Sports facilities					•									•	•	•						
	Public buildings		•						•				•	•	•	•	•	•					
	Kindergartens														•	•							
	Urban parks		•		•				•	•				•	•	•		•					•

Table 9. Parametric analysis of the sustainability criteria according to the One Planet Living framework.

6.2. What would be an efficient sustainability assessment tool for Southern European eco-neighborhoods?

The short analysis of the eco-neighbourhood design tools has shown that they have different orientations in terms of their focus on environmental, social or economic strategy. Some of the tools seem to be more design oriented while others have a social resilience focus. The Eco-Neighborhood is more than a mere buzzword or local marketing tagline. The neighborhoods have ambitious targets that go well beyond load reduction. They draw upon new and often complex practices, from urban project design to construction, use and assessment and it is often complicated for the local authorities to implement these practices as new methodologies need to be addressed and many of them recoil at doing so, from the very start of the project.

The above described experiences in Southern Europe show that an assessment tool for eco-neighborhoods cannot be efficient if nor directly linked with the valorisation of the territorial potential and the territorial capital (OECD, 2001), at local level. We have therefore proceeded to the formation of a new tool, largely inspired by the territorial capital and the territorial cohesion concepts, which gives emphasis to the holistic approach of the city, its neighborhoods and its relationship and interdependence with its region. The so-called "SDMed eco-neighborhood" tool is structured around the main determinants of the territorial capital (geographical, cultural, political, material, social and intellectual capitals) and is analysed in a series of criteria capable to manage the complexity and diversity of the Mediterranean urban phenomenon (Sinou &Kyvelou, 2006)

The concept of territorial capital was first proposed in a regional policy context by the OECD in its Territorial Outlook (OECD, 2001), and it has been later on reiterated by DG Regio of the European Commission: "Each Region has a specific 'territorial capital' that is distinct from that of other areas and generates a higher return for specific kinds of investments than for others, since these are better suited to the area and use its assets and potential more effectively. Territorial development policies (policies with a territorial approach to development) should first and foremost help areas to develop their territorial capital" (European Commission, 2005).Territorial capital is referring to the following elements: a/ a system of localised externalities, both pecuniary and technological; b/ a system of localised production activities, traditions, skills and know-how; c/ a system of localised proximity relationships which constitute a 'capital' – of a social psychological and political nature – in that they enhance the static and dynamic productivity of local factors, d/ a system of cultural elements and values which attribute sense and meaning to local practices and structures and define local identities; they acquire an economic value whenever they can be transformed into marketable products – goods, services and assets – or they boost the internal capacity to exploit local potentials; e/ a system of rules and practices defining a local governance model. Accordingly, the OECD has rightly drawn up a long list of factors acting as the determinants of territorial capital, and which range from traditional material assets to more recent immaterial ones. All the above have been used to draft the SDMed Eco-neighborhood tool.

Territorial capital	Territorial cohesion components
Geographic	Emmissions
	Landscape resources
Cultural	Cultural heritage
Political	Governance
Material	Economic growth-wealth
	Resources
	Sustainable transports-mobility
Social	Health and safety
	External accessibility
	Internal connectivity
Intellectual	Creativity

Table 10. The conceptual elements behind the SDMed Eco-neighborhood tool.

7. The SDMed Eco-Neighborhood tool

As it was mentioned in the introduction the SDMed Eco-Neighbourhood tool was based on:

1. the research and parametric analysis between the existing eco neighborhood tools;
2. the SDMed building performance assessment tool (Sinou & Kyvelou, 2006);
3. the concept of territorial capital (OECD, 2001);
4. The current economic and financial crisis and the consequent limitation of public funds.

By its definition, the concept of territorial capital can be divided in geographic, cultural, political, material, social and intellectual capital, while approaching territorial cohesion we can categorise actions in emissions, local resources, cultural heritage, governance economic growth, resources, sustainable transport-mobility, health and safety, external accessibility, internal connectivity and creativity (Kyvelou, 2010).

Thus, the structure of the new tool is constituted by the main subdivisions of territorial capital, eleven objectives-targets linked to territorial cohesion and in 39 accordingly sub-targets (Table 11). These sub-targets are further divided in more criteria in order to include all the parameters that can influence sustainable neighborhood. The final depiction of the tool is under study; however a first attempt is presented in the diagram of Table 10.

SDMed ECO NEIGHBORHOOD tool				
Territorial capital	11 Targets territorial cohesion	39 Sub-Targets		
Geographic	Emissions	1.1	Emissions (CO2)	Reduction of CO2 emissions
		1.2	Water waste	Management of water waste
		1.3	Production of solid waste	Management of domestic waste Management of construction waste Management of solid waste infrastructure
		1.4	Management of litter and waste	Network of sewage
		1.5	Urban heat island effect	Measures to reduce heat island effect
		1.6	Night-time light pollution	Reduction of light pollution
	Landscape resources	2.1	Natural and technological risks	Local management of natural risks (Earthquake, heat wave, tsunami) Local management of technological risks
		2.2	Influence to the urban form	Optimisation of land use Compact growth – Density Intelligent planning Integration of environmental issues in urban planning
		2.3	Adaptive opportunity	
		2.4	Quality of public spaces	Tree-lined and shaded streets Cohesion and linkage among urban spaces Design with bioclimatic criteria
		2.5	Comfort (thermal, visual, acoustic)	Nuisance linked to the neighborhood Noise pollution in the neighborhood from vehicles or activities Minimisation of construction nuisance

SDMed ECO NEIGHBORHOOD tool				
Territorial capital	11 Targets territorial cohesion	39 Sub-Targets		
				Visual quality of natural environment - view Visual quality of built environment Preferred locations Water efficient landscape Steep slope protection
		2.6	Air quality	Interior air quality Exterior air quality
		2.7	Spatial comfort and comfort of activity	Quality of building Quality of housing Variety in housing Satisfaction of users and residents Support of architectural quality Innovation and exemplary performance
Cultural	Cultural heritage	3.1	Maintenance of natural heritage - biodiversity	Maintenance of wetlands and natural beauty landscapes Protection of rural land Extinction species and ecological communities
		3.2	Maintenance of cultural heritage	Conservation and re-use of cultural heritage
		3.3	Maintenance of built environment	Use of existing buildings Preservation of historical resources and adaptive reuse Brownfield redevelopment
Political	Governance	4.1	Functionality and possibility of services control	
		4.2	Adaptability and flexibility of services	
		4.3	Robustness and maintenance of	

SDMed ECO NEIGHBORHOOD tool				
Territorial capital	11 Targets territorial cohesion	39 Sub-Targets		
			services	
		4.4	Community involvement	
		4.5	Ownership of land and buildings	
		4.6	Public Private Partnerships	
Material	Economic growth-wealth	5.1	Cost of land and construction cost	Local economic dynamic Creation of social economy
		5.2	Cost of of life cycle (€/year) (maintenance, exploitation et deconstruction)	
		5.3	Cost of waste management and cost of emissions (€/year)	
		5.4	Support of local economy	Presence economic activities Presence of retail trade Local food production Mixed use neighborhood centers
	Resources	6.1	Effect in the energy resources	Improvement of energy efficiency for heating, cooling and electricity (buildings and infrastructures) Use of renewable energy (locally) Orientation Heating and cooling of district Certified green buildings
		6.2	Exhaustion of raw material	Integration of recycled and reused materials, constructions and process of demolition in buildings and the public spaces

SDMed ECO NEIGHBORHOOD tool				
Territorial capital	11 Targets territorial cohesion	39 Sub-Targets		
		6.3	Use and management of water	Consumption of potable water Use and management of rain water of Stormwater management
	Sustainable transport-mobility	7.1	Transportation cost	Development ICT Proximity of housing and job
		7.2	Improvement of transportation and mobility	Improvement of transport Safe and convenient paths for pedestrians and bicycles Areas with decreased dependence automotive Transit centre Network bicycle path and storage Reduction of parking footprint Street network
Social	Health and safety	8.1	Health and productivity	Improvement of cleanliness in the neighborhood and communal spaces Right and acess to care and health
		8.2	Safety of users	Improvement of safety of people and goods Improvement of road safety
	External accessibility	9.1	Accessibility for people with special needs	Accessibility
		9.2	Accessibility in open spaces	Access in public spaces Access in recreation facilities
	Internal connectivity	10.1	Involvement of users	Involvement of residents and users in the process of sustainable urban development Participation of residents in decision making and projects related to the community Strengthening of community

SDMed ECO NEIGHBORHOOD tool				
Territorial capital	11 Targets territorial cohesion	39 Sub-Targets		
				Promotion and participation of community
		10.2	Creation of work places	
		10.3	Social diversity	
Intellectual	Creativity	11.1	Support of education - Levels of education and professional skills	Diversity in age distribution Mixed income communities Connected and open community Collaborations Promotion of academic success Reinforcement of the role of school in the community International cultural contacts - connectivity

Table 11. The new SDMed ECO-NEIGHBOURHOOD tool

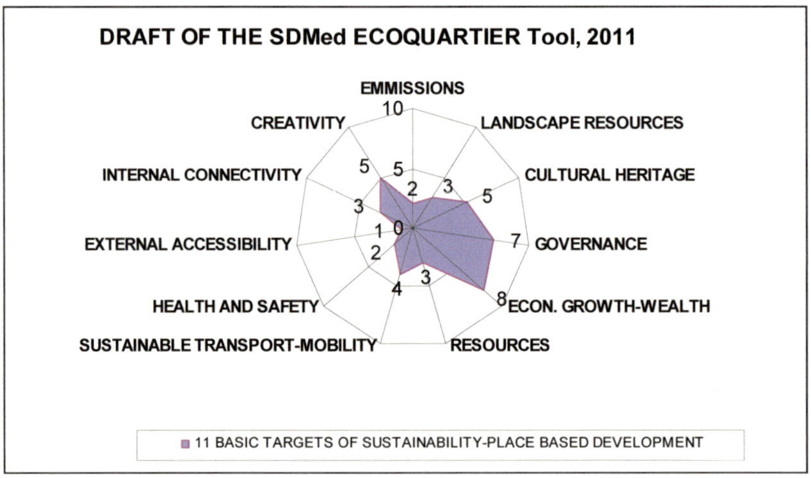

Figure 2. Draft illustration of the SDMed Eco-neighbourhood tool.

8. Conclusions

The climatic, social, cultural, governance and spatial planning related specificities of the Mediterranean region and the delay of eco-neighborhood development in the Mediterranean countries combined with current need of economic and social regeneration lead to the necessity to develop a tool adapted in these specific needs and particularities. The comparative research showed that the assessment tools for eco-neighborhoods have been structured around different principles and ideas. There is an abundance of tools and their evaluation constituted an aid for the creation of a new proposal. The parametric analysis of tools that were selected created a base for the creation of the new tool that has to be enriched in order to face the particular needs of the Mediterranean region, with sensitivity and taking into account current ecomonic and financial crisis.

Furthermore, the investigation of existing Mediterranean examples gives important information with regard to the sustainability criteria that are used in most of the cases.

Finally, the concept of place-based development and contemporary planning criteria like the one of territorial capital have been explored in order to form a proposal for a new methodological tool based on the need of place-making and urban regeneration under economic crisis and lack of public funds. Sustainability in the scale of a neighborhood cannot be a static process, has to do with innovation and has to ensure the creation of wealth at local level and the stimulation of local economy dynamics. Planning a contemporary eco-neighborhood is a complex procedure and addresses both economic, social, environmental and governance related challenges. Achieving an eco-neighborhood assumes the development and implementation of a process of project management and an action plan involving local actors at the different phases of a project. Furthermore, it involves setting objectives and implementing actions to achieve environmental, economic and social performances.

Author details

Stella Kyvelou
Panteion University of Social ad Political Sciences of Athens , Greece

Maria Sinou
Technological Educational Institution, Greece

Isabelle Baer
Syndicat National des professionels de l'aménagement et de lôtissement, France

Toni Papadopoulos
Workers' Housing Organisation, Greece

9. References

Barton H., (Ed.). (2000). *The Potential for Eco Neighbourhoods*, Earthscan Publications Ltd, ISBN 1 85383513 7, London.

Barton H.; Grant M. & Guise R. (2010). *Shaping Neighbourhoods, for local health and global sustainability*, Routledge, ISBN10: 0-415-49549-0; ISBN13: 978-0-415-49549-13, London and New York.

Chiotinis, N. (2006). The request of sustainability and architecture as cultural paradigm, *Management of Environmental Quality : An International Journal*, Vol. 17, No. 5, Emerald Group publishing Limited, pp.593-598, ISSN 1477-7835.

Chouvet, C. (2007). *Les quartiers durables: un exemple de démarche intégrée et participative*, Comité 21/Angenius, pp.4–20, http://www.comite21.org/docs/territoires-durables/ville-durable/les-quartiers-durables.pdf

Kyvelou, S. & Karaiskou, E. (2006). Urban development through PPPs in the Euro-Mediterranean region, *Management of Environmental Quality: An International Journal*, Vol. 17, No. 5, Emerald Group Publishing Limited, pp.599–610, ISSN 1477-7835.

Kyvelou, S., Hetzel, J., Sinou, M. & Iwamura, K. (2007). *L'application du développement durable au cadre bâti dans l'espace Méditerranéen: La démarche SD-MED*, Pulim (Ed.), pp.121, ISBN : 978-2-84287-425-4, France.

Kyvelou, S. (2010). *From spatial planning to spatial management: the concepts of strategic spatial planning and territorial cohesion in Europe*, KRITIKI (Eds.), ISBN 978-960-218-671-8, Athens.

Kyvelou, S. & Papadopoulos, A. (2010). *'Sustainable neighborhoods: lessons from Northern Europe – issues arising from a Mediterranean paradigm'*, in: I. Beriatos and M. Papageorgiou (Eds.): Spatial Planning-Urban Planning-Environment in the 21st Century, pp.315–326, University Editions of Thessaly, Volos, Greece, Mediterranean.

Kyvelou, S. & Papadopoulos T. (2010). Exploring a South-European eco-neighbourhood model: Planning forms, constraints of implementation and emerging resilience practices, *International Journal of Environment and Sustainable Development*, Vol. 14, Nos. 1/2, 2011, pp. 77-94, ISSN(print) 0960-1406.

Kyvelou, S., Sapounaki-Dracaki, L. & Papadopoulos, A. (2010). *«Eco-quartiers en Europe: leçons obtenues des pays du Nord, perspectives et politiques dans les pays de l'Europe du Sud»*, oral presentation at the 10th International EAUH Conference, Gand, August-September 2010.

Kyvelou, S., Marava, N. & Kokkoni, G. (2011). Perspectives of local public-private partnerships towards urban sustainability in Greece, *International Journal of Environment and Sustainable Development*, Vol. 14, Nos. 1/2, 2011, pp.95-111, ISSN(print)0960-1406.

Kyvelou, S., Dracaki, L., Sinou, M., & Papadopoulos A. (2011). Planning and building a South-European eco-neighborhood: from concepts and strategies to practice and assessment tools, *Review of decentralization, local government and regional development*, Vol.66, Athens, 2011, ISSN 1106-91-71.

Lefèvre, P. & Sabard, M. (2009). *Les écoquartiers*, Editions Apogée, p.261, ISBN: 978-2-84398-325-2, Paris.

Nonni, E. & Laghi, S. (2008). *Un eco quartiere mediterraneo: Il Quartiere Residenziale S. Rocco a Faenza'*, Comune di Faenza. http://www.avecnet.com/publications/files/page51_15.pdf

OECD Terrirorial Outlook (2001), ISBN: 9789264189911, Paris, p.300

Sinou, M. & Kyvelou, S. (2006). Present and future of building performance assessment tools, *Management of Environmental Quality: An International Journal*, Vol. 17 Iss: 5, pp.570 – 586, ISSN 1477-7835.

SNAL (2011), *"Ecoquartiers"*, ISBN 978-2-912683-84-7, Editions PC.

Souami, T. (2009). *Éco-quartiers, secrets de fabrication: analyse critique d'exemples européens*, Les Carnets de l'info (Eds.), ISBN 978-2-9166-2844-8, Paris.